Undergraduate
Chemistry
Volume–I

Undergraduate Chemistry

Volume–I

(Semester–I Course, Including Practicals)

S. C. DATTA

Ph.D. (Delhi), DIC (London),
Chartered Chemist, FRSC (London)

Former Associate Professor of Organic Chemistry
Hans Raj College, Delhi University, Delhi

Namita K. Johar

Ph.D. (Delhi)

Lecturer, Maharaja Agrasen Institute of Technology,
Guru Gobind Singh Indraprastha University, Delhi

Ane Books Pvt. Ltd.

New Delhi ♦ Chennai

Undergraduate Chemistry **(Volume–I)**

S.C. Datta and Namita K. Johar

First Edition : 2011

Reprint : 2018, **2020, 2023**

Published by

Ane Books Pvt. Ltd.
4821, Parwana Bhawan, 1st Floor, 24 Ansari Road, Darya Ganj,
New Delhi - 110 002, India, Tel.: +91(011) 23276843-44, Fax: +91(011) 23276863
e-mail: kapoor@anebooks.com, Website: www.anebooks.com

Branch Office:

- Avantika Niwas, 1st Floor, 19 Doraiswamy Road, T. Nagar,
  Chennai - 600 017, Tel.: +91(044) 28141554, 28141209
  e-mail: anebooks_tn@airtelmail.in

ISBN : 978-93-8061-888-3

Printed at : Repro Knowledgecast Limited, Thane

S. C. Datta

Dedicates to his ***Wife***

&

Namita K. Johar

Dedicates to her ***Husband***

Acknowledgements

I am thankful to my family members who supported this work by sparing me from lot of routine family responsibility for a long time and to my wife, Mrs. Pammi Datta, Principal, Kulachi Hans Raj Model School, in particular, who did many of my jobs, besides her busy administrative responsibilities.

Constant help and support from everyone at Ane Books Pvt. Ltd. is greatly appreciated.

S. C. Datta

I express my sincere gratitude to Dr. Nand Kishore Garg, Chairman, Maharaja Agrasen Institute of Technology, for his constant inspiration and encouragement without which this work would not have been possible. I am thankful to Prof. M. L. Goyal, Director, and Prof. O. P. Grover, Dean, Maharaja Agrasen Institute of Technology for their appreciation of work at the manuscipt stage, and to Prof. Gurmeet Singh, Proctor Delhi University, Delhi, for constant flow of ideas to enrich my life.

I am also thankful to all my colleagues who made helpful suggestions.

I feel short of words to express thanks to all my family members for their utmost patience, encouragement and help in sparing me from my responsibilities at home.

Constant help and support from everyone at Ane Books Pvt. Ltd. is greatly appreciated.

Namita K. Johar

Preface

The book is the outcome of the interactions of my students during my teaching career, extending over a period of forty one years at Hans Raj College, University of Delhi, Delhi and at some other Universities in UK and USA.

This book entitled 'Undergraduate Chemistry, Vol. I, of the book is divided into three units, comprising of twelve chapters, including inorganic, organic and practical chemistry course. Inorganic chemistry units cover the basic foundation course, like atomic structure and chemical bonding. It clarifies the structure of molecules in inorganic chemistry section. Organic section deals with basic concepts like molecular architecture, electronic effects (inductive effect, mesomeric effect and resonance), properties of molecules (dipole moments, molecular forces, aromaticity and some aspects of reactive species). Stereochemistry has been dealt with in fairly details, that helps to understand the basic behaviour of molecules in three dimensions. Chemistry and mechanism of reactions of alkanes, alkenes and alkynes has been included. These compounds are the back bone of whole of organic chemistry. Their knowledge is essential at this stage to form a complete foundation course. All the topics covered are essential for learning organic chemistry in an easy and systematic way.

It has been written keeping in mind with the needs of students preparing for first semester of B.Sc. course of Delhi University and other universities with similar course - a kind of foundation course. The book has emerged from the long teaching experience of the authors in the topics considered. Lectures delivered during the past many years, has given us the opportunity to understand the basic requirement of students in assimilating the concepts of Chemistry. The book has been written in a systematic manner using simple, necessary and clear language to enable even an average student for self study with minimal guidance. Each topic has been covered in a comprehensive way, so as to instill in students, a proper grasp of the subject. Solved examples in the chapters will reinforce the theoretical concepts covered in the chapter. Questions given at the end of each chapter will ensure sufficient revision and mastery of the subject.

To highlight some of the salient features of the book, we wish to point out the following:

1. Figures have been illustrated wherever required and are brought out to nearly the original placement of bonds and angles between different atoms.
2. Mathematical treatment of the derivations and difficult numericals have been done step-wise in detail for a better understanding.
3. The book contains a glossary of different terms often used in chemistry.
4. In the organic section, attempt has been made to give figures clearly explaining the 3D concepts of molecules.
5. In the appendix, periodic table that carries some of the properties of elements has been included.
6. To remind students that stereo-placement of a bond is differentiated by wedge diagram bond as in (◀ , ''''''||||) or (||||''''' , ▬), rather than simple (------ , ———) bonds.

Volume II shall involve physical chemistry concepts and more of organic chemistry, that shall completely lay the foundation of chemistry in the first year of any undergraduate course.

Suggestions and criticism are welcome as there is always a scope of improvement.

S. C. Datta
Namita K. Johar

Syllabus for
Delhi University
Undergraduate Chemistry (Volume–I)
Semester–I
Paper No. 101 : Chemistry–I

Section A: Inorganic Chemistry-1 **(30 Lectures)**

Unit 1: Atomic Structure: *Recapitulation of: Bohr's theory de-Broglie's relation, Heisenberg Uncertainty principle. Need of a new approach to Atomic structure.* Time independent Schrodinger equation ($H\Psi = E\Psi$). Significance of Ψ and Ψ^2, Schrödinger equation for hydrogen atom. Transformation of Cartesian coordinates (x, y, z) into polar coordinates $(r, \theta\phi)$. Radial and angular parts of the hydogenic wavefunctions (atomic orbitals) and their variations for 1*s*, 2*s*, 2*p*, 3*s*, 3*p* and 3*d* orbitals. (Only graphical representation), Radial and angular nodes and their significance. Radial distribution functions (1*s* and 2*s* atomic orbitals). Significance of quantum numbers, orbital angular momentum and quantum numbers *ml* and m_s. Shapes of *s*, *p* and *d* atomic orbitals, nodal planes. Discovery of spin, spin quantum number (*s*) and magnetic spin quantum number (m_s). Electronic configurations of the atoms. Concept of exchange energy. Relative energies of atomic orbitals, Anomalous electronic configurations.

Unit 2: Chemical Bonding and Molecular Structure

Ionic Bonding: Energy considerations in ionic bonding, lattice energy and solvation energy and their importance in the context of stability and solubility of ionic compounds. Born-Lande equation for calculation of lattice energy, Born-Haber cycle and its applications, polarizing power and polarizability. Fajan's rules, bond moment, dipole moment and percentage ionic character.

Covalent bonding: VB Approach: Concept of hybridization and VSEPR theory . Resonance and resonance energy: study of some inorganic and organic compounds.

Molecular Orbital Approach: LCAO method, bonding and antibonding MOs and their characteristics for *s-s*, *s-p* and *p-p* combination of atomic orbitals, non- bonding combination of orbitals ,MO treatment of homonuclear diatomic molecules of 1st and 2nd periods (including idea of *s-p* mixing)\ and heteronuclear diatomic molecules such as CO, NO and NO^+. Comparison of VB and MO approaches.

Section B: Organic Chemistry-1 (30 Lectures)

Unit 3: Fundamentals of Organic Chemistry

Physical Effects, Electronic Displacements: Inductive Effect, Electromeric Effect, Resonance and Hyperconjugation. Cleavage of Bonds: Homolysis and Heterolysis. Structure, shape and reactivity of organic molecules: Nucleophiles and electrophiles. Reactive Intermediates: Carbocations, Carbanions free radicals. Strength of organic acids and bases: Comparative study with emphasis on factors affecting *pK* values. Aromaticity: Benzenoids and Huckel's rule.

Unit 4: Stereochemistry

Conformations w.r.t. ethane, butane and cyclohexane. Interconversion of Wedge Formula, Newman, Sawhorse and Fischer representations. Concept of chirality (upto two carbon atoms). Configuration: Geometrical and Optical isomerism; Enantiomerism, Diastereomerism and Meso compounds). Threo and erythro; D and L; *cis* - *trans* nomenclature; CIP Rules: R/ S (for upto 2 chiral carbon atoms) and E / Z Nomenclature (for upto two C=C systems).

Unit 5: Aliphatic Hydrocarbons

Functional group approach for the following reactions (preparations and reactions) to be studied in context to their structure.

Alkanes: (Upto 5 Carbons)

Preparation: Catalytic hydrogenation, Wurtz reaction, Kolbe's synthesis, from Grignard reagent.

Reactions: Free radical Substitution: Halogenation.

Alkenes: (Upto 5 Carbons) *Preparation*: Elimination reactions: Dehydration of alkenes and dehydrohalogenation of alkyl halides (Saytzeff's rule); *cis* alkenes (Partial catalytic hydrogenation) and *trans* alkenes (Birch reduction). *Reactions*: *cis*-addition (alk. $KMnO_4$) and *trans*-addition (bromine). Addition of HX (Markownikoff's and anti-Markownikoff's addition). Hydration, Ozonolysis, oxymercuration-demercuration, hydroboration-oxidation

Alkynes: (Upto 5 Carbons) *Preparation*: Acetylene from CaC_2 and conversion into higher alkynes; by dehalogenation of tetra halides, dehydrohalogenation of vicinal-dihalides.

Reactions: formation of metal acetylides, addition of bromine and alkaline $KMnO_4$, ozonolysis and oxidation with hot alk. $KMnO_4$.

Contents

INORGANIC CHEMISTRY UNITS

1. Atomic Structure 1–54

2. Chemical Bonding 55–132

ORGANIC CHEMISTRY UNITS

3. Structure and Shape of Organic Molecules 135–152

4. Electronic Displacements and Reactive Intermediates 153–178

5. Strength of Acids and Bases 179–196

6. Aromaticity 197–210

7. Stereochemistry 211–266

PRACTICALS UNITS

INORGANIC CHEMISTRY

UNITS

1

Atomic Structure

John Dalton (1766–1844)

John Dalton was born on 6 September 1766 England. Around 1790 Dalton seems to have considered taking up law or medicine, but his projects were not met with encouragement from his relatives.Dalton's first publication was Meteorological Observations and Essays (1793), which contained the seeds of several of his later discoveries.

In 1794, shortly after his arrival in Manchester, Dalton was elected a member of the Manchester Literary and Philosophical Society.

In 1800, presented an important series of papers, entitled "Experimental Essays" on the constitution of mixed gases; on the pressure of steam and other vapours at different temperatures, both in a vacuum and in air; on evaporation; and on the thermal expansion of gases. The first theory of atomic structure of matter was announced by John Dalton. His reputation and continuous view of matter stalled the development of the atomic theory between 1810–60. These four essays were published in the Memoirs of the Lit and Phil in 1802.

He died on 27 July 1844 (aged 77)

IN THIS CHAPTER

1

Atomic Structure

Introduction

Immediately, after the start of the Big Bang, space expanded incredibly, quickly, which lasted for the minutest fraction of a second, and was called INFLATION. When, expansion began to slow down and different kinds of subatomic particles including electrons made their appearance. Just one millionth of a second after the birth of the universe, various particles had clumped together to form new particles called protons and neutrons. After a hundred seconds or so, some of the protons and nearly all of the neutrons gathered into bunches, consisting of two protons and two neutrons. Eventually, each bunch, or atomic nucleus, captured two electrons to form a helium atom, and each remaining proton captured a single electron to form a hydrogen atom. That's how the first building blocks of matter were born.

1.1 Atom

John Dalton, enunciated matter to be composed of indivisible particles called atoms. According to mordern researches, atoms are the smallest indivisible particles which cannot be further subdivided into smaller particles. However, it has been broken into smaller fragments like electrons, protons and neutron.

He could not explain the formation of chemical bond which actually led to the detailed study of the inner structure of atom. Various atomic models were discussed, detailing the step wise discovery of an atom and its fundamental particles.

1.2 Thomson Atomic Model or Plum Pudding Model

According to this model which was proposed by J.J. Thomson, in 1906 the atom is composed of corpuscles *i.e.*, the negative charge is surrounded by a soup or a cloud of the positive charge so as to balance the negative charge. It appears as if the negatively charge of plums are surrounded by the positively charged pudding. Therefore, it was called as plum pudding model of an atom as shown in Fig. 1.1.

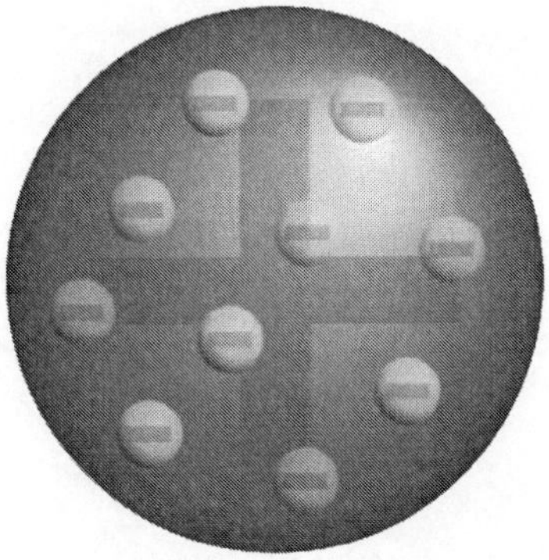

Fig. 1.1 A schematic representation of the plum pudding model of the atom.

1.3 Rutherford Atomic Model

The credit for giving the first successful model of the arrangement of electrons, protons and neutrons goes to Rutherford in the year 1909. His model was based on alpha-ray scattering experiment.

The model described the atom as a tiny, dense, positively charged core called a *nucleus*, in which nearly all the mass is concentrated, around which the light, negative constituents, called electrons, circulate at some distance, much like planets revolving around the Sun.

1.3.1 Description of the Alpha Particle Scattering Experiment

Alpha particles were produced from a radioactive element, placed in a lead block and allowed them to strike a thin gold foil. A movable screen, coated with ZnS placed on the other side of the foil helped in detecting the direction of the movement of the alpha particles on striking the foil.

Now, if the atoms were actually arranged according to Thomson, then the alpha particles would go undeflected. But during the experiment, it was found that though most of the alpha particles went right through the thin foil, a few alpha particles rebounded almost **directly backwards suggesting that the nucleus is positively charged, very dense, hard and rigid.**

From the above observations, he concluded that **large scattering angles** could only occur if most of the mass of the atom, and all of its positive charge reside in a very small, extremely dense region called the atomic nucleus.

Most of the alpha particles passed through the gold foil, without undergoing any deflection suggesting that most of the space around the nucleus is empty. So, this was the extranuclear part, where electrons revolve at very high speed in fixed path called orbits.

Thus, Rutherford established a new model of the atom that was similar to the structure of **planets in orbit around the sun**. Rutherford's model consisted of a dense positive core at the center of the atom with electrons orbiting around it. This is called as **Planetary model of atom**. Electrons are revolving around the nucleus, with high speed. Since the atom is electrically neutral, the number of electrons is equal to the number of protons.

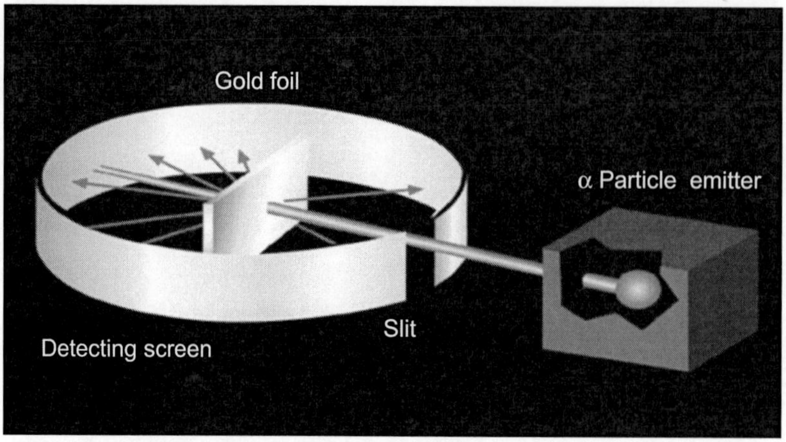

Fig. 1.2 Rutherford alpha particle scattering experiment.

1.3.2 Drawbacks of the Rutherford Model

In this model, the electrons are revolving around the nucleus with a fast speed. An electron while revolving around the nucleus loses energy and its speed is decreased, and ultimately the electron should fall into the nucleus, thereby making the atom unstable, which is not the case. Thus, the Rutherford model could not explain the stability of an atom. The drawbacks of the Rutherford model were explained by Bohr, that are given later in this chapter.

1.4 Three Fundamental Particles of an Atom

An atom is comprised of three fundamental particles namely, electrons, protons and neutrons.

- Electrons, have a negative charge and very small size which is not measurable. These are the least heavy of the three.
- Protons, have a positive charge, and are about 1836 times more heavy than electrons; and
- Neutrons, have no charge and are the same size and mass as that of protons.

Protons and neutrons make up a dense, massive atomic nucleus and are collectively called **nucleons**. The electrons form the much larger electron cloud that surrounds the nucleus. Electrons however, revolve around the nucleus in fixed orbits.

- Of all the fundamental particles present in an atom, neutron is the heaviest and the least stable particle, as it breaks down into electron, proton and neutrino.

In addition to the above fundamental particles some uncommon sub-atomic particles have also been found as mentioned below.

1. **Neutrinos and Antineutrinos:** These are the particles of small mass (approximately zero) and zero charge and were postulated by **Fermi** in 1934.
2. **Positrons:** In 1932, these were discovered by **Carl D. Anderson**, who gave its name, the positron. These are the positive counterparts of electron. The positron has an electric charge of +1, a spin of 1/2, and the same mass as that of an electron.

3. ***Pi*- mesons:** These are the particles having a mass intermediate between that of an electron and a proton. The Nobel Prize in Physics was awarded to *Yukawa* in 1949 (for predicting the existence of mesons) and to *Powell* in 1950 (for developing the technique of particle detection using photo-emulsions).

1.5 Atomic Number and Mass Number

The number of protons present in an atom is known as Atomic Number. It is denoted by letter 'Z' and written on the bottom of the symbol of the atom.

The entire mass of the atom lies in its proton and neutrons which are present in the nucleus. The sum of number of protons and neutrons is called the mass number of the atom. Mass number is generally denoted by the letter 'A' and written on the top of the symbol of the atom.

Note: *Mass number is the sum of the number of protons and neutrons in the nucleus, is always a whole number. On the other hand, atomic mass is not necessarily a whole number because it is the average mass of all the existing isotopes.*

A symbol for an atom of an element (*X*) indicating its Mass Number (*A*) and Atomic Number (*Z*) is given below:

$$_{Z}X^{A}$$

1.6 Electromagnetic Radiations

A disturbance which originates from some vibratory source and travels outwards as a continuous sequence of alternating crests and troughs is a wave.

Electromagnetic radiations are a form of energy transported in the form of waves. These are associated with both electric and magnetic fields. All forms of electromagnetic radiations travel in space in the form of waves with the velocity of light.

Every wave is associated with five characteristics *i.e.*, wavelength, frequency, velocity, wave number and amplitude.

1. **Wavelength (λ):** The distance between two neighboring troughs and crests is known as wavelength. It is denoted by λ and is expressed in cm, nanometers (*nm*) or Angstrom (Å) units.

2. **Frequency (ν):** The number of times a wave passes through a given point in one second is called a fequency of the wave. It is denoted by ν and is expressed in cycles per second or *Hertz*.The frequency of the wave is inversely proportional to its wavelength.

$$\nu = \frac{1}{\lambda}$$

3. **Velocity (*c*):** The distance travelled by the wave in one second is called its velocity. It is denoted by *c*. Therefore,

$$c = \nu\lambda$$

All types of electromagnetic radiations have velocity equal to the velocity of light which is equal to 3×10^{-10}cm/sec. or 3×10^{-8} m/sec.

4. **Wave number ($\bar{\nu}$):** It is defined as the number of wavelengths per cm, and is equal to the inverse of wavelength expressed in centimeters, $(1/\lambda)$

5. **Amplitude (α):** It is the height of the crest or depth of trough of a wave and is denoted by α. It determines the intensity or brightness of the beam of light.

1.7 Electromagnetic Spectrum

Electromagnetic radiations have wave character and do not require any medium for their propagation. Ordinary light rays, X-rays are all electromagnetic radiations. A radiation, may be mono or polychromatic.

A radiation of one wavelength is called **monochromatic radiation**, while that consisting of different wavelengths is known as **polychromatic radiation**.

When polychromatic radiation is passed through a prism it splits into seven different colours *i.e.* VIBGYOR (V = Violet, I = Indigo, B = Blue, G = Green, Y = Yellow, O = Orange and R = Red), all of different wavelengths. This process is called *dispersion* and the pattern of bands obtained is called *spectrum* or *visible region.*

The spectrum obtained by dispersion of ordinary light is continuous spectra and has seven colours merging into one another. There is no separation between different colours and hence spectrum is called a *continuous spectrum.*

The *emission spectrum* is obtained by passing radiations emitted by the excited atoms through the prism. It consists of bright lines against a dark background. It is characteristic of electronic environment of atoms and hence is known as *finger prints of atoms.*

Electromagnetic spectrum is the pattern of arrangement of different types of electromagnetic radiations in the order of increasing wavelength (or decreasing frequency).The arrangement of various radiations in the order of their wavelength or decreasing order of the frequencies is as follows:

Cosmic Rays < γ-Rays < X-Rays < Ultraviolet Rays < Visible Light < Infrared Radiations < Microwaves < Radiowaves.

The wavelengths of different radiations of the spectrum are : Cosmic rays = 0 to 10^{-2} Å, Gamma rays = (10^{-2} to 10^{-1} Å) X-Rays = (10^{-1} to 150Å) UV rays = (190 to 3800Å) Visible = (3800 to 7600 Å) Infrared (IR) = (7600 to 10^{6}Å) Microwaves = (10^{6} to 10^{9}Å) and Radiowaves= (10^{7} to 10^{14}Å).

1.8 Maxwell's Electromagnetic Wave Theory

Clark Maxwell in 1856, gave some postulates of Electromagnetic Theory which are as follows:

1. Light travels as waves of very short wavelength.
2. Energy is emitted or absorbed by a body continuously.

3. A Black Body is an object that absorbs all electromagnetic radiation that fall onto it. No radiation passes through it and none is reflected, and it emits radiation of one wavelength.
4. The energy of radiation is proportional to the intensity of radiation.

This theory could explain the mode of propagation of electromagnetic radiation and optical phenonmena like diffraction, interference, refraction etc. Therefore, came the Planck's Quantum Theory of Radiation.

1.9 Planck's Quantum Theory of Radiation

Maxwell could not explain the black body radiation and photoelectric effect. Therefore, **Quantum Theory** was proposed by Max Planck to explain black body radiations and photoelectric effect.

The main postulates of the Quantum Theory are:

1. The emission and absorption of energy does not take place continuously but in the form of small packets of energy called *quanta*. One quantum of light is known as photon which is defined as the smallest packet energy, which can exist independently and has no mass.
2. Energy of radiation is directly proportional to the frequency *i.e.*

$$E \propto \nu$$

Therefore, $E = h\,\nu$

Where, h is called Planck's constant, $h = 6.625 \times 10^{-27}$ ergs $= 6.625 \times 10^{-34}$ Js

3. A body emits or absorbs whole number multiple of quantum but never in fractions *i.e.* $E = nh\nu$, where n is 1, 2, 3 ... but it is never 0.8, 1.3 etc.

This condition is called Quantization of energy.

Example 1.1: Calculate and compare the energies of two radiations, one with wavelength of 400Å and the other with 800Å.

Solution: $E = h\nu = hc/\lambda$

$$E_{400} = 6.6 \times 10^{-34} \text{ Js} \times 3 \times 10^{8} / 400 \times 10^{-10} \text{ m}$$
$$= 4.95 \times 10^{-19}\text{J}$$
$$E_{600} = 6.6 \times 10^{-34} \text{ Js} \times 3 \times 10^{8} \text{ msec}^{-1} / 800 \times 10^{-10} \text{ m}$$
$$= 2.48 \times 10^{-19}\text{J}$$

Hence, $E_{400}/E_{800} = 4.95 \times 10^{-19} / 2.48 \times 10^{-19}$
$$= 2$$

Therefore, $E_{400} = 2E_{800}$

1.10 Photoelectric Effect

In 1887, **Heinrich Hertz** observed the photoelectric effect and the production and reception of Electromagnetic (EM) Waves. The effect is also termed the **Hertz Effect**, due to its discovery by **Heinrich Rudolf Hertz**, although the term has generally fallen out of use.

When a beam of light of sufficiently high frequency is allowed to strike a metal surface in vacuum, electrons are ejected from the metal surface. This phenomenon is known as *Photoelectric Effect* and the ejected electrons are called *Photoelectrons*.

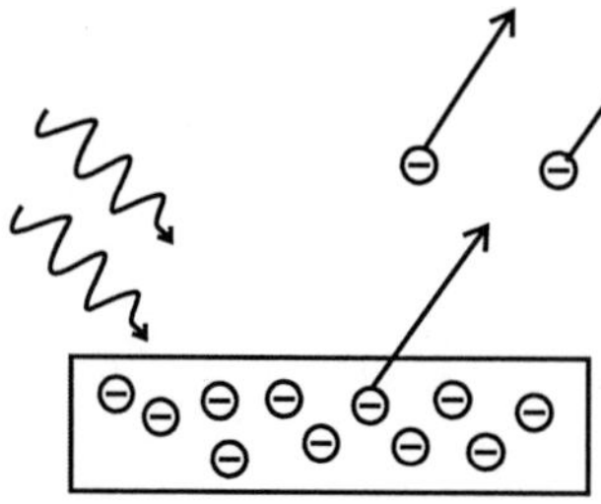

Fig. 1.3 Illustration of the photoelectric effect.

The energy of photoelectrons is proportional to the frequency of the radiations falling upon the metal surface and the number of photoelectrons emitted per second is proportional to the intensity of the incident radiation.

Important points about photoelectric effect

- There is certain minimum frequency, called threshold frequency, which can cause the emission of electrons.
- The kinetic energy of the photoelectrons increases linearly with the frequency of the incident radiation. If there is decrease in frequency below a certain critical value, there would not be any emission of electrons.
- An increase in intensity only increases the rate of emission of electrons.
- The time lag between the incidence of radiation and the emission of a photoelectron is very small even less than 10^{-9} seconds.

The photoelectric effect can be explained on the basis of **Quantum Theory**.

When the photon of light strikes the metal surface, it imparts its entire energy to the electron. This energy helps in overcoming the attraction between the electrons and the nucleus, if the frequency is less than the threshold frequency, there would be no emission of electrons from the surface. If, the frequency of light is more than the threshold frequency, some of the energy will be required to remove the electron from the atom and the extra energy gives electron the kinetic energy.

Therefore, Energy of photon = Energy needed to remove an electron + Kinetic energy of the emitted electron.

Algebraically :

$$h\nu = w + \tfrac{1}{2} mv^2$$

where, h is Planck's constant,

v is the frequency of the incident photon,

w ($h\nu_0$) is the work function, the minimum energy required to remove a delocalised electron from the surface of any given metal.

$\frac{1}{2} mv^2$ is the maximum kinetic energy of ejected electrons,

v_0 is the threshold frequency for the photoelectric effect to occur,

v is the velocity of the ejected electron.

1.11 Bohr's Atomic Model

To overcome the drawbacks of the Rutherford's model and to explain the origin of the production of spectrum of H and hydrogen like species (e.g., He^+, Li^{2+} etc), Niel Bohr in 1913 utilised the concept of quantisation and proposed a new model of an atom.

The various postulates of the model are as follows:

1. The electrons will not slowly loose energy as they travel, and hence will remain in stable, non-decaying orbits.
2. The absorption or emission of energy can occur only by transition of electron from lower to higher energy level or vice-versa respectively.
3. The electrons can move in only those circular orbits where the angular momentum ($mv\ r$) is a whole number multiple of $h/2\pi$ or it is quantised *i.e.* the magnitude of angular momentum is always a whole number and not in fractions.This is called the *Principle of Quantisation of Angular Momentum.*

   Therefore, $mvr = nh/2\pi$

Where,

m = mass of the electron,

v = velocity of the electron,

r = radius of the orbit in which the electron is moving,

n = Principal quantum number and

h = Planck's constant.

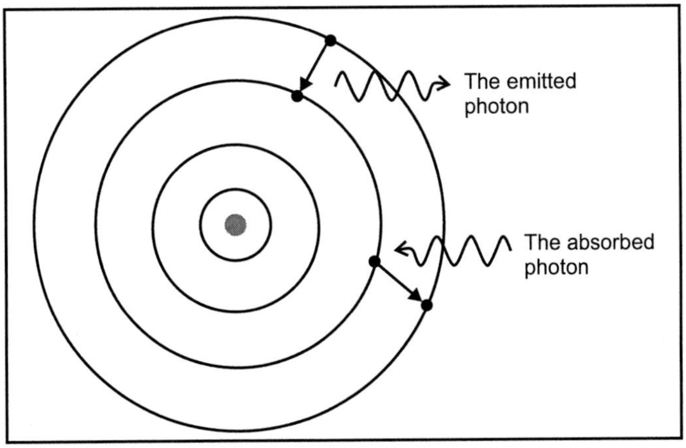

Fig. 1.4 The electron emits or absorbs the energy changing the orbits.

1.11.1 Bohr Theory as Applied to Hydrogen Atom and Hydrogen like Species

Bohr, worked out number of expressions for one electron species like H, He^+, Li^{2+} etc. He derived the expressions for the radius of the *n*th orbit, velocity and energy of an electron revolving in the *n*th orbit, frequency, wavelength (λ) of the spectra line emitted when an electron falls from the n_2 orbit to n_1 orbit.

A. Calculation of the radius of the *n*th orbit of hydrogen atom and hydrogen like species.

Let m be the mass of an electron, v is the velocity of an electron, r be the radius of the orbit, E is the energy of the electron, Z is the atomic number, e is the charge on the electron, h is the Planck's constant and k is the permitivity constant.

Charge on the nucleus = + Ze

Charge on electron = $-e$

The revolving electron is simultaneously effected upon by two forces (a) electrostatic force of attraction *i.e.*, the centripetal force (b) The force of repulsion *i.e.*, the centrifugal force.

Electrostatic force of attraction is the force which exists between the nucleus and the revolving electron and is responsible for the attraction of electron towards the nucleus.

Force of repulsion or the centrifugal force is the force which acts away from the nucleus and tries to keep the revolving electron away from its orbit.

Electrostatic force of attraction between the nucleus and electron is given by:

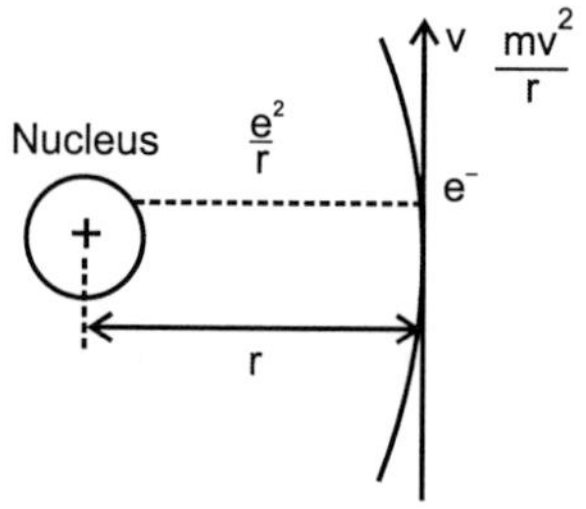

Fig. 1.5 Electron - Nucleus attraction.

$$F_e = \frac{kZe^2}{r^2} \quad ...(1)$$

k is a constant and is given by

$$k = \frac{1}{4\pi \in_0} \text{ where } \in_0 \text{ is the permittivity of free space}$$

Therefore, $$k = \frac{1}{4\pi \in_0} = 9 \times 10^9 \text{ Nm}^2/\text{C}^2$$

The centrifugal force is given by

$$F_c = \frac{mv^2}{r} \quad ...(2)$$

For a stable electron, the electrostatic force balances the centrifugal force.

Therefore, $F_c = F_e$

$$\frac{mv^2}{r} = \frac{kZe^2}{r^2} \quad ...(3)$$

$$v^2 = \frac{kZe^2}{mr} \quad ...(4)$$

According to Bohr's atomic model,

$$mvr = n\left(\frac{h}{2\pi}\right) \quad ...(5)$$

Solving for v, we get

$$v = \frac{nh}{2\pi mr} \quad ...(6)$$

$$v^2 = \frac{n^2h^2}{4\pi^2m^2r^2}$$

Substituting the value of v^2 from equation (4) we get,

$$\frac{kZe^2}{mr} = \frac{n^2h^2}{4\pi^2m^2r^2} \quad ...(7)$$

Solving for r, we get

$$r = \frac{n^2h^2}{4\pi^2mkZe^2} \quad ...(8)$$

For hydrogen atom $Z = 1$,

the above relation becomes $r = \frac{n^2h^2}{4\pi^2me^2k}$

Also, for the radius of the first orbit where $n = 1$

$$r = \frac{h^2}{4\pi^2me^2k}$$

Substituting the value we get,

$$r = \frac{(6.625\times10^{-34})^2}{4 \times (3.14)^2 \times 9 \times 10^{-31} \times (1.6 \times 10^{-19})^2 \times 9 \times 10^9}$$

$$= 5.29 \times 10^{-11} \text{ m}$$

$$= 0.529 \text{ Å}$$

Therefore, the radius of the first orbit of hydrogen atom is 0.529Å.

The radius of the first orbit of hydrogen like species is represented by r_0 and is referred to as Bohr's radius.

Since, radius of an orbit is directly proportional to square of n, the radius of the nth orbit can be calculated as:

$$R_n = 0.529\text{Å} \times n^2$$

Example 1.2: (*a*) Calculate the radius of Bohr's first orbit for hydrogen atom and velocity of electron in this orbit.

(*b*) Calculate the Bohr's radius for the 5th orbit of the hydrogen atom.

Solution (*a*)

$$r_n = \frac{n^2h^2}{4\pi^2me^2k}$$

Where,

$$h = 6.626 \times 10^{-34} \text{ J sec}$$

$$m = 9.1 \times 10^{-31} \text{ kg}$$

$$e = 1.6 \times 10^{-19} \text{ C}$$

Substituting these values we get,

$$m = 9.1 \times 10^{-31} \text{ kg}$$

$$r_1 = \frac{(1)^2 \times (6.626 \times 10^{-34})^2}{4 \times (3.14)^2 \times 9.1 \times 10^{-31} \times (1.6 \times 10^{-19})^2 \times 9 \times 10^9}$$

$$= 0.529 \times 10^{-10} \text{ m}$$

$$= 0.529\text{Å}$$

$$\text{Velocity } (v) = \frac{nh}{2\pi mr} = \frac{1 \times 6.626 \times 10^{-34}}{2 \times 3.14 \times 9.1 \times 10^{-31} \times 0.529 \times 10^{-10}}$$

$$= 2.188 \times 10^6 \text{ m/sec.}$$

(*b*)

$$r_n = r_0 \times n^2$$

$$r_5 = (0.529 \times 10^{-10}\text{m}) \times (5)^2$$

$$= 13.225 \times 10^{-10}\text{m}$$

$$= 13.225\text{Å}$$

Example 1.3: Calculate the distance of separation between second and first orbit of hydrogen atom.

Solution: The radius of the *n*th orbit is given by

$$r_n = r_0 \times n^2$$

$$= 0.529 \times n^2$$

$$r_2 - r_1 = (0.529 \times 2^2) - (0.529 \times 1^2)$$

$$= 0.529 \times (4 - 1)$$

$$= 0.529 \times 3$$

$$= 1.59 \text{ Å}$$

B. Calculation of Velocity of electron in an orbit

Velocity with which electron is revolving in an orbit can be calculated by the following relation:

where velocity $v = \dfrac{nh}{2\pi mr}$

Substituting the value of r, $v = \dfrac{nh}{2\pi\ m} \times \dfrac{4\pi^2 me^2 k}{n^2 h^2}$

$$v = \frac{2\pi e^2 k}{nh}$$

Substituting the values of π, e, h we get

$$v = \frac{2.19 \times 10^8}{n} \text{ cm sec}^{-1}$$

Also, we can calculate the number of revolutions per second made by an electron round the nucleus *i.e.*

$$= \frac{\text{velocity of electron}}{\text{circumference of the orbit}}$$

C. Calculation of energy of an electron

The total energy of an electron revolving in a particular orbit is calculated by adding its potential and kinetic energies

Potential energy of an electron, $V = \dfrac{-kZe^2}{r}$

Kinetic energy $= \dfrac{1}{2}mv^2$

Therefore, the total energy $= \dfrac{mv^2}{2} - \dfrac{kZe^2}{r}$ ...(9)

From eq. (4), we know that

$$mv^2 = \frac{kZe^2}{r} \quad \text{...(10)}$$

Substituting the value of mv^2 in eqn (9).

$$E = \frac{kZe^2}{2r} - \frac{kZe^2}{r} = \frac{-kZe^2}{2r} \quad \text{...(11)}$$

Substituting the value of r from eqn (8) in eqn (11) we get,

$$E = \frac{-kZe^2}{2} \times \frac{4\pi^2 Ze^2 mk}{n^2 h^2}$$

$$= \frac{-2\pi^2 Z^2 e^4 mk^2}{n^2 h^2}$$

Therefore, the total energy of the electron in an orbit of hydrogen atom is given by

$$E_n = \frac{-2\pi^2 e^4 Z^2 mk^2}{n^2 h^2}$$

The magnitude of total energy obtained is in C.G.S. units.
Substituting the values of constant

$$\pi = 3.1416$$
$$k = 9 \times 10^2 \text{ Nm}^2/\text{C}^2$$
$$e = 1.6 \times 10^{-19} \text{ C}$$
$$m = 9.1 \times 10^{-31} \text{ kg}$$
$$h = 6.625 \times 10^{-34} \text{ Joules sec.}$$

When, $Z = 1$,

i.e., the value of E_n of an electron revolving in the nth orbit of one electron system.

$$E = \frac{-21.79 \times 10^{-19}}{n^2} \text{ Joules atom}^{-1}$$

$$= \frac{-21.79 \times 10^{-19}}{n^2} \times 6.2419 \times 10^{18} eV$$

($\because$ 1 Joule = 6.2419×10^{18} eV)

$$= \frac{-13.6}{n^2} eV = \frac{-13.6}{n^2} \times 23.053 \text{ kcal}$$ ($\because$ $1eV = 23.053$ kcal)

$$= \frac{-313.52}{n^2} \text{ kcal/mole}$$

Therefore, $$E_n = \frac{-313.52}{n^2} \text{ kcal/mole}$$

$$E_n = \frac{-2.17 \times 10^{-11} \times Z^2}{n^2} \text{ erg/atom}$$

Example 1.4: Calculate the wavelength in Angstroms of the photon that is emitted when an electron in Bohr's orbit $n = 2$ returns to the orbit $n = 1$ in the hydrogen atom. The ionization potential of the ground state of hydrogen atom is 2.17×10^{-11} ergs per atom.

Solution: The energy of the electron in orbit $n = 1$ is

$$E_1 = -2.17 \times 10^{-11} \text{ ergs}$$

The energy of the electron in orbit $n = 2$

$$E_2 = \frac{-2.17\times10^{-11}}{2^2} \text{ ergs}$$

$$E_2 - E_1 = -2.17\times10^{-11}\left(\frac{1}{4}-1\right) \text{ ergs}$$

$$= \frac{3\times2.17\times10^{-11}}{4} \text{ ergs}$$

If the frequency of the emitted photon is v, then

$$E_2 - E_1 = hv = \frac{hc}{\lambda}$$

$$\lambda = \frac{hc}{E_2 - E_1}$$

$$\lambda = \frac{4\times6.62\times10^{-27}\times3.0\times10^{10}}{3\times2.17\times10^{-11}}$$

∴ Therefore $\lambda = \frac{79.44}{6.51}\times10^{-6}$ cm

$$= 1.22 \times 10^{-5} \text{ cm}$$

$$= 1220 \text{ Å}$$

Example 1.5: To which orbit the electron in the hydrogen atom will jump on absorbing 12.16 eV of energy?

Solution: Energy of an electron in the nth orbit of hydrogen atom is

$$E_n = \frac{-13.6}{n^2} eV$$

$$E_1 = -13.6 \; eV$$

After absorbing 12.16 eV, the energy will be

$$= -13.61 + 12.16 = 1.5eV$$

$$n^2 = 9 \text{ or } n = 3$$

1.11.2 Achievements of Bohr's Atomic Model

- Bohr's concept of stationary orbits of electrons explains the emission and absorption spectra of hydrogen like species.

- When an element is heated, its electrons absorb energy and jump to higher energy levels. Electrons are unstable in the higher energy levels and jumps back to the lower energy levels, thereby releasing energy. The energy, thus released, gives the emission spectrum. Similarly, an electromagnetic radiation when passed through the solution of a substance, some of its wavelengths are absorbed by the substance, and causes the excitation of its electrons. The radiation absorbed by the substance, appear as dark lines in the spectrum and corresponds to the *absorption spectrum.*
- According to him, the electron does not lose energy, as long as it revolves in a particular orbit. Thus, we say that atom is stable.
- The experimental values of radii and energy of the permissible orbits in hydrogen atom are in good agreement with that the calculated on the basis of the Bohr's theory.
- The experimental values of the frequencies of the spectral lines of the hydrogen spectrum are in close agreement with the frequencies of the same lines calculated with the help of the Bohr's theory.

1.11.3 Emission Spectrum of Hydrogen

When an electric discharge is passed through hydrogen gas filled in a discharge tube at a very low pressure, bluish glow is obtained. On analyzing this glow in the prism, a series of spectra lines were obtained in the visible and ultra–violet region called the emission spectrum of Hydrogen. These lines are grouped into five series named after the name of their discoveres as given below:

1. **Lyman Series:** The first line in the ultraviolet spectrum of the Lyman series was discovered in 1906 by Harvard physicist, Theodore Lyman. The **Lyman series** is the series of transitions and resulting emission lines of the hydrogen atom as an electron goes from $n \geq 2$ to $n = 1$ (where n is the principal quantum number referring to the energy level of the electron).
2. **Balmer Series:** The Balmer series is characterized by the electron transitioning from $n \geq 3$ to $n = 2$ in the visible region, where n refers to the principal quantum number of the electron.
3. **Paschen Series:** The **Paschen series** (also called **Ritz-Paschen series**) is the series of transitions and resulting emission lines of the hydrogen atom appears as an electron goes from $n \geq 4$ to $n = 3$ in the near infrared region, where n refers to the principal quantum number of the electron.
4. **Brackett Series:** The lines appear in emission when the electron of an hydrogen atom descends to the 4th energy level from a higher level. These lines lie in the infrared part of the electromagnetic spectrum.
5. **Pfund Series:** The lines were experimentally discovered in 1924 by *August Herman Pfund*, and correspond to the electron jumping between the 5th and higher energy levels of the hydrogen atom.

Calculation of the frequency and wavelength of a spectra line emitted when an electron of H atom falls from n_2 orbit to n_1 orbit.

When an electron moves from an energy level to another, a photon is emitted. Using the derived formula for the different energy levels of hydrogen, one may determine the wavelengths of light that a hydrogen atom can emit.

The energy of a photon emitted by a hydrogen atom is given by the difference of two hydrogen energy levels.

$$E_{n_2} - E_{n_1} = h\nu = \frac{hc}{\lambda} \qquad (\because \nu = \frac{c}{\lambda})$$

$$\lambda = \frac{hc}{E_{n_2} - E_{n_1}}$$

Where, E_{n_2} and E_{n_1} are the energies of higher and lower energy levels

We know, $$E_n = \frac{-2\pi^2 mZ^2e^4k^2}{h^2n^2}$$

Therefore, $$\Delta E = \frac{-2\pi^2 mZ^2e^4k^2}{h^2n_2^2} - \left(\frac{-2\pi^2 mZ^2e^4k^2}{h^2n_1^2}\right)$$

$$\Delta E = h\nu$$

$$\nu = \frac{2\pi^2 mZ^2e^4k^2}{h^3}\left(\frac{1}{n_1^2} - \frac{1}{n_2^2}\right)$$

Substituting the values of m, e, h, k, Z, π ($Z = 1$, in case of hydrogen) we get,

$$\nu = 3.289 \times 10^{15}\left(\frac{1}{n_1^2} - \frac{1}{n_2^2}\right) \text{ cycles/sec or Hz}$$

We also know that, the wave number

$$\bar{\nu} = \frac{\nu}{c} = \frac{2\pi^2 mZ^2e^4k^2}{h^3c}\left\{\frac{1}{n_1^2} - \frac{1}{n_2^2}\right\}$$

$$\nu = {}^*R_H\ \frac{1}{n_1^2} - \frac{1}{n_2^2}$$

Where R_H is the Rydberg constant and its calculated value for hydrogen comes out to be 1.097×10^7 /m

(Experimental value is 1.09677×10^7/m)

The above is known as the Rydberg formula. This formula was known in the 19th century to scientists studying spectroscopy, but there was no theoretical justification for the value itself, until Bohr derived them, more or less along the lines above. In fact, Bohr's derivation of the empiric Rydberg constant to high accuracy, in terms of other known physical constants, was an early reason for the acceptance of Bohr's model as an advance in understanding.

---

*See page 52 for description of R, R_H, R_∞ and R_C.

When an electron jumps from nth level to the ground level, the number of spectral lines in the spectrum will be given by:

$$\frac{n(n-1)}{2}$$

In general, the intensities of these spectral lines decreases with increase in the value of n *i.e.* $2 \rightarrow 1 > 3 \rightarrow 1 > 4 \rightarrow 1 > 5 \rightarrow 1$.

Example 1.6: Calculate the wavelength of radiation emitted, producing a line in Lyman series, when an electron falls from the third stationary state in hydrogen atom.

Solution:

$$\tilde{\nu} = R_H\left[\frac{1}{n_1^2}-\frac{1}{n_2^2}\right]$$

$$\frac{1}{\lambda} = R_H\left[\frac{1}{n_1^2}-\frac{1}{n_2^2}\right] \qquad (\therefore R_H = 1.1 \times 10^7)$$

$$\frac{1}{\lambda} = 1.1 \times 10^7\left[\frac{1}{1^2}-\frac{1}{3^2}\right]$$

$$\frac{1}{\lambda} = 1.1 \times 10^7\left[\frac{1}{1}-\frac{1}{9}\right]$$

$$= 1.1 \times 10^7 \times 8/9$$

$$\lambda = \frac{9}{8}\times\frac{1}{1.1\times10^{-7}}$$

$$\lambda = 1.02 \times 10^{-7}\ m$$

1.11.4 Limitations of the Bohr's Atomic Model

- It explains the spectra of species having only one electron.
- No justification was given for the principle of quantisation of angular momentum *i.e.*, for the assumption that the electron can rotate only in those orbits in which the angular momentum of the electron, *i.e.*, mvr, is a whole number multiple of $\left(\frac{h}{2\pi}\right)$.
- It does not explain the splitting of spectral lines under the influence of magnetic field (Zeeman Effect) or electric field (Stark's Effect).
- Bohr's theory predicts the origin of only one spectra line from an electron between any two given energy states. But, under the spectroscope of strong resolution, a single line was found to split in a number of very closely related lines. Therefore, Bohr, could not

explain this multiple or fine structure of spectral lines. These multiple lines indicate the energy levels of similar energy for each Principal quantum number, *n*. Due to this, the concept of quantum numbers came into existence.

- Bohr assumed that an electron in an atom is located at a definite distance from the nucleus and is revolving with a definite velocity around it. This is against the Heisenberg's uncertainty principle, which has been explained in the later part of the chapter.

1.12 The Dual Nature of Electrons–Particles and Wave

Wave particle duality is a conceptualisation that all objects in our universe exhibit properties of both wave (such as non-locality) and of particle (such as quantisation of some of their properties).

To overcome the inadequacies of Bohr's model, the modern structure of atom was developed in the light of observations by de-Broglie and Heisenberg Uncertainty Principle.

In 1924, Louis-Victor de-Broglie formulated the de-Broglie hypothesis, claiming that *all* matter, not just light, has a character-wave and particle;

Differentiation between a particle and a wave

Particle	Wave
1. A particle is localised in space as it occupies a well-defined position in space. 2. When a number of particles are present in a region of space, the total value is equal to the sum of the individual particles.	1. A wave is delocalised in space as it spreads out in space. 2. If the number of waves are present in a given region of space, due to interference, the resultant wave can be small or large depending upon the fact, whether the wave is destructive or constructive.

He related wavelength (denoted as λ), and momentum (denoted as *p*):

$$\lambda = \frac{h}{p} = \frac{h}{mv}$$

where, *h* is the Planck's constant or 6.626×10^{-34} kg m^2 s^{-1}, *m* is the mass of the object (kg), *v* is the velocity and *p* is the momentum.

1.12.1 Derivation of de-Broglie's Equation

Energy of a photon is given by: $E = h\nu = hc/\lambda$ (Planck's equation) ...(*i*)

According to Einstein equation, $E = mc^2$ ...(*ii*)

Where ν is the frequency, c is the velocity of light, λ is the wavelength and m is the mass of the photon.

Combining (*i*) and (*ii*) we get,

$$hc/\lambda = mc^2$$

therefore, $h/\lambda = mc$

or $\lambda = h/mc$

or $\lambda = h/p$, where p is the momentum of the particle.

We can also calculate the de-Broglie's wavelength of the electron from the potential applied to accelerate it.

If a potential V is applied to accelerate the electron e, the kinetic energy acquired by the electron is given by eV.

Thus, eV = Charge on the electron in coulombs × Potential applied in volts.

This energy is the kinetic energy of an electron, $\frac{1}{2}mv^2$

Therefore, $\frac{1}{2}mv^2 = eV$

$$v^2 = \frac{2eV}{m}$$

We know that, $\lambda = h/mv$

$$\lambda = \frac{h}{m} \times \sqrt{\frac{m}{2eV}}$$

$$\lambda = \frac{h}{\sqrt{2meV}}$$

Substituting the values of various constants, we get,

$$\lambda = \frac{1.2261 \times 10^{-9}}{\sqrt{V}}$$

1.12.2 Experimental Verification of Wave Nature of Electrons by Davisson and Germer

In physics, the **Davisson-Germer experiment**, provided a critically important confirmation of the de-Broglie hypothesis that particles, such as electrons, could behave as waves. Davisson and Germer designed and built a vacuum apparatus for the purpose of measuring the energy of electrons scattered from a metal surface. Electrons from a heated filament were accelerated by a voltage, and allowed to strike the surface of nickel metal.

The electrons were scattered by the crystal and were obtained on the photographic plate on which electrons produced the diffraction rings.These rings were found to be same as those produced by the X-rays when diffracted by the nickel crystal suggesting that electrons like X-rays have wave character.

Significance of the de-Broglie equation

- The wave character is significant only in case of small particles like atoms, electrons, protons etc., having small masses.The wave character is of no significance in case of large macroscopic objects like cricket ball, a car, a train etc., having larger masses. The associated wavelength of the waves is too small to be observed, as $\lambda \propto \frac{1}{m}$ (if m is large, λ has to be small)
- Construction of electron microscope and study of surface structure of solids by electron diffraction are the important applications of de-Broglie concept.

Example 1.7: Calculate the wavelength of 1000 kg rocket moving with a velocity of 300 km per hour.

Solution: Mass of the rocket = 1000 kg

$$\text{Velocity} = 300 \text{ km per hour} = \frac{300 \times 1000}{60 \times 60}$$

$$= 83.33\ ms^{-1}$$

From de-Broglie equation

$$\lambda = h/mv$$

$$\lambda = 6.625 \times 10^{-34} \text{ J}/1000 \text{ kg} \times 83.33\ m \text{ sec}^{-1}$$

$$\lambda = 7.95 \times 10^{-39}\ m$$

1.13 Heisenberg Uncertainty Principle

In 1926, Wernel Heisenberg, suggested that simultaneous measurement of the position and momentum of microscopic or sub-atomic particle like electron with complete accuracy is not possible.

In other words, the product of uncertainty in momentum (Δp) and that in position (Δx) almost remains constant.

Mathematically, this principle can be expressed as

$$\Delta x.\Delta p \geq \frac{h}{4\pi}$$

We can also write as, $\Delta x.m\Delta v \geq h/4\pi$

Uncertainty in measurement is not due to lack of any experimental technique but it is due to the nature of subatomic particle itself.

1.13.1 Illustration of the Uncertainty Principle

Let us consider, an electron *e*, moving in a circular path. When colloided with a photon of wavelength say λ, the electron gains energy from this photon and hence changes its path. Due to the changed path, the velocity of electron is also changed and the determination of the exact velocity of the electron becomes impossible. Although, the measurement of the position of the electron can still be determined with the help of a microscope.

Whereas, when a photon of higher wavelength collides with the electron, the electron would not change the path due to the low energy of the electron and hence the velocity is not changed, but the exact position cannot be determined with accuracy.

Therefore, we say that the exact determination of the position and velocity cannot be done simultaneously.

The value of Δp calculated for a given value of Δx is the minimum value, similarly, the value of Δx for a given value of Δp is the minimum value.

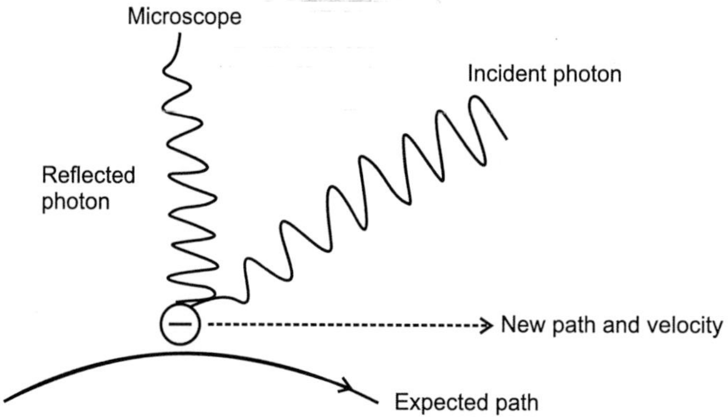

Fig.1.6 Change in the path and velocity of a moving electron when collided with the photon.

Example 1.8: A dust particle having mass equal to 10^{-11}g and velocity 10^{-4} cm sec^{-1}. The error in the measurement of velocity is 0.1%. Calculate the uncertainty in its position.

Solution:

$$v = 10^{-4} \text{ cm sec}^{-1}$$

$$\Delta v = 0.1 \times 10^{-4}/100$$

$$= 1 \times 10^{-7} \text{ cm sec}^{-1}$$

$$\Delta v \times \Delta x = h/4\pi\text{m} \qquad (\therefore m.\Delta v \times \Delta x = \frac{h}{4\pi})$$

$$\Delta x = 6.625 \times 10^{-34}/ 4 \times 3.14 \times 10^{-11} \times 10^{-7}$$

$$= 0.527 \times 10^{9} \text{ cm}$$

Wave Mechanical Concept of Atomic Structure

Introduction

In physics, especially Quantum Mechanics, the Schrödinger equation is an equation that describes how the Quantum state of a physical system changes in time. It is as central to Quantum Mechanics as Newton's Laws are to Classical Mechanics. The Schrödinger equation tells us the behaviour of ψ, which was successfully interpreted by Max Born in 1926, as a probability amplitude. Solutions to schiödinger's equation describe atomic and subatomic systems, electrons and atoms, also the Macrosopic systems, possibly even the whole universe.

Erwin Schrödinger

1.14 Postulates of the Wave Mechanical Model

A completely different model of the atom was proposed, named as **Wave Mechanical Model of Atom** which was based on the following postulates:

- Electron wave is compared to a stationary or standing wave which is formed by a vibrating string fixed between two points.
- Energy values of an electron were quantised by Bohr.
- The wavelength of the electron which is associated with a wave is given by the De-Broglie relation as discussed in the previous section.

$$\lambda = h/mv$$

- The term probability distribution or mathematical wave function is used to describe the motion of an electron.

1.15 Schrödinger Wave Equation

The **Schrödinger equation** was proposed by the Austrian Physicist Erwin Schrödinger in 1926. He said that since the electron behaves as a material particle and a wave, there must be a wave equation which should describe the wave motion of the electron.Therefore, the equation was **Schrödinger Wave Equation.**

This equation describes the wave motion of an electron-wave propagating in three dimensions (x, y and z-axis) in space. This equation can be written in many forms, one of which is given below:

$$\frac{d^2\psi}{dx^2} + \frac{d^2\psi}{dy^2} + \frac{d^2\psi}{dz^2} + \frac{8\pi^2 m}{h^2}(E - V) = 0$$

Here x, y, z are three space coordinates, m = mass of electron, h = Planck's constant, E = total energy, V = potential energy and ψ = wave function of electron wave.

ψ represents the amplitude of the electron wave at various points surrounding the nucleus or we can say, that it is the measure of the particle being at a given point in space at a given time.

The above expression can also be expressed as:–

$$\Delta^2\psi + \frac{8\pi^2 m}{h^2}(E - V)\psi = 0$$

Here, Δ^2 is the Laplacian vector.

The permitted solutions of Schrödinger equation are known as wave function, which denotes the definite energy state which is a orbital. In simple words, this equation may be interpreted by a body of mass m, velocity v, potential energy E, has wave like characteristics associated with it, with an amplitude given by wave function Ψ.

The total energy of the electron is the sum of the K.E. and P.E., so the total energy operator. $\hat{H}$ is called the Hamiltonian operator, which is the sum of the K. E (E) and P.E (V).

i.e., $$\hat{H} = \hat{E} + V$$

In Quantum mechanics, to every corresponding property, there is an operator, which is represented by symbol ^ over that property.

Therefore, Schrödinger wave equation can also be represented in the form

$$\hat{H}\Psi = E\Psi$$

or

$$(\hat{E} + \hat{V})\Psi = E\Psi$$

This equation can be solved to get the values of E and Ψ for the system.

While deriving the equation, Shrödinger assumed that the nucleus is surrounded by a vibrating electron wave which are produced by the vibrations of a stretched string.

He considered if x is the displacement of the electron wave of amplitude ψ and wavelength λ, then the value of ψ in one dimension is given by the sine wave equation,

$$\psi = A \sin\frac{2\pi x}{\lambda} \qquad ...(1)$$

On differentiating (1) w. r. t. x, we get,

$$\frac{d\psi}{dx} = \frac{A2\pi}{\lambda}.\frac{(\cos 2\pi x)}{\lambda} \qquad ...(2)$$

On differentiating (2) w.r.t. *x*, we get,

$$\frac{d^2\psi}{dx^2} = \frac{A2\pi}{\lambda}.\frac{2\pi}{\lambda}\frac{(-\sin 2\pi x)}{\lambda}$$

$$= \frac{-4\pi^2}{\lambda^2} A \sin\frac{2x\pi}{\lambda} \qquad ...(3)$$

Substituting (1) in (3), we get

$$\frac{d^2\psi}{dx^2} = \frac{-4\pi^2\psi}{\lambda^2}(\psi = A \sin\frac{2\pi x}{\lambda}) \qquad ...(4)$$

This is the wave equation for one dimension wave vibrating along *x*-axis only. Therefore, eqn. (4) in 3-dimensions is expressed as:

$$\frac{d^2\psi}{dx^2}+\frac{d^2\psi}{dy^2}+\frac{d^2\psi}{dz^2} = \frac{-4\pi^2}{\lambda^2}\psi \qquad ...(5)$$

i.e.,

$$\frac{d^2\psi}{dx^2}+\frac{d^2\psi}{dy^2}+\frac{d^2\psi}{dz^2}+\frac{4\pi^2\psi}{\lambda^2} = 0 \qquad ...(6)$$

Since, Ψ is a function of Cartesian co-ordinates *x*, *y*, *z*

Equation (6) may be written as

$$\Delta^2\psi+\frac{4\pi^2}{\lambda^2}\psi = 0, \text{ where } \Delta^2 = \frac{d^2\psi}{dx^2}+\frac{d^2\psi}{dy^2}+\frac{d^2\psi}{dz^2} \qquad ...(7)$$

Incorporating, de-Broglie's relationship in equation (7) we get,

$$\Delta^2\psi+\frac{4\pi^2m^2v^2}{h^2}\psi = 0 \qquad \text{where, } \lambda = \frac{h}{mv} \qquad ...(8)$$

Since, total energy is the sum of its Kinetic Energy (½ mv^2) and potential energy (*V*)

Thus, E = ½ $mv^2 + V$

or

½ $mv^2 = (E - V)$

or

$$v^2 = \frac{2(E-V)}{m} \qquad ...(9)$$

Substituting (9) in (8) we get,

$$\Delta^2\psi + \frac{4\pi^2 m^2\psi}{h^2} \times \frac{2(E-V)}{m} = 0$$

i.e.,
$$\Delta^2\psi + \frac{8\pi^2 m}{h^2}(E-V)\psi = 0$$

Therefore, the equation becomes,

$$\frac{d^2\psi}{dx^2} + \frac{d^2\psi}{dy^2} + \frac{d^2\psi}{dz^2} + \frac{8\pi^2 m}{h^2}(E-V)\psi = 0$$

or

$$\left(\frac{d^2}{dx^2} + \frac{d^2}{dy^2} + \frac{d^2}{dz^2}\right)\psi + \frac{8\pi^2 m}{h^2}(E-V)\psi = 0$$

or

$$\Delta^2\psi + \frac{8\pi^2 m}{h^2}(E-V)\psi = 0 \qquad ...(10)$$

This is Schrödinger Wave Equation.

1.16 Physical Significance of Ψ and Ψ^2

Ψ is a mathematical function which is called wave function; it represents the amplitude of the electron wave at various points surrounding the nucleus. Since it is an amplitude function of certain space coordinates *x*, *y*, *z*, it should be written as Ψ (*x*, *y*, *z*). In an atom with a single electron, such as hydrogen, the wave function of the electron provides a complete description of how the electron behaves. For this single electron, it can occupy any point between a certain distance from the center of the atom to a large distance.

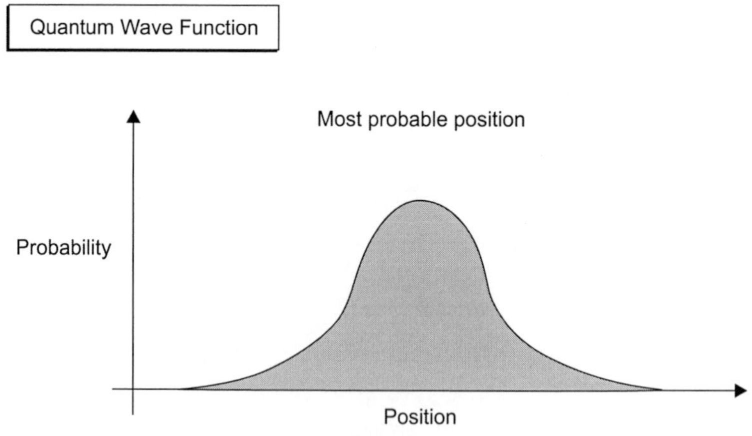

Fig. 1.7 Illustration of the wave function.

In general terms, Ψ is the measure of the probability of an event to take place or we can say, that it is related to the likelihood of the particle being at a given point in space at a given time.

Since wave function is a complex number, containing imaginary numbers, therefore, it cannot represent such a probability since this would mean that Ψ should always be real valued function, but this is not true.

Therefore, the square of the wavefunction is the property which has physical significance.

Ψ^2 represents the probability of finding the electron at the given point.

In order to avoid imaginary values, $\Psi\ \Psi^*$ is used for Ψ^2. Ψ^* is the complex conjugate of Ψ. If Ψ extends over a finite volume dV, then $\Psi\ \Psi^*$ dV gives the probability of finding the electron in the volume dV.

The values of Ψ are obtained by solving Schrödinger wave equation, only those values are accepted which obey the following conditions:

- Ψ must be single valued, *i.e.* it must have only one definite value at a particular point in space.
- Ψ must be continuous at all points in space.
- Ψ must have a finite value over the space dV that the electron can occupy.
- Ψ must be normalised, *i.e.* the probability of finding the electron in the space under the consideration must be unity.
- Ψ must be zero at infinity.

Therefore, these acceptable values of the wave function Ψ are called **eigen (acceptable) wave functions**. The value of *E* given by a particular eigen (acceptable) wave function is called **eigen value**. Thus, every value of wave function is not an eigen (acceptable) wave function.

Therefore, we can summarise important points as :

- The wave function of an electron (ψ) of nucleus of an atom is called atomic orbital, which is a three dimensional amplitude of an electron wave.
- The wave function of an electron (ψ) of nucleus of an atom is called atomic orbital, which is a three dimensional amplitude of an electron wave.
- ψ^2 dV is the probability of finding the electron in a small volume element dV, surrounding the nucleus of an atom.
- The particular value of ψ is called eigen function and the values of energies which correspond to it are called eigen values.
- On solving the equations, there are regions in space which is positive or negative. Where ψ^2, is the probability density which is always positive.

1.16.1 Normalised and Orthogonal Wave Functions

The wave functions ψ and ψ^* are said to be **orthogonal functions**, if the product of the two sets between the appropriate integration boundaries is zero, *i.e.*,

$$\int \psi\psi^*\, d\,\psi = 0$$

Whereas these functions would be called **normalised functions** if the product of the two is equal to one, *i.e.*,

$$\int \psi\psi^* \, d\,\psi = 1$$

Wave functions which are both normalized and orthogonal are called **orthonormal functions**.

1.17 Schrödinger Wave Equation for Hydrogen Atom in Cartesian and Polar Co-ordinates

Schrödinger wave equation for a single particle of mass m is given by :

$$\Delta^2\psi + \frac{8\pi^2 m}{h^2}(E - V)\psi = 0$$

$$\frac{1}{m}\Delta^2\psi + \frac{8\pi^2}{h^2}(E - V)\psi = 0$$

There are two particles in the hydrogen atom, *i.e.*, the electron and the nucleus, its motion is described by the following equation:

$$\frac{1}{m_1}\Delta^2\psi + \frac{1}{m_2}\Delta^2\psi + \frac{8\pi^2}{h^2}(E - V)\psi = 0$$

where m_1 is the mass of an electron and m_2 is the mass of the nucleus.

1.17.1 Schrödinger Wave Equation for Hydrogen Atom in the Cartesian Co-ordinates

In H-atom, electron has charge equal to $-e$, which moves around the nucleus with charge equal to $+e$. Therefore, Potential energy term becomes $-e^2/r$.
Thus, Schrödinger Wave Equation becomes,

$$\frac{d}{dx^2} + \frac{d}{dy^2} + \frac{d}{dz^2}\psi + \frac{8\pi^2 m}{h^2}\left(E + \frac{e^2}{r}\right)\psi = 0 \qquad ...(1)$$

1.17.2 Schrödinger Wave Equation for Hydrogen Atom in Polar Spherical Co-ordinates

Cartesian co-ordinates are converted into polar co-ordinates, with the help of following relations:

$$x = r\cos\theta$$

$$y = r\sin\theta\sin\phi$$

$$z = r\sin\theta\cos\phi$$

Eqn. (1) takes the shape of the following equation for H-atom in polar co-ordinates

$$\left[\frac{\partial}{\partial r}\left(r^2\frac{\partial}{\partial r}\right) + \frac{1}{\sin\theta}\frac{\partial}{\partial\theta}\left(\sin\theta\frac{\partial}{\partial\theta}\right) + \frac{1}{\sin^2\theta}\frac{\partial^2}{\partial\phi^2}\right]\psi + \frac{8\pi^2 mr^2}{h^2}\left[E + \frac{e^2}{r}\right]\psi = 0 \qquad ...(2)$$

The following figure correlates the cartesian co-ordinates and the polar co-ordinates and makes the above transformation more clear.

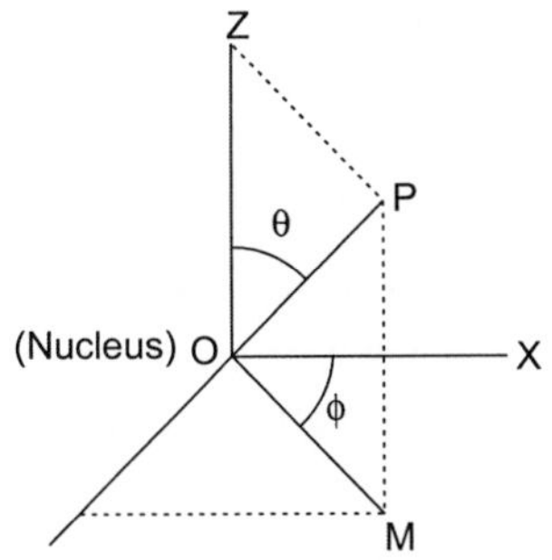

Fig. 1.8 Correlation of Cartesian and Polar co-ordinates.

The point P in the spherical polar co-ordinates is relative to point O in the Cartesian co-ordinates.

1.18 Concept of Electron Charge Cloud and Atomic Orbital

The electron revolves around the nucleus and the nucleus remains fixed at its position. The probability of finding the electron is maximum nearest to the nucleus and goes on decreasing as the distance of the electron from the nucleus increases.

The region in three dimensional space in which there is probability of finding the electron is called electron charge cloud.

The region in three dimensional space around the nucleus of an atom where the probability of finding the electron is maximum is called an atomic orbital.

1.19 Quantum Numbers

In general, it can be said that the **address of the electrons is given by the Quantum Numbers.**

We can say, that the set of four integers required to define the state of electron in an atom are called Quantum Numbers. The set of Quantum Numbers are:

1. **Principal Quantum Number (*n*):** It was proposed by *Bohr* and determines the main energy level, the average distance of the electron from the nucleus, the magnitude of energy of the electron and describes the size of the orbital.

   It can have integral values excluding zero such as 1, 2, 3 etc.

   The maximum number of electrons in any principal shell is given by $2n^2$, where n is the principal quantum number.

2. **Azimuthal (or Subsidiary) Quantum Number (*l*)**: It was proposed by *Sommerfeld.* It is also known as orbital or angular momentum Quantum Number. The **angular quantum number** (l) describes the shape of the orbital. Orbitals have shapes that are best described as spherical ($l = 0$), dumb-bell ($l = 1$), or double dumb-bell ($l = 2$). They can even take on more complex shapes as the value of the angular quantum number becomes larger. The l values depend upon the Principal quantum number n. For a given n value, l can have values starting from 0 to $n–1$, a total of n values.

The maximum number of orbitals in a sub shell is given by $(2l + 1)$ and the number of orbitals in n^{th} shell = n^2.

3. **Magnetic Quantum Number (m):** It was proposed by *Lande*. It describes the orientation in space of a particular orbital. For a given value of l, there can be $2l + 1$ values for m ranging from –1 to + 1 including zero. For a sub shell with $l = 0$, there is only one orientation of the orbital, for a sub shell with $l = 1$, there can be 3 orientations of the orbital.

4. **Spin Quantum Number (s):** It was proposed by *Uhlenbeck and Goudsmith*. Spin Quantum Number gives us an idea about the electron spinning on its axis. Each spinning electron can have two values of spin quantum numbers *i.e.* +1/2 (clockwise spin) and –1/2 (anticlockwise spin).

   It is the only quantum number which can have non-integral values.

   Spin Quantum angular momentum has a magnitude

$$S = h/2\pi \, [s\,(s + 1)]^{1/2}$$

Rules Governing the Allowed Combinations of Quantum Numbers

- The three quantum numbers (n, l, and m) that describe an orbital are integers: 0, 1, 2, 3, and so on.
- The principal quantum number (n) cannot be zero. The allowed values of n are therefore 1, 2, 3, 4, and so on.
- The angular quantum number (l) can be any integer between 0 and $n - 1$. If $n = 3$, for example, l can be either 0, 1, or 2.
- The magnetic quantum number (m) can be any integer between $-l$ and $+ l$. If $l = 2$, m can be either –2, –1, 0, +1, or +2.

Example 1.9: Is the following sets of quantum numbers permissible or not, if not why?

(*i*) $n = 3, l = 0, m = -1, m_s = -1/2$

(*ii*) $n = 2, l = 2, m = 0, m_s = ½$

Solution: (*i*) It is not allowed because value of m is governed by the values of l (+l..., 0...., –l). If $l = 0$, m cannot have values other than zero.

(*ii*) It is not allowed because values of l are governed by n. For $n = 2$, l can have values 0, 1 only. So, if $n = 2$, l cannot be equal to 2.

1.20 Radial and Angular Wave Function

1.20.1 Radial Wave Function

The electronic wave function, as we have discussed above, describes the distribution of the electron positions in terms of the distance of the electron from the nucleus, **r**, and the orientation of the electron relative to the nucleus. We can separate the wave function into an orientationally dependent part, known as the angular wave function, and an orientationally independent part, which is known as the **radial wave function.**

$$\psi(r, \theta, \phi) = R_{n,\,l}(r)\, \Theta_{l,\,m}(\theta)\, \Phi_m(\phi) \qquad ...(1)$$

The Radial Wave function, $R_{nl}(r)$, depends only on the distance of the electron from the nucleus, and is characterised by the values of the principal and orbital angular momentum quantum numbers, whereas the angular wave function, depends on the angles of the electron from the nucleus, and is characterised by the values of the orbital angular momentum and magnetic quantum numbers.

Radial wave function is given by the exponential term:

$$e^{-Zr/na_0} \qquad ...(2)$$

where, e = base of natural logarithm

Z = atomic number, $Z = 1$ for hydrogen atom

a_0 = Bohr radius = 0.529 Å for 1s in hydrogen atom

n = radial principal quantum number

The radial wave function may be represented in 3 ways:

1. The mathematical expressions for the radial part of the wave functions of 1s, 2*s* and 2*p* orbital of hydrogen atom are given below:

1*s*-orbital:

Here $n = 1$, $l = 0$

∴ the radial part of the wave function is

$$R(r) = \left(\frac{Z}{a_0}\right)^{3/2} e^{-Zr/a_0} \qquad ...(3)$$

$R(r)$ of the 1s orbital depends primarily on the exponential term :

$$e^{-Zr/a_0}$$

i.e., $R(r)$ should decrease with increase in the value of r as shown in Fig. 1.9.

2*s*-orbital:

Here, $n = 2$, $l = 0$, the radial part of the wavefunction is given by

$$R(r) = \left(\frac{Z}{2a_0}\right)^{3/2} \left(2 - \frac{Zr}{a_0}\right) e^{-Zr/2a_0}$$

∴ the radial wave function for 2*s* orbital is primarily governed by the term.

$$\left(2 - \frac{Zr}{a_0}\right) e^{-Zr/2a_0}$$

As r increases, the value of $R(r)$ decreases and becomes zero at $r = \frac{2a_0}{Z}$ as shown in Fig. 1.10.

$$\left[2-\frac{Zr}{a_0}\times\frac{2a_0}{Z}\right]e^{\left[-Z\times\frac{2a_0}{Z}\times\frac{1}{2a_0}\right]}$$

i.e., $$R(r) = (2-2)\,e^{-1} = 0$$

The point at which $R(r) = 0$ is called the node, this node is a spherical surface on which $R(r) = 0$. *i.e.* the point where the probability density of electronic charge is zero. Now, **the radial wave function** can be represented by plotting $R(r)$, which is r dependent part of the wave function Ψ, against distance r from the nucleus as shown for $1s$, $2s$, $2p$, $3s$ and $3p$ orbitals for the hydrogen atom in Figs. 1.9 – 1.12.

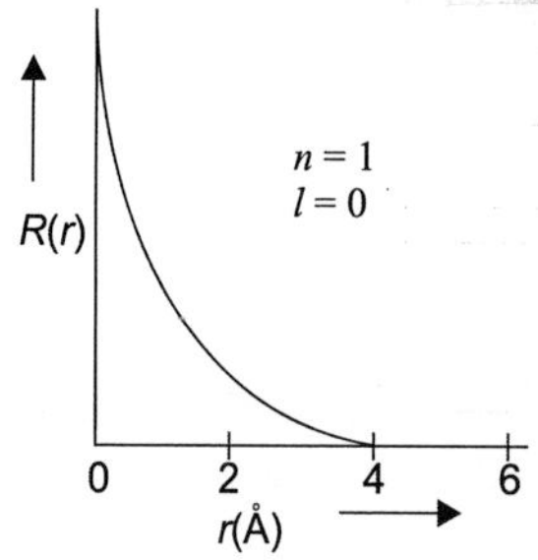

Fig 1.9

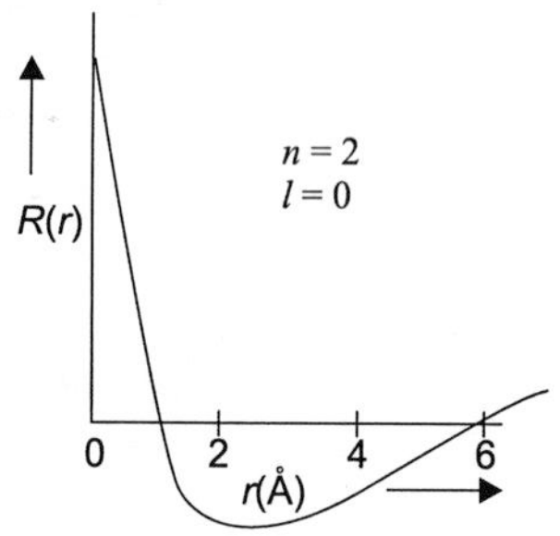

Fig 1.10

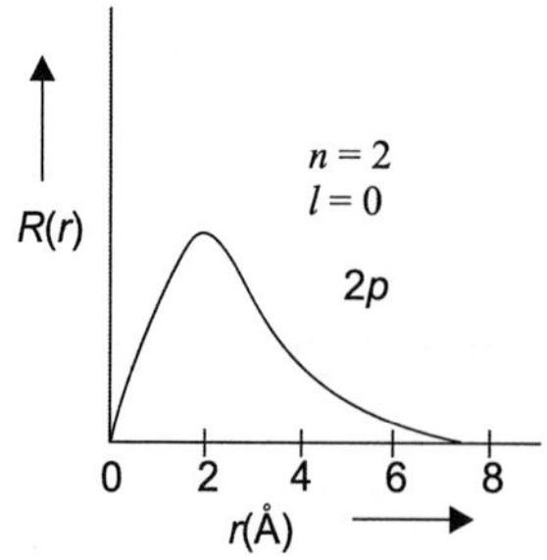

Fig 1.11

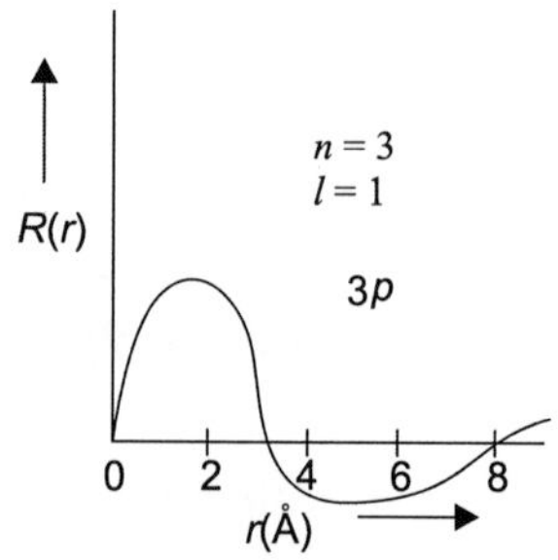

Fig 1.12

2*p*-orbital :

Here $n = 2$, $l = 1$, the radial part of the wave function is given by the following expression.

$$R(r) = \frac{1}{2\sqrt{6}}\left(\frac{Z}{a_0}\right)^{3/2}\left(\frac{Zr}{a_0}\right)e^{-Zr/2a_0}$$

Therefore, the radial wave function for 2*p* orbital of H –atom depends mainly on the term

$$(r/a_0)e^{-r/2a_0}$$

The above expression, does not become zero except where $r = 0$, hence, this curve does not have any nodes as shown in Fig. 1.11.

As r increases, the radial wave function $R(r)$ increases to a maximum and then decreases slowly, before becoming infinitesimally small.

Similarly, the radial wave function for 3*p* is shown in Fig. 1.12, where the wave function increases in the beginning and after attaining the maximum value, falls down and becomes zero.

For all the plots shown, the value of $R_{n,\,l}$ is maximum at $r = 0$, this can't be true as the probability of finding the electron at the nucleus is always zero. Therefore, we move on to another way of representing these radial wave functions. Also, $R_{n,l}$ is showing positive as well as negative values in the probability curves for 2*s* and 3*s* orbitals. The negative values of probability density is not possible.

2. $R^2_{n,l}$, plotted against *r* from the nucleus

Plots of $R^2_{n,\,l}$ versus r for the electron present in 1*s*, 2*s* and 3*s* orbitals are shown in Figs. 1.13, 1.14, 1.15 respectively.

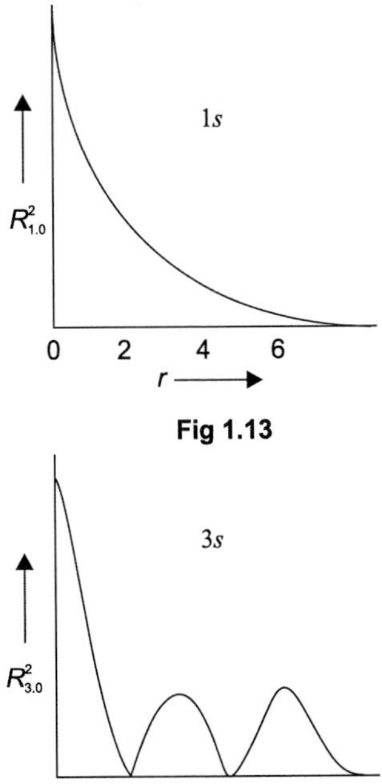

Fig 1.13

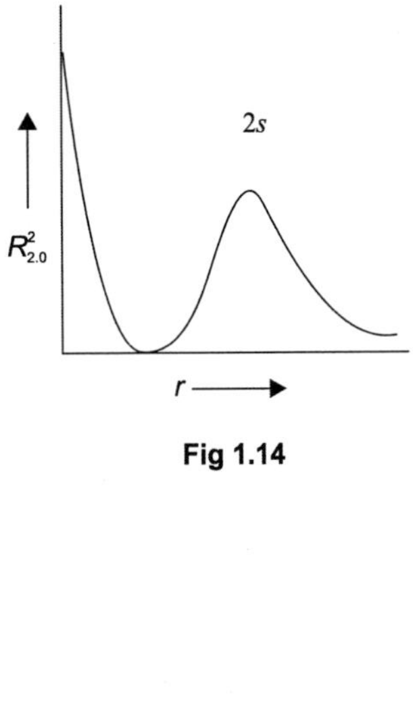

Fig 1.14

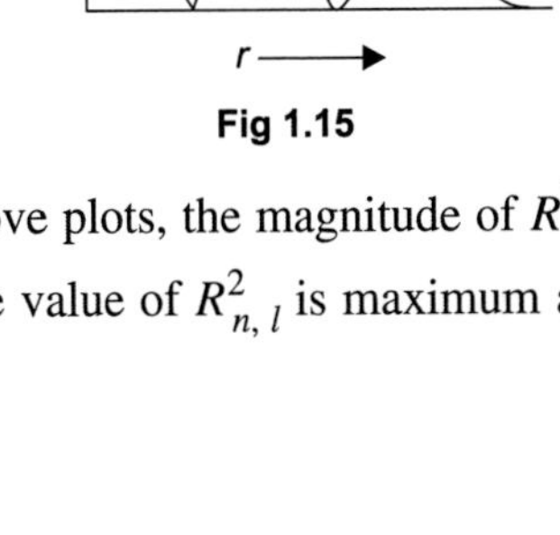

Fig 1.15

In all the above plots, the magnitude of $R^2_{n,\,l}$ is decreasing with increase in the value of *r.*

Secondly, the value of $R^2_{n,\,l}$ is maximum at $r = 0$, which is not possible.

Therefore, even these plots cannot be a satisfactory way of representing the radial wave functions. Therefore, we move on to the Radial Probability Distribution Curves.

3. By Plotting Radial Probability Distribution Curves

It is more beneficial to study the probability of finding the electron in a spherical shell of thickness *dr* at a distance *r* from the nucleus *i.e.*, the volume between the spherical shells of radii *r* and *r* + *dr* is equal to *dv*.

$$dv = \frac{4}{3}\pi(r + dr)^3 - \frac{4}{3}\pi r^3$$

$$= \frac{4}{3}\pi(r^3 + dr^3 + 3r^2dr + 3rdr^2 - r^3)$$

$$= \frac{4}{3}\pi(dr^3 + 3r^2dr + 3rdr^2)$$

Since, dr^3 and dr^2 are very small, these can be taken as zero.

Therefore, the above equation becomes,

$$dv = \frac{4}{3}\pi \times 3r^2dr = 4\pi r^2dr$$

The radial probability of finding the electron in a small radial shell of thickness *dr* around the nucleus is given by

$$= R^2_{n,\,l} \times dv = R^2_{n,\,l} \times 4\pi r^2dr$$

Therefore, the radial distribution between $r = 0$ and $r = r$ would be

$$= \int_{r=r}^{r=0} R^2_{n,\,l}\ 4\pi r^2\, dr$$

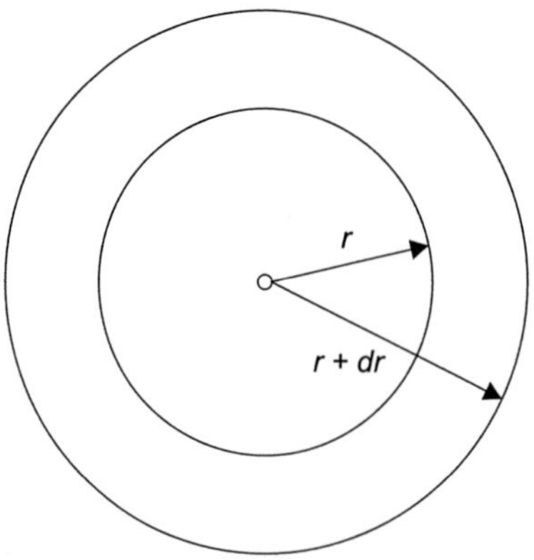

Fig. 1.16 Spherical shell of thickness *dr*.

For 1*s* orbital

The radial probability distribution of 1*s* is zero at the nucleus and increases to a maximum at $r = r_0$ which is equal to 52.9 pm and is referred to as Bohr's radius for a hydrogen atom. The distance r_0 is the radius of maximum probability and is shown in Fig. 1.17.

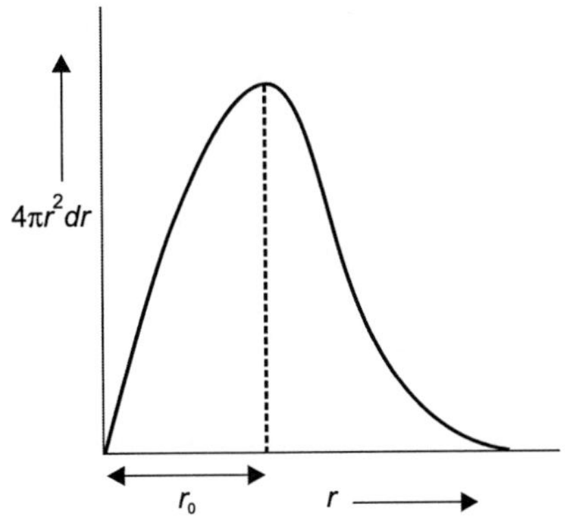

Fig. 1.17

For 2*s* orbital

The radial probability plot for 2*s* has two regions. The point where the probability of finding the electron is maximum, is separated by a node as shown in Fig. 1.18

Note: *A node is a region, where the probability of finding the electron is zero or we can say that the number of minima which appear in a particular curve gives the number of radial nodes or the orbital. The number of radial nodes is equal to (n – l – 1).*

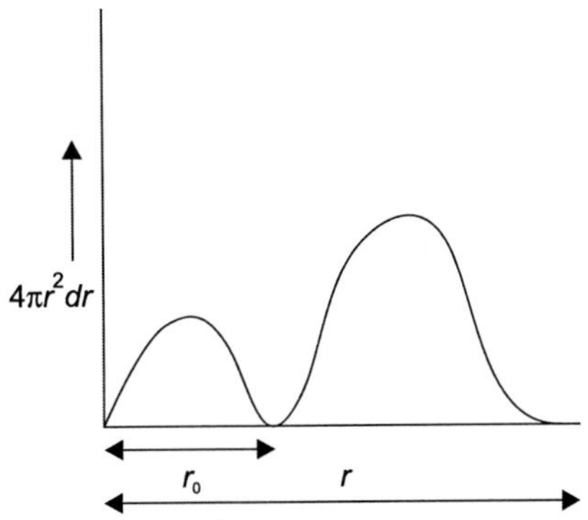

Fig. 1.18

2*p* orbital

For $n = 2$, $l = 1$, there are three *p* orbitals corresponding to $m = +1, 0, -1$

Radial probability increases as the *r* increases and reaches a maximum at r_{max}.

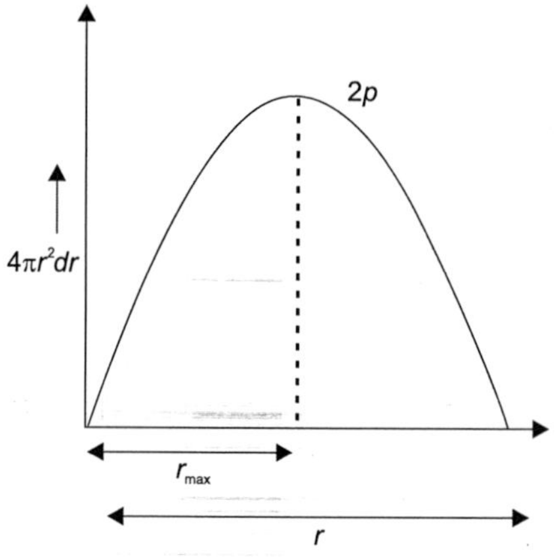

Fig. 1.19

For 1*s*, 2*s*, 2*p* 3*s*, 3*p* and 3*d* orbitals, the number of nodal points is 0, 1, 0, 2, 1, 0.

1.20.2 Angular Wave Function

Angular part of the wave function depends on *m* and *l* values.The shape of the orbitals depend upon angles θ and φ which determine the geometry of the atom.

The Schrödinger wave equation in polar coordinates is written as :

$$\left[\frac{d}{dr}\left(r^2\frac{\partial}{\partial r}\right)+\frac{1}{\sin\theta}\frac{\partial}{\partial\theta}\left(\sin\theta\frac{\partial}{\partial\theta}\right)+\frac{1}{\sin^2\theta}\frac{\partial^2}{\partial\phi^2}\right]\psi+\frac{8\pi^2mr^2}{h^2}\left(E+\frac{e^2}{r}\right)\psi=0 \qquad ...(1)$$

This equation (1) is obtained by writing

$$\psi(r, \theta\ \phi) = R\,(r)\ \Theta\,(\theta)\ \Phi\,(\phi)$$

This particular equation can be split into the following three equations (*i*), (*ii*), (*iii*)

(*i*) $$\frac{1}{R}\cdot\frac{\partial}{\partial r}\left(r^2\frac{\partial R}{\partial r}\right)+\frac{8\pi^2mr^2}{h^2}\left(E+\frac{e^2}{r}\right) = l(l+1)$$

(*ii*) $$\frac{\sin\theta}{\Theta}\frac{\partial}{\partial\theta}\left(\sin\theta\frac{\partial\Theta}{\partial\theta}\right)+l(l+1)\sin^2\theta=m^2$$

(*iii*) $$\frac{1}{\phi}\cdot\frac{\partial^2\Phi}{\partial^2\phi}=m^2$$

Equation (*ii*) is a function of θ only, and on solution gives the value of Θ (θ).

If the values of *l* and *m* are known for a given orbital, the value of angular wave function $\Theta_{l,m}$ can be obtained as shown below:

(*i*) $\Theta_{0,0}(\ell = 0, m = 0) = \frac{\sqrt{2}}{2} = \frac{1}{\sqrt{2}} (s - \text{orbital})$

(*ii*) *p*-orbitals

$$\Theta_{1,0}(\ell = 1, m = 0) = \sqrt{\frac{3}{2}} \cos\theta$$

$$\Theta_{1,\pm1}(\ell = 1, m = \pm 1) = \sqrt{\frac{3}{2}} \sin\theta$$

(*iii*) *d*-orbitals

$$\Theta_{2,0}\ (\ell = 2, m = 0) = \sqrt{\frac{5}{8}} (3\cos^2\theta - 1)$$

$$\Theta_{2,\pm1}\ (\ell = 2, m = \pm 1) = \sqrt{\frac{15}{4}} \sin\theta \cos\theta$$

$$\Theta_{2,\pm2}\ (\ell = 2, m = \pm 2) = \sqrt{\frac{15}{16}} \sin^2\theta$$

Equation (*iii*) is a function of *m* only and on solution, gives the value of Φ(ϕ). The values of Φ_m for the various orbitals are written below:

$$\Phi_0(m = 0) = \frac{1}{\sqrt{2\pi}} e^0 = \frac{1}{\sqrt{2\pi}}$$

$$\Phi_{+1}(m = +1) = \frac{1}{\sqrt{2\pi}} e^{+i\phi} = \frac{1}{\sqrt{2\pi}} (\cos\phi + i \sin\phi)$$

$$\Phi_{-1}(m = -1) = \frac{1}{\sqrt{2\pi}} e^{-i\phi} = \frac{1}{\sqrt{2\pi}} (\cos\phi - i \sin\phi)$$

$$\Phi_{+2}(m = +2) = \frac{1}{\sqrt{2\pi}} e^{+2i\phi} = \frac{1}{\sqrt{2\pi}} (\cos 2\phi + i \sin\phi)$$

$$\Phi_{-2}(m = -2) = \frac{1}{\sqrt{2\pi}} e^{-2i\phi} = \frac{1}{\sqrt{2\pi}} (\cos 2\phi + i \sin\phi)$$

The expression for the *s*-orbital has no angular (θ and φ) dependence and hence is spherically symmetrical. The physical interpretation of angular wave function is in the form of shape of orbitals shown in the later part.

The angular wave function of an orbital is the product of $\theta_{l,\,m}$ and ϕ_m where l and m are the azimuthal and magnetic quantum numbers for the orbital.

The $\Theta_{l,\,m} \times \Phi_m$ values for all the orbitals are shown below:

***s*-orbital**

$$\Theta_{0,\,0} \times \Phi_0 = \frac{1}{\sqrt{2}} \times \frac{1}{\sqrt{2\pi}} = \frac{1}{2\sqrt{\pi}}$$

This equation is independent of the angles θ and φ, hence *s*-orbital is symmetrical all over.

***p*-orbital**

For *p*-orbitals, $\ell = 1, m = 0, +1, -1$

Angular wave function for $\ell = 1, m = 0$

$$\Theta_{1,0} \times \Phi_0 = \frac{\sqrt{3}}{2}\cos\theta\frac{1}{\sqrt{2\pi}}$$

$$= \sqrt{\frac{3}{4\pi}}\cos\theta$$

For $\ell = 1, m = \pm 1$

$$\Theta_{1,\,+1} \times \Phi_{+1} = \sqrt{\frac{3}{4}}\sin\theta \times \frac{1}{\sqrt{2\pi}}(\cos\varphi + i\sin\varphi)$$

$$\Theta_{1,\,+1} \times \Phi_{+1} = \sqrt{\frac{3}{8\pi}}\sin\theta(\cos\phi + i\sin\phi)$$

$$\Theta_{1,\,-1} \times \Phi_{-1} = \sqrt{\frac{3}{4}}\sin\theta \times \frac{1}{\sqrt{2\pi}}(\cos\theta - i\sin\phi)$$

or $$\Theta_{1,\,-1} \times \Phi_{-1} = \sqrt{\frac{3}{8\pi}} \times \sin\theta \times \left(\cos\varphi - i\sin\varphi\right)$$

***d*-orbitals**

For *d*-orbitals, $\ell = 2$ and $m = 0, \pm 1, \pm 2$

Angular wave function for $\ell = 2, m = 0$

$$\Theta_{2,\,0} \times \Phi_0 = \sqrt{\frac{5}{8}}(3\cos^2\theta - 1) \times \frac{1}{\sqrt{2\pi}}$$

$$= \sqrt{\frac{5}{16\pi}} \times (3\cos^2\theta - 1)$$

For $\ell = 2, m = \pm 1$

$$\Theta_{2,+1} \times \Phi_{+1} = \sqrt{\frac{15}{4}} \sin\theta\cos\theta \times \frac{1}{\sqrt{2\pi}} e^{+i\phi}$$

$$= \sqrt{\frac{15}{8\pi}} \sin\theta\cos\theta \times e^{-i\phi}$$

$$\Theta_{2,-1} \times \Phi_{-1} = \sqrt{\frac{15}{4}} \sin\theta\cos\theta \times \frac{1}{\sqrt{2}\ \pi} e^{-i\varphi}$$

$$= \sqrt{\frac{15}{8\pi}} \sin\theta\cos\theta \times e^{-i\phi}$$

For $\ell = 2, m = \pm 2$

$$\Theta_{2,+2} \times \Phi_{+2} = \sqrt{\frac{15}{16}} \sin^2\theta \times \frac{1}{\sqrt{2\pi}} e^{+2i\phi}$$

$$= \sqrt{\frac{15}{32\pi}} \sin^2\theta \times e^{+2i\phi}$$

$$\Theta_{2,-2} \times \Phi_{-2} = \sqrt{\frac{15}{16}} \sin^2\theta \times \frac{1}{\sqrt{2\pi}} e^{-2i\phi}$$

$$= \sqrt{\frac{15}{32\pi}} \sin^2\theta e^{-2i\phi}$$

1.21 Concept and Shapes of Atomic Orbitals

An **atomic orbital** is the region in which an electron may be found around a single atom. Specifically, atomic orbitals are the quantum states of the individual electrons in the electron cloud around a single atom. Classically, the electrons were thought to orbit the atomic nucleus, much like the planets around the Sun.

Shape of *s*-orbital

The *s*-orbitals is spherical in shape and have symmetrical orientation. For all *s*-orbitals there are $(n - 1)$ number of spherical nodes. *s*-orbitals have no angular dependance as discussed in the angular wave function part and hence these are spherically symmetrical.

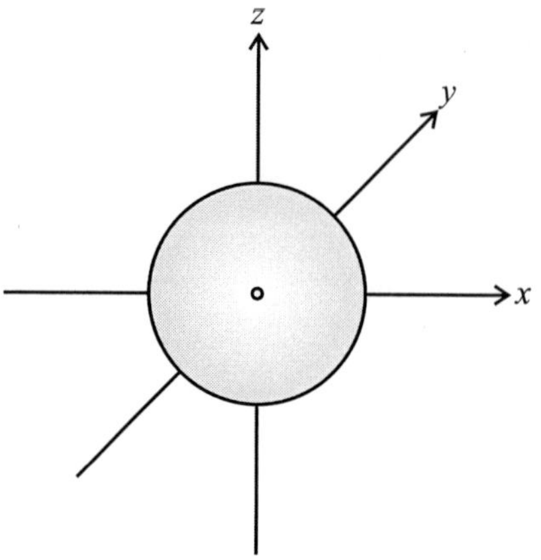

Fig. 1.20 Spherical *s*–orbital

Shape of the *p*-orbital

For the *p*-orbital, $l = 1$, m has three values *i.e.*, +1, 0, –1 representing three orientations of *p*-orbitals, p_x, p_y and p_z Here comes the concept of degenerate orbitals, degenerate orbitals for electrons in an atomic subshell are orbitals at identical energy levels. *e.g.* 2*p* orbitals, *i.e.* $2p_x$, $2p_y$, $2p_z$ are degenerate orbitals.

The Angular wave function expression for the p_z orbital is $(3/4\pi)^{1/2}$ cos θ gives two tangent dumb-bells symmetrical about *z*-axis.We can say, that the angular part of the probability distribution is concentrated along *z*-axis and the resulting surface of p_z electron has the form of a dumb-bell along the *z*-axis.The dumb-bell shaped lobes of high probability are separated by a nodal surface in *xy* plane.The probability of finding the electron in both the lobes is equal. Both the lobes are seperated by a nodal plane which passes through the nucleus.

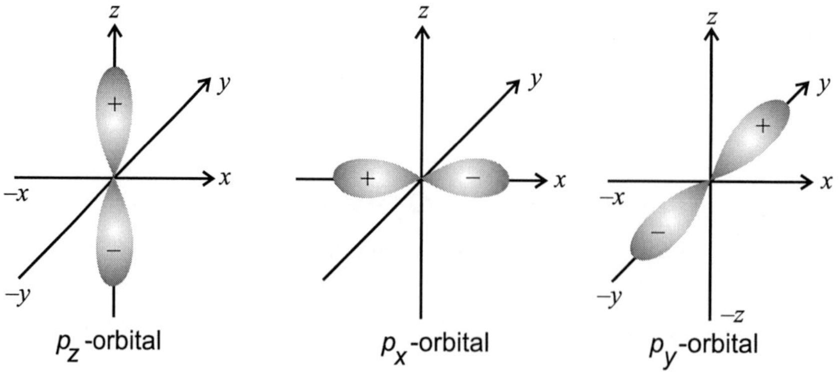

Fig. 1.21 Dumb-bell shaped *p*-orbitals

Shape of the *d*-orbitals

Mathematical expressions for the angular wave functions for the *d*-orbitals are complex and already mentioned in the previous section.

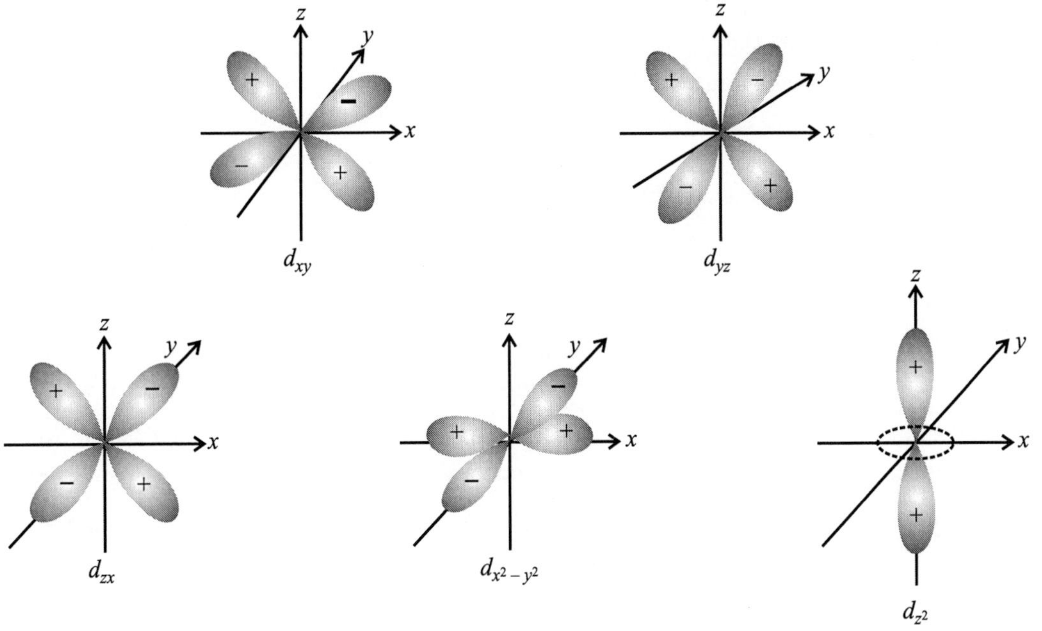

Fig. 1.22 Double Dumb-bell shaped *d*-orbitals

For *d*-orbitals l = 2, and m = 0, +1, –1, +2, –2 representing the five orientations of *d*-orbitals designated as d_{z^2}, $d_{x^2-y^2}$, d_{xy}, d_{yz}, d_{zx}.

d_{xy}, d_{yz}, d_{zx}, each of these orbitals are double - dumb-bell shaped, *i.e.*, these have four lobes of high electron density in *xy*, *yz* and *zx* plane as shown in Fig. 1.22 $d_{x^2-y^2}$ also has a dumb-bell shape with four lobes.These four lobes lie in *xy* plane and are concentrated along *x* and *y* plane. d_{z^2} has a dumb-bell symmetrical shape with two lobes along the *z*-axis and a concentric ring around the nucleus in the *xy* plane.

1.22 Pauli's Exclusion Principle

The **Pauli exclusion principle** is a quantum mechanical principle formulated by *Wolfgang Pauli* in 1925.

For electrons in a single atom, it states **"that no two electrons can have the same four quantum numbers"**, that is, if *n*, *l*, and *m* may be same but m_s must be different such that the electrons have opposite spins since the state with opposite spins is the state of lower energy and higher stability, whereas the state with same spins is the state of higher energy and lower stability.

In other words, it can be said that **"In an atom, a maximum of two electrons can occupy the same orbital".**

For example, consider a neutral *helium* atom, which has two bound electrons. Both of these electrons can occupy the lowest-energy (*1s*) states by acquiring opposite spin. This does not violate the

Pauli principle, because spin is part of the quantum state of the electron, so the two electrons are occupying different quantum states (both the electrons differ in the value of spin quantum number *i.e.*, (+1/2 and –1/2). In a *lithium* atom, which contains three bound electrons, the third electron cannot fit into a 1*s* state, and has to occupy one of the higher-energy 2*s* states instead. Similarly, successive elements produce successively higher-energy shells. The chemical properties of an element largely depend on the number of electrons in the outermost shell, which gives rise to the *periodic table of the elements*.

1.23 Hund's Rule of Maximum Multiplicity

Friedrich Hund in 1925 discovered this rule and others referred to collectively as *Hund's rules* for atomic energy levels. They are important for *Spectroscopy* and *Quantum Chemistry.*

It states that pairing of the electron in degenerate orbitals, belonging to a particular sub-shell (*i.e.*, *p*, *d*, *f*) does not take place till the orbital is occupied by the single electron with parallel spin.

Pairing of electrons in the same orbital should be made in such a way that the two electrons of the pair have opposite spins (↑↓). According to the rule, the arrangement of the three electrons in three degenerate orbitals *i.e.* p_x, p_y, p_z should be as $\underline{\uparrow}\ \underline{\uparrow}\ \underline{\uparrow}$ and not as $\underline{\uparrow\downarrow}\ \underline{\uparrow}\ \underline{\ \ }$.

Half-filled and fully filled electronic configurations are stable since they have more symmetry and exchange energy.

Exchange energy is the energy associated with maximum number of exchange pair of electrons with the same spin. Two electrons placed in the same orbital leads to higher repulsion. More the number of exchange pair of electrons, more is the exchange energy, lower is the energy of the system and hence higher is the stability. Similarly, symmetry also plays an important role in establishing the stability of the system as half-filled and fully filled orbitals have a symmetrical electronic charge distribution.

1.24 The Aufbau's Principle

The principle takes its name from the *German Aufbauprinzip*, "**building-up principle**", rather than being named for a scientist. In fact, it was formulated by the Danish physicist *Niels Bohr circa* in 1920.

The **Aufbau principle**, is used to determine the *electron configuration* of an *atom*, *molecule* or *ion*. The principle postulates a hypothetical process in which an atom is "built up" by progressively adding electrons. According to the principle, *electrons* fill orbitals starting at the lowest available (possible) *energy states* before filling higher states (*e.g.*, 1*s* before 2*s*).

In other words, we can say that the electrons are fed in the orbitals in order of increasing energy and the energy of the orbitals is governed by $(n + l)$ rule.

The order of energies of the orbitals is given below:

$1s < 2s < 2p < 3s < 3p < 4s < 3d < 4p < 5s < 4d < 5p < 6s < 4f < 5d < 6p$

Limitations of Aufbau's Principle:

According, to Aufbau's Principle, the electronic configuration of Cr (Z = 34), should be $3d^4\ 4s^2$, but the experimentally supported configuration is $3d^5\ 4s^1$. Similarly, Pd (Z = 46), should have configuration, $5s^2 4d^8$, but the observed configuration is $5s^0 4d^{10}$. The reasons for these deviations are due to extra stability of half-filled or completely-filled subshell.

Example 1.10: Why half-filled and completely-filled sub shells are more stable?

Solution: This can be explained by taking the example of Cr (Z = 34). As, already told, the actual configuration is $3d^5 4s^1$. This is so, because in this configuration, the distribution of electronic charge is symmetrical than in $3d^4 4s^2$.

The concept of exchange energy, comes into consideration. It is the energy associated with maximum number of exchange pair of electrons with the same spin. Placing two electrons in the same orbital will lead to higher repulsion, than if they occupy the different orbital.More the number of exchange pair of electrons, more is the exchange energy. Higher exchange energy lowers the energy of the system and hence increases the stability. Since, $3d^5 4s^1$ configuration has five exchange pair of electrons with the same spin, therefore, it is a stable configuration than $3d^4 4s^2$.

Similarly, symmetry also plays an important role in establishing the stability of the system as half-filled and fully filled orbitals have a symmetrical electronic charge distribution.

1.25 Screening Effect and Effective Nuclear Charge

The **shielding effect** describes the decrease in attraction between an *electron* and the *nucleus* in any atom with more than one electron shell. It is also referred to as the **Screening Effect** or **Atomic Shielding**.

The electrons between the nucleus and the valence shell offers as a screen thereby decreasing the force of attraction between the nucleus and the electrons in the outer most shell thereby decreasing the effective nuclear charge *i.e.* is the net positive charge experienced by an *electron* in a multielectron *atom* by the quantity σ which is called the screening constant. Therefore, the effective nuclear charge is given by

$$Z_{eff} = Z - \sigma$$

where, Z_{eff} is the effective nuclear charge

Z is the nuclear charge

σ is the screening constant

s-electrons tend to shield the nucleus more effectively than the *p*-electrons which in turn, will shield the nucleus more effectively than *d*-electrons.This, because of the fact that the *s*-electron penetrates more towards the nucleus than the *p* and *d* electrons.

Therefore, the order of penetration of electrons is given by:

$$s > p > d > f$$

When we move down the group, the number of intervening electrons increases and hence the magnitude of σ also increases which in turn decreases the effective nuclear charge which actually lead to the increase in the atomic radii.

As we move from left to right in the periodic table, effective nuclear charge increases which leads to the decrease in the atomic radii.

1.26 Slater Rules

A set of rules were proposed after the name of scientist *John C. Slater* in the 1930 to calculate the screening constant (σ) and effective nuclear charge (Z_{eff}).

Rules are divided into two categories:

Category 1. Rules for calculating the value of σ for an electron residing in an *ns* or in one of the *np* orbitals of the *n*th shell of an atom or an ion is the sum of the following contributions:

1. Each of the remaining electron in the *n*th shell makes a contribution of 0.35 to the value of σ.
2. Each of the electrons in $(n - 1)$th shell makes a contribution of 0.85.
3. Each of the electrons in the next inner shells makes a contribution of 1.0.
4. There will be no contribution to the value of σ by the electrons residing in orbitals having higher value of their principal quantum number than the orbital containing the electron for which σ is being calculated.
5. If σ is being calculated for an electron of 1*s* orbital, there will be a contribution of 0.30 from other single electron in 1*s* orbital.

Therefore, the value of σ for an electron residing in *ns* or *np* orbital = 0.35 × [Number of remaining electrons in the *n*th shell] + 0.85 × [Number of electrons in the $(n–1)$th shell] + 1.0 × [Number of electrons in the inner shells]

Calculation of screening constant can be well explained by taking the following examples:

Example 1.11: Calculate σ and Z_{eff} for 4*s* electron in Cu (Z = 29).

Solution: $Cu_{29} = \underbrace{1s^2 2s^2 2p^6}_{(10)} \underbrace{3s^2 p^6 d^{10}}_{(18)} \underbrace{4s^1}_{(1)}$

σ for $4s^1$ electron is = 0.35 × [No. of the remaining Cu atom electrons in 4s orbital] + 0.85 × [No. of electrons in 3*s*, 3*p* and 3*d* orbitals] + 1.0 × [No. of electrons in the inner shells]

$= 0.35 \times 0 + 0.85 \times 18 + 1.0 \times 10 = 25.30$

Z_{eff} experienced by $4s^1$ electron = 29 – 25.30 = 3.70

Example 1.12: Calculate the Z_{eff} for the last electron in (*i*) K atom (19) and (*ii*) K^+ ion

Solution: (*i*) K atom (19) = $\underbrace{1s^2}_{2}\ \underbrace{2s^2p^6}_{8}\ \underbrace{3s^2p^6}_{8}\ \underbrace{4s^1}_{1}$

σ for $4s^1$ electron of K atom

$$= 0.35 \times 0 + 0.85 \times 8 + 1.0 \times 10 = 16.80$$

$$Z_{eff} = 19 - 16.80 = 2.20$$

(*ii*) K^+ ion

$$K^+ \text{ ion } (18) = \underbrace{1s^2}_{2}\ \underbrace{2s^2p^6}_{8}\ \underbrace{3s^2p^6}_{8}$$

$$\sigma = 0.35 \times 7 + 0.85 \times 8 + 2 \times 1.0$$

$$\sigma = 11.25$$

$$Z_{eff} = 15 - 11.25 = 3.75$$

Example 1.13: Calculate the Z_{eff} experienced by the last electron in phosphorous atom.

Solution: $Z = 15 = \underbrace{1s^2}_{2}\ \underbrace{2s^2p^6}_{8}\ \underbrace{3s^2p^3}_{5}$

$$\sigma = 0.35 \times 4 + 0.85 \times 8 + 1.0 \times 2$$

$$\sigma = 10.20$$

$$Z_{eff} = 15 - 10.20 = 4.80$$

Example 1.14: Calculate the effective nuclear charge for 1*s* electron in nitrogen atom.

Solution: $N\ (Z = 7)\ \underbrace{1s^2}_{2}\ \underbrace{2s^2p^3}_{5}$

There will be no contribution from the five electrons present in 2*s* and 2*p* orbitals to σ

Thus, $\sigma = 5 \times 0 + 1 \times 0.30 = 0.30$

$$Z_{eff} = 7 - 0.30 = 6.70$$

Category 2: The value of σ for an electron residing in the (*n* – 1) d orbitals of (*n* – 1)th shell of an atom or an ion is the sum of the following contributions

1. There will be no contribution to the value of σ by the electrons residing in the *ns* orbital.
2. Each of the remaining electrons present in (*n*–1)*d* orbitals, makes a contribution of 0.35.
3. Each of the electrons present in the (*n*–1)*s*, (*n*–1)*p* orbitals and the inner shells makes a contribution of 1.0.

Therefore, σ = 0.35 × [No. of the electrons in (*n*–1)*d* orbitals] + 1.0 × [No. of electrons in (*n*–1)*s*, (*n*–1)*p* orbitals and the inner shells].

Example 1.15: Calculate σ and Z_{eff} for 3*d* electron in Mn (Z = 25)

Solution: $4s^2$ electron will not contribute to the value of σ for 3*d* electron.

Therefore, = 0.35 × {No. of remaining electrons in 3*d* orbitals} + 1.0 × [No. of electrons in 1*s*, 2*s*, 2*p*, 3*s*, 3*p* orbitals]

$$= 0.35 \times 4 + 1.0 \times 18 = 19.40$$

$$Z_{eff} = 25 - 19.40 = 5.60$$

Effective nuclear charge can also be calculated at the periphery of an atom or ion.

Z_{eff} felt at the periphery of an atom or ion is equal to the atomic number of the element minus the screening constant (σ) for all the electrons present in the electronic configuration of the atom or ion. For example, Z_{eff} can be calculated at the periphery of the Cr-atom.

$$\text{Cr} = \underbrace{1s^2 2s^2 p^6}_{10}\ \underbrace{3s^2 p^6 d^5}_{13}\ \underbrace{4s^1}_{1}$$

σ for all the electrons in Cr-atom= 0.35 × 1 + 0.85 × 13 + 1.0 × 10 = 21.40

$$Z_{eff} \text{ at the periphery} = 24 - 21.40 = 2.60.$$

1.27 Electronegativity

Electronegativity, is a chemical property which describes the power of an atom to attract electrons towards itself.

Methods of Calculation of Electronegativity

A. Pauling's Method

Pauling, first proposed the concept of electronegativity in 1932 as an explanation of the fact that the covalent bond between two different atoms (A–B) is stronger than would be expected by taking the average of the strengths of the A–A and B–B bonds.

The difference in electronegativity between atoms A and B is given by:

$$x_a - x_b = \text{B.E. (A–B)} - [\text{ B.E. }(A_2) \times \text{B.E. }(B_2)]^{1/2}$$

where, B.E. (A–B) is the bond energy of A–B, B.E. (A_2) is the bond energy of A–A and B.E. (B_2) is the bond energy of B–B.

B. Mulliken Scale

Mulliken in 1934, proposed that the arithmetic mean of the first ionization energy and the electron affinity should be a measure of the tendency of an atom to attract electrons.

$$\text{Electronegativity of an atom} = \frac{\text{I.E.} + \text{E.A.}}{2}$$

For ionization energies and electron affinities in Electron Volts is given by:

$$x = 0.187\ (\text{I.E.} + \text{E.A.}) + 0.17$$

and for energies in kilojoules per mole

$$x = (1.97 \times 10^{-3})\ (\text{I.E.} + \text{E.A.}) + 0.19$$ *x is the electronegativity*

Example 1.16: Calculate the electronegativity of fluorine atom, given that ionisation energy of F-atom = 17.41 *eV* and electron affinity of F-atom = 3.45 *eV*.

Solution: According to the equation given by Mulliken.

$$x = 0.187\ [\text{I.E.} + \text{E.A.}] + 0.17$$

$$= 0.187\ [17.41 + 3.45] + 0.17$$

$$x_{\text{Fluorine}} = 4.07$$

Allred and Rochow Scale

Allred and *Rochow* in 1958, considered that electronegativity should be related to the charge experienced by an electron on the "surface" of an atom. Higher the charge per unit area of atomic surface, the greater is the tendency of that atom to attract electrons. The effective nuclear charge, Z_{eff}, experienced by *valence electrons* can be estimated using *Slater's rules*, while the surface area of an atom in a molecule can be taken to be proportional to the square of the *covalent radius*, r_{cov}. Where, r_{cov} is expressed in Å

$$x = 0.359\ Z_{eff}/r^2_{\text{cov}} + 0.744$$

Table 1.1: Electronegativity value of different elements according to Pauling and Allred–Rochow Scale

Element	Pauling	Allred-Rochow
C	2.50	2.48
N	3.04	3.06
O	3.44	3.50
F	3.95	4.10
I	2.65	2.20
Cl	3.16	2.83

Factors Affecting the Electronegativity

1. **Oxidation State:** The higher the oxidation state, more is the attraction for electrons and higher is the electronegativity. For example, the electronegativity of S in SF_6 is higher than that in SF_4, because the former molecule is in higher oxidation state of S.

2. **Hybridisation:** Hybridisation affects the value of electronegativity to a great extent as *s*-orbitals has higher penetration effect as it is held more tightly than other orbitals to the nucleus. This increases its tendency to attract electrons towards itself and hence has higher electronegativity.

Therefore,

Electronegativity of Carbon	CH_4	<	$CH_2{=}CH_2$	<	$CH{\equiv}CH$
Hybridisation of carbon	sp^3		sp^2		sp
s-character	25%		33%		50%

Points to Remember

- Atom can be defined as the simplest particle of a matter that takes part in chemical reactions.
- Atoms can be photographed using a technique known as Scanning Tunneling Microscopy (STM).
- Nucleus is a small positively charged part of an atom, which is situated at the centre and carries almost entire mass of the atom.
- Atom is electrically neutral, so every atom has the same number of electrons as protons.
- Electrons, protons and neutrons are the most important fundamental particles of atoms of all elements (except hydrogen).
- Electrons are also named as Cathode Rays.
- Neutrinos and Antineutrinos are the particles of small mass and zero charge.
- Positrons are the positive counterparts of electron.
- Pi-mesons are the particles having a mass intermediate between that of the electron and the protons.
- The credit for the first successful model about the arrangement of electrons, protons and the neutrons goes to Rutherford whose model was based on the Alpha-Ray scattering experiment.
- An atom of an element (*X*) indicating the Mass Number (*A*) and the Atomic Number (*Z*) is represented by $_ZX^A$.
- The arrangement of the radiations in the order of their increasing wavelengths and the decreasing frequencies is as follows:

Cosmic rays < γ-rays < *X*-rays < *U.V* rays < Visible light < *IR* radiations < Microwaves < Radiowaves

- The emission and absorption of energy dosen't take place continuously but in the form of small packets of energy called Quanta.
- The energy of each photon is given by Planck's relationship

  $E = hv$, where h is the Planck's constant
- When a beam of light of sufficiently high frequency is allowed to strike a metal surface, electrons are ejected from the surface of the metal. This phenomenon is Photoelectric Effect.
- The radius of the *n*th orbit for the hydrogen atom can be obtained by

  $$R_n = 0.529 \times n^2$$
- The velocity of an electron revolving in an orbit can be calculated by

  $$v = 2.19 \times 10^8/n \text{ cm/sec}$$
- The energy of an electron can be expressed by the following expression:

  $$E = -\frac{13.6}{n^2} eV$$
- The frequency of the emitted radiation when an electron jumps from n_2 orbit to n_1 orbit is given by

  $$v = R_H \left(\frac{1}{n_1^2} - \frac{1}{n_2^2} \right)$$
- Bohr's model of circular orbits was further extended by Sommerfeld who introduced the concept of elliptical orbits, with the nucleus situated at one of the focii of the ellipse.
- The principal Quantum number *n*, used by Bohr and Azimuthal Quantum number 1 used by Sommerfeld are related to one another by the following expression:

  n/l = Length of major axis of the elliptical orbit/Length of minor axis of the elliptical orbit.

  When $n = 1$, orbit becomes circular.
- But when $l \neq 0$ because in that case minor axis is zero, *i.e.* the motion of the electron passing through the nucleus is linear.
- According to de-Broglie principle, "the momentum of a particle in motion is inversely proportional to the wavelength of the waves associated with it".

- de-Broglie concept is applicable in the construction of electron microscope used for measurement of objects of very small size.
- According to Heisenberg's Uncertainty Principle it is impossible to determine at any given moment both the momentum and position of an electron.
- The mathematical expression for the Heisenberg's Uncertainty principle is given as

$$\Delta x.\Delta p = \frac{h}{4\pi}$$

- Uncertainty Principle is not on account of any limitation of the measuring instrument.
- The branch of science based on Newton's law of motion is called Classical Mechanics.
- Schrödinger Wave Equation is

$$\frac{d^2\psi}{dx^2} + \frac{d^2\psi}{dy^2} + \frac{d^2\psi}{dz^2} + \frac{8\pi^2 m}{h^2}(E - V)\psi = 0$$

where x, y, z are the space co-ordinates

h is the Planck's constant

m is the mass of the electron

E is the total energy of the electron

V is the Potential Energy of the electron

ψ is the amplitude of the electron wave, called wave function.

This system is valid for an atom or a molecule whose energy does not change with time.

- Schrödinger Wave Equation cannot be exactly solved for a multielectron atom.
- ψ represents the amplitude of the electron wave at various points surrounding the nucleus.
- The square of the amplitude ψ^2 of the wave at any point tells us about the electron probability at that point.
- Quantum Numbers were discovered by different scientists given below:

  Principle by Bohr, Azimuthal by Sommerfeld, Magnetic by Lande and Spin by Uhlenbeck and Goldsmith.
- Number of sub shells in the nth shell = n

  Number of orbitals in a sub shell = $2l + 1$

  Number of orbitals in nth shell = n^2

  Maximum number of electrons in nth shell = $2n^2$

- The region of space around the nucleus where the probability of finding the electron is maximum is an orbital.
- For an equation,

  The part $R(r)$ is called radial part whereas the product is called the angular part.
- In the plots of radial probability versus distance from the nucleus, the number of peaks, *i.e.* the regions of maximum probability = $n - 1$.
- A plane passing through the nucleus on which the probability of finding the electron is zero is called Nodal Plane.
- According to Pauli's Exclusion Principle, **"No two electrons in an atom can have the same set of four Quantum Numbers"**
- The angular momentum of the electron

  $$= l\,(l + 1)\,h/2\pi,$$ it depends on l and not on m
- Half-filled and fully filled electronic configurations are more stable due to symmetry and exchange energy.
- The screening constant (σ) and effective nuclear charge (Z_{eff}) can be calculated by a set of rules, known as Slater Rules.
- The value of σ for an electron residing in *ns* or *np* orbital can be obtained by the following expression:

  = 0.35 × (No. of remaining electrons in the *n*th shell) + 0.85 × (No. of electrons in the $(n-1)^{th}$ shell) + 1.0 (No. of electrons in the inner shells).

The value of symbol for Rydeberg constant depends on its use, as given below:

The Rydeberg constant, symbol $R_{infinity}$, named after the Swedish physicist Johannes Rydeberg, is a physical constant relating to atomic spectra in the science of spectroscopy. It is a constant which governs the relationship of the spectral line features of an atom through the Rydeberg formula. Rydeberg initially determined its value empirically from spectroscopy, but it was later found that its value could be calculated from more fundamental constants by using quantum mechanics.

When we write *R*, as Rydeberg constant, it should not be confused with Gas constant, R.

The Rydberg constant, R_H (for H atom), R_M (for an atom with one electron) , R_C (c stands for speed of light), $R_{infinity}$ (is the normal symbol for Rydeberg constant)

R_H (Rydeberg constant for H atom) = 109677 cm^{-1}

$R_M = 1.0973731568549 \times 10^7 m^{-1}$

$R_C = 3.289841960368 \times 10^{15}$ Hz

$R_{infinity} = 109737.311 \pm 0.012\ cm^{-1}$ or $109737.31573\ cm^{-1}$

$Rhc = 2.17987190 \times 10^{-18}$ J , where c = speed of light [ms^{-1}] , h = Planck's constant [Js]

R (Gas constant) = 8.314 K^{-1} mol^{-1}.

Exercises

1. State the important characteristics of the three fundamental particles in an atom.
2. Explain briefly the Rutherford atomic model.
3. Differentiate between electromagnetic waves and matter waves.
4. State the postulates of Planck's Quantum theory.
5. Explain Photoelectric effect with suitable examples.
6. What is the relationship between wavelength and momentum of a particle? Explain
7. State de-Broglie's Principle.
8. Give the mathematical expression for Heisenberg Uncertainty Principle.
9. Calculate the uncertainty in the position of a dust particle with mass equal to 2mg if the uncertainty in velocity is 6.5×10^{-20} m/sec.
10. The threshold energy for photo-electric emission of electrons from a metal is 3.056×10^{-15} Joule. If the light of 3000Å wavelength is used, will the electrons be ejected or not?
11. Derive an expression for the velocity of an electron in an orbit of hydrogen atom.
12. Derive a expression for the energy of an electron in an orbit of hydrogen atom.
13. Discuss the main postulates of Bohr's atomic model.
14. Write the limitations of Bohr's atomic model.
15. Explain the significance of the Quantum Numbers.
16. What are the values for n, l and m Quantum Numbers for
    (*i*) 3*s* (*ii*) 2*p* (*iii*) 5*d* orbitals
17. What is the difference between an orbit and the orbital ?
18. What is the physical significance of ψ and ψ^2 ?
19. Derive Schrödinger Wave Equation.
20. Write the Schrödinger wave equation for hydrogen atom in cartesian and polar co-ordinates.
21. Draw the plots of $R(r)$ versus r for 1*s*, 2*s*, 2*p* and 3*p* orbitals for the hydrogen atom.
22. What do you understand by radial probability distribution curve?
23. Explain why 1*s* orbital is spherically symmetrical.
24. Which Quantum Number does not follow from the Schrodinger wave equation and why?
25. Write a short note on Slater Rules.

26. What do you understand by shielding effect and screening constant?

27. Explain Effective nuclear charge with suitable examples.

28. Calculate the effective nuclear charge experienced by an electron residing in K and L shell in Be_4, O_8 and F_9 atoms ?

29. Draw the shapes of the following orbitals.

(*i*) $2p_x$ (*ii*) $3d_{z^2}$ (*iii*) $3d_{x^2-y^2}$

30. What is Pauli's Exclusion Principle ?

31. State Aufbau's Principle and what are the limitations related to it?

32. Why half-filled and fully filled orbitals are more stable. Explain?

2

Chemical Bonding

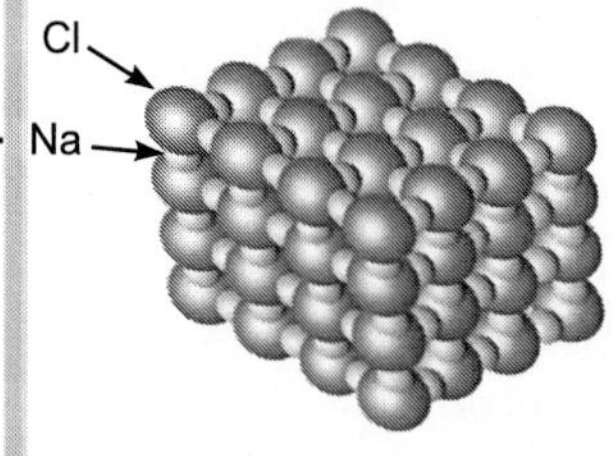

Sodium Chloride Crystal Structure

Sodium chloride, also known as common salt, table salt, or halite, is an ionic compound with the formula NaCl. It has chemical bonding between sodium cation and chloride anion. The properties of sodium chloride are as below :

Molecular mass = 58.443 g/mol, Appearance-Colorless/white crystalline solid, Odor-Odorless, Density 2.165 g/cm^3, Melting point 801 °C (1074 K), Boiling point, 1465 °C (1738 K), Solubility in water, 35.6 g/100 mL (0 °C), Refractive index (n_D)-1.5442 (589 nm). Structure- Crystal structure Cubic cF8 Space group, Coordination geometry-Octahedral (Na^+), Octahedral (Cl^-)

It is found in ocean and in the extracellular fluid of many multicellular organisms. natural rocks of salt also found at many places. As the major ingredient in edible salt, it is commonly used as a condiment and food preservative.

In 2002, world production was estimated at 210 million metric tonnes, the top five producers (in million tones) being the United States (40.3), China (32.9), Germany (17.7), India (14.5).

As well as the familiar uses of salt in cooking, manufacturing pulp and paper, to setting dyes in textiles and fabric, to producing soaps, detergents. It is the major source of industrial chlorine and sodium hydroxide.

Sodium chloride is sometimes used as a cheap and safe desiccant because it appears to have hygroscopic properties, making salting an effective method of food preservation historically.

IN THIS CHAPTER

CHAPTER **2**

Chemical Bonding

Introduction

In 1704, Issac Newton outlined his atomic bonding thereby saying that atoms attach to each other by some force, which was further modified as "Particles attract one another by some force, which has immediate contact exceedingly strong, at small distances and performs the chemical operations, and reaches not far from the particles with any sensible effect".

In 1819, Jakob Berzelius, developed a theory of chemical combination stressing the electronegative and electropositive character of the combining atoms. Finally, in 1916, chemist Gilbert N. Lewis developed the concept of the electron-pair bond, in which two atoms may share one to six electrons thus forming the single, double or a triple bond.

2.1 Chemical Bond

Atoms combine to form molecules which are more stable and have lower energy than the individual atoms. Thus, atoms combine with each other, to lower their energy and hence gain extra stability.

Therefore, **Chemical Bond is defined as the force of attraction which binds the atoms together within a molecule.**

2.1.1 Need of Chemical Combination

Chemical combination takes place due to following reasons:

A. Tendency to acquire the noble gas configuration

The inertness of noble gases is due to eight electrons in the valence shell. Hence, the electronic arrangement of (s^2p^6) is of maximum stability and minimum energy. All other elements, other than noble gases, tend to acquire stable gas configuration, by losing, gaining or sharing of electrons with other elements.

This led to conclusion, that elements participate in bond formation in order to complete their octet.

The outermost shell configuration of the atoms of noble gases is a stable configuration of 8 electrons, is called an octet.

B. Tendency to acquire a state of minimum energy

Individual atoms have high energy and less stability, when atoms approach each other, they form a bond, thereby, acquiring the state of minimum energy by redistribution of electrons in their valence shell.

Therefore, it can be said that:

If Force of Attraction > Force of Repulsion
Energy is lowered and a bond is formed
If Force of Attraction < Force of Repulsion
A bond is not formed.

The above fact becomes more clear by the following diagram:

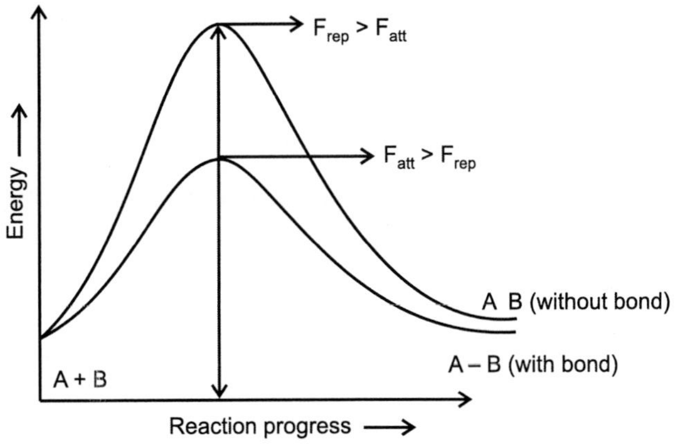

Fig. 2.1 Energy considerations during formation of a bond.

2.1.2 Formation of a Chemical Bond

We know, that the electrons in the outermost shell are actually responsible for the formation of a chemical bond. It can be either by gaining, losing or sharing of electron from the outermost shell, which consequently leads to formation of a bond.

The participation of the outermost electrons only, in the bond formation is because of the following reasons:

(*a*) The outermost electrons are away from the attractive influence of the nucleus and hence can be easily removed.

(*b*) The outermost electrons are closest to any neighbouring atom which may approach it and thus are attracted to another atom.

2.1.3 Types of Chemical Bond

There are different types of chemical bonds depending upon the different ways of combination of atoms.

(*a*) Ionic or Electrovalent Bond
(*b*) Covalent Bond
(*c*) Co-ordinate Bond
(*d*) Hydrogen Bond
(*e*) Metallic Bond
(*f*) Van der Waal forces of attraction.

2.2 Electrovalent/Ionic Bond

Ionic bond is formed by the complete transference of electrons between the participating atoms. In Ionic bond formation, a metal donates an electron, due to low ionization energy (low electronegativity) to form a positive ion or cation. The non metal atom has electronic configuration just short of a noble gas structure and has high electro-negativity and so readily gains electron to form negative ions or anions.

In the formation of an ionic bond or ionic compound, energy is released. The cations and anions obtained by the loss or gain of electrons attain the nearest noble gas configuration.

Often Ionic Bonds are formed between metals and non metals.

The number of electrons lost or gained is called ***Electrovalency***.

We can also, say, that electrovalency of an element is equal to the number of positive or negative charges acquired by the atom of an element in forming cation or anion respectively. There are conditions for formation of an Ionic Bond *i.e.*,

1. **Low Ionisation Energy of the element forming cation:** If Ionisation Energy is low, electron can easily be lost from an atom to form a cation.
2. **High Electron Affinity of an element forming an anion:** Electron Affinity is the amount of energy released when an atom accepts the electron to form a gaseous anion, *i.e.*,

$$X\,(g) + \text{electron} \rightarrow X^- + \text{Electron Affinity}$$

It is expressed in kJ/mole

More easily it will accept electron, more easily it will form an anion, and more amount of energy is released.

Example of ionic bond:

The most common example is formation of NaCl (Sodium Chloride)

Electronic configuration of Sodium atom = $1s^2\ 2s^2\ 2p^6\ 3s^1$

Electronic configuration of Chlorine atom = $1s^2\ 2s^2\ 2p^6\ 3s^2\ 3p^5$

By losing one electron from its outermost shell, sodium acquires the inert gas configuration and changes into a cation.

$$\underset{2,8,1}{Na} \rightarrow \underset{2,\,8}{Na^+} + e$$

On the other hand, chlorine atom accepts one electron, released by sodium to complete its octet and gets converted to chloride ion by attaining stable configuration of argon.

$$\underset{2,\,8,\,7}{Cl} + e \rightarrow \underset{2,\,8,\,8}{Cl^-}$$

When these two oppositely charged ions approach each other, they are held together by strong electrostatic forces of attraction and form an ionic compound, Sodium Chloride as shown below:

$$\underset{2,\,8,\,1}{Na} + \underset{2,\,8,\,7}{Cl} \rightarrow \underset{\text{Sodium ion}}{[Na]^+}\quad \underset{\text{Chloride ion}}{[Cl]^-}$$

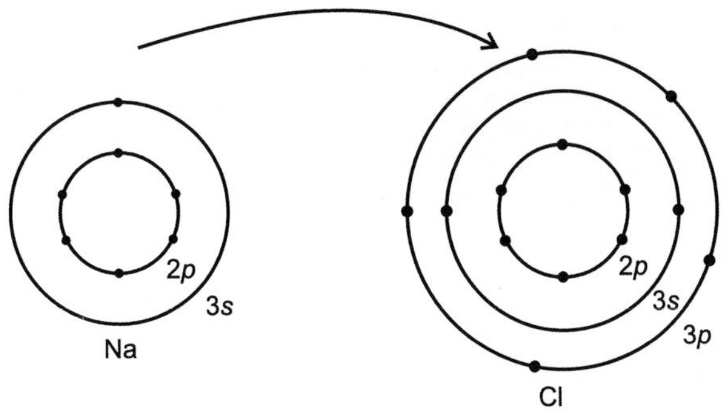

Fig. 2.2 Formation of NaCl Molecule.

2.2.1 General Characteristics of Ionic Bonds

A. High Melting and Boiling point

Ionic compounds have high melting point and boiling point due to electronic cation-anion interaction, extending throughout the crystal lattice. Since, the forces of attraction are very strong, therefore, considerable amount of energy is needed to break up their crystal lattice. When crystal lattice is broken on melting, ions are still present. Due to this, boiling point is also high. The melting points of various sodium halides are given below:

Table 2.1 Melting point of Sodium Halides

Compound	NaF	NaCl	NaBr	NaI
M.P. (°C)	995	800	750	651

B. Hardness

The existence of electrostatic forces of attraction extending throughout the crystal lattice makes ionic compounds hard. Hardness increases with decrease in the inter-ionic distances. Also, increase in the charge of the ions brings the oppositely charged ions closer to each other, making the solid harder.

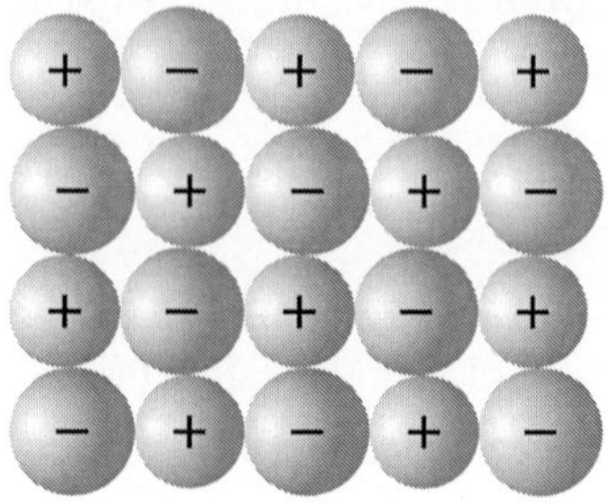

Fig. 2.3 Crystalline Lattice of Sodium Chloride where (+), (−) represents Na^+ and Cl^- ions respectively.

C. Crystalline Nature

Ionic compounds are usually crystalline in nature. X-ray diffraction has proved that constituent units in a crystal are ions and not molecules. These ions are arranged in a regular pattern to form a lattice.

D. Solubility

Ionic compounds are soluble in water but insoluble in benzene or other organic solvents. Water and other polar solvents have high dielectric constants, decreases the electrostatic force of attraction and ions get separated or we say that, the compound is dissolved.

E. Conductance

Ionic compounds do not conduct electricity in the solid state. In the solid state, the ions occupy fixed positions and hence are not free to move. But the compound conducts electricity when present in fused state or in aqueous solution. This is because, in fused state or in aqueous solution, the ions are free to move and hence are able to conduct electricity. NaCl(s) is an insulator but NaCl (fused) conducts electricity.

F. Brittle in Solid State

Ionic compounds are brittle in solid state.

This is because when a force is applied, the relative positions of cations and anions change with respect to each other. The cation-cation and anion-anion repulsions make the crystal brittle or cause it to break apart.

Therefore, inspite of the hardness, ionic compounds are brittle and can be powdered easily.

G. Isomorphism

Ionic compounds show Isomorphism *i.e.* the crystals of different ionic compounds having equal number of atoms and similar arrangement of atoms in crystalline state thus resulting in same geometry. It is known as isomorphism and crystals become isomorphs to each other *e.g.*, NaF and MgO show isomorphism as they have identical electronic structure of the ions.

Example 2.1: Out of CuCl and CuBr, which is hard and why?

Solution: Charge on the cation and anion remains same but the size increases from Cl to Br. Since, Cu and Cl are compatible in sizes, therefore, the force of attraction between Cu and Cl is making it more hard.

2.2.2 Structure of Ionic Solids

Ionic compounds in the solid state form a continuous ionic lattice structure in an ionic crystal. The constituent particles (atoms, ions or molecules) of a crystalline solid are arranged in a regular order. The position of these particles in the crystalline solid are referred to as ***Lattice Points***. The regular arrangement of infinite set of these lattice points is called crystal lattice or space lattice or simply crystal.

The crystal lattice is composed of an infinite number of small units, each of these small units is called a **unit cell**. When a large number of unit cells of the crystal combine together and arrange themselves in a regular order, the entire crystal lattice is obtained. A unit cell of a crystal lattice has all those properties which the crystal lattice has. The simplest form of ionic crystal is a **simple cubic** wherein all the atoms are placed at the corner of a cube as shown:

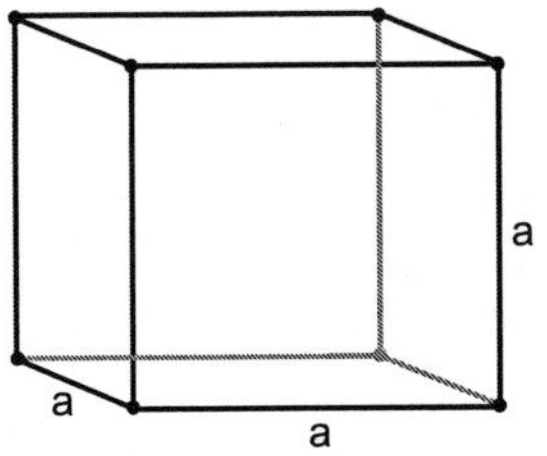

Fig. 2.4 Simple Cubic Structure.

The number of atoms per unit cell is one. Each atom at the lattice points is then shared equally between eight adjacent cubes, and the unit cell, therefore contains in total, one atom *i.e.* (1/8 × 8).

When all the ions are approximately the same size, they can form a different structure called a **face centered cubic structure.**

In this lattice, atoms are present at all the 8 corners and at the centre of each face as shown:

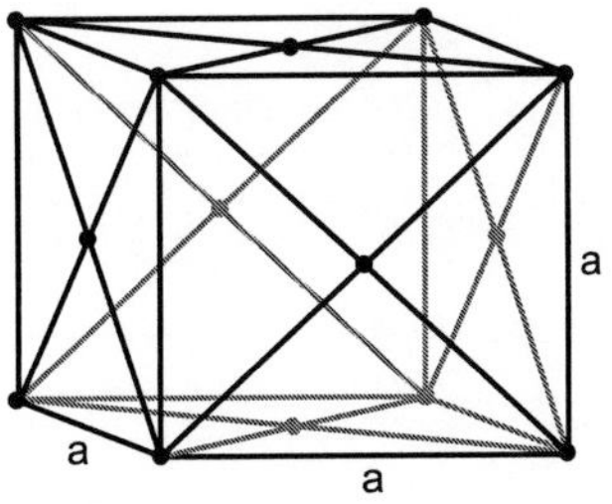

Fig. 2.5 Face centered cubic structure.

The **face centered** cube has atoms on the faces of the cube of which each unit cube gets exactly one half contribution, giving a total of 4 atoms per unit cell $\left(\frac{1}{8}\times 8+\frac{1}{2}\times 6=4\right)$. When the ions are of different sizes, the structure is often **body centered cubic**. The atoms are present at all the corners and at the centre of the unit cell.

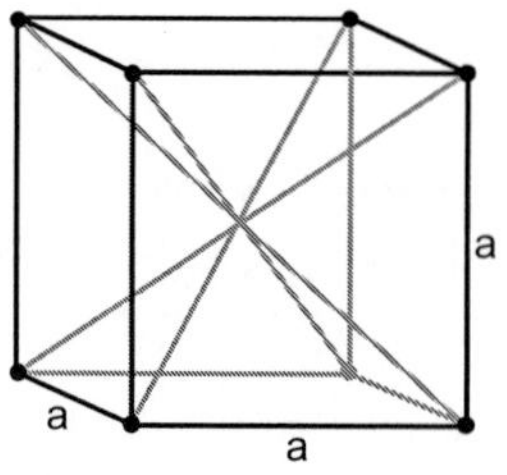

Fig. 2.6 Body centered cubic structure.

The **body centered** cubic system has one atom in the center of the unit cell in addition to the eight corner points. It has a contribution of 2 atoms per cell $\left(\frac{1}{8}\times 8+1=2\right)$.

2.2.3 Band Theory of Solids

Band structure of a solid describes ranges of energy that an electron is forbidden or allowed to have.

Any solid has a large number of bands. Fig 2.7 shows a simplified picture of the bands in a solid that allow the three major types of materials, *i.e.* metals, semiconductors, and insulators.

Band gap is one of the most useful aspects of the band structure, as it strongly influences the electrical and optical properties of the material.

Valence band is the highest range of electron energies where electrons are normally present at absolute zero. Conduction band is the range of electron energy higher than the valence band, sufficient to make the electrons accelerate under the influence of an applied electric field.

In insulators, the electrons in the valence band are separated by a large gap from the conduction band, in conductors like metals, the valence band almost overlaps with the conduction band. In semiconductors, there is a small gap in between the valence and the conduction band.

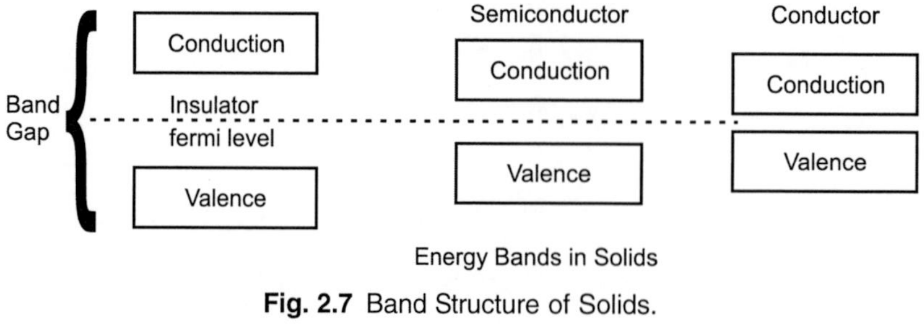

Fig. 2.7 Band Structure of Solids.

To make this point more clear, some of the elements with band energy gaps in kJ/mol is given below:

Table 2.2 Band Energy Gaps of Some Elements

C (Diamond)	*Si*	*Ge*	*Sn* (grey)
502	104	60	10

Metals are electrical conductors with low resistance to the flow of electrons because of their lattice arrangement *i.e.*, they have mobile or delocalised electrons. At the other extreme are the solid non-metals with covalent bonds and are termed as insulators. Diamond is an insulator. In between them, are the semi-conductors.

For Ge, conduction electrons will be released by the normal thermal excitation.These elements are called Intrinsic Semiconductors.

Promotion of an electron leaves behind a hole in the valence band. This means, there is a space for another electron.

The conducting properties of a semiconductor are greatly increased by controlled addition of certain impurities. If Phosphorous is introduced, to Si lattice, 4 of the valence electrons of the Phosphorous will be accommodated in the lattice but the 5th electron is excited to higher level thereby decreasing the energy gap to 1.3 kJ/mol and consequently increasing the conductance.

Here, conduction is stimulated by electrons.

If Boron is introduced to Si lattice, a hole will be created and now the energy gap is 0.98 kJ/mole.

Here, conduction is stimulated by positive charge.

In the fig 2.7 there is one more term used *i.e.* Fermi Level, which is the highest occupied energy level at absolute zero which means that all the energy levels up to the Fermi Level are occupied by the electrons.

2.3 Lattice Energy

The lattice energy or lattice enthalpy of an ionic solid is a measure of the strength of bonds in that ionic compound. It is given by the symbol U and is equivalent to the amount of energy required to separate a solid ionic compound into gaseous ions. Lattice energy can also be considered as the energy released in the formation of 1gm mole of a crystal from the gaseous ions.

Greater the value of lattice energy more is the stability.

The lattice energy of sodium chloride is –774 kJ /mol.

The negative sign, indicates that the process is exothermic and the energy of the system is lowered as the solid is formed.

Lattice energies can be determined by using:

(*a*) **Born Lande Equation:** This equation is used for theoretical calculation of Lattice energy on the basis of the coulombic interactions (attractive and repulsive interactions) between the ions present in the crystal.

(*b*) **Born Haber cycle:** Experimental values of lattice energy are obtained from thermodynamic data using Born Haber cycle.

2.3.1 Born Lande Equation

Consider, the two oppositely charged ions M^{z+} and X^{z-}, as point charges separated by a distance '*r*', z^+ and z^- are the number of charges on cations and anions respectively.

The attractive and repulsive interactions would take place between the ions.

The electrostatic energy of attraction is given by:

$$E = \frac{-z^+z^-e^2}{r} \qquad ...(1)$$

Where, *e* is the electronic charge (1.6×10^{-19} C). The value of energy is negative, because energy is given out when a crystal is formed.

For one mole, the attractive energy is

$$E = \frac{-N_0 z^+z^-e^2}{r} \qquad ...(2)$$

Where N_0, is the Avogadro's Number. Ionic crystals are not composed of simple ion pairs but of ionic clusters in which each cation is surrounded by a definite number of anions and vice-versa. In order to calculate the coulombic energy of any single ion, we must know the number of, and arrangement of neighbouring ions. Because of electrostatic interactions with surrounding ions, the coulombic energy of a single ion is greater than indicated by the above equation, by a factor typical of each crystal structure. This factor is designated as A and is known as Madelung Constant. Madelung Constant depends entirely on the geometry of the crystal.Thus, the equation becomes

$$E = \frac{-N_0 A z^+z^-e^2}{r} \qquad ...(3)$$

where *A* is the Madelung Constant

Table 2.3 Madelung constant for some typical crystal structures

Crystal Structure	Coordination No. of ions	Madelung constants
NaCl	6 : 6	1.7475
CsCl	8 : 8	1.7627
ZnS (Zinc Sulphide)	4 : 4	1.6381
CaF_2	8 : 4	5.0388

When the cation and anion approach close to one another, repulsive force comes into play due to the interaction between the electron clouds on the two atoms or ions.

The repulsive forces increase rapidly as r decreases. The repulsive force is given by B/r^n where B is a constant known as *Repulsion Coefficient* and n is also a constant called the ***Born Exponent***.

$$E \text{ (repulsion)} = B/r^n \quad ...(4)$$

For one mole,

$$E = \frac{N_0 B}{r^n}$$

Therefore, the total energy holding the crystals together is U, the lattice energy. This is the sum of the attractive and repulsive forces as given below:

$$U = -\frac{N_0 A z^+ z^- e^2}{r} + \frac{N_0 B}{r^n} \quad ...(5)$$

At equilibrium distance, $r = 0$, the attractive and repulsive forces exactly balance each other. At equilibrium distance, U is minimum and hence $dU/dr = 0$.

Differentiating equation (5) with respect to r, we get

$$dU/dr = -N_0 A\, z^+ z^- e^2 (-1)\, r^{-2} + BN_0 (-n)\, r^{(-n-1)}$$

$$= \frac{N_0 A z^+ z^- e^2}{r^2} - \frac{nN_0 B}{r^{n+1}} \quad ...(6)$$

At equilibrium, $r = r_0$ and $\frac{dU}{dr} = 0$

Equation (6) becomes,

$$\left.\frac{dU}{dr}\right|_{r=r_0} = 0$$

$$\frac{N_0 A z^+ z^- e^2}{r_0} - \frac{nN_0 B}{r_0^{n+1}} = 0$$

$$\frac{N_0 A z^+ z^- e^2}{r_0^2} = \frac{nN_0 B}{r_0^{n+1}}$$

Therefore,

$$B = \frac{A z^+ z^- e^2 r_0^{n-1}}{n} \quad ...(7)$$

Substituting the value of B in eqn. (5) we get,

$$U = -\frac{N_0 A z^+ z^- e^2}{r_0} + \frac{N_0 A z^+ z^- e^2 r_0^{n-1}}{r^n}$$

$$U = -\frac{N_0Az^+z^-e^2}{r_0} + \frac{N_0Az^+z^-e^2}{nr_0}$$

$$= -\frac{N_0Az^+z^-e^2}{r_0}\left[1-\frac{1}{n}\right]$$

The equation is known as the **Born Lande Equation.**

Example 2.2: Calculate the lattice energy of sodium chloride (in kJ/mol) from the following data:

$$A = 1.75,\ r_0 = 2.8\text{Å},\ n = 9$$

$$N_0 = 6.023 \times 10^{23},\ e = 4.8 \times 10^{-10} \text{ e.s.u}$$

Solution:

$$U = \frac{-N_0Az^+z^-e^2}{r_0}\left[1-\frac{1}{n}\right]$$

$$= \frac{-6.023\times10^{23}\times1.75\times(4.8\times10^{-10})^2\times1^2}{2.8\times10^{-8}}\left[1-\frac{1}{9}\right]$$

$$= -770.56 \times 10^{10} \text{ ergs/mol}$$

Since 10^7 ergs = 1 Joule

10^{10} = 1kJ

Therefore, $U = -770.56$ kJ/mol

Conclusions Drawn from Born Lande's Equation

(1) Lattice Energy depends upon the values of r, smaller the value of r (internuclear distance) more is the Lattice Energy as $U \propto \frac{1}{r}$. Therefore, the Lattice Energy of the following ionic solids increases in the following order:

LiI < Li Br < Li Cl < Li F

The internuclear distance is minimum in case of LiF because of smaller atomic radii of Li and F, leading to maximum value of Lattice Energy.

(2) Higher the charge on cation and anion, greater would be the magnitude of lattice energy and therefore, greater would be the stability of the crystal.

For example, the order of lattice energy for the ionic solids is

Ba S	>	**Ca F_2**	>	**Li F**
$Ba^{2+}S^{2-}$		Ca^{2+} F^{-}		Li^{+} F^{-}
(Bi- bivalent)		(Bi-univalent)		(Uni-univalent)

(3) The lattice energy depends on the value of the Madelung constant, which depends on the electron number and the geometric pattern of the ions in the crystal lattice.

2.3.2 Born-Haber Cycle

The Born-Haber cycle is an approach to analyze reaction energies. It was named after and developed by the two German scientists **Max Born and Fritz Haber**.

The Born-Haber cycle involves the formation of an ionic compound from the reaction of a metal with a non metal. Born-Haber cycle is used primarily as a means of calculating lattice enthalpies, which otherwise cannot be measured directly.

The lattice enthalpy is the enthalpy change involved in formation of the ionic compound from gaseous ions. Born-Haber cycle calculates the lattice enthalpy by comparing the standard enthalpy change of formation of the ionic compound to the enthalpy required to make gaseous ions from the elements. This is an application of Hess's Law.

According to which "The enthalpy of formation of a compound remains the same whether the reaction proceeds in one step or in several steps".

To make gaseous ions from elements, it is necessary to atomise the elements and then to ionise the atoms. If the element is normally a molecule then we have to consider its bond dissociation energy as well.

The energy involved in removing electrons to make a cation is called the **Ionisation Energy**.

The enthalpy of adding electrons to an atom to make it an anion is called the **Electron Affinity**.

Let us study the Born Haber cycle for the standard enthalpy change of formation of lithium fluoride.

Step 1 : Sublimation Energy, (SE) of metal (In case of Lithium)

Li (s) $\rightarrow$ Li (g) SE

(Sublimation energy is the energy required for the conversion of one mole of metallic lithium into gaseous lithium atoms)

Step 2 : Ionisation enthalpy, (IE) of metal

Li(g) $\rightarrow$ Li^{+}(g) + IE

Step 3 : Dissociation energy, (DE) of the non metal

F_2 $\rightarrow$ F ½ DE

(The energy required for the formation of one mole of gaseous fluorine atoms = D/2)

Step 4 : Electron affinity, (EA) of the non metal

$F + e \rightarrow F^{-} - EA$

Step 5 : Lattice enthalpy, (LE) of LiF

$Li^{+} + F^{-} \rightarrow LiF - LE$

According to Hess law, the enthalpy of formation of MX should be the same whether the reaction proceeds in one step or several steps.

Hence, the enthalpy of formation, H_f for lithium fluoride is equal to

$$H_f = SE + ½ DE + IE + EA + U$$

Using positive sign for energy absorbed and negative sign for energy released, we get

$$-H_f = SE + ½ DE + IE - EA - U.$$

where, H_f is the heat of formation of LiF

Example 2.3: Calculate the lattice energy of NaCl crystal from the following data by the use of Born Haber Cycle.

Ionisation Energy of Na (g)	=	489.5 kJ/mol
Electron Affinity for Cl (g)	=	– 351.4 kJ/mol
Sublimation Energy	=	108.7 kJ/mol
Dissociation Energy for Cl_2	=	225.9 kJ/mol
Heat of formation of NaCl (H_f)	=	– 414.2 kJ/mol

Solution: Substitute the values in the expression

$$H_f = SE + ½ DE + IE + EA + U$$

$$-414.2 = 108.7 + ½(225.9) + 489.5 - 351.4 + U$$

$$U = -773.95 \text{ kJ/mole}$$

There is another way of calculating Lattice Energy in the form of **Kapustinskii equation** which is experimentally difficult to determine. It is named after Anatoli Kapustinskii. The equation given by him to calculate the lattice energy is:

$$U = -1202.5 n z^{+} z^{-} \left[1 - \frac{0.345}{r_{+} + r_{-}}\right]$$

In this equation, n is the number of ions in the empirical formula, z is the anionic and cationic charge, and r is the radius of the anion/cation. The units of the radii in this equation are Ångström, and the lattice energy is given in kJ/mol. The calculated lattice energy gives a good estimation although the real value differs in most cases by less than 5%.

Furthermore, when the lattice energy is known one is able to determine the atomic radii using the Kapustinskii equation. This is useful for rather complex ions like sulphate (SO_4^{2-}) or phosphate (PO_4^{3-}).

2.3.3 Role of Lattice Energy (LE)

Lattice Energy helps us in predicting the following information about the ionic solid.

A. Stability of Ionic Solid

Higher the value of lattice energy, more energy is required to separate the oppositely charged ions and hence higher is the stability of an ionic solid. For example,

Oxides	BeO	MgO	CaO	SrO	BaO
LE (kJ/mol)	4541	3895	3520	3325	3108

Therefore, the order of stability

BeO > MgO > CaO > SrO > BaO

B. Melting Point and Boiling Points of Ionic Solids

Higher the value of lattice energy, the electrostatic forces of attraction between the oppositely charged ions are strong, leading to higher value of melting points and boiling points of ionic solids. *e.g.*,

Halides of Li	LiF	LiCl	LiBr	LiI
(Lattice energy) kJ/mol	1034	840.1	781.2	718.2
Melting Points (°C)	870	613	597	446

C. Solubility of the Ionic Solids

Whether a given ionic solid is soluble or insoluble in polar or a non polar solvent, depends on the two factors *viz.,*

1. **Lattice Energy**
2. **Hydration energy** of the ionic solid which is the energy released when the ions of the ionic solid obtained by dissolution of ionic solid in ionic solvent are hydrated and form hydrated ions.

   Therefore, higher the value of lattice energy, the ions are not separated easily in the ionic crystal and are not hydrated, leading to lesser solubility. This can be very well understood by comparing the lattice energies of the halides of lithium.

Thus,

Halides of Li	LiF	LiCl	LiBr	LiI
LE	1034	840.1	781.1	718.
Solubility	**LiF**	**< LiCl**	**< LiBr**	**< LiI**

Since the lattice energy of LiF is higher, its hydration energy is less which decreases its solubility in water.

Example 2.4: Why $PbSO_4$ is sparingly soluble in water?

Solution: Since Pb^{2+} and SO_4^{2-} ions are comparable in sizes, the force of attraction between the two are good enough that makes the lattice energy higher than the value of hydration energy, leading to lower solubility in water.

2.4 Covalent Character in Ionic Compounds

No ionic bond is 100% ionic. Ionic bond has some covalent character due to interaction between the cation and an anion, as the positive charge exerts an attractive pull on the electrons in an anion, leading to the concept of polarisation or distortion. The phenomenon in which the symmetrical shape of the anion gets deformed by the approach of the cation to it is called polarisation. Similarly, covalent bond acquires some ionic character due to the greater differences in electro negativities of the two atoms.

2.5 Polarisation and Polarisability

When the cation approaches the anion, the net positive charge on the cation attracts the electron cloud of the anion and repels the positively charged nucleus of the anion. Therefore, the electron cloud of the anion gets distorted or bulged towards the cation. This is called **polarisation of the anion.** The electron charge cloud of the anion no longer remains spherical but gets distorted, as shown:

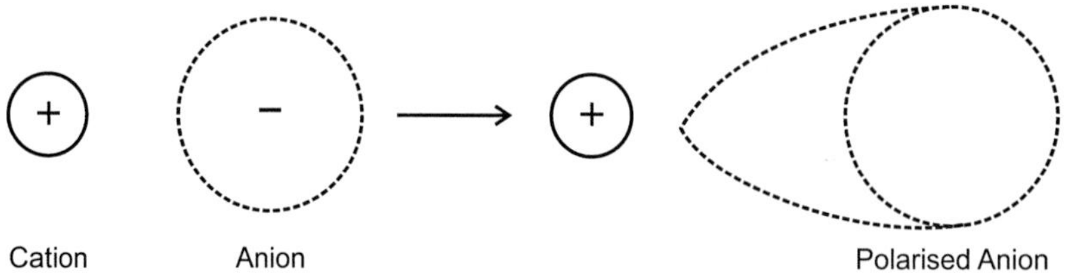

Fig 2.8 Polarisation of the anion.

High electron density results between the two nuclei due to polarisation of ions leading to an appreciable covalent character.

The power of the cation to distort the anion is known as **polarising power** of the cation and the tendency of an anion to get polarised is called **polarisability** of the anion.

Depending upon the polarising power of the cation, the ionic bond between the cation and anion becomes either partially covalent or purely covalent. Greater is the polarising power of a cation, greater will be the amount of covalent character produced in the ionic molecule.

2.6 Fajan's Rule

In Inorganic chemistry, Fajan's rule, formulated by Kasimir Fajans was used to predict whether a chemical bond will be covalent or ionic, which depends on the charge on the cation and the relative sizes of the cation and anion. These are summarised in the Table 2.4.

Table 2.4: Charge and size of ionic and covalent bond

Ionic	Covalent
Low positive charge	High positive charge
Large cation	Small cation
Small anion	Large anion

Factors affecting the extent of Polarisation:

(*a*) **Small Positive Ions:** Small positive ions have high polarising power because positive charge is concentrated over a small surface area (*i.e.* the positive field of the nucleus is strong) and thus it can distort the electron cloud of the anion efficiently.

(*b*) **Large Size of the Anion:** The large size of the anion would keep its orbitals away from the positive field of the nucleus and hence such anions will get polarised easily.

(*c*) **High Charge on the Cation and Anion:** High positive charge on the cation would attract the electron cloud of the anion more easily and hence distort it. High negative charge on the anion would lead to more number of orbitals or we can say electrons, which can be easily polarised by the cation.

Even the electronic configuration of the cation affects the polarisation in ionic compounds. The molecules having cations with 18 electron valence shell configuration are more covalent than those having 8 electron valence shell configurations. This is because of the d-electrons of the 18 electrons which shield the nuclear charge less effectively than s and p electrons. Hence, a cation with 18 electron shell has larger effective nuclear charge than the cations with 8 electrons in the outermost shell, even though the two may have same size and same charge. For e.g. AgCl is more covalent than NaCl even though the size of Ag^+ and Na^+ is almost the same.

Example 2.5: *Why Sodium Chloride is ionic and Aluminium iodide is covalent*?

Solution: Sodium Chloride has a low positive charge (+1), a fairly large cation and small anion thereby making it ionic; but Aluminium iodide (AlI_3) has a high positive charge (+3) and a large anion,which causes the distortion of the anion thereby giving it a covalent character.

Example 2.6: *$AlCl_3$ (anhydrous) is covalent and $AlCl_3.6H_2O$ is ionic. Explain*

Solution: Al^{3+} is small in size and has high positive charge, whereas chloride is larger in size and is easily polarised, giving $AlCl_3$ a covalent character.

$AlCl_3.6H_2O$ in solution exists as $[Al(H_2O)_6]^{3+} . 3Cl^-$. Due to its existence in the form of ions, it is considered as an ionic compound.

2.7 Covalent Bond

Gilbert N Lewis, in 1916 gave the concept of covalent bonding. Covalent bond is a form of chemical bond that is characterised by the sharing of pairs of electrons between atoms. Therefore, the bond formed between the two atoms by mutual sharing of electrons between them so as to complete their octets is a covalent bond. The formation of a covalent bond between the two atoms A and B is favoured by two factors:

(*a*) Both the combining atoms should have higher values of Ionisation Energies, so that none of them should be able to lose electrons to form a cation.

(*b*) Both the combining atoms should have comparable electronegativities, so that symmetrical distribution of the electron pair between the two atoms is obtained leading to purely covalent bond.

Gilbert Newton Lewis gave the Concept of covalent bond

Examples of Covalent Bond:

A. Formation of hydrogen molecule

The electronic configuration of H is $1s^1$

Hydrogen atom has one electron in its valence shell. In the formation of hydrogen molecule, both the atoms contribute one electron each for mutual sharing.

$$H\cdot\cdot H \longrightarrow H{:}\,H \ \text{ or } \ H{-}H$$

B. Formation of NH_3

N– atom has 5 outer electrons and H has only 1 $\overline{e}$ and by sharing its electrons with the 3 electrons of nitrogen, it attains a stable configuration.

$$\cdot\ddot{N}\cdot + 3[H.]^{-} \longrightarrow H\cdot\cdot\ddot{\underset{\cdot\cdot}{N}}\cdot\cdot H$$
$$\underset{}{H}$$

The shared pairs of electron present between the atoms are **bond pairs** and the valence electron not involved in bonding are **non bonding pairs or lone pairs.**

This concept was called lewis concept named after the name of the scientist. Although this concept gave a new sense to bonding but had certain limitations.

Limitations of Lewis concept:

1. It couldn't explain how the atoms are held together in the molecules like H_2, Cl_2 etc., as no electrostatic forces are operating between them.
2. It couldn't explain the shapes of molecules containing covalent bonds.
3. It could not explain the paramagnetic character of the oxygen molecule.
4. It couldn't explain the formation of molecules which violate the octet rule which is the tendency of the atom to have eight electrons in their outer most shell. *e.g.*, PCl_5, SF_6, $AlCl_3$ etc.

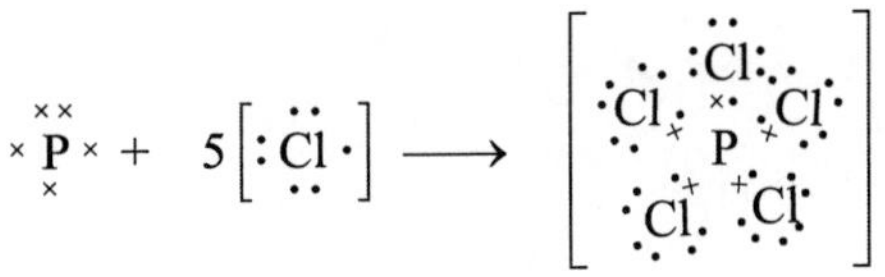

This is not clearly understandable so it is the limitation.

(5) It could not explain the formation of the molecules which have odd number of electrons such as NO and ClO_2. In these molecules, the two atoms are linked with each other by means of a bond that has odd number (generally three) of electrons and such bonds are 3–e bonds. For example,

S in SF_6 has 12 electrons around them as shown

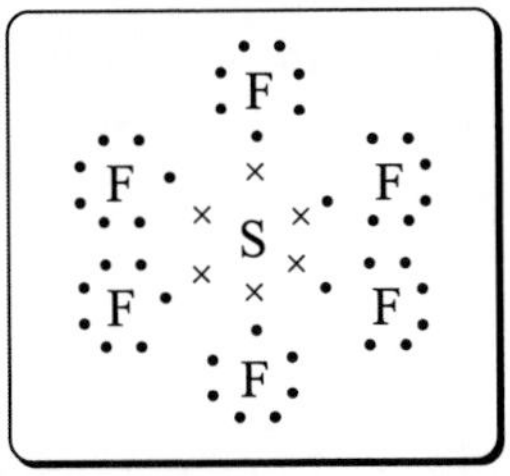

The reason for this failure can be explained on the basis that in some molecules, the atoms are linked not only by covalent bonds but also by an other type of linkage known as singlet linkage or we can say half bond, wherein one sided sharing of only one electron takes place between two atoms and is represented by a half arrow as shown in SF_6 molecule.

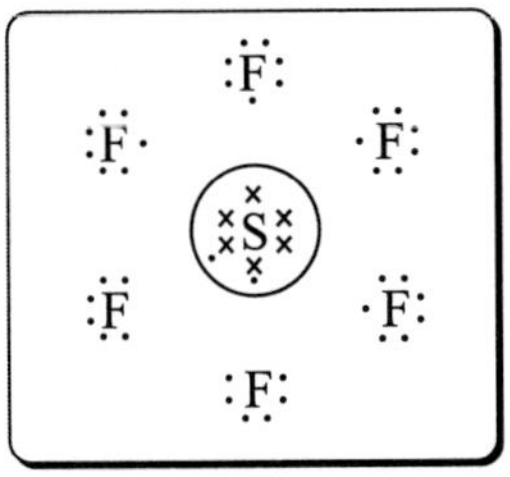

There are 8 electrons around S-atom and F-atoms. Therefore, there are 4 singlet linkages and 2 covalent bonds formed.

Therefore, the molecule is represented as

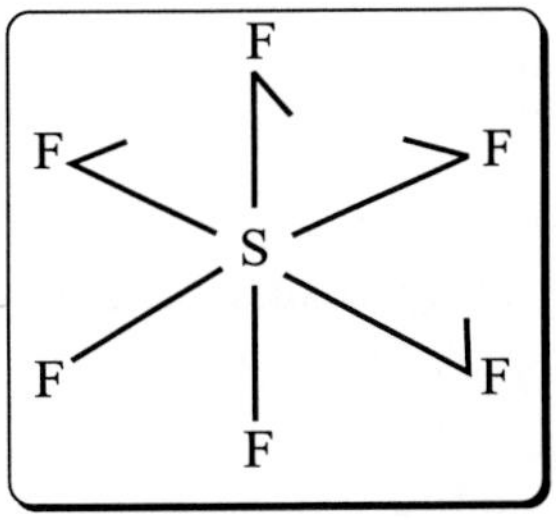

Covalent bond can be **polar** or **non-polar** covalent bond depending upon the electronegativity difference between the atoms.

If a covalent bond is formed between similar atoms, the shared pair of electrons lies almost in the centre and electron cloud is uniformly distributed around the two atoms, such a covalent bond is called **Non-Polar covalent bond**.

For example: H–H, Cl–Cl.

Examples of other non-polar compounds include fats, oil and petrol. Most non-polar molecules are water insoluble (hydrophobic) at room temperature. However, many non-polar organic solvents, such as turpentine, are able to dissolve non-polar substances - *like dissolves like*.

If a covalent bond is formed between the different atoms, then the shared pair of electrons is displaced towards the more electronegative atom causing greater concentration of electron density around the more electronegative atom developing partial positive and negative charges, such a bond is a **Polar covalent bond**.

For example: H$\rightarrow$ Cl. Here the electron density is more displaced towards the Chlorine atom.

Examples of other polar molecules also include ammonia and sugar . Polar molecules are generally able to dissolve in water (hydrophilic) due to the polar nature of water *i.e. like dissolves like*.

The following values of electronegativity difference between the two atoms forming a bond decides about the nature of the bond formed are cited below in Table 2.5

Table 2.5 Electronegativity scale for different bonds

Nature of bond	Electronegativity difference
Non-polar covalent bond	0.0 – 0.4
Slightly polar bond	0.5 – 0.9
Moderately polar bond	1 – 1.3
Highly polar bond	1.4 – 1.7
Slightly ionic bond	1.8 – 2.2
Ionic Bond	2.3 +

Example 2.7: HCl in the gaseous state is covalent while in the aqueous solution, it is ionic, explain.

Solution: H and Cl atoms, acquire a partial positive and negative charges due to electronegativity difference between the two and exist as $H^{\delta+} - Cl^{\delta-}$ .Therefore, H–Cl becomes polar covalent bond. But, when HCl molecule is dissolved in H_2O, it surrounds the water molecule and splits $H^{\delta+} - Cl^{\delta-}$ into $H^{\delta+}$ and $Cl^{\delta-}$ ions. Therefore, the aqueous solution of H–Cl is ionic.

2.7.1 Properties of Covalent Compounds

1. **Physical state:** Generally covalent compounds are gases or liquids at room temperature because of weak inter molecular forces of attraction between the molecules.
2. Covalent compounds have low melting and boiling points because only weak Van der Waals forces exist except for diamond, silica, which have covalent giant lattices and requires large amount of energy to break the lattice.
3. Covalent compounds are soluble in non polar solvents like benzene and chloroform etc.
4. Covalent compounds neither have ions or mobile electrons, and hence do not conduct electricity in the solid state.
5. Covalent compounds are soft and waxy.
6. Covalent bonds are directional and rigid as these are directed in a particular direction in space and have definite geometry.

Characteristic Properties of Covalent Bond

A. Bond Length
B. Bond Energy
C. Bond Angle

A. Bond Length

It is the average distance between the centers of nuclei of two bonded atoms and is expressed in Å (10^{-8}cm) or Pico meters (10^{-12} m).

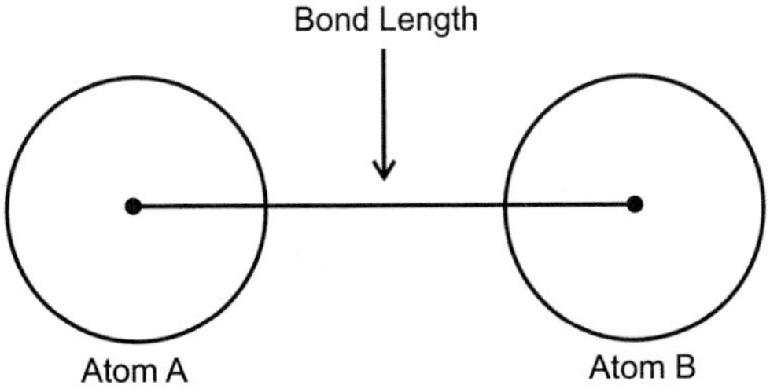

Fig. 2.9 Illustration of the bond length between the two atoms A and B.

Some important points to remember

(1) For a given atom the bond length increases with increase in the size of the other atom bonded to it *e.g.*,

$$HI > HBr > HCl > HF$$

(2) Bond length is related to bond order *i.e.* Bond length $\propto \frac{1}{Bond\ order}$. Bond order is also given as $\frac{1}{2}[N_b - N_a]$ where N_b, N_a is the no. of electrons in the bonding and antibonding

orbitals. The bond would get shorter, when more electrons participate in the bond formation. Bond length is also related to bond strength and the bond dissociation energy, since a stronger bond is also a shorter bond. In short, we can say, that lower the bond length more is the bond order.

(3) Bond length decreases with the increase in *s*-character since an *s*-orbital is smaller because of its greater penetration towards the nucleus. *e.g.*,

$$sp^3\ C-H = 110pm > sp^2\ C\text{–}H = 109\ pm > sp\ C\text{–}H = 108\ pm$$

B. Bond Energy

It is a measure of bond strength in a chemical bond. For example, the bond energy of C – H bond in methane is the enthalpy change involved in breaking up one molecule of methane into a carbon atom and 4 hydrogen atoms divided by 4. Bond energy should not be confused with bond dissociation energy.

Another example: O – H bond of water molecule (H–O–H) has 493.4 kJ/mol of bond dissociation energy and 424.4 kJ/mol is needed to cleave the remaining O–H bonds.

Important features among bond energies

1. Bond energy increases with increase in the electronegativity difference between the atoms because of increase in the bond polarity as the electronegative atom acquires a partial negative charge and less electronegative atom has a partial positive charge. For example, the following is the order of bond energies of different hydrogen halides.

$$H—F > H—Cl > H—Br > H—I$$

2. The bond energy increases with the increase in the percentage of *s*-character in a hybrid orbital as more is the s-character, more is the penetration of electrons towards the nucleus. Thus, bond energy increases in the following order :

	sp^3	<	sp^2	<	sp
% of *s*-character	25		33.3		50

3. Bond energy increase from a single bond to a triple bond *i.e.*,

$$C—C < C=C < C\equiv C$$

4. The bond energy decreases with the increase in the number of lone pairs on the bonded atom, which is due to the presence of electrostatic repulsion between lone pairs of electrons on the two bonded atoms as the order of repulsions is *lp – lp* > *lp – bp* > *bp – bp*. For example,

	C — C	>	N — N	>	O — O	>	F — F
B. E (kcal/mol)	81.6		39		34.2		33.3

C. Bond Angle

The average angle between the bonded orbitals (*i.e.*, between the two covalent bonds) is known as bond angle, **denoted by θ.**

Important features among bond angles:

1. It has been observed that as the *s*-character of the hybrid bond increases, bond angle increases.

Bond Type	sp^3	sp^2	sp
Bond Angle	109°28′	120°	180°

Depending upon the kind of repulsions operating in the molecule, there may be deviations from the expected bond angles. NH_3 is sp^3 hybridised, the expected bond angle is 109°28′ but due to repulsion between lone pair on Nitrogen and bond pair between N and H, the bond angle is 107.5°.

Similarly, H_2O is also sp^3 hybridised but instead of 109°.28′, the bond angle is 104.5° due to repulsions between lone pairs present on oxygen.

2. If the electronegativity of the central atom decreases, bond angle decreases. For example,

Molecule:	H_2O	>	H_2S	>	H_2Se	>	H_2Te
Bond angle:	104.5°	>	92.2°	>	91°	>	89.5°

Since sulphur is less electronegative than oxygen, the bond pairs in H_2S are more away from the sulphur atom than in water where O is a more electronegative atom with the result, the repulsive forces between bond pairs are smaller producing smaller bond angles.

2.8 Calculation of Ionic Character from Electronegativity

The percentage of ionic character in a polar covalent bond can be calculated from electronegativity values. Electronegativity is the tendency of an atom in a molecule to attract the shared pair of electrons towards itself. It depends upon :

(*i*) Charge on the nucleus

(*ii*) Size of the atom

(*iii*) Shielding effect.

In polar covalent molecules, greater is the difference in the electronegativities of the bonded atom, greater is the polarity in the molecule leading to higher percentage of ionic character. Easy development of charge increases the magnitude of ionic character.

Various empirical relationships are also given to calculate the percent ionic character of a bond of which the simplest and the most acceptable relationship was driven by Hanny and Smith in 1946.

It is given by the following expression:

$$\% \text{ I.C.} = 16(\chi_a - \chi_b) + 3.5\,(\chi_a - \chi_b)^2$$

Where I.C. is the ionic character, χ_a and χ_b are the electronegativites of the bonded atoms A and B.

It has been proposed by Pauling that "if the difference in the electronegativities of the bonded atoms is 1.9 then the bond is 50% ionic and 50% covalent in character."

If it is less than 1.9, the bond is predominantly covalent and if it is more than 1.9 then the bond is predominantly ionic.

Example 2.8: The electronegativities of hydrogen and fluorine atoms are 2.1 and 4.0, Calculate the percentage ionic character of HF.

Solution: $\% \text{ I.C} = 16\,(\chi_a - \chi_b) + 3.5\,(\chi_a - \chi_b)^2$

$= 16\,(4.0 - 2.1) + 3.5\,(4.0 - 2.1)^2$

$= 16\,(1.9) + 3.5\,(1.9)^2$

$= 30.4 + 12.6 = 43\ \%$

Therefore, HF is 43% ionic in nature.

2.9 Co-ordinate Bond

A co-ordinate bond is a description of covalent bonding between two atoms in which both electrons shared in the bond come from the same atom. Oncc the bonds have been formed using this, its strength and description is no different from that of other polar covalent bonds.

Co-ordinate covalent bonds are involved when a Lewis base (an electron donor) donates a pair of electrons to a lewis acid (an electron acceptor) to give an adduct. The process of forming a dative bond is called co-ordination.

The electron donor acquires a positive formal charge as it becomes deficient of electrons while the electron acceptor acquires a negative formal charge, as it gains electrons.

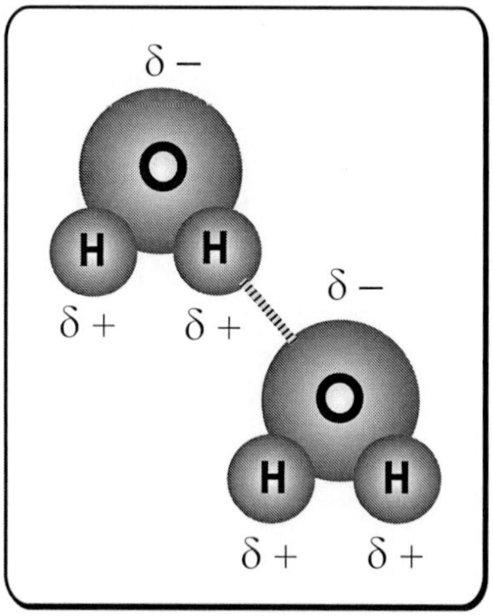

Hydrogen bonding in water

An example of co-ordinate covalent bonding in the ammonium ion, is shown below:

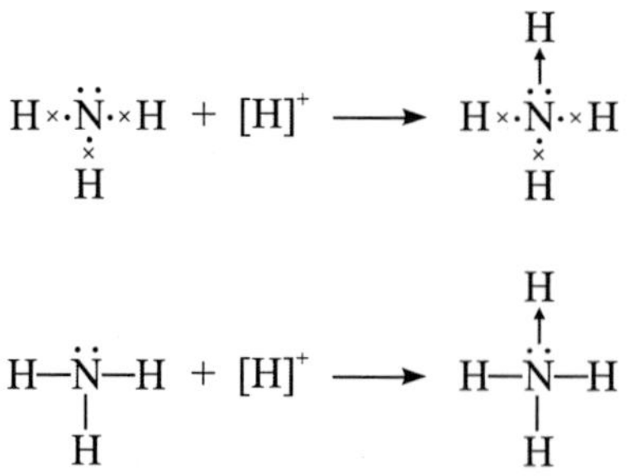

The orbital concept of co-ordinate bond can be explained by taking an example of H_3O^+ ion.

According to this concept, when an orbital containing a lone pair of electrons overlaps with an empty orbital present in the other atom, a co-ordinate bond is formed.

This can be explained by considering the formation of H_3O^+ ion.

The O-atom in water has four orbitals $2s^2$, $2p_x^{\ 2}$, $2p_y^{\ 1}$, $2p_z^{\ 1}$

Two of which are used to form covalent bond with H-atom, while the other two contains lone pair of electrons. The 's' orbital of H^+ ion is empty and overlaps with one of the lone pairs in the *p*-orbital of *O*-atom to form H_3O^+ ion.

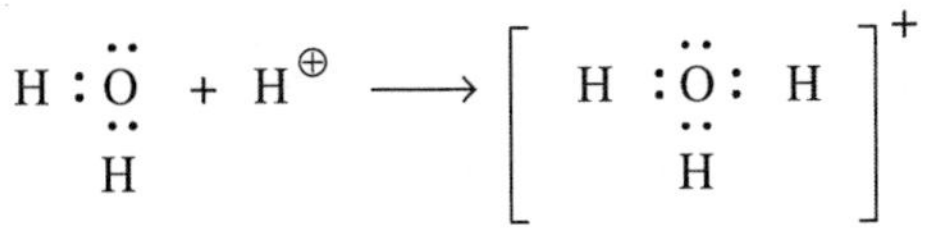

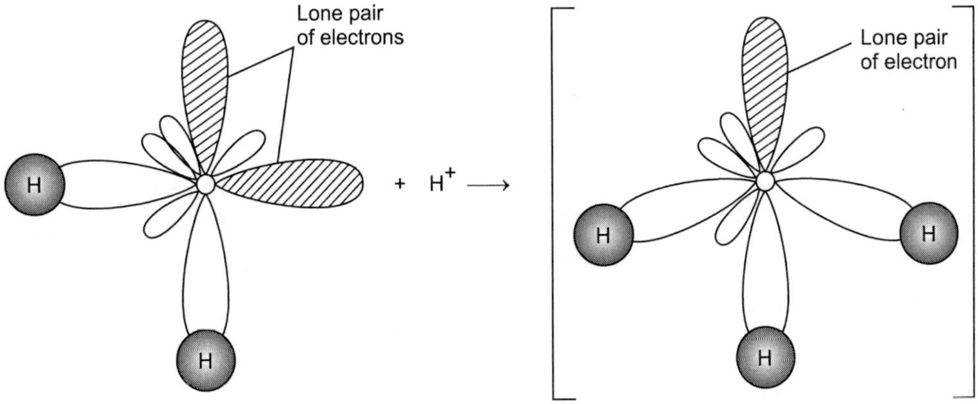

2.9.1 Characteristics of Co-ordinate Bond

1. These are usually insoluble in water, but soluble in non-polar solvents.
2. Co-ordinate compounds do not conduct electricity as these do not dissociate into ions when dissolved in a solvent.
3. The melting and boiling point of these compounds are higher than covalent compound, but lesser than the ionic compounds.
4. Co-ordinate bond is directional, therefore, it has definite geometry and exhibits isomerism.

2.10 Hydrogen Bond

A hydrogen bond is a special type of attractive interaction that exists between an electronegative atom and a hydrogen atom bonded to another electronegative atom.

Hydrogen bond is defined as an electrostatic attractive force between covalently bonded hydrogen atom of one molecule and an electronegative atom of another molecule.

Hydorgen bonds can occur between molecules or within different parts of a single molecule. The hydrogen bond is stronger than Van der Waals forces, but weaker than covalent or ionic bonds.

Hydrogen bond
↑
H——F-------- H——F

Hydrogen bond is represented by a dotted line.

Fig. 2.10 Illustration of Hydrogen Bond

2.10.1 Conditions for Formation of a Hydrogen Bond

The following conditions should be fulfilled for the formation of hydrogen bond:

1. The molecule must possess a higher electronegative atom such as F, O or N, directly linked to hydrogen atom.
2. The size of the electronegative atom should be small.

The electronegative atom is usually fluorine, oxygen or nitrogen. Thus, hydrogen bonds are usually found in hydrogen fluoride (HF), water (H_2O), ammonia (NH_3), alcohol (ROH), carboxylic acids (RCOOH), and amines ($RNII_2$).

Hydrogen bonds can vary in strength from very weak (1–2 kJ/mol) to extremely strong (40 kJ /mol) as in the ion HF_2^-. For example,

O ------- H ------- N (29 kJ/mol)

O ------- H ------- O (21 kJ/mol)

N ------- H ------- N (13 kJ/mol)

N ------- H ------- O (8 kJ /mol)

The bond strength is dependent on temperature, pressure, bond angle, and environment. Liquids that display hydrogen bonding are also called as **associated liquids.**

2.10.2 Types of Hydrogen Bond

The hydrogen bonds are classified into two types:

(*a*) **Intermolecular Hydrogen Bonding:** The bond formed between H-atom of one molecule with electronegative atom of the neighboring molecule.

For example, in Ammonia, two or more than two molecules of NH_3 combine together to give a polymeric aggregate. This is intermolecular hydrogen bond

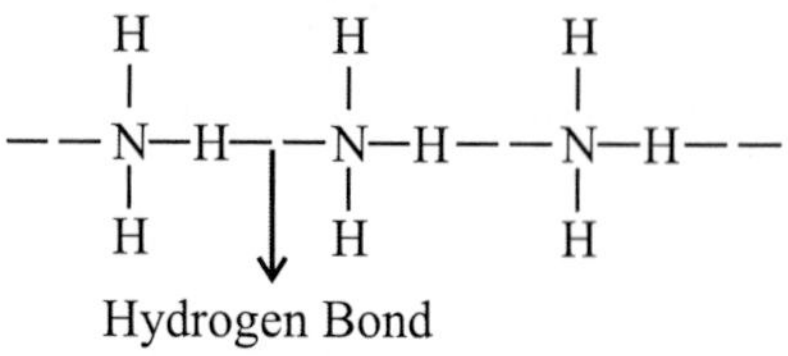

(*b*) **Intramolecular Hydrogen Bonding:** The bond is formed within two atoms of the same molecule. This type of hydrogen bonding is commonly known as Chelation and frequently occurs in organic compounds.

For example:

(*i*) *o*-nitrophenol

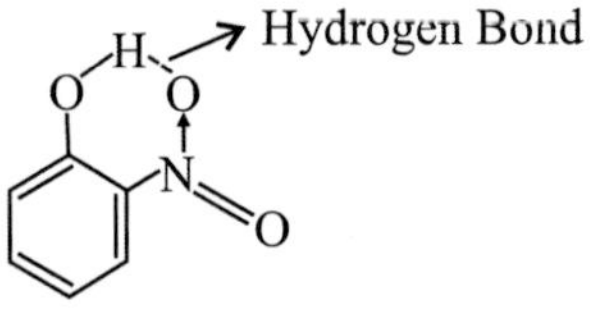

(*ii*) *o*-chlorophenol

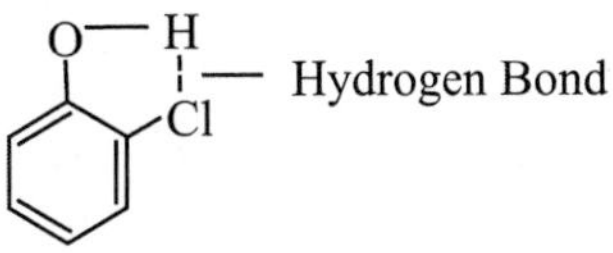

Such type of bonds are Intramolecular hydrogen bonds.

Example 2.9: Why KHF_2 exists but $KHCl_2$ does not?

Solution: Due to hydrogen bonding in HF, we have H — F....H — F....H — F.... This can dissociate to give HF_2^- ion and hence KHF_2 exists but there is no hydrogen bonding in H – Cl as the size of Cl is large enough as compared to F. So, HCl_2^- ion does not exist and hence $KHCl_2$ also does not exist.

Example 2.10: Why glucose, fructose, sucrose etc., are soluble in water though they are covalent compounds?

Solution: These compounds contain —OH groups which can form H—bonds with water, which makes them soluble in it.

2.10.3 Effect of Hydrogen Bonding on the Properties of the Compounds

(*a*) **Boiling Point:** Due to H bonding, there is a greater tendency of association with the result greater energy is required to break the bond, thereby increasing the boiling point. Therefore, intermolecular hydrogen bonding is responsible for the high boiling point of water (100°C).

This is explained on the basis, that water molecules associate through hydrogen bonding and require more energy to get separated for vaporization.

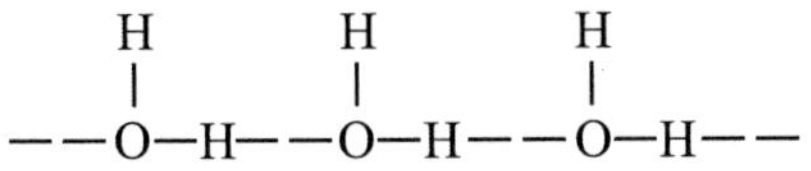

Hydrogen Bonding in Water

1. The boiling point of water is higher than ethyl alcohol because of greater degree of hydrogen bond formation in water than alcohol.
2. There are two unusual properties of water, namely (*i*) low density in the solid state, than that in the liquid state as shown and, (*ii*) Contraction when heated between 0°C and 4°C. This is due to H bonding in the solid state as shown in the Fig. below an open cage like structure is formed as a result of which mass per unit volume decreases and hence the density, also decreases.

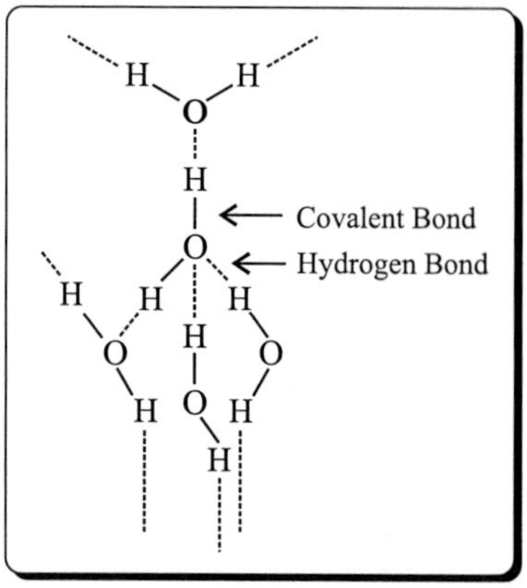

The tetrahedral open cage like structure of ice.

The plot given below in Fig. 2.11 shows how the volume of water varies with the temperature; the large increase (about 9%) on freezing shows why ice floats on water and why pipes burst when they freeze. The expansion between –4°C and 0°C is due to the formation of larger clusters. Above 4°C, thermal expansion sets in as thermal vibrations of the O—H bonds becomes more vigorous, tending to move the molecules farther apart.

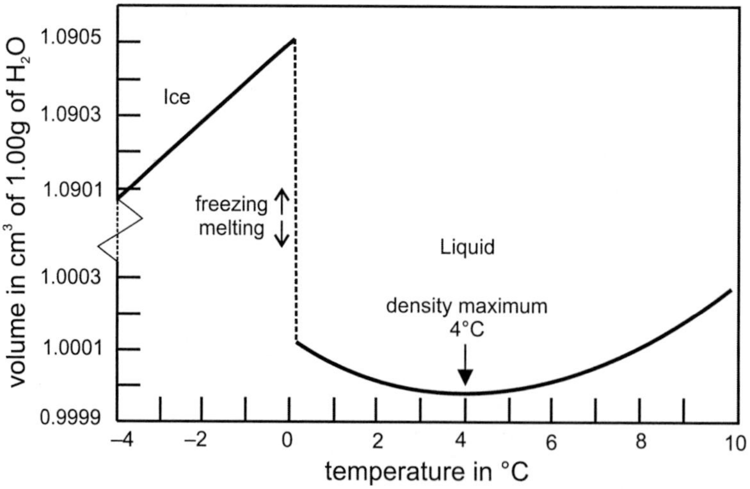

Fig. 2.11 Variation of volume of water with temperature

3. The molecules having intra molecular H bonding are found to have lower boiling point than the molecules having intermolecular H bonding. Therefore, ortho substituted compounds have lower melting point and boiling point than that of meta and para isomers. This is because of the fact that intramolecular hydrogen bonding takes place between the two ortho substituted groups and thus prevents association of the

molecules which would have raised the melting and boiling points. Whereas, intra molecular H-bonding is not possible in *m*- and *p*- isomers. Thus, here the intermolecular hydrogen bonding takes place which causes some degree of association thereby resulting the *m*- and *p*- isomers to melt and boil at higher temperatures.

For example, *o*-hydroxy benzoic acid

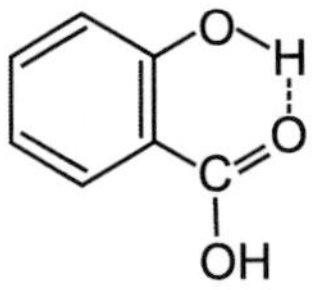

(*b*) **Water Solubility:** A substance is said to be soluble in water if it is capable of forming hydrogen bonding with water molecules. Thus, organic compounds like alkanes, alkenes and others which lack the formation of hydrogen bonds with water are insoluble in water.

Whereas, alcohols and acids are soluble in water since they form H-bond with water easily as shown below:

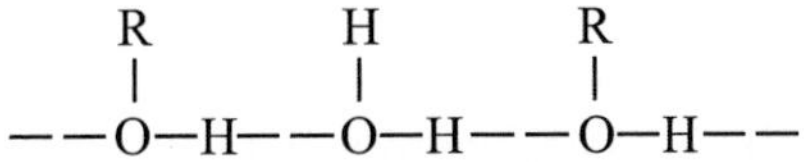

Fig. 2.12 Hydrogen bonding between alcohol and water.

Alcohols having alkyl groups with three carbon atoms are not freely soluble in H_2O. With increase in the number of C-atoms, the solubility decreases.

(*c*) **Strength of Acids:** The strength of acids is greatly influenced by the type of H-bond present in the molecule. For example, let us consider the dissociation constants of the hydroxyl derivatives of benzoic acid.

Hydroxy benzoic acid	Ortho	Meta	Para
$HO.C_6H_4\,COOH$	105×10^5	8.3×10^5	2.9×10^5

The dissociation constant for the ortho derivative was found to be higher since its anion is stabilised by chelation due to intramolecular H-bonding which increases the acidity of the o-isomer.

Similarly, Maleic acid is more acidic than fumaric acid which is again due to stabilisation of maleate ion by chelation which is not possible in case of fumarate ion because hydrogen and oxygen are on the opposite sides to each other, to form a hydrogen bond.

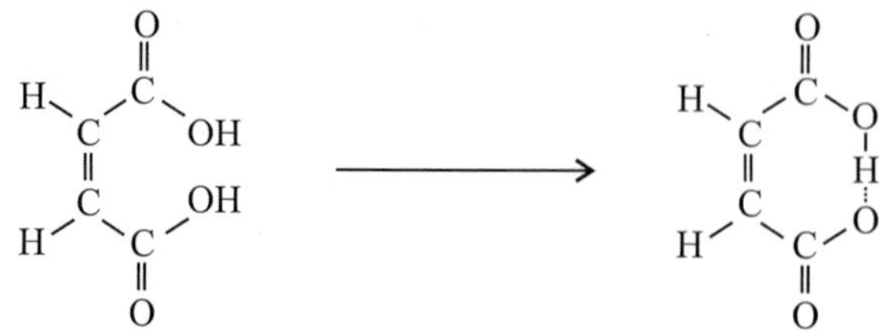

Maleic acid (Chelation not possible) — Maleate ion (Chelation possible)

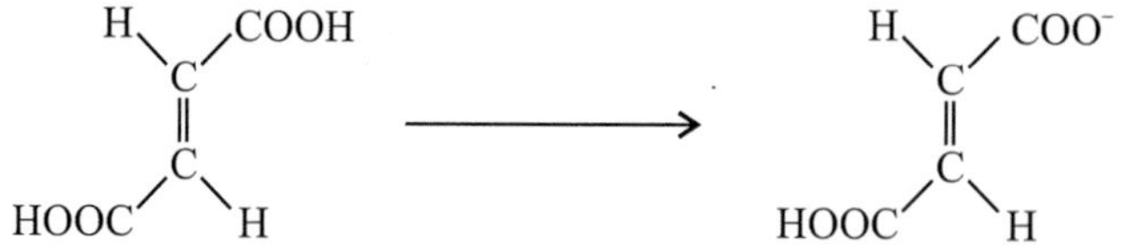

Fumaric acid (Chelation not possible) — Fumarate ion (Chelation not possible)

(*d*) **Viscosity of Liquids:** Viscosity of a liquid can be defined as the resistance that one part of the liquid offers to another part of the liquid or simply we can say, that, it is the resistance to the flow of liquid. Due to intermolecular hydrogen bonding, the association between the molecules of hydrogen bonded liquids is increased, this increase in the association, decreases the tendency of the liquid to flow smoothly.

For example, glycerol (CH_2OH — CHOH — CH_2OH) forms more number of H-bonds than ethyl alcohol (C_2H_5OH) as glycerol has three –OH groups as compared to C_2H_5OH which has one –OH group. Therefore, the viscosity of glycerol is greater than that of C_2H_5OH.

Example 2.11: How many H-bonds are formed by each H_2O molecule and how many water molecules are attached to each water molecule and in what direction?

Solution: Each H_2O molecule forms four H-bonds, two with O-atom and two with H-atoms. Further, each H_2O molecule is linked to four H_2O molecules through H-bonds tetrahedrally.

2.10.4 Importance of Hydrogen Bonding in Biological Systems

More than 50% of the weight of plants and animals is constituted by water. Water is attached to the proteins of these plants and animals by hydrogen bond.

Another important significance of hydrogen bonding is that it stabilizes the structure of proteins and nucleic acids in biological systems.

2.11 Dipole Moment

A diatomic molecule with charged ends is said to be a Dipole.

The term Dipole moment was used to express the extent of polarity in a molecule.

The dipole moment (μ) for a pair of opposite charges of magnitude q is defined as the magnitude of the charge times the distance between them and the defined direction is from positive end to negative end.

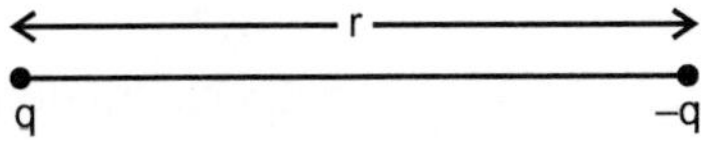

Fig. 2.13 Illustration of the dipole moment.

That is,

$$\mu = q \times r$$

where, μ = Dipole moment

q = charge on an atom

r = distance between the atoms

μ is a vector quantity. It is a useful concept in atoms and molecules where the effects of charge separation are measurable; it is also a useful concept in dielectric and other applications in solid and liquid materials.

The unit of Dipole moment is **Debye.** in C.G.S. units.

i.e., 10^{-18} esu.cm = 1 D

Therefore, $\mu = 4.8\text{D}$

Dipole moment in SI units is expressed in Coulomb meter (C.m).

$$1 \text{ C.m} = 2.996 \times 10^{29}\text{D}$$

Non-polar diatomic molecules have a zero Dipole moment. While diatomic molecules formed by atoms of different electro negativity possesses dipole moments.

Whereas, in case of Dipole moment of polyatomic molecules is taken as the resultant of all the bonds. Each bond in a given molecule has a definite value of its bond moment which, is called bond moment.

(*a*) Molecules with regular geometry have a zero dipole moment, while

(*b*) Molecules with irregular geometry (having bond pairs and lone pairs) possess a definite dipole moment.

(*c*) Molecules which are non-polar in nature like A – A, have zero dipole moment, as both the atoms have same electronegativity values.

(*d*) Molecules which are polar in nature (A – B), have a definite value for dipole moment, as both the atoms have different values of electronegativity.

Dipole moment values are quite beneficial as it has following applications.

(*a*) The percentage Ionic character can be calculated.

(*b*) Molecular structures of various molecules can be predicted.

2.11.1 Calculation of Ionic Character from the Dipole Moment

The percentage ionic character can be calculated from the dipole moment data also as it was calculated by electro-negativity values. For example:

(*i*) A diatomic molecule having a dipole moment of 1.92 Debye and bond length of 2.0Å shall have 20% ionic character. This can be calculated by another method which is

$$p = q \times r$$

or 1.9×10^{-18} esu cm = $q \times 2 \times 10^{-8}$ cm

or charge (q) = $1.92 \times 10^{-18}/ 2 \times 10^{-8} = 0.96 \times 10^{-10}$ esu

Since charge on isolated ion = 4.8×10^{-10} esu

Therefore % ionic character = $0.96 \times 10^{-10} / 4.8 \times 10^{-10} \times 100 = 20\%$

The percentage ionic character can also be calculated by another method which is as follows:

$$\% \text{ ionic character} = \frac{\text{Experimental value of dipole moment}}{\text{Theoretical value of dipole moment}} \times 100$$

For example, the dipole moment of HX molecule is 1.92D and bond distance is 1.2Å. Calculate the ionic character of HX.

Theoretical value of dipole moment (μ) when ionic character is 100% = $q \times d$

$$= 4.8\times10^{-10}\ esu \times 1.2\times10^{-8}\ \text{cm} \qquad (\text{Because } 1(\text{Å}) = 10^{-8}\text{cm})$$

$$= 5.76 \times 10^{-18}\ \text{esu cm}$$

$$\mu = 5.76\ \text{D}$$

We know that,

$$\% \text{ ionic character} = \frac{\text{Experimental value of dipole moment}}{\text{Theoretical value of dipole moment}} \times 100$$

$$= 1.92/5.76 \times 100 = 33.3\%$$

Dipole moment values can be used to distinguish between the *cis*- and *trans*- isomers. Usually *cis*- isomers have higher dipole moment and are more polar than the *trans*- isomers. Since the substituent groups in *cis*- lie on the same side. Therefore, the total dipole moment is obtained by adding the individual dipole moment values of the bonds, whereas in *trans*- the substituent groups lie on the opposite sides which cancels the individual dipole moment values of the bonds leading to zero dipole moment value. *e.g.*,1,2 Dichloroethylene

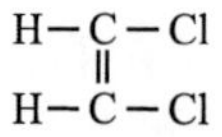

cis-1,2 Dichloroethylene
(μ = 1.9D)

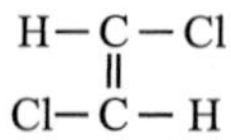

trans-1,2 Dichloroethylene
(μ = 0.0)

2.11.2 Dipole Moment and Molecular Structure

By illustrating the molecular structures of various molecules, the dipole moment values can be predicted.

Water has two very polar covalent bonds between oxygen and hydrogen. Because, water is bent (the oxygen atom is sp^3 hybridised), the two bond dipole moments add up to give water a relatively larger molecular dipole.

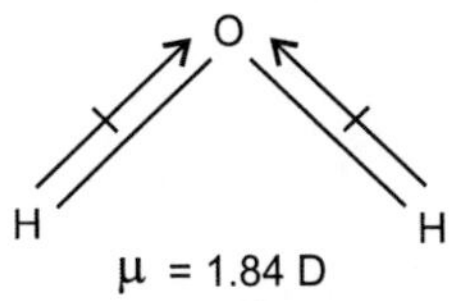

Fig. 2.14 Illustration of water molecule possessing dipole moment.

The measured magnitude of this dipole moment is $\mu = 6.2 \times 10^{-3}$ cm

1. Carbon dioxide has two polar covalent C – O bonds. However, the bond dipole moments exactly cancel each other since they are pointing in exactly opposite directions. Thus, CO_2 has no molecular dipole moment.

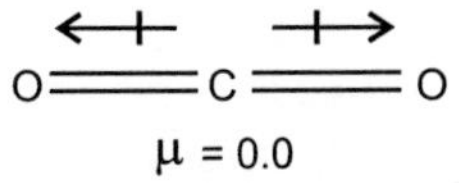

Fig. 2.15 Illustration of carbon dioxide molecule showing zero dipole moment.

2. Difluoromethane (CH_2F_2) has two polar covalent C – F bonds. The entire molecule has a molecular dipole moment resulting from the vector sum of the two C – F bond dipole moments.

3. Tetrafluoromethane (CF_4) has four polar covalent bonds. Because, they are arranged in a symmetrical tetrahedral way, all of the bond dipole moment vectors exactly cancel, leaving no resultant molecular dipole moment for CF_4.

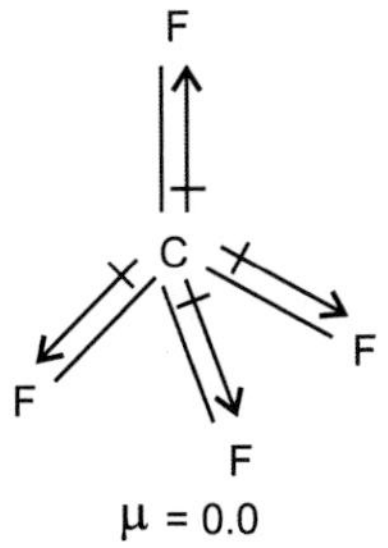

Fig. 2.16 CF_4 Molecule possessing dipole moment.

4. Among the *ortho-*, *meta-* and *para-* isomers, dipole moment is greatest for the *ortho-* isomer and zero for the *para-* isomer as the direction of the bond moment vectors cancel each other, when the substituents are same but the same is not true in case of *p*-dihydroxybenzene because the two hydroxyl groups are not symmetrical about the C–O–H that may not lie in the same plane as that of benzene ring.

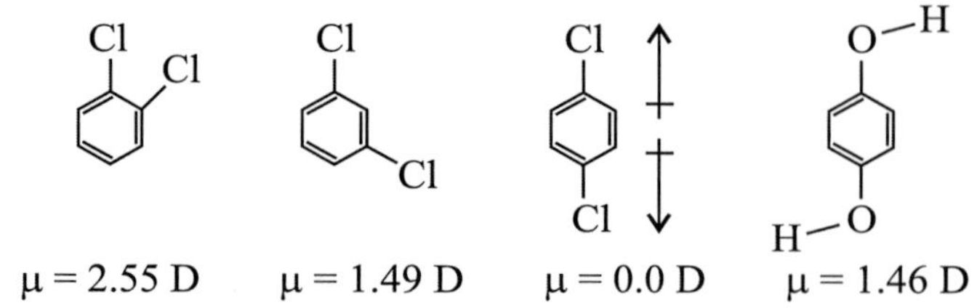

5. The dipole moment of BF_3 molecule is zero even if the B – F bonds are polar because of the planar triangular structure (symmetrical) in which the effect of different bond moments cancels each other.

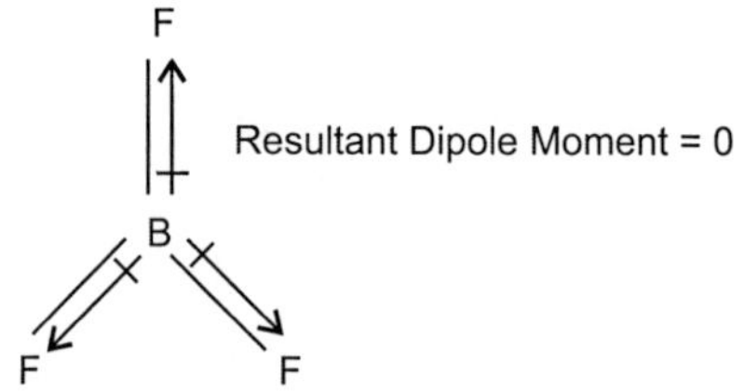

Fig. 2.17 Illustration of Dipole moment of BF_3 Molecule.

6. The dipole moment of NH_3 is greater than that of NF_3, as the lone pair moment adds on the resultant of the N – H moment, giving NH_3 a higher value of dipole moment. But in case of NF_3 molecule, the lone pair moment on nitrogen is opposite to the N – F bond moments. It results in lesser value of dipole moment than NH_3.

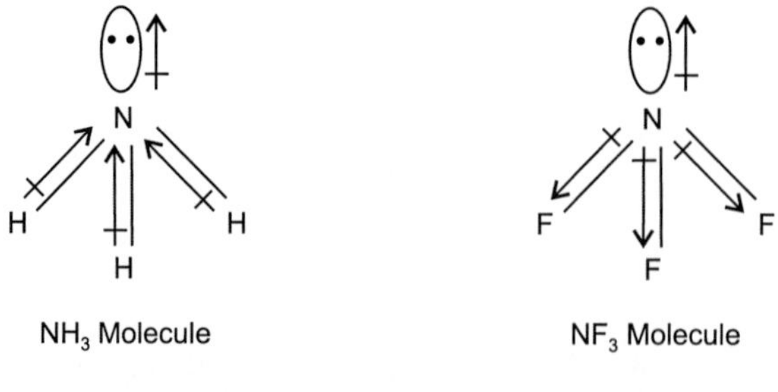

Fig. 2.18 Illustration of the dipole moments of NH_3 and NF_3 molecules.

Example 2.12: Out of CH_3OH and CH_3F, which has a higher value of dipole moment and why?

Solution: CH_3F has a higher value of dipole moment than CH_3OH because, F is more electronegative than oxygen.

Example 2.13: Out of 1-Butyne and 1-Butene, which has a larger value of dipole moment and why?

Solution: 1-Butyne involves sp-sp hybridised bond which is much more polarised bond than $sp^2 - sp^2$ hybridised bond in 1-butene, which leads to higher value of dipole moment of 1-Butyene, as $\mu = q \times r$. More polarisation would increase the magnitude of q leading to higher value of μ.

2.12 Van der Waals Forces

Van der Waals forces are used as synonym for non-covalent or intermolecular forces that are weak compared to those appearing in covalent bonding.

Therefore, non-bonded attraction and repulsive forces between molecules are called **Van der Waals forces**.

Or

We can say, that these forces are due to the strong force of attraction of nucleus of an atom of the molecule and valence electron of an atom present in the adjacent molecules.

Strength of the Van der Waals forces depends upon the size and increases with the number of electrons and surface area.

These forces are short range forces of the order of 10^{-7} cm.

For example, Water molecules in liquid water are attracted to each other by electrostatic forces, and these forces have been described as Van der Waals forces.

Van der Waals forces are present in molecules of solid CO_2, solid iodine and paraffins etc. The magnitude of these forces is maximum in case of solids and decreases as we move from Solid → Liquid → Gas.

2.12.1 Types of Van der Waals Forces

A. Dipole – Dipole Interactions
B. Dipole – Induced dipole Interaction
C. Induced Dipole – Induced dipole Interaction.

A. Dipole – Dipole Interactions:

These are called keesom interactions after the name of William Hendrik Keesom, are the forces that occur between two molecules with permanent dipoles. These dipoles orient themselves in such a way that the positive end of the molecule will be attracted to the negative end of the other molecule, when these are in close proximity.

For example,

$$\overset{\delta+}{H}—\overset{\delta-}{Cl}\cdots\cdots\overset{\delta+}{H}—\overset{\delta-}{Cl}$$

B. Dipole – Induced Dipole Interaction:

The interaction between permanent multipole or polar molecules with an induced multipole on another. This interaction is called after Debye or simply called as dipole-induced dipole interaction.

The attraction between the two will largely depend upon the polarising power of the permanent dipole and polarisability of the non-polar molecule.

For example, noble gases get polarised in presence of H – Cl which is a polar molecule.

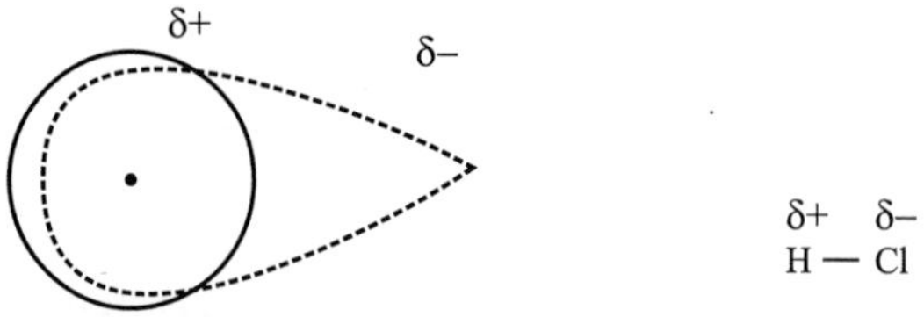

Non-Polar Molecule Polar Molecule

C. Induced dipole – Induced Dipole Interactions:

Also called London forces, these involve the attraction between temporarily induced dipoles in non-polar molecules which often disappear within an instant.

A non-polar molecule may be considered as a positive centre surrounded by symmetrical electron cloud. But electron cloud keeps on oscillating with respect to nuclei of the atoms. Therefore, at a given instant the electron cloud may be concentrated on one side and a temporary dipole exists in a molecule *i.e.*, a non-polar molecule may become temporarily polarised. This self polarised molecule may induce a dipole moment in a neighbouring molecule. The weak forces of attraction between these temporarily polarised non polar molecules are induced – dipole induced dipole interactions.
For example, Chlorine dissolving in water

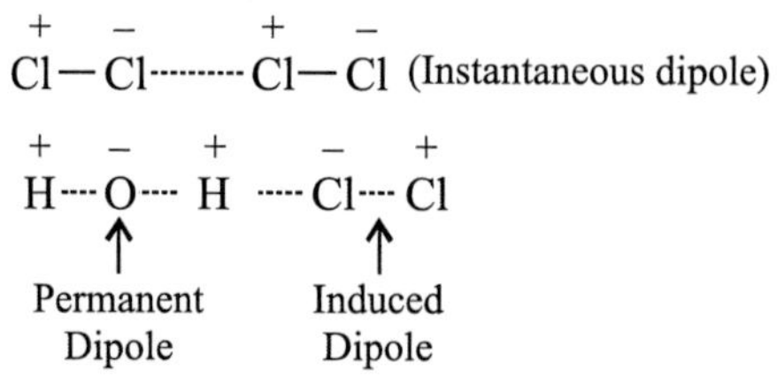

2.13 Metallic Bond

Metallic bonding was first discovered by Frederick Louise in 1922.

Metallic bonding is the bonding between atoms within metal. It involves the delocalised sharing of free electrons among a lattice of metal atoms. Therefore, metallic bonding is the electrostatic attraction between the metal atoms or ions and the delocalised electrons, also called conductor electrons. Greater the number of valence electrons, greater is the attraction by the positively charged species (kernels) and hence stronger the bond. Similarly, smaller the size of the kernel, greater is the attraction on the valence electrons and hence stronger the bond. Alkali metals are soft, because they have only one valence electron and large size kernel. *i.e.*, positively charged metal ion.

That is why atoms or layers are allowed to slide past each other, resulting in properties of malleability and ductility. The electrons and the positive ions (Kernels) in the metal have a strong attractive force between them. This means that more energy is required to negate these forces. Therefore, metals have high melting or boiling points. Metallic bonding is non-polar because for pure metals or even for alloys, there is no electro negativity difference among the atoms participating in the bonding reaction. The metallic bond accounts for many physical characteristics of metals, such as strength, malleability, conduction of heat and electricity, and lustre. In a metallic bond, electrons move independently of the positive ions in a sea of negative charge. Metal gains some electrical conductivity. It quickly possess the energy through the electrons generating a current.

The conductivity of the metal decreases with the increase in temperature because with the rise in temperature, the positively charged kernels start vibrating to and for disturbing the flow of electrons thereby decreasing the electrical conductivity.

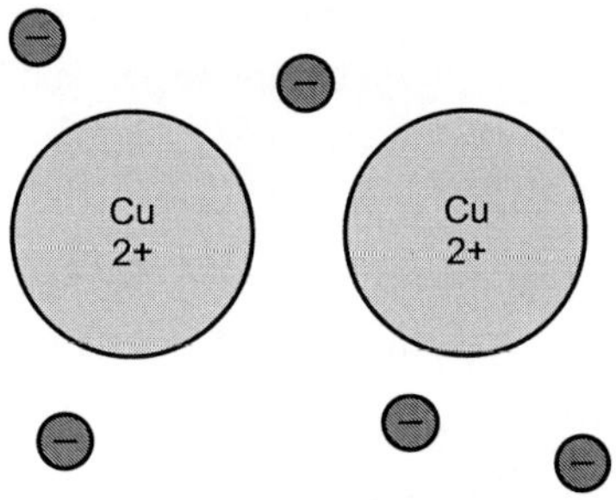

Fig. 2.19 Metallic bonds are found in metals like Copper.

Here, the force of attraction between the Cu^{2+} ion and the electron is the **Metallic Bond.**

Comparison between Covalent and Metallic Bond

Covalent Bond	Metallic Bond
1. In covalent bond, the valence electrons in them are strongly attracted by the positively charged nuclei, making it strong.	1. In metallic bond, the valence electrons are mobile and are weakly attracted by the positively charged kernels, making it a weaker bond.
2. It is directional, as the valence electrons in them are strongly attracted by the nucleus of an atom.	2. It is non-directional, as the valence electrons in them are spread all over the crystal with less uniformity.

Valence Shell Electron Pair Repulsion Theory and Molecular Orbital Theory

Introduction

VSEPR and MO theory: The word orbital was introduced by Mulliken in 1932. In 1933, the Molecular Orbital Theory was accepted as a valid and useful theory. In 1938, Accurate calculation of a Molecular Orbital wave function was made by **Charles Coulson** on the hydrogen molecule. By 1950, molecular orbitals were completely defined as eigen functions of the self-consistent field Hamiltonian and at this point, Molecular orbital theory became fully rigorous and consistent.

2.14 An Atomic Orbital

An atomic orbital is the region in which an electron may be found around a single atom or we can say, that atomic orbitals are the quantum states that electrons surrounding an atom may exist in.

The orbital names (*s*, *p*, *d*, *f*, *g*, *h*) are derived from the quality of their spectroscopic lines: sharp, principal, diffuse and fundamental. The quantum number n determines the size and energy of the orbital, l determines the orbitals shape, and m, its orientation.

The single *s*-orbitals ($l = 0$) are spherically symmetrical. For $n = 1$, the sphere is solid (it is most dense at the centre and fades exponentially outwardly), but for $n = 2$ or more, each single *s*-orbital is composed of spherically symmetric surfaces which are nested shells. The *s*-orbitals for all n numbers are the only orbitals with no node.

The three *p*-orbitals have the form of two ellipsoids with a point of tangency at the nucleus and are referred to as dumb-bell shaped. The three–orbitals have the form of two ellipsoids with a point of tangency at the nucleus. The three *p*-orbitals in each shell are oriented at right angles to each other, as determined by their respective values of m_l.

Four of the 5 *d* - orbitals look similar, each of the 4 pear shaped balls, each ball tangent to two others, and the centres of all four lying in one plane, between a pair of axes thereby giving them a double-dumb-bell shaped.Three of these planes are the *xy*, *xz* and *yz* planes, and the fourth has the centres on the *x* and *y* axes. The fifth and the final *d*-orbital consists of three regions of high probability density, *i.e.*, a torus with two pear shaped regions placed symmetrically on its *z* axis. These are seven *f*-orbitals each with shapes more complex than those of *d*-orbitals. The detailed explanation about the atomic orbital has been given later in this chapter.

2.15 Pauling Slater's Theory : Orbital Overlap Theory of Covalent Bond

The idea that atoms form covalent bonds by sharing pairs of electrons was first proposed by G. N Lewis in 1902, however, **Walter Heitler** and **Fritz London** in 1927 showed how the sharing of pairs of electron holds a covalent molecule together. The last major step in the evolution of this theory was by **Pauling** and **Slater** in 1931. Pauling Slater theory is the modified form of Heitler London Theory.

The main postulates of the theory are:

1. An atom must possess one or more unpaired electrons to enter into a chemical combination. The atomic orbitals of the valence shell which have paired electrons, do not contribute to the bond formation. The electron pairs in these orbitals are called non-bonding electron pairs or lone pairs of electrons.
2. Atoms with unpaired electrons tend to combine with other atoms which also have unpaired electrons.
3. The new orbital which is formed due to overlapping of atomic orbitals is more stable.
4. The strength of the bond is determined by the extent of overlapping of two half filled orbitals. More is the overlapping, more is the decrease in energy and stronger is the bond formed.
5. The relative strength of the covalent bonds obtained by *s–s*, *p–s* and *p–p* overlaps is as *s–s* < *p–s* < *p–p*. Covalent bond formed by *s–s* overlap is weak, because these have spherical distribution of electron density whereas the *p–s* overlap is comparitively better as *p*-orbitals are concentrated in a particular direction and their lobes are longer than the radius of the corresponding *s*-orbital.
6. The atoms which unite to form a molecule completely retain their identities in the resulting molecules.
7. The shape of the molecule is determined by the directions in which orbitals point.
8. According to this theory, an electron moves under the influence of only one nucleus of an atom.
9. The number of bonds formed by an atom is equal to the number of unpaired electrons it has.

The overlapping of atomic orbitals take place in different ways.

For example,

1. *s–s* overlap:

This can be explained by taking up the example of H_2 molecule. H has 1 electron in its valence shell *i.e.*, $1s^1$

H_2 molecule is formed when two *s*-orbitals overlap along their axis. This bond is called as *s-s* bond as it is formed by the overlapping of two *s*-orbitals as shown:

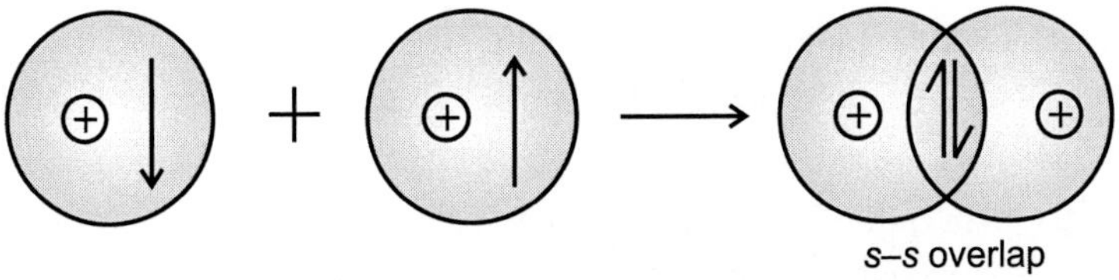

Fig. 2.20 Overlapping of the *s-s* orbitals.

2. *p – p* overlap :

This can be explained by taking up the example of F_2 molecule.

Electronic configuration of fluorine is $1s^2\ 2s^2\ 2p_x^{\ 2}\ 2p_y^{\ 2}\ 2p_z^{\ 1}$.

Its $2p_z$ orbital has 1 unpaired electron which would overlap with the electron in $2p_z$ orbital of another *F* atom to form F_2 molecule. Such type of overlap is *p – p* overlap as shown:

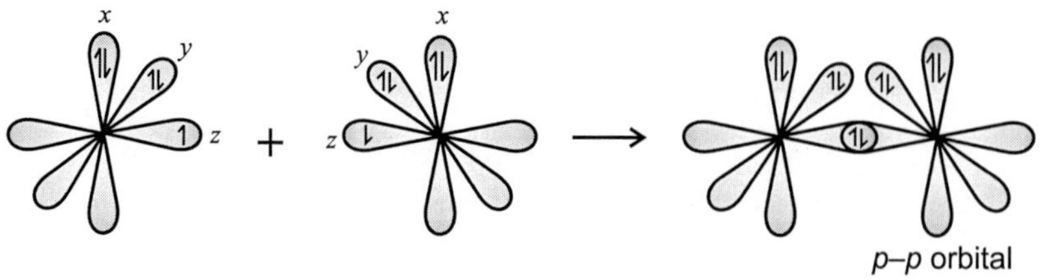

Fig. 2.21 Overlapping of the *p–p* orbital.

3. *s-p* overlap :

This can be explained by taking up the example of HF molecule. In the formation of HF molecule, half filled $2p_z$ orbital of fluorine overlaps with half filled 1*s* orbital of hydrogen atom resulting in *s-p* overlap as shown:

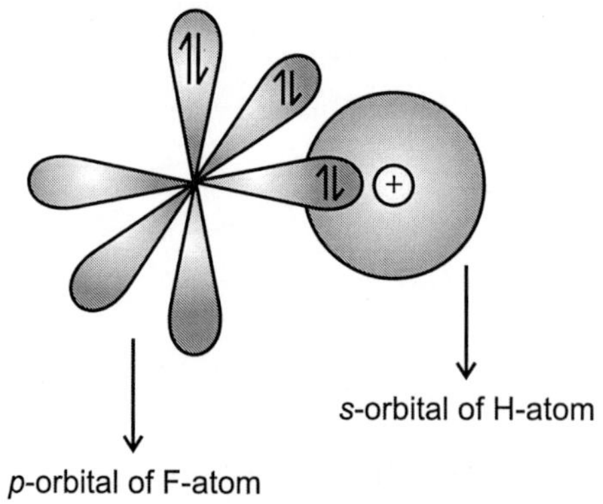

Fig. 2.22 Overlapping of the *s-p* orbitals.

Whenever the two orbitals overlap with each other, it results in the formation of bonds.

Limitations of the Theory

1. This Theory is unable to explain the paramagnetic character of B_2 and O_2 molecules as according to their structures. :B – B: and :Ö = Ö:, these have all the electrons in the paired state and hence should be diamagnetic.
2. According to this theory, the formation of bonds formed by the overlap of pure atomic orbital of the atoms, does not explain the properties like bond lengths, bond angles etc. For example, according to Pauling-Slater's theory/VB theory, all the four C – H bonds in CH_4 molecule should be of unequal length, but actually these have been found to be of equal length.

2.15.1 Sigma Bond

Sigma bond (σ) is defined for diatomic molecules using the language and tools of symmetry groups. Therefore, the bond formed by the overlap of atomic orbitals along the internuclear axis is a **sigma bond**. In this formal approach, sigma bond (σ) is symmetrical with respect to rotation about the bond axis. By this definition, common forms of sigma bonds are $s + s$, $p_z + p_z$ and $s + p_z$ and $d_z^2 + d_z^2$ (where z is the inter nuclear axis). A sigma bond is also produced when two hybrid orbitals overlap together along the inter nuclear axis.

For example, Formation of H_2 molecule

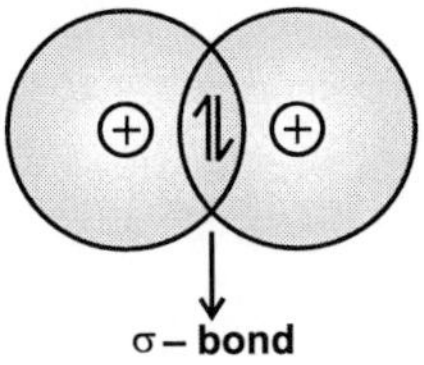

Sigma bonds are the strongest type of covalent bonds. Electrons in sigma bonds are referred to as sigma electrons.

2.15.2 *Pi*-Bond

The bonds formed by the sidewise overlapping or lateral overlapping of the atomic orbitals. The orbital symmetry of the pi- bond is same as that of the *p*-orbital when seen down the bond axes. *Pi*- bonds are usually weaker than sigma bonds, this bond weakness is explained by significantly less overlap between the component *p*-orbitals due to their parallel orientation.

Formation of N_2 molecule

Electronic configuration of N is *G.S.* $1s^2\ 2s^2\ 2p^3$

E.S. $1s^2\ 2s^2\ 2p_x^1\ 2p_y^1\ 2p_z^1$

$2p_z$ orbital of one nitrogen atom overlaps with the $2p_z$ of another nitrogen atom forming a sigma bond. $2p_x^1$ and $2\ p_y^1$ of one *N* atom overlaps with $2p_x^1$ and $2p_y^1$ orbitals of other nitrogen atom laterally to form two **Pi-bonds**.

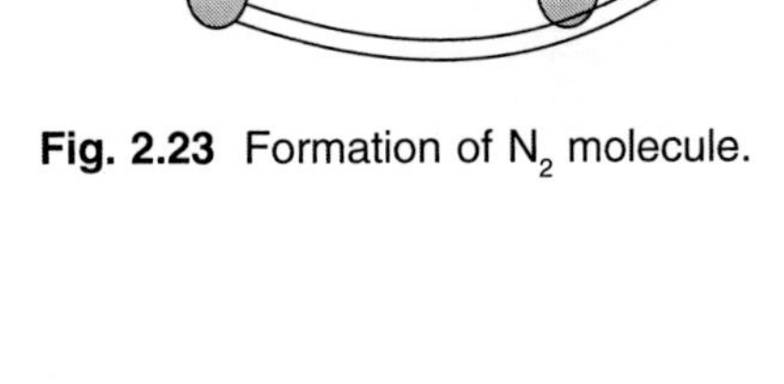

Fig. 2.23 Formation of N_2 molecule.

2.15.3 Comparison between Sigma (σ) and π-bonds

Sigma Bond	π-Bond
1. It results from the end to end overlapping of two *p*-orbitals or one *s* and one *p*-orbital.	1. It results from the sideways overlapping of two *p*-orbitals.
2. The bond is strong as the overlapping is along the axis.	2. The bond is weak as the overlapping of orbitals is along the sideways, which is only partial.
3. Sigma electrons are localized	3. π-electrons are mobile.
4. Sigma bonds are less reactive.	4. π-bonds are more reactive.
5. The shape of the molecule is determined by the σ bonds present in the molecule.	5. π-bonding does not affect the shape of the molecule in which it occurs.
6. σ bonds can have independent existence.	6. π-bond always exists along with a σ bonds.

2.16 Resonance

The concept of resonance was introduced by **Linus Pauling** in 1928.

Resonance is a tool used to represent and model certain types of non-classical molecular structures.

Therefore, we can say that resonance involves representing or modeling the structure of a molecule as an intermediate, average (a resonance hybrid) between several simpler but incorrect structures.

For example, Benzene, the two carbons can be connected by one or two covalent bonds, but in the observed benzene molecule, the bond lengths are longer than the double bonds yet shorter than the single bonds. More importantly they are all equivalent.

Therefore, one calls the two Lewis structures, canonical, contributing or resonating structures and the real molecule is considered to be the average, called a **resonance hybrid.**

So, we can say, that, when a molecule is represented by more than one Lewis structure, none of which is able to explain the properties of the molecule and the true structure is a resonating structure. This phenomenon is called **Resonance**.

Resonance structures of the same molecule are connected with a double headed arrow.

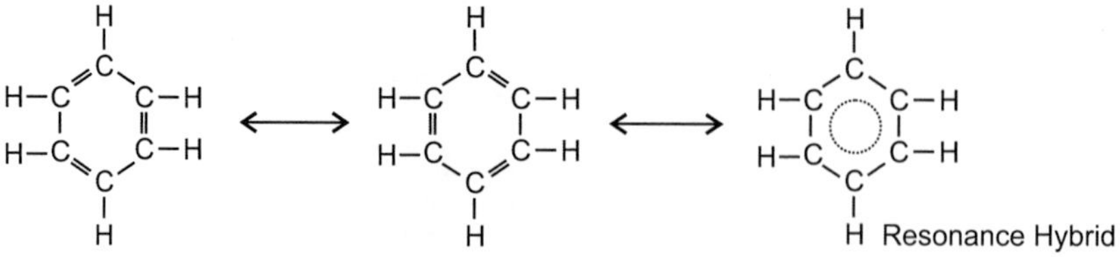

Fig. 2.24 Resonating structures of Benzene Molecule.

2.16.1 Characteristics of Resonating Structures

1. Position of nuclei must be the same in all structures; otherwise they would be isomers with real existence. They should differ only in the position of electrons. *e.g.*,

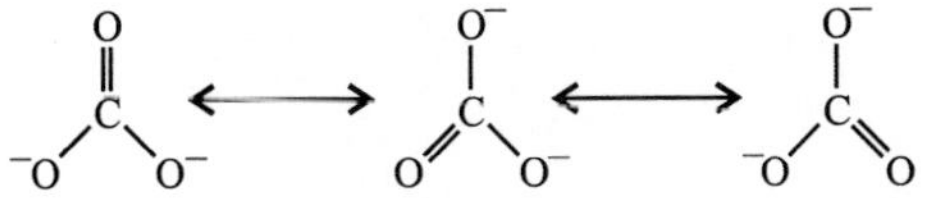

2. Total number of electrons and thus total charge must be constant or we can say, that the number of unpaired electrons if any, in the molecule must remain the same in all the forms.

   The resonating structures of O_2 can be represented as:

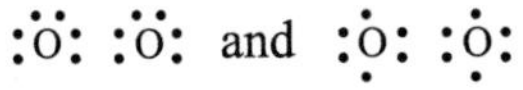

(I) (II)

   (I) and (II) are not the resonating structures of oxygen because the number of unpaired electrons in the two structures are different.

3. When separating charge (giving rise to ions), usually structures where negative charges are on less electronegative elements have little contribution. Contributing structures should be such where the negative charge resides on the most electronegative atom and positive charge on the most electropositive atom *e.g.*,

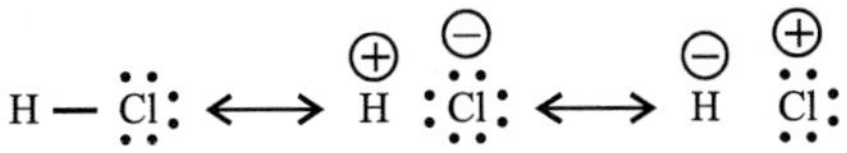

(I) (II) (III)

   The resonating structures can be (I) and (II) but practically there is no contribution from (III) as in this structure positive charge resides on the most electronegative atom.

4. In contributing structures, like charges should not reside on the adjacent atom, otherwise the structure will be unstable due to electrostatic repulsion. For example, HN_3 (Hydrazoic acid)

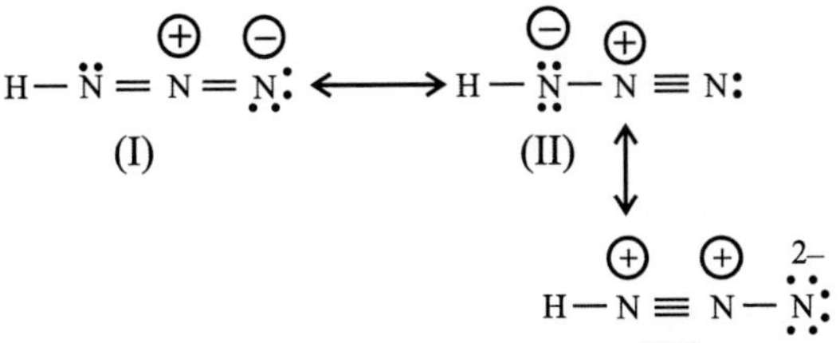

(I) (II) (III)

   (I) and (II) are the resonating structures of each other while (III) is not, because it is unstable as both the adjacent N atoms have similar charge leading to electrostatic repulsions.

5. The larger is the number of contributing structures, the greater is the stability of the molecule.

2.16.2 Resonance Energy

Resonance hybrids (true structure) are always more stable than any of the contributing structure (which are not existent). Resonance hybrid is more stable because of delocalisation of electrons which lowers the energy and imparts stability. The gain in stability of the resonance hybrid over the most stable (non-existent) contributing structures is called the **Resonance Energy**.

or

The difference in energy of the actual molecule and the most stable contributing structures is called **Resonance Energy.**

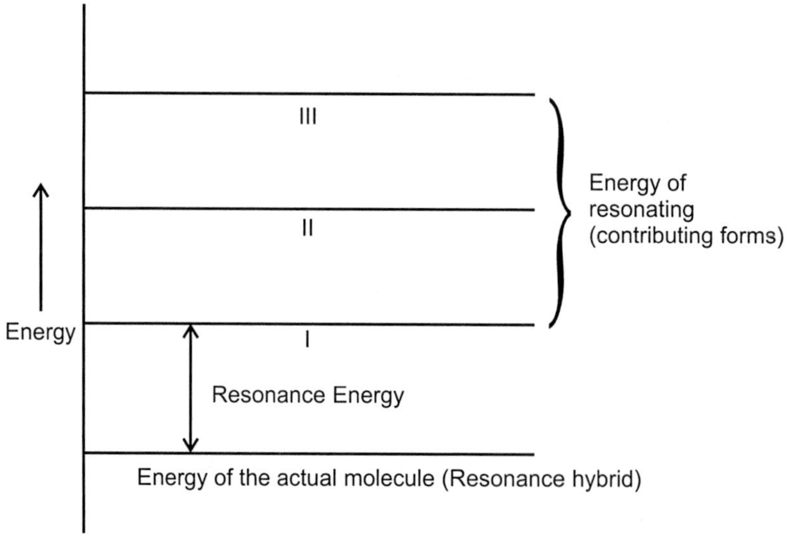

Fig. 2.25 Concept of Resonance Energy.

Resonance energy of a conjugated system can be measured by heat of hydrogenation of the molecule. Consider the example of benzene, the energy required to hydrogenate an isolated pi-bond is around 28.6 kJ/mole. So, the complete hydrogenation of benzene should require 85.8 kJ/mole. However, the experimental heat of hydrogenation of benzene is around 49.8 kJ/mole. The difference of 36 kJ/mole is looked upon as a measure of resonance energy. Even though, resonance structures have no physical existence, and the term "**resonance energy**" is quite meaningless, if offers an insight into how different the valence bond picture of a molecule is from the actual molecule itself.

2.17 Valence Shell Electron Pair Repulsion Theory (VSEPR Theory)

Valence shell electron pair repulsion theory (VSEPR) was given in 1957, is a model in chemistry, which is used for the representation of shapes of individual molecules, based upon the extent of electron pair electrostatic repulsion. The theory is also called the **Gillespie – Nyholm** theory after the two main developers.

Postulates of Valence Shell Electron Pair Repulsion Theory:

1. The geometry of a molecule or ion depends upon the number of electron pairs (bonded or not) in the valence shell of its central atom.
2. The electron pairs try to stay as far away as possible, so that the repulsive forces between electron pairs are minimum and the geometry is not distorted.
3. The geometry of the molecule is determined by the total number of electron pairs, bond pairs as well as lone pairs, surrounding the central metal atom.
4. When the central atom in a molecule is surrounded by bond pairs only, the molecule will have a regular shape or geometry as shown in the Table 2.6.

Table 2.6 Geometries of Different Molecules

No. of electron pairs	Shape	Bond Angle (Degree)
2	Linear	180°
3	Plane Triangular	120°
4	Tetrahedral	109° 28′
5	Trigonal Bipyramid	120° and 90°
6	Octahedral	90°
7	Pentagonal Bipyramid	72° and 90°

5. When the central atom in a molecule is surrounded by bond pairs as well as lone pairs, the geometry is distorted or irregular. This is because lone pair is close towards the nucleus, whereas bond pair is under the influence of two nuclei. Consequently, the presence of lone pairs will alter the bond angle to a small extent and disturbs the regular geometry.
6. It is found that the repulsion between the lone pair electrons is greater than the repulsion between bond pair electrons. The order of decreasing repulsion is

   Lone pair – Lone pair > Lone pair – Bond pair > Bond pair – Bond pair

The reason for this is that the bond pair electrons are held between the two atoms linked together by covalent bond while a lone pair of electrons is held by only one atom. Therefore, the orbital containing a lone pair is more spread out in space as compared to bonding electron pair. As a result, a l *p* will exert more repulsion on the bonding electron pair as compared to repulsion exerted by a bonding electron pair on the other bonding electron pair.

The following table shows the VSEPR geometries, where E is the central atom and X as the surrounding atoms.

Table 2.7 VSEPR Geometries of different Molecules

No. of Bonding electrons	Basic Geometry	1 Lone Pair	2 Lone Pair	3 Lone Pair
2		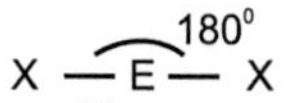 Linear		
3	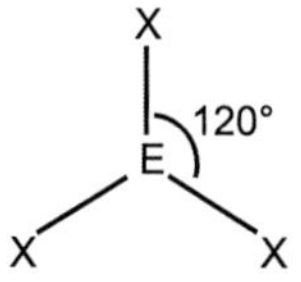 Trigonal Planar	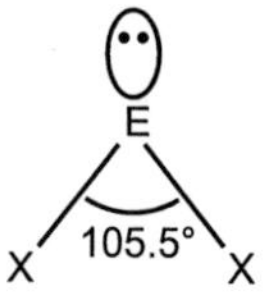 Bent or angular		
4	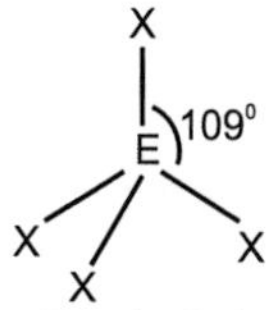 Tetrahedral	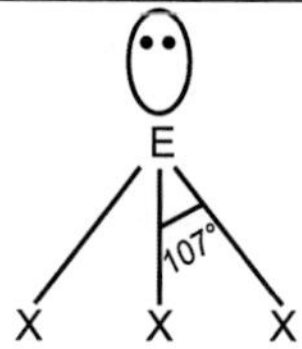 Trigonal Pyramid	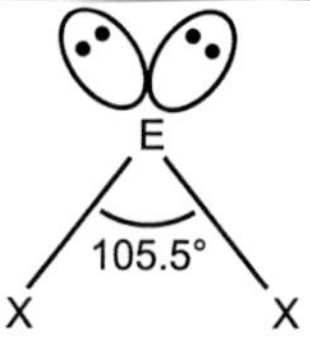 Bent or angular	
5	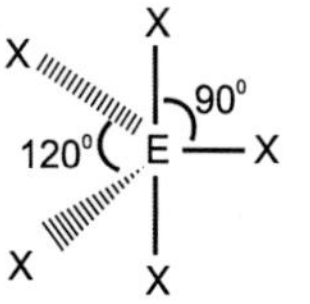 Trigonal Bipyramid	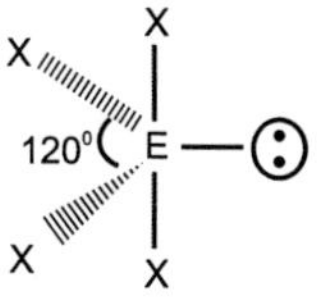 Sawhorse	 T-shape	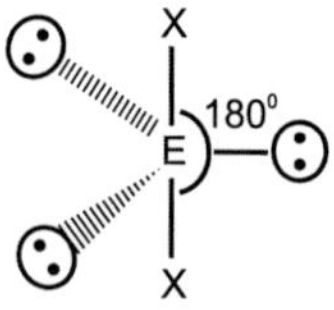 Linear
6	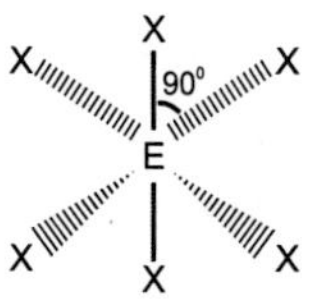 Octahedral	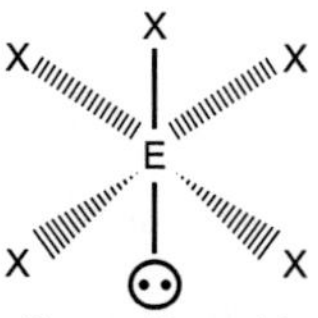 Square Pyramid	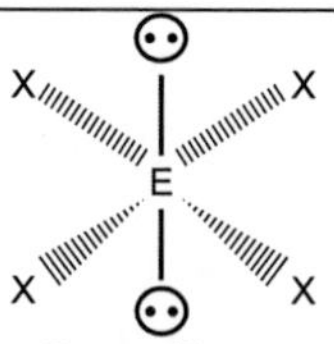 Square Planar	

Let us discuss few examples:

1. **Shape of BeF_2 molecule:** In BeF_2, the central atom Be has 2 valence electrons and forms 2 covalent bonds with the 2 fluorine atoms. According to VSEPR theory, these bond pairs arrange themselves at an angle of 180° and the molecule will be linear.

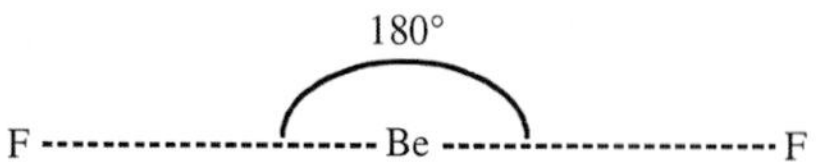

2. **Shape of BF_3:** Boron, the central metal atom has 3 electrons in the valence shell and forms 3 covalent bonds with 3 fluorine atoms which are directed towards the corners of a regular equilateral triangle leading to a bond angle of 120°.

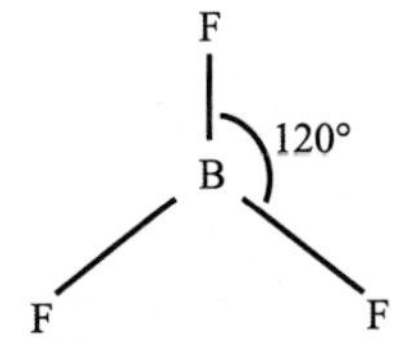

3. **Shape of CH_4:** Carbon atom has 4 valence electrons which form 4 covalent bonds with the 4 hydrogen atoms leading to a regular tetrahedral shape, with a bond angle of 109°. 28′.

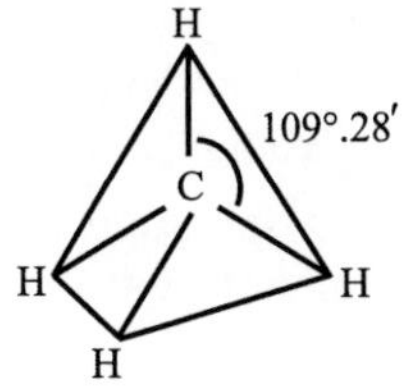

4. **Shape of PCl_5:** Phosphorous has 5 valence electrons which form 5 covalent bonds with 5 chlorine atoms to form 5 bond pairs. Hence, the geometry is Trigonal Bipyramidal with bond angles of 120° and 90°.

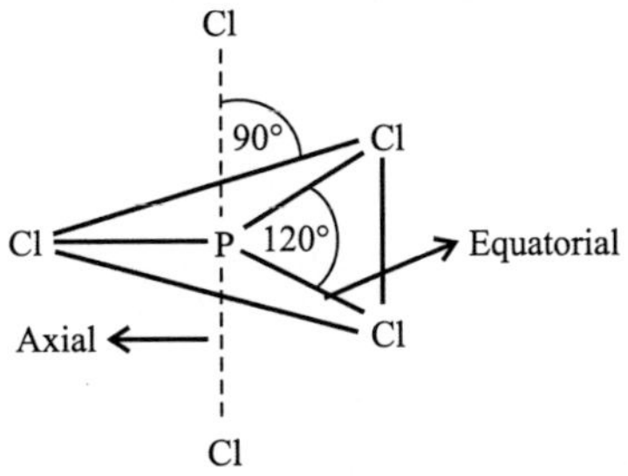

5. **Shape of SF_6:** Sulphur has 6 valence electrons and forms six covalent bonds with fluorine atoms. The geometry of the molecule is octahedral, with a bond angle of 90°.

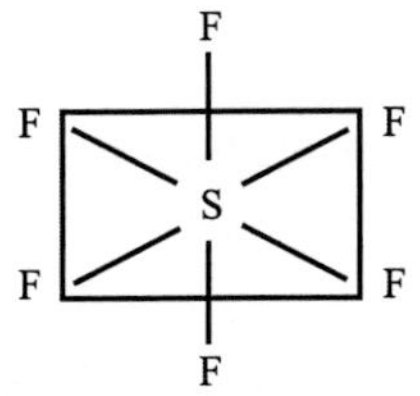

6. **Shape of NH_3:** Nitrogen has 5 valence electrons. The nitrogen atom to which the three hydrogen atoms in ammonia are attached also bears an unshared pair of electrons in a

p-orbital. This lone pair forms the fourth corner on the tetrahedron. Due to bond pair-lone pair repulsion, the resulting shape is not a perfect tetrahedron as in methane but is Triangular Pyramidal with a bond angle of 107°.48′ as shown:

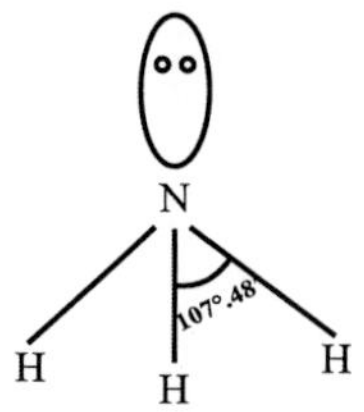

7. **Shape of H_2O Molecule:** In water molecule H_2O, the single electron of each H is shared with 2 of the six valence electrons of oxygen leaving 4 electrons organised into 2 Non-bonding (Lone) pairs. Since, these 2 lone pairs are closer to the oxygen atom, these exert a stronger repulsion against the two covalent bonding pairs, effectively pushing the two hydrogen atoms closer together resulting in a distorted tetrahedral arrangement *i.e.*, bent or angular shape in which the H—O—H angle is 104.5°.

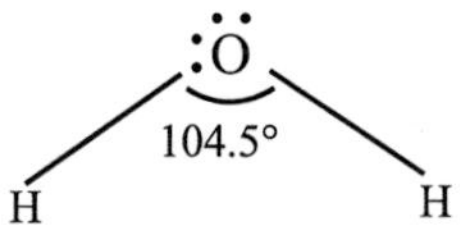

8. **Shape of ClF_3 Molecule:** Chlorine has 7 valence electrons in its valence shell. Three electrons form bond to fluorine atoms and four electrons do not take part in bonding.Therefore, there are five electron pairs, three bond pairs and two lone pairs. Hence, the structure is Trigonal Bipyramid.

Theoretically three different arrangements are possible as shown below:

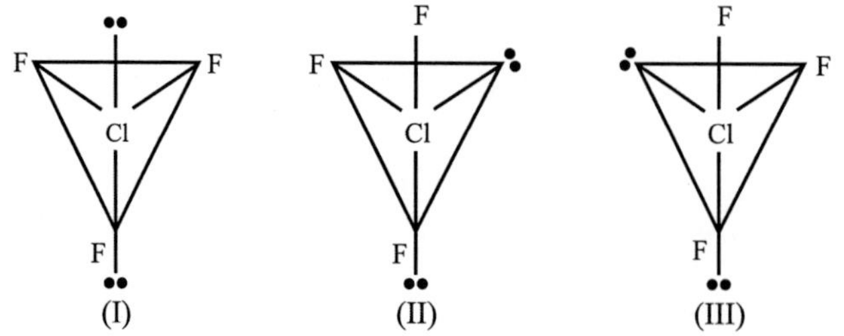

Structure I has 6 lone-bond pair repulsions, no *lp-lp* repulsion and 2 *bp-bp* repulsion.

Structure II has one lone pair-lone pair repulsion and 3 lone pair-bond pair repulsion at 90° and there are 2 *bp-bp* repulsions

Structure III has 4 lone pair-bond pair repulsions, no *lp-lp* repulsion and 2 *bp-bp* repulsions.

Structure II has one lone pair-lone pair repulsion and hence least stable.

Out of structures I and III, III is more stable as the number of lone pair-bond pair repulsion is less as compared to structure I.

9. **Shape of SF_4 Molecule:** Sulphur has 6 valence electrons with *S* at the centre with one lone pair and four fluorine atoms attached. The lone pair occupies the equatorial position to minimize the repulsion and F atoms occupy the other four corners. Due to lone pair-bond pair repulsion, the shape changes from Trigonal Bipyramidal to see saw. Although there are two geometries.

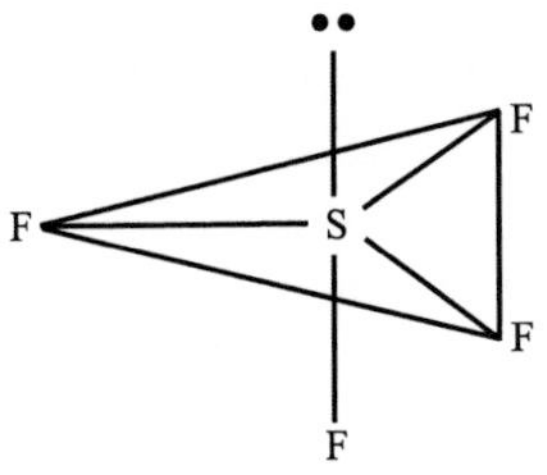

(I)

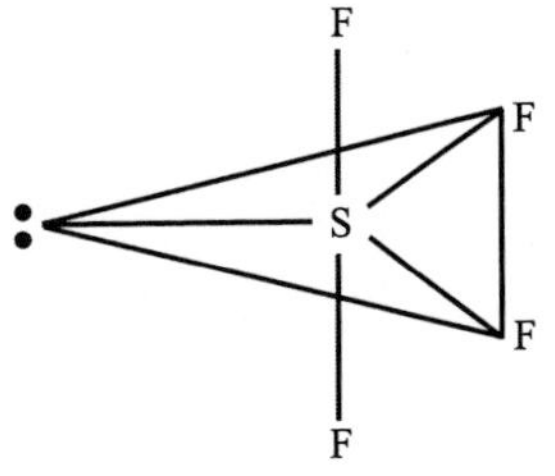

(II)

In geometry (I) *lp* is at an axial position there are 3 *lp-bp* repulsions and 3 *bp-bp* repulsions whereas, in geometry (II) *lp* is at an equatorial position where there are 2 *lp-bp* repulsions and 4 *bp-bp* repulsions. Therefore, the correct geometry is (II).

Example 2.14: BeF_2 is linear but H_2O is angular although both are triatomic.

Solution: BeF_2 is linear because the two bonded pair of electrons in the valence shell of Be arrange themselves as far apart as possible, that is 180° apart. As there is no lone pair on Be, therefore, the bond pairs do not face any repulsion and hence the bond angle is 180°.

Whereas, in H_2O, the central atom O has two bond pairs and 2 lone pairs which causes lone pair-lone pair repulsions leading to angular shape.

2.18 Hybridisation

The hybridisation was promoted by chemist **Linus Pauling**, in order to explain the structure of molecules such as methane. Historically, this concept was developed for simple chemical systems but the approach was later applied more widely, and today it is considered an effective heuristic for rationalising the structures of organic compounds.

The number of half filled atomic orbitals of the element is referred to as **covalency** of the element. But this concept, could not explain the formation of beryllium, boron and carbon compounds.

The electronic configuration of these elements is represented as:

Element	At. No.	1*s*	2*s*	2*p*
Beryllium	4	[↑↓]	[↑↓]	
Boron	5	[↑↓]	[↑↓]	[↑][ ][ ]
Carbon	6	[↑↓]	[↑↓]	[↑][↑][ ]

But, on the basis of electronic configuration of these elements we say, that beryllium behaves as noble gas whereas boron and carbon have covalency of one and two respectively, which is not correct.

In order to explain this point, it was assumed that the paired electrons become unpaired to participate in bonding. For example, in carbon one of the electron from 2*s* orbital jumps to 2*p* orbital.

Carbon	2*s*	2*p*
Carbon (Ground State)	[↑↓]	[↑][↑][ ]
Carbon (Excited State)	[↑]	[↑][↑][↑]

Hence, total of 4 unpaired electrons are available for bonding. But, according to the electronic configuration, the three electrons in the *p*-orbital would be oriented at right angles and the electron in the *s*-orbital will be spherically symmetrical.

Thus, the bonds formed by the overlap of *p*-orbital of carbon atom will be different from the bond formed by the *s*-orbital of carbon as *s* and *p*-orbitals are of different energies.

Experimentally, it was shown that the four carbon bonds are of equal strength *i.e.*, they are equivalent.

So, in order to explain the formation of four equivalent bonds, the concept of hybridisation was introduced.

Hybridisation is the concept of mixing atomic orbitals to form new hybrid orbitals suitable for the qualitative description of atomic bonding properties.

Hybridised orbitals are very useful in the explanation of the shape of molecular orbitals for molecules.

Hybrid orbitals are named by considering the type and the number of atomic orbitals from which they arose.

For understanding the concept of hybridisation, following points should be remembered:

1. Hybridisation is not a real physical process, it is hypothetical which has been introduced to explain some structural properties which could not be explained by simple valence bond theory.
2. Only orbitals of similar energies and belonging to the same atom undergo hybridisation.
3. The number of hybrid orbitals produced is equal to the number of pure orbitals mixed during hybridisation. *e.g.*, four hybrid orbitals are produced by mixing one *s* and 3*p* (p_x, p_y, p_z) pure orbitals.
4. The hybrid orbitals formed out of a particular hybridisation are similar in all respects.
5. Hybrid orbitals are much stronger than the non-hybrid ones.
6. Hybrid orbitals follow Hund's rule and Pauli's exclusion principle
7. Types of hybridisation indicate the shapes of the molecule.
8. A hybrid orbital, can accomodate a maximum of two electrons with opposite spins.
9. Hybrid orbitals generally form either σ-bonds or contain lone pair of electrons but not π-bonds.

2.18.1 Types of Hybridisation

Various types of hybridisations have been discussed below:

A. *sp* hybridisation

The combination of an s and a p orbital leads to two hybrid orbitals known as *sp*-orbitals. *sp* hybridisation can be very well explained by taking an example of BeF_2 molecule.

The electronic configuration of Be is $1s^2\ 2s^2$

The electronic configuration of these elements is represented as

Berrylium	1*s*	2*s*	2*p*
Be (Ground State)	↑↓	↑↓	[ ] [ ] [ ]
Be (Excited State)	↑↓	↑	↑ [ ] [ ]
Hybridisation	↑↓	↑	↑ [ ] [ ]
		sp- hybridisation	

Lobes of *s* and *p* orbitals combine with each other in 2 different ways.

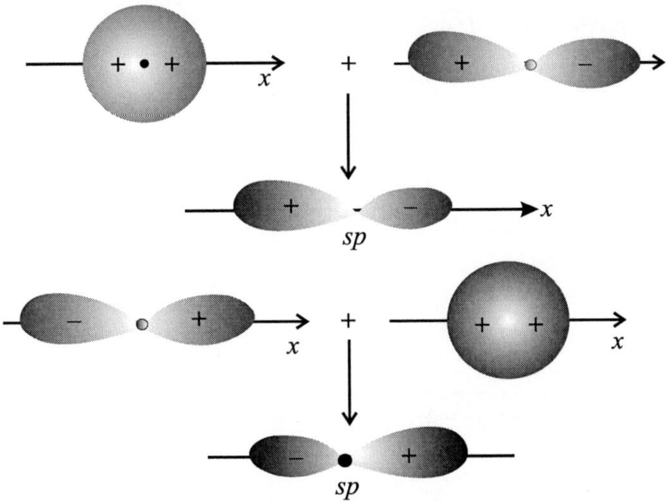

Here, two *sp* hybrid orbitals are equivalent in energy and are in the opposite direction, *i.e.*, these hybrid orbitals are inclined to each other at an angle of 180° as shown:

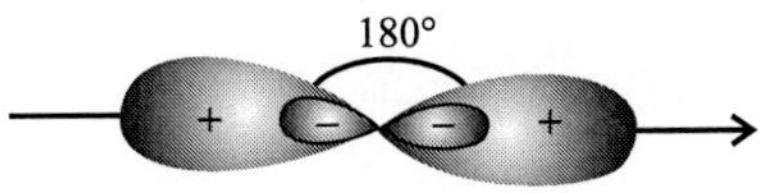

Fig. 2.26 Formation of two *sp* hybrid orbitals in case of Be atom.

$2s$ and $2p$ orbitals of beryllium mix to form two sp hybridised orbitals having equal energies. These hybrid sp orbitals overlap with the $2p$ orbitals of fluorine atoms to form two bonds *i.e.*, (Be – F) sigma bonds. The molecule is linear with a bond angle of 180° as shown:

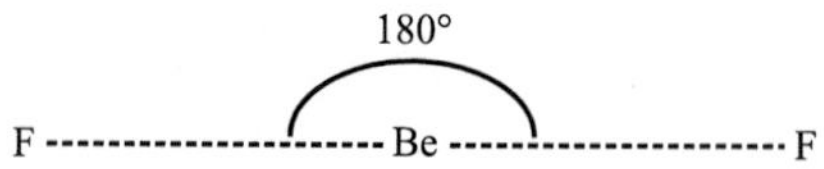

B. sp^2 hybridisation

The combination of an s and 2 p-orbitals leads to three hybrid orbitals, known as sp^2 orbitals. This can be explained by taking an example of BF_3 molecule.

Boron	1s	2s	2p
Boron atom (G. S.)	↑↓	↑↓	↑ | |
Boron atom (E. S.)	↑↓	↑	↑ | ↑ |
Hybridisation	↑↓	↑ (encircled with 2p)	↑ | ↑ | (encircled)
		sp^2-hybridisation	

The electronic configuration of B atom is $1s^2\ 2s^2\ 2p_x^{\ 1}$.

Boron in the excited state has 3 unpaired electrons, *i.e.*, 1 electron from $2s$ orbital of Boron jumps to $2p$ orbital as shown above.

One s and two p orbitals hybridise to give 3 sp^2 hybridised orbitals which are directed at an angle of 120° to each other. Each of the sp^2 hybrid orbitals contain unpaired electron which overlaps with the p orbitals of 3 fluorine atoms to form BF_3 molecule.

Fluorine	1s	2s	2p
Fluorine atom (G. S.)	↑↓	↑↓	↑↓ | ↑↓ | ↑

One s and any two p-orbitals of the valence shell of the central metal atom combines to form a set of 3 equivalent hybrid orbitals.

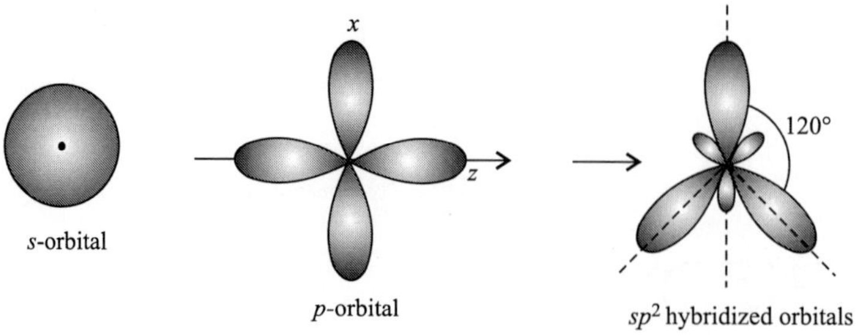

C. sp^3 hybridisation

The combination of one *s* and 3*p* (p_x, p_y, p_z) leads to formation of 4sp^3 hybridised orbitals. CH_4 molecule exhibits sp^3 hybridisation.

The electronic configuration of carbon atom is $1s^2\ 2s^2\ 2p^2$

Carbon atom (G. S.)	↑↓	↑↓	↑ ↑ □
Carbon atom (E. S.)	↑↓	↑	↑ ↑ ↑
		↓ sp^3-hybridisation	

Each of the hybridised orbital contains an unpaired electron which overlaps with the 1*s* orbital of hydrogen to form four bonds in CH_4 molecule with a bond angle of 109° 28′.

Atomic orbitals having lone pairs of electrons (completely filled orbitals) are also involved in the hybridisation, if the central atom of the given molecule has one or more lone pairs of electrons. This can be very well explained by taking an example of NH_3 and H_2O molecule.

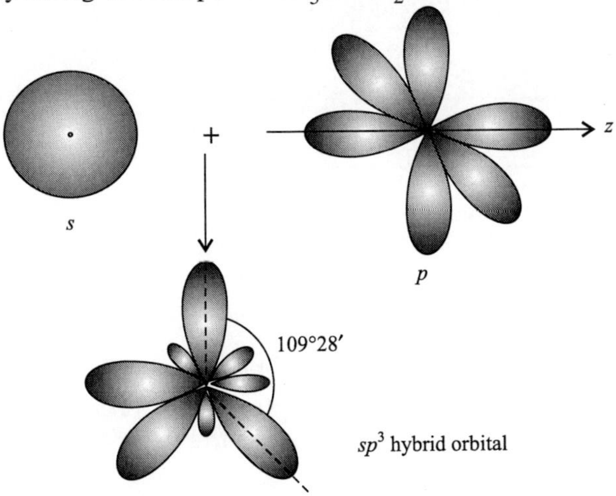

Hybridisation in NH_3 molecule:

The electronic configuration of nitrogen is $1s^2\ 2s^2\ 2p_x^{\ 1}\ 2p_y^{\ 1}\ 2p_z^{\ 1}$

Nitrogen	1*s*	2*s*	2*p*
Nitrogen	↑↓	↑↓	↑ ↑ ↑
Hybridisation	↑↓	↑↓	↑ ↑ ↑
		↓ sp^3-hybridisation	

One *s* orbital which is completely filled and three half filled *p*-orbitals hybridise to form four sp^3 hybridised orbitals. Out of the four hybridised orbitals, three orbitals are used for bonding with three hydrogen atoms, while fourth orbital contains a lone pair. Since, the lone pair-bond pair repulsion is greater than bond pair-bond pair repulsion, the shape of the molecule is distorted from tetrahedral to pyramidal with a bond angle of 107°.

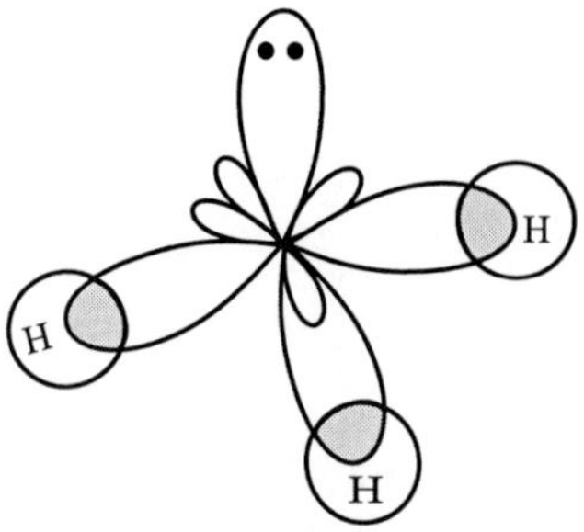

Fig. 2.27 Pyramidal shape of NH_3 molecule.

Hybridisation in H_2O molecule

The electronic configuration of oxygen atom is $1s^2$, $2s^2$, $2px^2$, $2py^1$, $2pz^1$

Oxygen	1*s*	2*s*	2*p*
Oxygen atom	⇅	⇅	⇅ ↑ ↑
Hybridisation	⇅	⇅	⇅ ↑ ↑
		sp^3-hybridisation	

Out of these four hybridised orbitals, two contain lone pairs while the other two are used for bond formation with two hydrogen atoms. Due to the presence of two lone pairs, the shape of the molecule is distorted from tetrahedral to bent with a bond angle of 104.5° as shown:

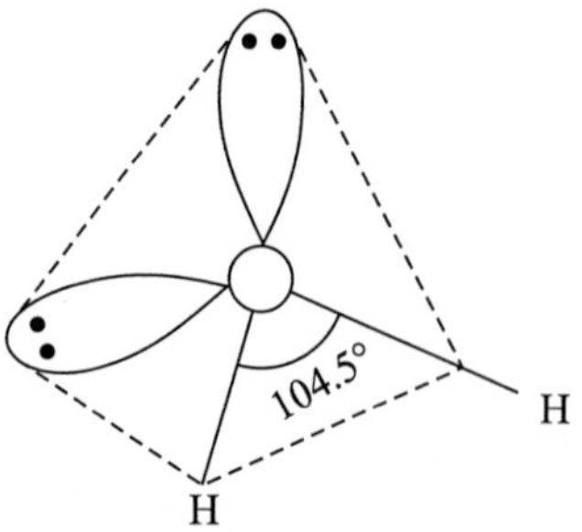

Fig. 2.28 Angular or Bent shape of H_2O molecule.

***sp*³*d* hybridization**

The combination of one *s*, 3 *p* (p_x, p_y, p_z) and one *d* orbital(d_{z^2}) leads to formation of five sp^3d hybrid orbitals of equal energy. This can be explained by taking up the example of PCl_5 molecule.

	3*s*	3*p*	3*d*
Phosphorous (G. S.)	↑↓	↑ | ↑ | ↑	| | | |
Phosphorous (E. S.)	↑	↑ | ↑ | ↑	↑ | | | |
		sp^3d-hybridisation	

5 of the sp^3d hybrid orbitals containing unpaired electrons overlaps with the 3*p* orbitals of Cl to form 5 bonds in PCl_5 molecule leading to Trigonal Bipyramidal shape of the molecule, the equatorial and axial angles being 120° and 90° respectively.

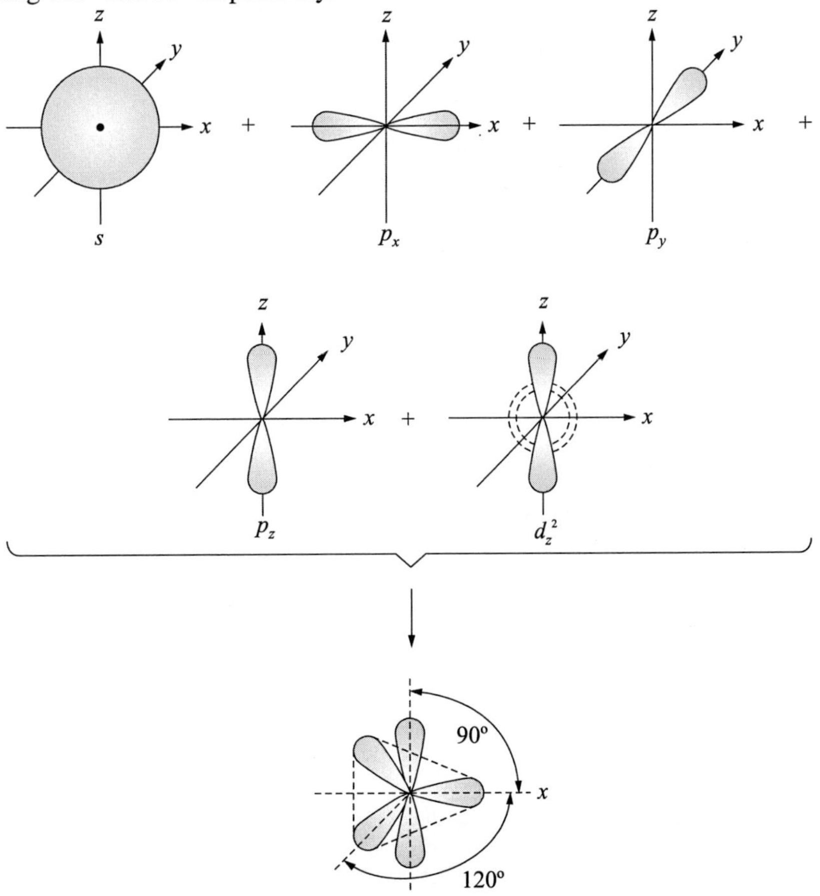

Fig. 2.29 Five *sp*³*d* hybrid orbitals

The trigonal bipramidal shape of PCl_5 is shown below:

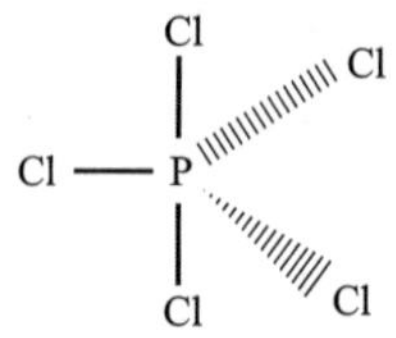

There are two P-Cl bonding environments in this molecule:

- **Each equatorial P – Cl bond makes two 90°** and two 120° bond angles with other bonds in the molecule.
- Each axial P-Cl bond makes three 90° and 180° bond angles with the other bonds in the molecule.

*sp*³*d*² hybridisation

The combination of one s, 3 p (p_x, p_y, p_z) and d_z^2, $d_x^2 - d_y^2$ leads to formation of 6 sp^3d^2 hybrid orbitals of equal energy. Let us take the example of SF_6 molecule.

Sulphur (G. S.)

Sulphur (E. S.)

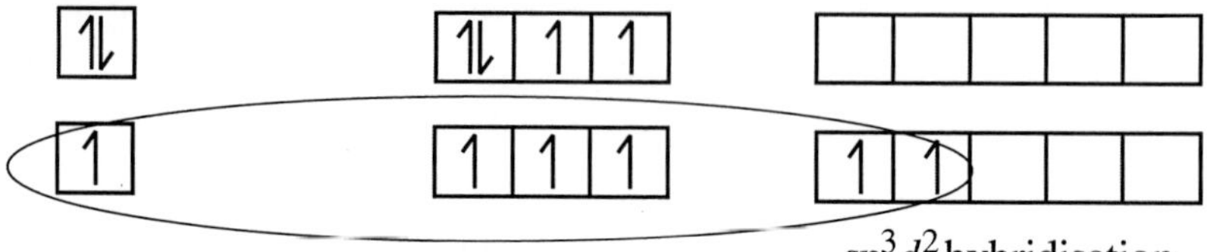

sp^3d^2 orbitals

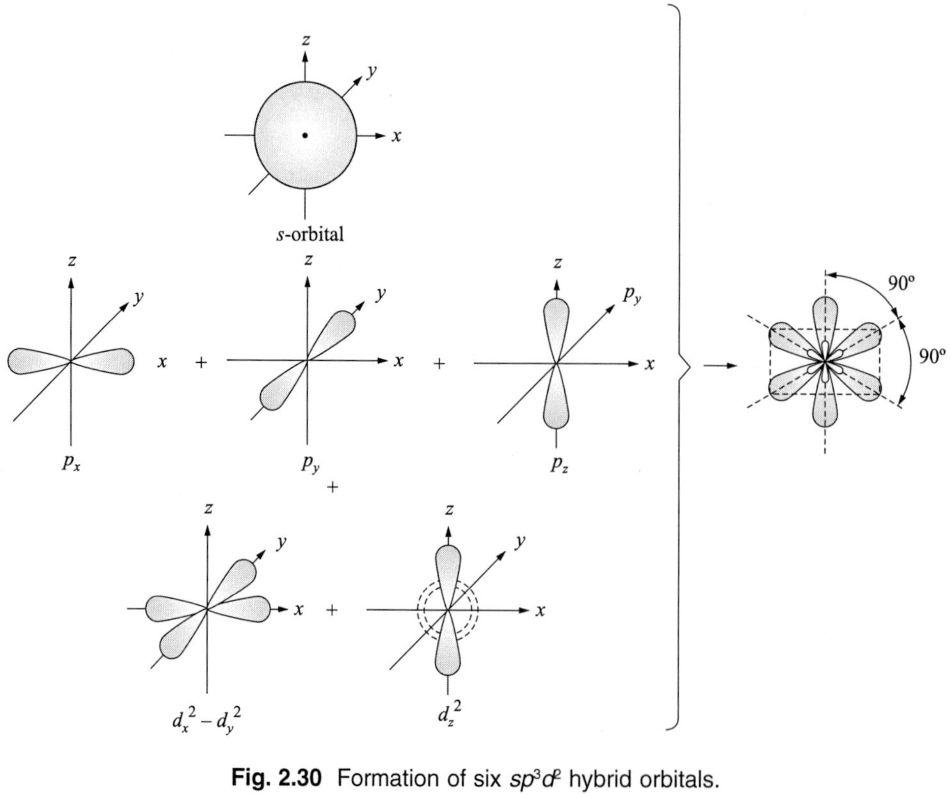

Fig. 2.30 Formation of six *sp*³*d*² hybrid orbitals.

The molecule has octahedral shape with bond angle of 90° as shown:

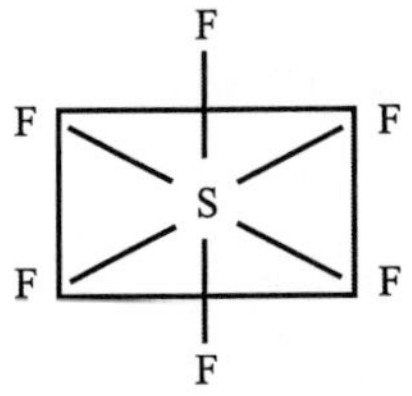

2.18.2 Structure and Geometries of Various Molecules

The ground state electronic configuration of Sn is [Kr] d^{10}, $5s^2$, $5p^2$

Sn (G. S.)

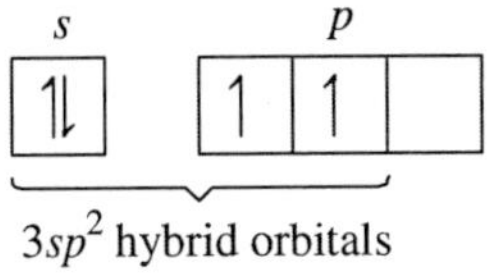

Since, Sn is bonded to two chlorine atoms, therefore, only two unpaired electrons are required. But tin atom gives 3 hybridised orbitals, two of which will be used for bonding with two chlorine atoms and the third hybridised orbital shall have lone pair of electrons on tin.

The two singly occupied sp^2 hybrid orbital would form σ-bond with the 3p-orbitals of chlorine atom whereas, the third sp^2 hybrid orbital is used to accomodate the lone pairs of electron as shown:

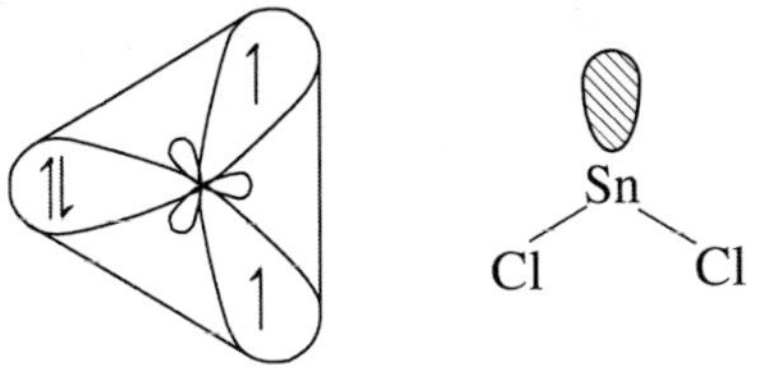

Therefore, $SnCl_2$ molecule is angular

Sulphate ion (SO_4^{2-})

The ground state electronic configuration of sulphur atom is [Ne] $3s^2$, $3p^4$

Sulphur atom [G. S]

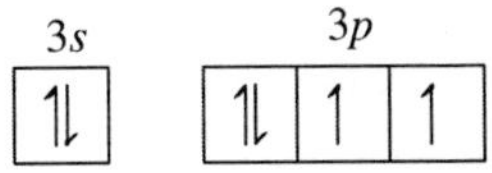

Sulphur atom [E. S]

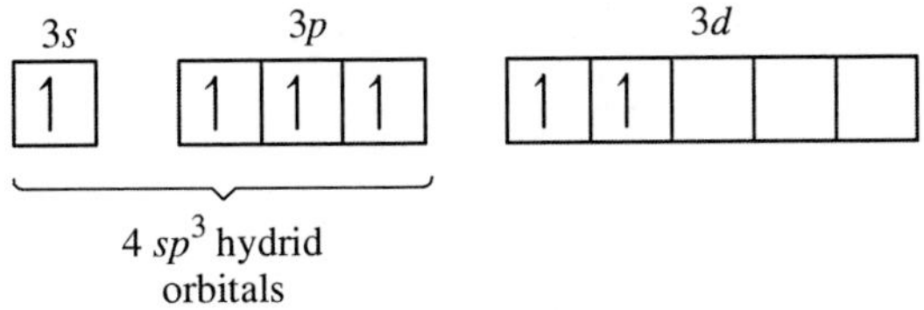

Sulphur-atom contains two unpaired electrons but it has to form four σ-bonds to oxygen atoms, so it requires four unpaired electrons, which is obtained by promoting one 3s electron and one 3p electron to the next higher energy vacant 3d orbitals.

One $3s$ and three $3p$ hybridise to produce four sp^3 hybrid orbitals which accounts for the tetrahedral shape of SO_4^{2-} ion.

The two electrons in the $3d$ orbitals are used to form 2π bonds to oxygen atoms.

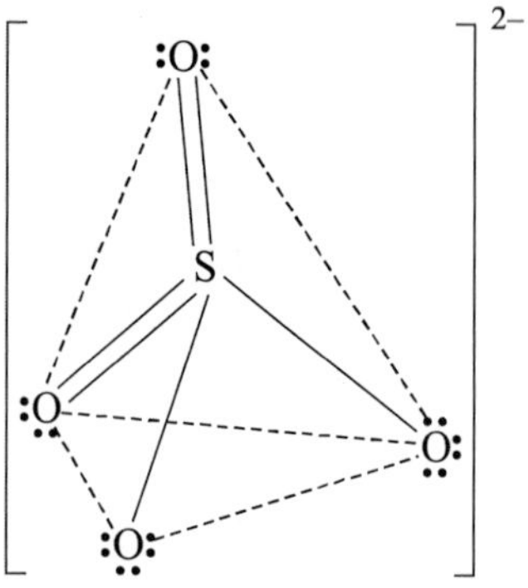

Xenon Tetrafluoride (XeF_4)

The ground state electronic configuration of xenon atom is $[Kr]\ 4d^{10}, 5s^2, 5p^6$

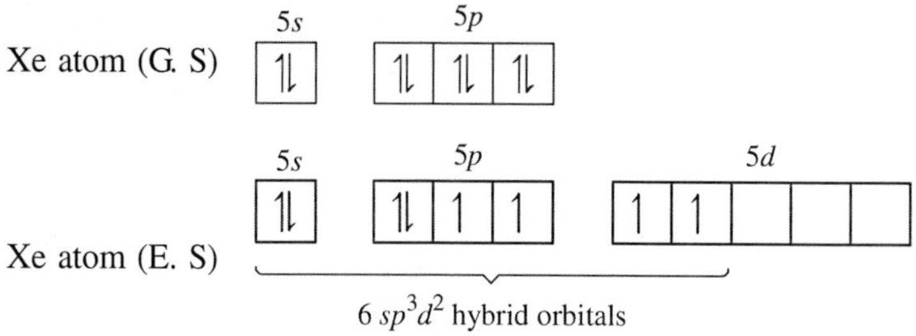

One $5s$, three $5p$ and two $5d$ orbitals hybridise to give 6 sp^3d^2 hybrid orbitals.

The four hybrid orbitals contain one electron each that is used to form bond with four fluorine atoms. The other two hybrid orbital contains lone pair of electrons. Therefore, XeF_4 has planar geometry as shown below:

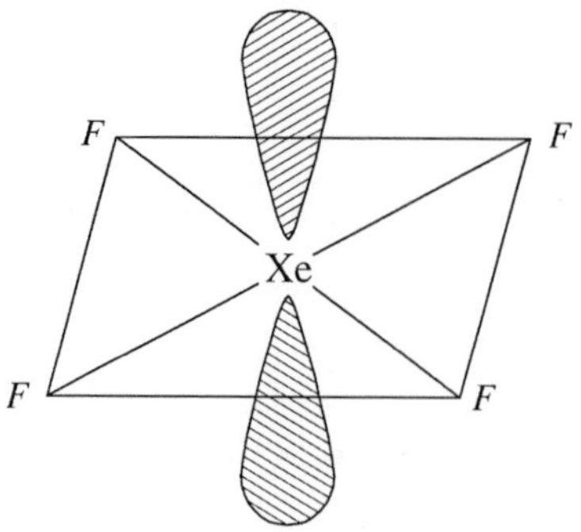

Example 2.15: ClO_3^- and ClO_4^- ions have same number of electron pairs around the central Chlorine atoms but their geometries are different. Why?

Solution: In ClO_3^-, one tetrahedral position is occupied by a lone pair , therefore, the ion has pyramidal shape, instead of being sp^3 hybridised.But, in ClO_4^-, as expected from sp^3 hybridisation, ClO_4^- achieves tetrahedral shape.

2.19 Molecular Orbital Theory

Molecular orbital theory was developed after Valence Bond Theory (1927) had been established, by the efforts of Friedrich Hund, Robert Mulliken, John C. Slater, and John Lennard-Jones, and was originally called the Hund-Mulliken theory. The word *orbital* was introduced by Mulliken in 1932. By 1933, the molecular orbital theory had become accepted as a valid and useful theory.

Before moving to molecular orbital theory we must refresh ourselves with the definition of atomic orbital.

"An atomic orbital is the 3-D space around the nucleus wherein the probability of finding the electron is maximum."

Depending upon the type of atomic orbitals involved in bond formation, different bonds are formed, i.e. σ- bond and π-bond, whose formation is already discussed in the previous part.

Molecular orbital theory (MO) uses a linear combination of atomic orbitals to form molecular orbitals which cover the whole molecule. These are often divided into bonding orbitals, anti-bonding orbitals, and non-bonding orbitals.

A **molecular orbital** is a region in which an electron may be found in a molecule. In general, **atomic orbitals combine to form molecular orbitals**.

Bonding Molecular orbitals are obtained when two atomic orbitals combine together in such a way that their lobes with + + sign overlap. Antibonding Molecular orbitals are obtained when two atomic orbitals combine together in such a way that their lobes with +– sign overlap.

Some of the important properties of Molecular Orbital theory are as follows:

- When nuclie of two atoms come close to each other, their atomic orbitals interact, to form molecular orbitals in which the identity of both the atomic orbitals is lost.
- The number of molecular orbitals is equal to the number the atomic orbitals included in the linear combination.
- The electrons in the atomic orbital are influenced by one nucleus, the electrons in the molecular orbitals are influenced by all the nuclie.
- Electrons are filled in the molecular orbital in accordance to Aufbau principle, Pauli's Exclusion Principle and Hund's rule of maximum multiplicity.

Molecular Orbital is formed by combination of two atomic orbital and this method is known as **Linear Combination of Atomic Orbitals**.

2.19.1 Linear Combination of Atomic Orbitals

An electron in an atom is described by a wavefunction ψ, as electrons are considered either as a particle or wave.

Let us consider two atoms *A* and *B* which have atomic orbitals described by the wavefunctions ψ_a and ψ_b respectively.The wavefunction of the molecular orbital ψ_{MO} can be obtained by the linear combination of ψ_a and ψ_b.

Therefore, $$\psi_{MO} = C_1 \psi_a + C_2 \psi_b$$

Where C_1 and C_2 are the coefficients chosen to give the lowest energy for the molecular orbital.

The probability of finding the electron in a volume of space is given by ψ^2, which can be obtained by

$$\psi^2_{MO} = (C_1 \psi_a + C_2 \psi_b)^2$$

$$\psi^2_{MO} = (C_1^2 \psi_a^2 + 2C_1C_2 \psi_a \psi_b + C_2^2 \psi_b^2)$$

The term $C_1^2 \psi_a^2$ gives the probability of finding an electron on atom A and the term $C_2^2 \psi_b^2$ gives the probability of finding an electron on *B*.

A. Combination of *s–s* orbitals

To produce the molecular orbitals for hydrogen, we add together the atomic wavefunctions ψ_a and ψ_b to produce the molecular orbitals for hydrogen. Each hydrogen atom in H_2 has only 1s orbital, so we add the two 1s wave functions. As we know, that atomic wave functions can have either plus or minus phases—this means the value of the wave function ψ is either positive or negative. There are two ways to add the wave functions, either both in-phase (either both plus or both minus) or out-of-phase (one plus and the other minus). Figure 2.31 shows how atomic wave functions can be added together to produce molecular orbitals.

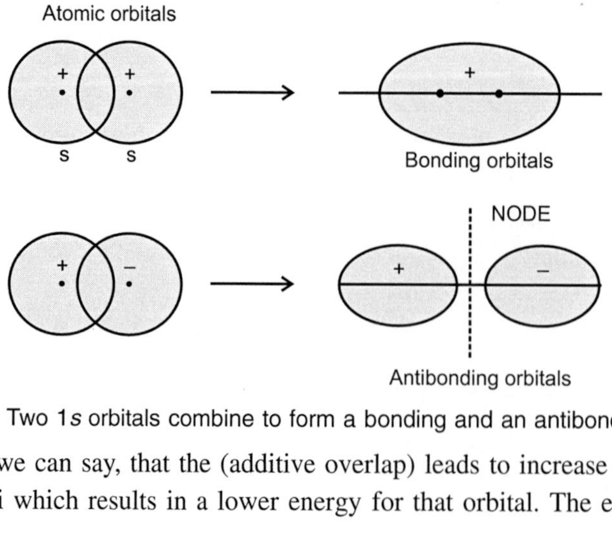

Fig. 2.31 Two 1*s* orbitals combine to form a bonding and an antibonding M.O.

The in-phase, or we can say, that the (additive overlap) leads to increase in the electron density between the two nuclei which results in a lower energy for that orbital. The electrons occupying the

ψ_{H-H} orbital represent the bonding pair of electrons of H_2 and is named as **bonding molecular orbital**. The other molecular orbital produced, ψ^*_{H-H} shows a decrease in electron density between the nuclei reaching a value of zero at the midpoint between the nuclei where there is a nodal plane i.e. at this point the probability of finding an electron is zero. Since the ψ^*_{H-H} orbital shows a decrease in electron density between the two nuclei, it is called an antibonding molecular orbital. Due to the decrease in electron density between the nuclei, the antibonding orbital is higher in energy than both the bonding orbital and the hydrogen 1*s* orbitals. In the molecule H_2, no electrons occupy the antibonding orbital.

For better understanding, an energy level diagram is shown below in figure 2.32.

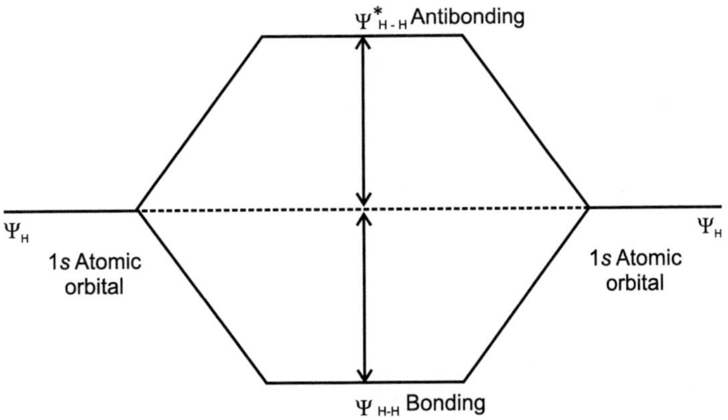

Fig. 2.32 An energy level diagram of H_2 molecule

B. Combination of *s–p* orbitals

An *s*-orbital combines with the *p*-orbital, which is directed along the internuclear axis. The combination of *s* and *p*-orbitals takes place in the following way:

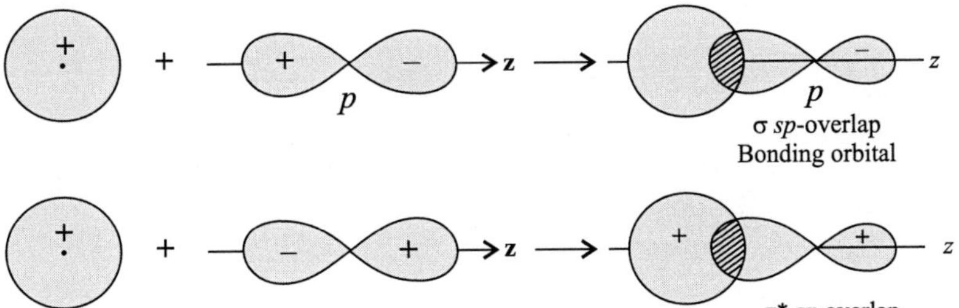

Combination of *s-p* orbitals

C. Combination of *p-p* orbitals

The overlap of two p orbitals along the axis joining the nuclei results in formation of σ_{2p} bonding and σ^*_{2p} antibonding orbitals. The combination is shown as below:

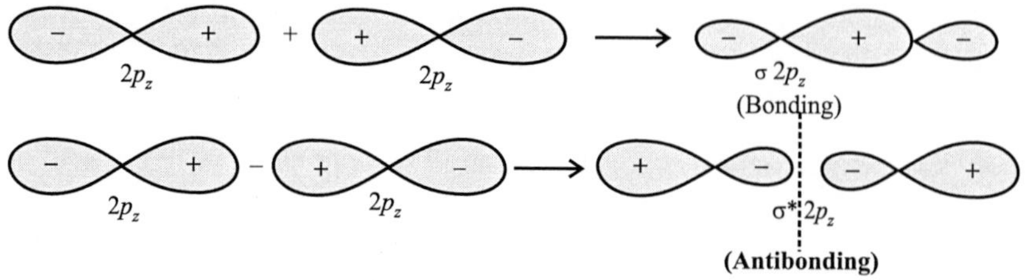

If the overlap of two *p*-orbitals, having lobes perpendicular to the axis joining the nuclei takes place, it would result in the formation of π_{2p_x} bonding and $\pi^*_{2p_x}$ antibonding orbitals as shown below:

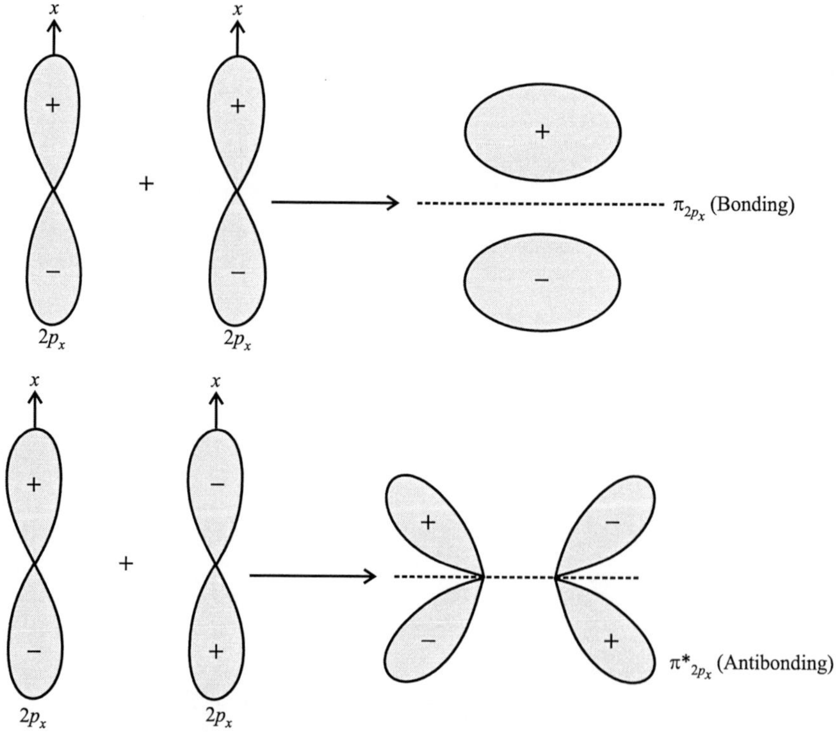

2.19.2 Energy Level Diagram for Molecular Orbitals

Different molecular orbitals have different energies. Antibonding molecular orbitals have higher energy than bonding molecular orbital. Therefore, the increasing order of energy of molecular orbitals is given as:

$$\sigma_{1s} < \sigma^*_{1s} < \sigma_{2s} < \sigma^*_{2s} < \sigma_{2p_z} < \pi_{2p_x} = \pi_{2p_y} < \pi^*_{2p_x} = \pi^*_{2p_y} < \sigma^*_{2p_z}$$

For molecules like Li_2, Be_2, B_2, C_2 and N_2, the energy of the orbitals is as follows:

$$\sigma_{1s} < \sigma^*_{1s} < \sigma_{2s} < \sigma^*_{2s} < \pi_{2p_x} = \pi_{2p_y} < \sigma_{2p_z} < \pi^*_{2p_x} = \pi^*_{2p_y} < \sigma^*_{2p_z}$$

The main reason for the difference in sequence, is the interaction of $2s$ and $2p_z$ atomic orbitals.For these molecules the difference in the energies of $2s$ and $2p_z$ orbitals is small and are of same symmetry, therefore,these interact with each other due to repulsive forces between the electrons. As a result of interaction, σ_{2s} and σ^*_{2s} molecular orbitals do not retain their pure s character and σ_{2p_z} and $\sigma^*_{2p_z}$ molecular orbitals do not retain their pure *p* character. All the molecular orbitals acquire a mixed *sp* character. Due to this mixing, σ_{2s} and σ^*_{2s}, which also contain some *p*-character, becomes more stable and thus lowered in energy. Similarly, σ_{2p_z} and $\sigma^*_{2p_z}$ also contain some *s*-character and becomes less stable and thus raised in energy. Since, $\pi 2p$ orbitals are not involved in mixing, the energies of these remain unchanged.

2.19.3 Bond Order

It is defined as half of the difference between the number of electrons present in the bonding and the antibonding orbitals. *i.e.*

Bond Order = ½ (No. of electrons in bonding orbitals – No. of electrons in anti-bonding orbitals).

A negative or zero bond order corresponds to an unstable molecule whereas positive bond order indicates a stable molecule.

Important points regarding bond order:

(*a*) Bond order of 1, 2 or 3 corresponds to a single, double or triple bond respectively.

(*b*) Bond order may be fractional or even zero.

(*c*) A zero bond order indicates the equal number of bonding and antibonding electrons. Zero value indicates that the species does not exist, while the fractional value indicates that the species is unstable.

2.19.3.1 Filling of Electrons in Molecular Orbitals of Homonuclear Diatomic Molecules

A. Hydrogen Molecule

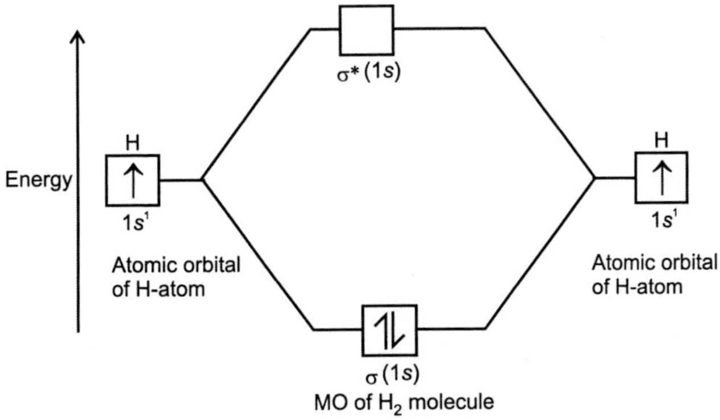

Fig. 2.33 MO diagram of H_2 Molecule.

Here, $N_b = 2$, $N_a = 0$

Therefore, Bond order= ½ $(N_b - N_a)$ = ½(2–0) =1

H_2 molecule is diamagnetic as it dosen't have any unpaired electron as shown in the MO. Diagram

B. Helium Molecule

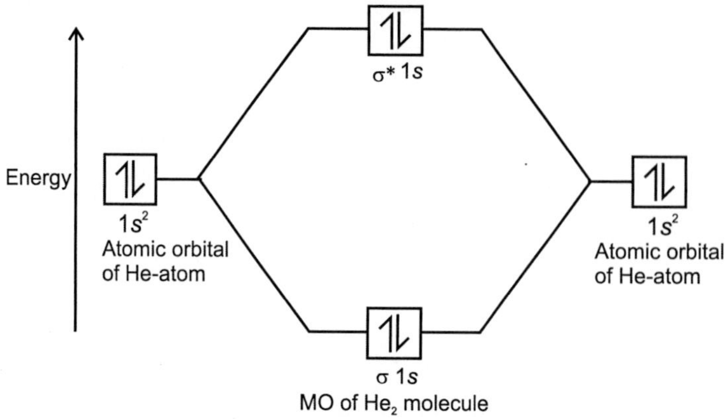

Fig. 2.34 MO diagram of He_2 Molecule

$N_b = 2$, $N_a = 2$. Therefore, Bond order = ½ $(N_b - N_a)$ = ½ (2 – 2) = 0

i.e. No bond is formed between 2 He atoms, therefore, He_2 does not exist. Although, He_2 does not exists but even if it existed it would have been a diamagnetic molecule.

C. Helium Molecule ion (He_2^+)

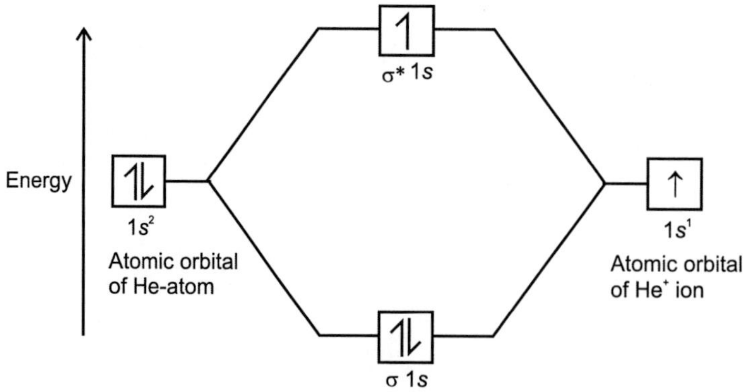

Fig. 2.35 MO Diagram of He_2^+ ion

Here, $N_b = 2$, $N_a = 1$, therefore, Bond order = ½ (2–1) = ½.

He_2^+ is a paramagnetic and not very stable molecule.

D. Boron Molecule B_2

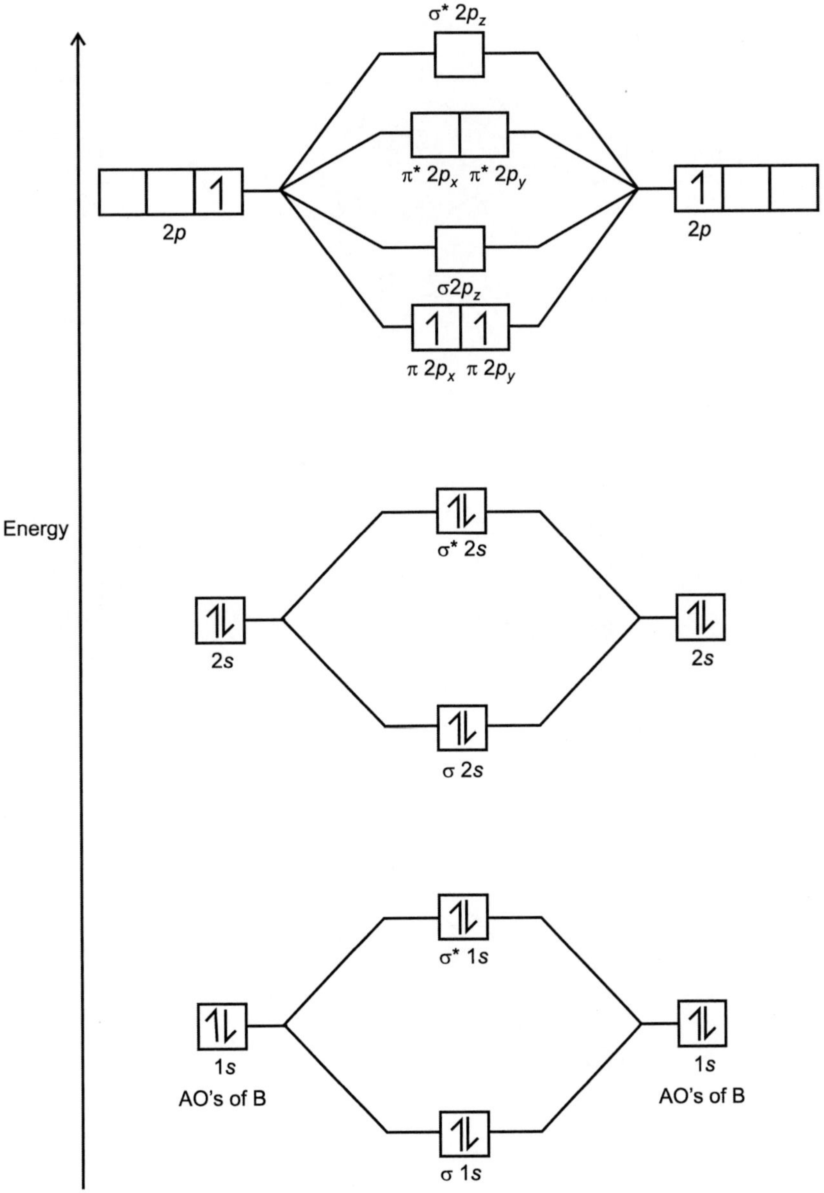

Fig. 2.36 MO diagram of B_2 Molecule.

Here, $N_b = 6$, $Na = 4$. Therefore, bond order $= \frac{1}{2}(6 - 4) = 1$, B_2 molecule is also a paramagnetic molecule.

E. Nitrogen Molecule N_2

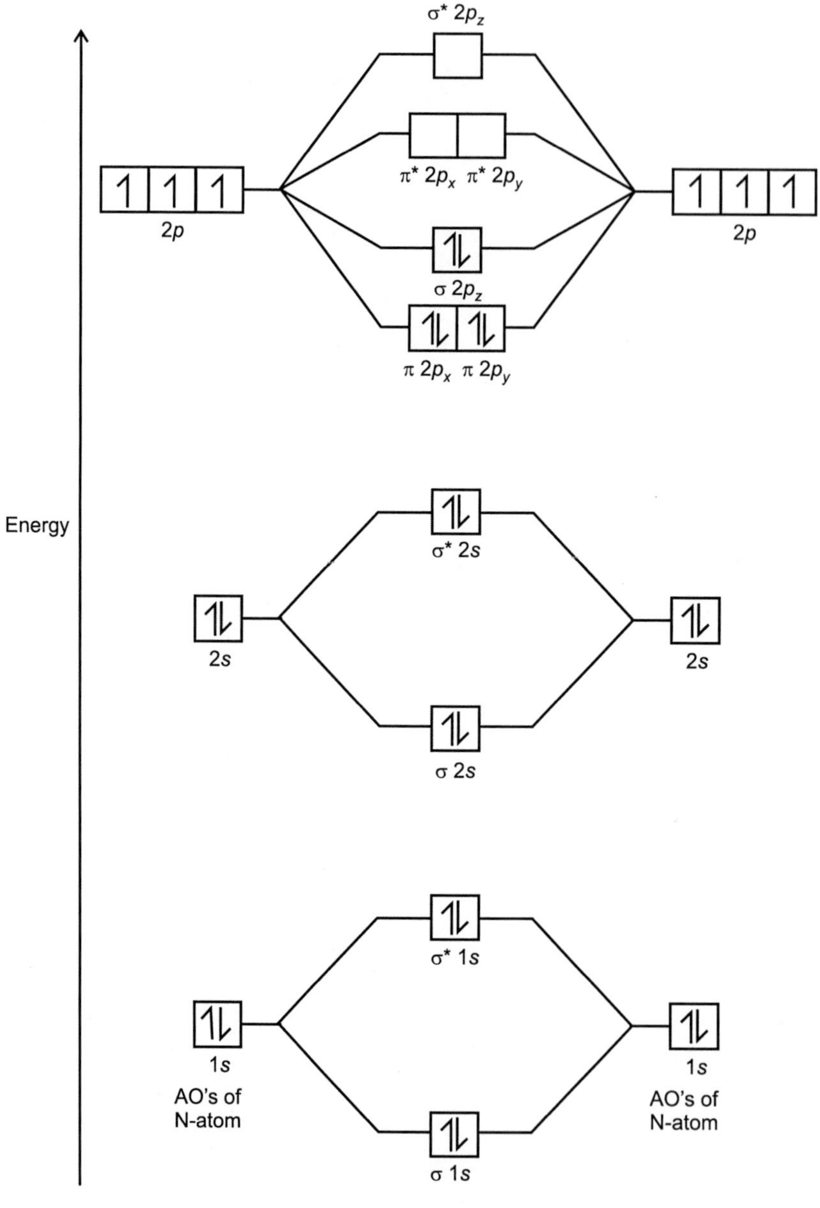

Fig. 2.37 MO diagram of N_2 Molecule.

Here, $N_b = 10$, $N_a = 4$. Therefore, bond order = ½ $(N_b - N_a)$ =½(10 – 4) = 3

This suggests that N_2 has a triple bond and it is a Diamagnetic Molecule.

2.19.3.2 Filling of electrons in Molecular Orbitals of Heteronuclear Diatomic Molecules

The filling up of orbitals of heteronuclear diatomic molecules is different from filling up of orbitals of homonuclear diatomic molecules. Since the electronegativities of the two atoms is different, the electrons in a bonding molecular orbital lies closer to the atom having high electronegativity whereas the electrons in an anti-bonding molecular orbital are closer to the less electronegative atom.

Let us now discuss the Molecular orbital diagrams for heteronuclear diatomic molecules:

A. Carbon Monoxide Molecule, CO

In case of CO molecule, oxygen is more electronegative than carbon; therefore, the atomic orbitals of oxygen are lower than those of carbon.

Due to this difference, the σ* 2*s* MO is slightly higher in energy than $\sigma\, 2p_z$, $\pi\, 2p_x$ and $\pi\, 2p_y$ MO's. Bond order = $\frac{8-2}{2} = 3$

Thus, CO molecule has a triple bond.

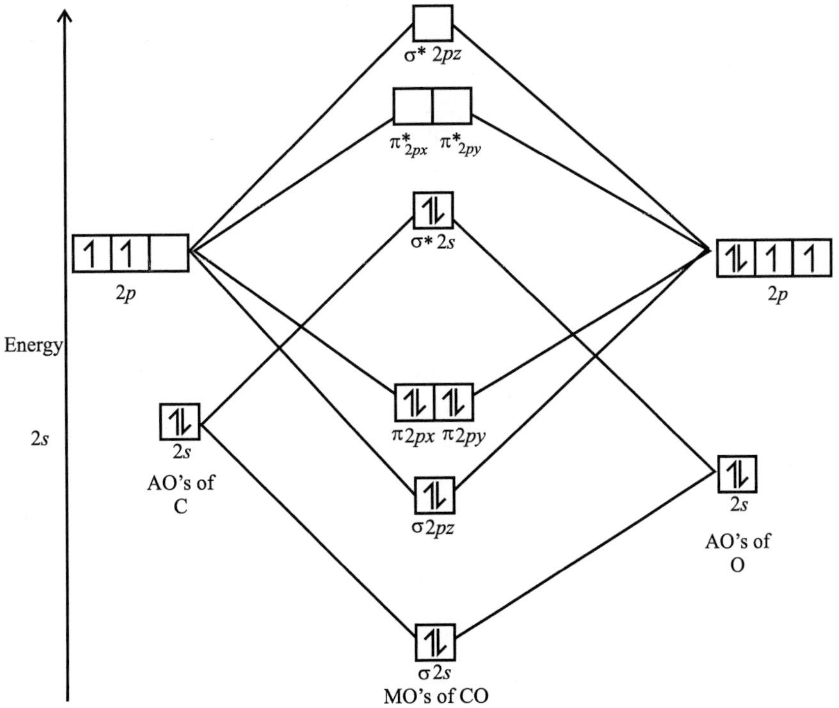

Fig. 2.38 MO diagram of CO molecule.

B. HCl Molecule

The energy of 3*s* and 3*p* atomic orbitals of the Cl atom is lower than that of 1*s* atomic orbital of the hydrogen atom.

The 3*s* atomic orbital does not interact with the 1s atomic orbital, because of the energy considerations.

1s atomic orbital of H-atom interacts with the $3p_z$ atomic orbital of Cl atom, because of appropriate energy and symmetry giving one sigma bonding and anti-bonding MO's.

$3p_x$ and $3p_y$ orbitals of Cl yield non-bonding MO.

The configuration becomes $(3s)^2$ $(\sigma 3p_z)^2$ $(3p_x)^2$ $(3p_y)^2$

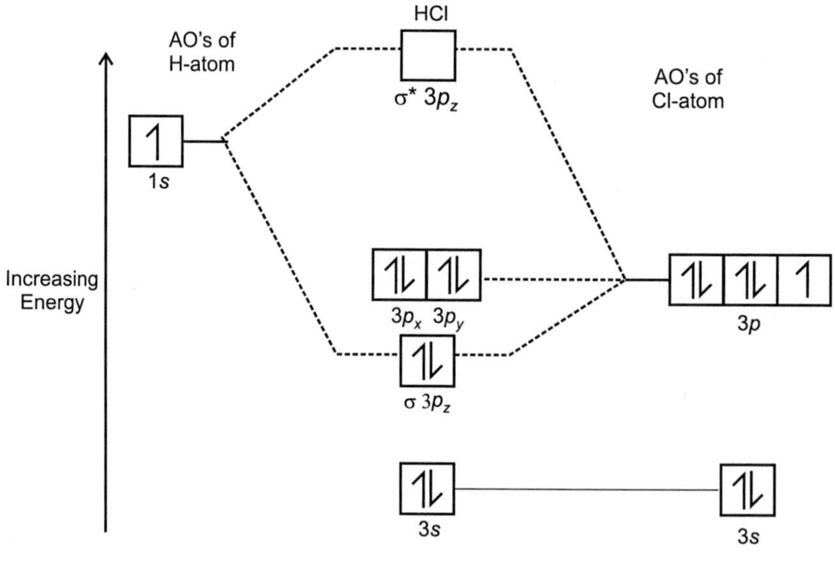

Fig. 2.39 MO diagram of HCl Molecule.

C. HF Molecule

The electronic configuration of H and F atoms are

$$H = 1s^1$$

$$F = 1s^2\ 2s^2\ 2p^5$$

HF molecule is formed by the linear combination of the atomic orbital of $H(1s^1)$ and the $\sigma 2p_z$ of *F*-atom.

The energies of $1s^2$ and $2s^2$ electrons of F are very low and do not take part in bonding, therefore, these are non-bonding orbitals.

Out of three $2p$ orbitals, only $2p_z$ combines with the s-orbital due to symmetry reasons.

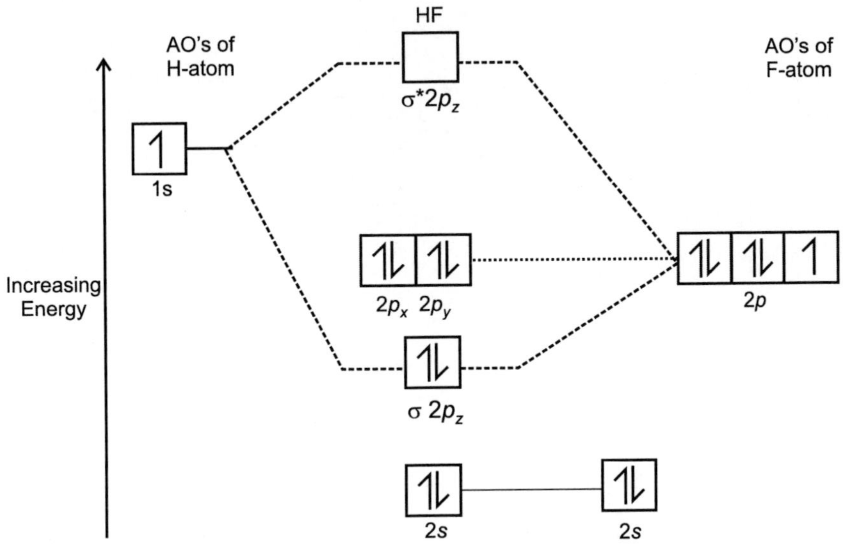

Fig. 2.40 MO Diagram of HF moleucle.

D. Nitrogen oxide molecule

Nitrogen oxide is paramagnetic molecule, and it has been asssumed that this molecule has two normal covalent bonds (one s- and one p- bonds) and one three electron bond. Therefore, the structure of NO molecule can be written as $[\ {}^{x}_{x}N \overset{x\bullet\bullet}{\equiv} O:]$. Also the total number of electrons in the valence shell of NO shall be 11. Therefore the bond order, of NO is 2.5 and hence it is paramagnetic in nature. It has been suggested that N=O has a triple bond and both the atoms have one lone pair of electrons each. Thus, the structure is $[{}^{x}_{x}N \overset{x\bullet\bullet}{\equiv} O:]$

$$N_b = 10,\ N_a = 5, \text{ thus Bond Order} = \frac{10-5}{2} = 2.5$$

The Molecular Orbital picture of this molecule is given in Fig.2.41.

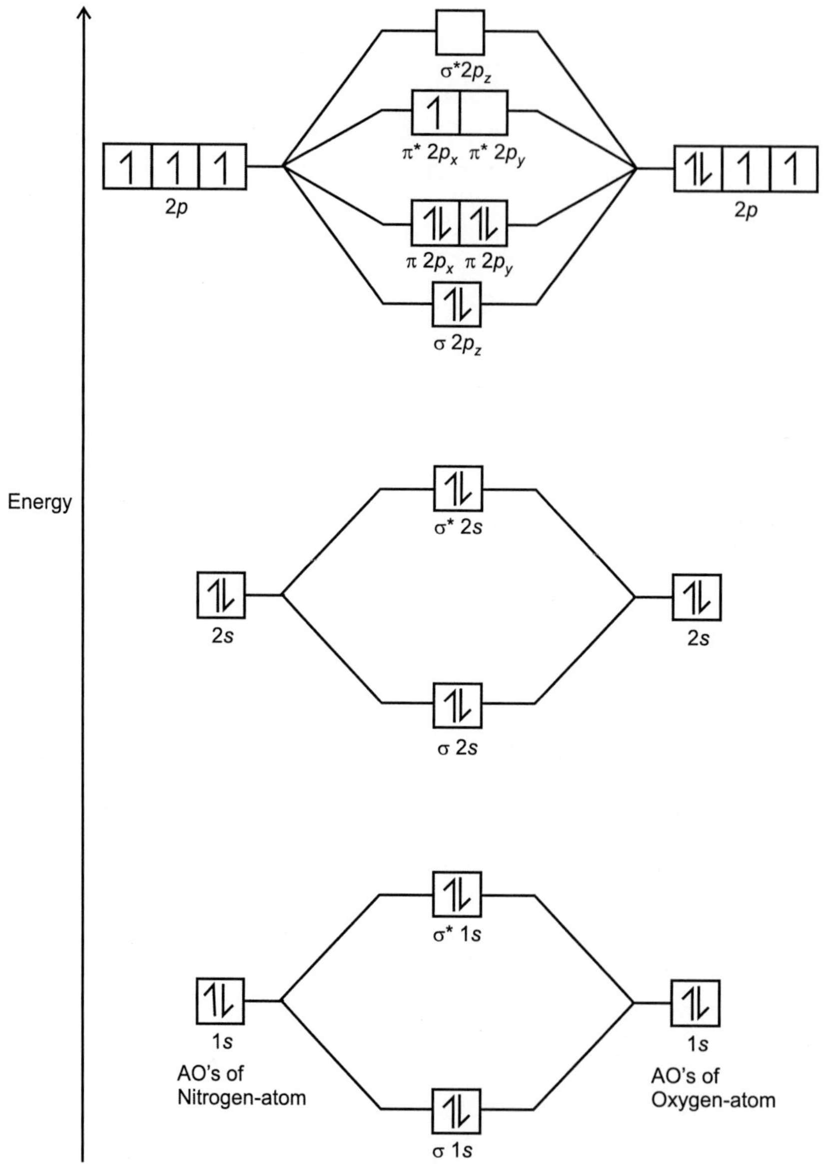

Fig. 2.41 Molecular Orbital diagram of Nitrogen oxide.

2.20 Comparison of the Valence Bond Theory and Molecular Orbital Theory

Valence Bond Theory	Molecular Orbital Theory
1. According to valence bond theory, a molecule is composed of atoms which retain their individual identity even after bond formation.	1. According to Molecular orbital theory, the combining atoms forming molecules loose their individual identity.
2. Atomic orbitals are represented as *s*, *p*, *d* and *f*.	2. Molecular orbitals are represented by $\sigma\sigma^*$ and $\pi\pi^*$ orbitals.
3. It does not involve the formation of bonding and anti-bonding orbitals.	3. It involves the formation of bonding and anti-bonding orbitals.
4. Electrons in molecules are localised as if they are in isolated atoms.	4. It treats the nuclei of the molecule as polycentric and then constructs one electron molecular orbitals.
5. It could not explain the paramagnetism of O_2.	5. It can easily explain the paramagnetism of O_2.

Points to Remember

- If the force of attraction is more than the force of repulsion, a bond is formed.
- If the force of attraction is less than the force of repulsion, a bond is not formed.
- **Low Ionisation Energy** of the element forming a cation and **High Electron Affinity** of an element forming an anion leads to formation of an Ionic Bond.
- The number of electrons lost or gained by the atom to form Ionic compound is known as the valency of the respective atom in that compound.
- Ionic Bonds are found in hydrides, halides, nitrides, sulphides, oxides and carbides of alkali and alkaline earth metals.
- The crystal lattice is composed of an infinite number of small units; each of these small units is called a unit cell.

- Band gap in the band structure strongly influences the electrical and optical properties of the material.
- The band gap is maximum for an insulator and minimum for a conductor.
- Fermi level is the highest occupied energy level at absolute zero.
- The energy released in the formation of 1gm mole of a crystal from the gaseous ions is called Lattice Energy of the crystal.
- Born Lande equation is given by the following expression :

$$U = \frac{-N_0 A z^+ Z^- e^2}{r_0}[1-1/n]$$

- The energy changes involved in the formation of ionic compounds from their constituent elements can be studied with the help of a thermochemical cycle called Born-Haber cycle.
- Madelung constant depends on the electron number and the geometric pattern of the ions in the crystal lattice
- Lattice Energy can also be calculated by Kapustinskii Equation,

$$U = 1202.5 n z^+ z^- \frac{(1-0.345)}{(r^+ + r^-)}$$

- The power of the cation to distort the anion is known as the polarising power of the cation and the tendency of an anion to get polarised is called polarisability of the anion.
- Covalent Bond is formed between atoms of comparable electronegativities, wherein equal number of electrons are shared between the two concerned atoms.
- If a covalent bond is formed between similar atoms, the shared pair of electrons lie almost in the centre, such a covalent bond is Non-Polar Covalent Bond.
- If a covalent bond is formed between the different atoms, the shared pair is displaced towards the more electronegative atom, such a covalent bond is a Polar Covalent Bond.
- The order of repulsion between the electrons is given as $lp - lp > lp - bp > bp - bp$
- The order of increase in the s-character is sp – H > sp^2C – H > sp^3C – H hybridised.
- Dipole Moment is defined as the product of the net positive or negative charge and the distance between the two charged ends.

- Percentage Ionic Character can be obtained from the electronegativity values by the following expression:

$$\% \text{ I.C} = 16\ (X_a - X_b) + 3.5\ (X_a - X_b)^2$$

where X_a and X_b are the electronegativities of the bonded atoms A and B.

- Percent Ionic character can also be calculated from dipole moment values *i.e.*

$$\% \text{ Ionic character} = \frac{\text{Experimental value of D.M}}{\text{Theoretical value of DM}} \times 100$$

- Dipole Moment is a vector quantity.
- By illustrating the structures of various molecules, their dipole moment values can be predicted satisfactorily.
- **Co-ordinate bond** is formed when both the electrons forming the bond are contributed only by one atom.
- NH_4Cl, $CuSO_4$, $[Cu(NH_3)_4]SO_4$, $K_4[Fe(CN)_6]$ are the compounds having all the ionic, covalent and dative bonds.
- An electrostatic attractive force between covalently bonded hydrogen atom of one molecule and the electronegative atom of another molecule is a **Hydrogen Bond.**
- An atomic orbital is the volume in space where an electron may be found around an atom.
- VSEPR was developed by *Gillespie and Nyholm* to predict the shape of the simple covalent bonds.
- The repulsive interactions between the electrons decreases in the order

$$lp - lp > lp - bp > bp - bp$$

- Resonance hybrid is more stable because of the delocalisation of electrons that lowers the energy and imparts stability.
- The difference in energy of the actual molecule and the most stable contributing structures is called Resonance Energy.
- Valence bond theory fails to explain the paramagnetic behaviour of substances and geometry of non-linear molecules.
- Hybridisation is the concept of mixing atomic orbitals to form new hybrid orbitals suitable for the qualitative description of atomic bonding properties.
- Hybrid orbitals are much stronger than the non-hybrid ones.

- Hybrid orbitals follow Hund's rule and Pauli's exclusion principle.
- BeF_2, $BeCl_2$, CO_2, C_2H_2, HCN exhibits *sp* hybridisation where the orientation of hybrid orbitals is linear.
- BF_3, SO_2, SO_3, C_2H_4 exhibits sp^2 hybridisation with Trigonal planar orientation.
- SiF_4, CCl_4, NH_3, H_3O^+, ClO_4^-, NF_3 exhibits sp^3 hybridisation with tetrahedral orientation of hybrid orbitals.
- ClF_3, SF_4, PCl_5 exhibits sp^3d hybridisation where the orientation of hybrid orbitals is Trigonal bipyramidal.
- XeF_4, SF_6, BrF_5, $XeOF_4$ exhibits sp^3d^2 hybridisation with octahedral orientation of hybrid orbitals.
- Molecular orbital is a region in which an electron may be found in a molecule.
- The number of covalent bonds formed in a molecule is called a Bond Order.
- Bond Order explains about the Bond Dissociation Energy, Bond length and Magnetic behaviour of molecules.
- Bond Order for H_2^+ and H_2^- is samc but H_2^- is less stable than H_2^+, since H_2^- has one electron in the anti-bonding orbital resulting in the repulsion.
- Order of stability and bond dissociation energies are

  (*i*) $O_2 > O_2^- > O_2^{2-}$

  (*ii*) $N_2 > N_2^+ > N_2^{2-}$
- In case of molecules having resonance, bond order can be calculated as follows:

$$\text{Bond order} = \frac{\text{Total number of bonds between two atoms in all the structures}}{\text{Total number of resonating structures}}$$

For Example,

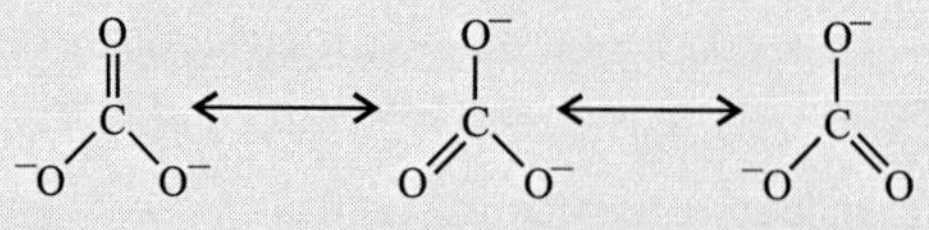

$$\text{Bond order} = \frac{\text{Double bond + Single bond + Single bond}}{3}$$

$$= \frac{2+1+1}{3} = 1.33$$

Exercises

1. What do you understand by electrovalency?
2. Name the factors which influence the formation of ionic compounds.
3. Explain why electrovalent compounds are mostly solids and have higher melting points than covalent compounds.
4. Solid NaCl, does not conduct electricity, whereas molten (fused) NaCl does, why?
5. Why NaCl is an electrolyte but CCl_4 is not?
6. What is Born Haber's cycle? Discuss its usefulness in explaining the stability of solids.
7. Derive an expression for the energy released when one mole of solid LiF is formed from its gaseous ions.
8. What is Madelung Constant?
9. Calculate the lattice energy of CaF_2.The madelung constant for fluorite structure is 2.519, ionic radii for Ca^{2+} and F^- are 0.99 and 1.36Å respectively, the electronic charge $e = 4.8 \times 10^{-10}$ e.s.u, Avagadro's Number = 6.023×10^{23} and repulsive constant $n = 7$.
10. Why ionic compounds melt at a higher temperature as compared to covalent compounds?
11. Ionic radius of Potassium ion is smaller than Potassium atom while that of the fluoride ion is greater than that of fluorine atom, why?
12. What is a covalent bond? Illustrate with examples?
13. Why covalent compounds are often gases, liquid or solids with low MPt.
14. Differentiate between Ionic and Covalent bond.
15. Describe the factors favouring the formation of covalent bond in reference to Fajan's rule.
16. Write short notes on :

    A. Fajan's rule of covalency

    B. Polarisation of covalent molecules.

    C. Polarisibility

    D. Polarisation

    E. Polarising power of ions.

17. Explain the formation of singlet linkages in chemical compounds giving suitable examples.

18. Define the term bond order. Give its significance.

19. How the concept of electronegativity been used to predict the nature of bond?

20. The electronegativity values of A and B are 2.0 and 3.0 respectively. Predict the nature of bond between the two atoms.

21. Write a short note on

A. Polar and Non-Polar covalent bond.

B. Percentage ionic character of covalent bonds.

22. Write down the electronic structures of

A. PCl_5

B. H_3PO_4

C. SF_4

23. Calculate the percentage ionic character of HCl molecule when the bond distance is 1.275 Å and the observed dipole moment is 1.03 D.

24. Explain why bond angles in CH_4, NH_3 and H_2O molecules decrease from 109.5° to 107.5° to 104.5°.

25. What is dipole moment? How it is useful in the determination of molecular structure?

26. Dipole moments of CO_2 and SO_2 molecules are zero and 1.60D respectively, why?

27. Explain why the dipole moment of NH_3 is greater than that of NF_3.

28. Arrange the following molecules in the increasing order of their dipole moment values

$$CH_4, CHCl_3, CCl_4, CH_2Cl_2, CH_3Cl$$

29. Boron has the tendency to form covalent compounds. Explain.

30. Why the polarising power of Cu^+ ion is greater than that of Na^+ ion?

31. Explain why beryllium compounds are covalent.

32. Write a note on metallic bond.

33. Write a note on co-ordinate bond.

34. Describe the basic ideas of VSEPR theory.

35. What are the limitations of VSEPR theory?

36. Explain the shape of ClF_3, PCl_5, SF_4 molecules on the basis of VSEPR theory.

37. Describe the structures of H_2O, NH_3 and CH_4 molecules on the basis of VSEPR theory. Explain why the bond angles are different in the three molecules.

38. What do you mean by linear combination of atomic orbitals?

39. What are the conditions necessary for the formation of molecular orbitals?

40. Differentiate between bonding and anti-bonding molecular orbitals.

41. What are sigma and Pi molecular orbitals. Explain with suitable examples.

42. Explain on the basis of MOT, why He_2 and Be_2 molecules do not exist?

43. Why O_2 molecule is paramagnetic while N_2 molecule is diamagnetic?

44. Draw molecular orbital configurations of O_2, O_2^+, O_2^- and O_2^{2-} ?

45. Draw molecular orbital energy level diagram for NO and find its bond order.

46. Draw the molecular orbital diagrams for H_2, H_2^+, He^+, He_2^+ ?

47. Write short notes on

(*a*) Resonance

(*b*) Resonance energy.

48. What is hybridisation? Explain with suitable examples.

49. State the rule which governs the hybridisation of orbitals.

50. Explain *sp*, sp^2 and sp^3 hybridisation with suitable examples.

51. Why hybrid orbitals form comparatively stronger bonds.

52. Explain the structure of CH_4, NH_3, H_2O on the basis of hybridisation theory.

53. Explain sp^3d and sp^3d^2 hybridisation with suitable examples.

❑❑❑

ORGANIC CHEMISTRY

UNITS

3

Structure and Shape of Organic Molecules

Jacobus Hendricus Van't Hoff (1852-1911)

IN THIS CHAPTER

He was born in Rotterdam, Netherlands in the year 1852, and was a student at the University of Utrecht. He along with Joseph Achille leBel in 1874, introduced the idea of tetrahedral valency of carbon that could explain the existence of stereo-isomers. He received his Ph.D. in synthetic organic chemistry. Shortly before, he submitted an only average doctoral thesis in synthetic organic chemistry to the University of Utrecht, the twenty-two-year-old van't Hoff had printed and distributed a twelve-page pamphlet at his own expense that, although ignored for many years, essentially outlined the foundation of modern stereochemistry. He acquired fame particularly by his epoch making publications.that gave the impetus to the develoment of stereochemistry. He was a founder of modern stereochemistry and physical chemistry. The concept of the asymmetric carbon atom, supplied an explanation of the occurrence of stereo-isomers, inexplicable by means of the then current structural formulae. At the same time, he pointed out the relationship between optical activity and the presence of an asymmetric centre in a molecule. Dutch chemist Jacobus Hendricus van't Hoff, received the 1901 Nobel Prize in chemistry, "in recognition of the extraordinary services he has rendered by the discovery of the laws of chemical dynamics and osmotic pressure in solutions.

As is the case with many creative minds, van't Hoff's career initially encountered many road blocks, mostly related to the ignorance of the scientific establishment. Unemployed for two years after receiving his doctorate, he eventually took a teaching job in physics at the veterinary school in Utrecht. Success finally came in 1877 when the newly founded University of Amsterdam offered him a lectureship; the next year he became a professor of chemistry at the same institute. Although his international reputation led to the formation of a new chemical laboratory at the Amsterdam school, which was completed in 1891, he moved to Berlin in 1896 as both a university professor and member of the prestigious Prussian Academy of Science.

Inspired by the earlier ideas of the German chemist Johannes Wislicenus to extend chemical structure theory from constitutional chemical formulas to representations in three-dimensional space, van't Hoff suggested a structural distinction between optical **isomers** that had been represented up until that time by the same formula. By screening known substances for optical activity (i.e., the rotation of the plane of polarized light), he found that all their constitutional formulas contained at least one carbon atom that combined with four different atomic groups. If the latter were placed at the corners of a tetrahedron in three-dimensional space, with the carbon atom at its center, there were exactly two possible tetrahedra with asymmetric carbon atoms, each being the mirror image of the other, that could account for the pairs of optical isomers. This structural theory was further supported by his observation that for every known chemical transformation, where the optical activity of a reactant disappeared, there was no more asymmetric carbon atom in the structural representation of the reaction product. Independently and virtually simultaneously, Le Bel arrived at the same theory in Paris.

Based on numerous measurements of organic reaction rates, he classified chemical reactions according to different orders and molecularities and formulated the temperature dependence of reaction rates, now known as the "Arrhenius equation". Following through on earlier ideas, he interpreted chemical equilibria as dynamic states where backward and forward reaction rates were equal, which he represented by a double arrow. He further formulated the temperature dependence of the equilibrium constant, known as the "van't Hoff isochore," and suggested what later came to be known as "Le Chatelier's principle." Most important, he established the foundation of chemical thermodynamics by expressing the relationship between what is now called the free energy of a reaction and its equilibrium constant.

During the late 1880s van't Hoff turned the seemingly exotic phenomenon of osmotic pressure into a crucial part of the new physical chemistry of solutions. He discovered that the osmotic pressure p increased with the concentration of the solute, $c = n / V$, and the absolute temperature T according to the equation which is equivalent to the ideal gas law, if one replaces the osmotic pressure p , with the gas pressure, p : with R being the gas constant in both cases. The surprising equivalence between the two, suggested that the osmotic pressure depended only on the number of solute molecules and not their chemical nature, like other so-called colligative properties such as vapor pressure lowering and freezing point depression. Van't Hoff did not simply advance all these phenomena on a common thermodynamic basis; he also successfully explained apparent anomalies with the Arrhenius theory of electrolytic dissociation.

In his later years, van't Hoff applied chemical thermodynamics to the formation of marine salt deposits, which made him one of the pioneers in the science of petrology. Van't Hoff had a generalizing and speculative mind which gave him insights into the newly developing field of physical chemistry. He died on March 1, 1911.

CHAPTER 3

Structure and Shape of Organic Molecules

Introduction

In the present chapter, the already described concepts of atomic structure in earlier chapters of inorganic chemistry are recapitulated. Some newer concepts about organic molecular structure are discussed.

3.1 The Electron as a Wave and Atomic Orbitals

We have learnt that electrons are located in orbits. Thus, to plot a path for electron (orbit of electron) you need to know exactly where it is at any given moment and where it is likely to be after an instant. To work out exactly where it's going to be after an instant later, we do need to know the direction and velocity exactly. To understand this we have de-Broglie equation ($\lambda = h/mv$) and Heisenberg's Uncertainty Principle.

Heisenberg, 1925 announced that it is not possible in case of submicro particles like electron, to know exactly or precisely its position and momentum simultaneously. It is known as ***Heisenberg Uncertainty Principle.*** During 1938-45 development of ***quantum mechanics*** took place. Soon it was realised that microparticles like electrons behave like waves. That is, energy and associated properties of an electron could be described by an equation, like that of a wave.

Schrödinger, in 1927 described the properties of an electron, or matter in general, by a quadratic equation called ***Schrödinger Wave Equation***. Each solution of the equation is a wave function (ψ) that describes the properties of an electron in reference.

$$\frac{\partial^2\psi}{\partial x^2} + \frac{\partial^2\psi}{\partial y^2} + \frac{\partial^2\psi}{\partial z^2} + \frac{8\pi^2 m}{h^2}(E - V)\psi = 0$$

Where

ψ = Wave function

m = mass of the electron

h = Planck's Constant

V = Potential energy

x, y, z = are the cartesian co-ordinates

The square of this wavefunction (ψ^2) best describes the probability of finding the electron in a certain volume. So instead of calculating the energy of an electron in an orbit, we look at the Probability density of electrons in a volume in space around the nucleus. There is a certain volume around the nucleus, where the probabiliity of finding the electron is maximum. This space around the nucleus is referred to as *Atomic Orbital* instead of orbit- the exact path in circle described earlier.

Atomic Orbital: It is also described as a mathematical solution of Schrödinger wave equation for an electron. It tells us the energy of electron and the volume in space having maximum probability of locating the electron.

3.2 Representation of Atomic orbitals of Carbon

Carbon has six electrons. Two of them are found in the 1*s* orbital that is closest to the nucleus. The next two electrons go to 2*s* orbital. The remaining two are in two 2*p* orbitals. This is in accordance to the Hund's rule. The shape of orbitals is given below:

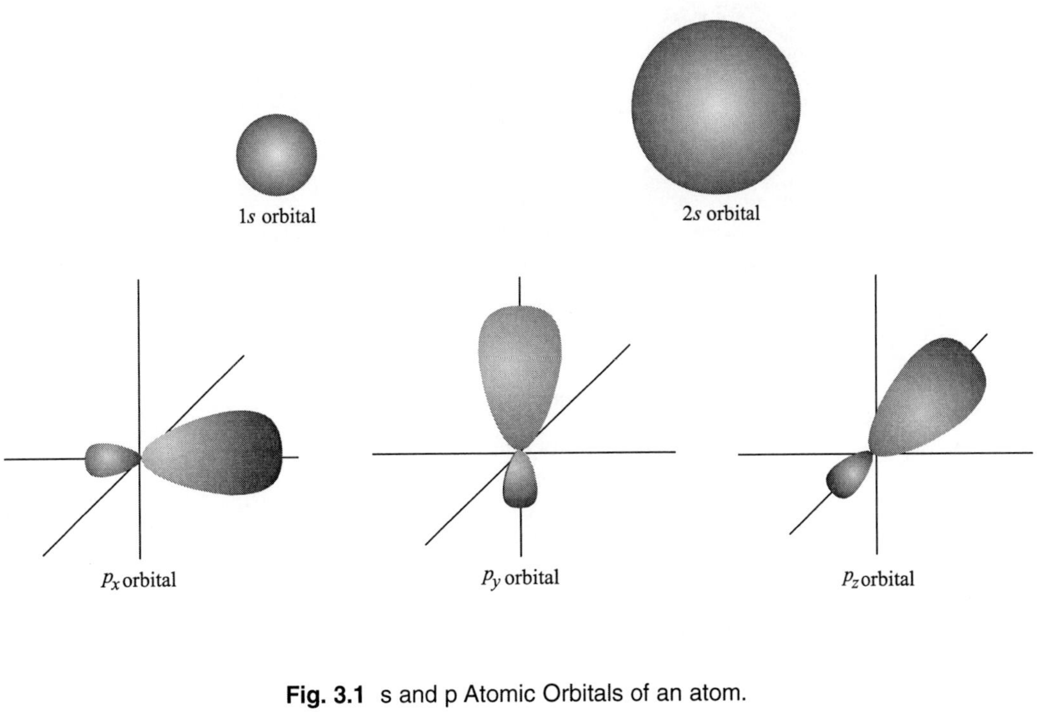

Fig. 3.1 s and p Atomic Orbitals of an atom.

Combining these orbitals, we get the orbital structure of carbon. Electronic configuration of carbon in ground and excited states, are written as $1s^2\ 2s^2\ 2p_x^{\ 1}\ 2p_y^{\ 1}\ 2p_z^{\ 0}$ and $1s^2\ 2s^1\ 2p_x^{\ 1}\ 2p_y^{\ 1}\ 2p_z^{\ 1}$ respectively. These are represented as shown ahead:

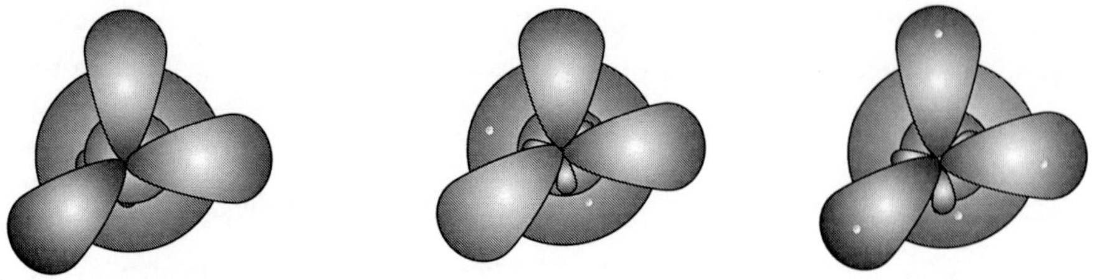

Vacant Orbitals $\quad 1s^2\ 2s^2\ 2p_x^1\ 2p_y^1\ 2p_z^0$ Ground state $\quad 1s^2\ 2s^1\ 2p_x^1\ 2p_y^1\ 2p_z^1$ Excited state

Fig. 3.2 The orbital picture of electrons in ground and excited state.

(In the Fig.3.2, electrons are shown by dots, which is not a correct representation of orbitals of an atom, because, an orbital is a volume in space and electron does not have a fixed position. The above representation is just for understanding the number of electrons.)

Remember

*The **Bohr model** was a one-dimensional model that used one quantum number to describe an electron in an atom. Only the size of the orbit was important, which was described by the Prinicple Quantum Number, n*

***Schrödinger** described an atomic model with electrons in three dimension. This model required three co-ordinates, or three quantum numbers to completely describe the electrons.*

The three coordinates from Schrödinger's wave equation are the principal (n), angular (l) and magnetic (m) quantum numbers. These quantum numbers describe the size, shape and orientation in space in an atom.

Summary of important facts

1. Planck's Law	$E = hv$
2. Pauli's Exclusion Principle	The values of four quantum of electrons in an atom are never same.
3. Hund's Rule	Pairing of electron spin shall start after all the degenerate singly filled.
4. de-Broglie Equation	$\lambda = \frac{h}{p} = \frac{h}{mv}$
5. Heisenberg's Uncertainty Principle	$\Delta x \times \Delta v \geq \frac{h}{4\pi.m}$
6. Schrödinger's Wave Equation	$\frac{\partial^2\psi}{\partial x^2} + \frac{\partial^2\psi}{\partial y^2} + \frac{\partial^2\psi}{\partial z^2} + \frac{8\pi^2 m}{h^2}(E - V)\psi = 0$
7. Probability Factor	ψ^2
8. Einstein's Law	$E = mc^2$

Max Planck (1858–1947)

Albert Einstein (1879–1955)

3.3 How a Covalent Bond is Formed?

Let us consider the formation of a carbon-carbon covalent bond from the electronic structure of carbon atom. There are number of theories of bond formation that depict the way, the formation of covalent bond takes place. **Linus Pauling** in1930, developed the ***Valence Bond Theory***. According to this, a covalent bond between two atoms is formed by overlap of atomic orbitals that are singly occupied and are of proper energy and proper symmetry. In the process of overlap, pairing of spins of electrons takes place. It looks to be a kind of extension of Lewis structures, where electrons come in between the atoms and share their spins to form a covalent bond. In case of orbital picture, we bring the orbitals close to or overlap each other to form a covalent bond. It requires a linear overlap of atomic orbitals to form a stable bond.

G. N. Lewis (1875–1946)

Linus Pauling (1901–1994)

3.3.1 Molecular Orbital Theory

The wave functions of an atom are called atomic orbitals. When atoms are brought closer to each other may be from infinity, their energy decreases to a minimum at a certain distance between each other. This distance is called bond distance. If atoms are brought closer than than this distance, energy increases again. A graph of energy and distance is given below. Bonding between atoms occurs, when atomic orbitals linearly overlap each other to reach a state of lower energy. The new orbital called Molecular orbital (MO, with wave function, ψ_{MO}, the molecular orbital wavefunction) is obtained. It is called MO theory. It was developed by R S Mulliken.

The mathematical solution of Schrödinger wave equation, gives value of wave (ψ) function. The wave equation being quadratic, has two solutions with same value but with opposite signs. Molecular orbital is considered to get formed by addition and subtraction of atomic orbitals. When addition occurs, electrons lie in between the orbitals and new orbital called bonding molecular orbital is formed. When subtraction occurs, electrons lie away from either of atomic orbitals, new orbital called antibonding molecular orbital is formed. This process of M.O. formation takes place by Linear Combination of Atomic Orbitals, and is called LCAO theory.

For example, the formation of covalent bond in hydrogen molecule in terms of orbitals is represented below. Two hydrogen atoms add up their A.O.'s or substract their A.O.'s to give bonding or antibonding M.O. Addition of A.O.'s, contrary to mathemtical addition term, leads to decrease in energy of M.O. and substraction of A.O. leads to increase in energy of M.O..

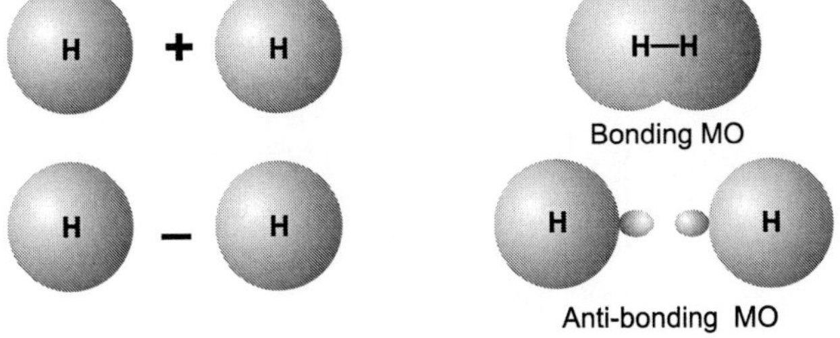

Energy of bonding orbital is lower than that of antibonding orbital as given in their energy diagram Fig. 3.3.

Mathematically, bonding orbitals are formed by addition of wave functions of electrons that lie in the region between two nuclei and have opposite spins. The lowering of energy leads to higher stability. **Antibonding** orbitals, formed by subtraction of wave functions, have electrons outside the region of two distinct ***nuclei***. The ***electrons*** present in each antibonding orbital are ***repulsive*** and act to ***destabilize*** the ***molecule***. The energy of the anti-bonding MO is higher than that of bonding MO.

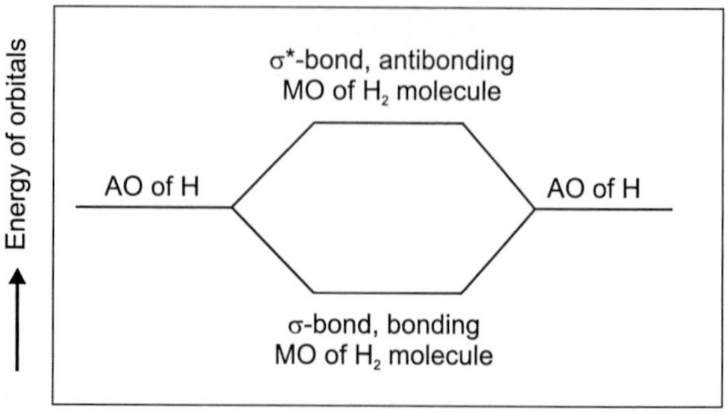

Fig. 3.3 1*s* orbital of two hydrogens overlap to form a covalent bond.

3.4 Why do We Hybridise the Atomic Orbitals?

Carbon in the unhybridised state does not give us the real shape of molecules, so hybridisation was adopted. The ***Electronic Configuration*** of carbon in the ground state (*i.e.*, the lowest energy state) is $1s^2 2s^2 2p^2$. If energy is supplied to raise one of the 2*s* electrons to a higher energy level, it fills the vacant 2*p* orbital. The electronic configuration of carbon in the excited state, would be $1s^2 2s^1 2p^3$. In this state, the carbon atom would be expected to produce three bonds of one sort (*i.e.*, from the three (*p*) orbitals) and one bond of another sort (*i.e.*, from the single (*s*) orbital).

Let us consider the formation of four bonds of methane between one carbon and four hydrogens atoms. In the process, four atomic orbitals of one carbon atom shall overlap with '*s*' orbital of each of the four hydrogens.

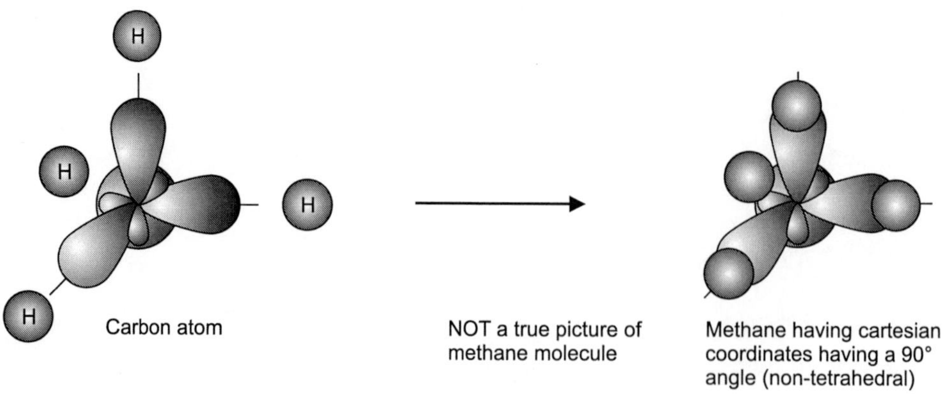

For ethane, one orbital of carbon overlaps with one orbital of another carbon to form a C–C bond. The three orbitals of each carbon then overlap with three hydrogen orbitals. From the structure of carbon orbitals (directed towards three cartesian coordinates), we can conclude that the shape of methane or ethane molecule shall be as depicted ahead:

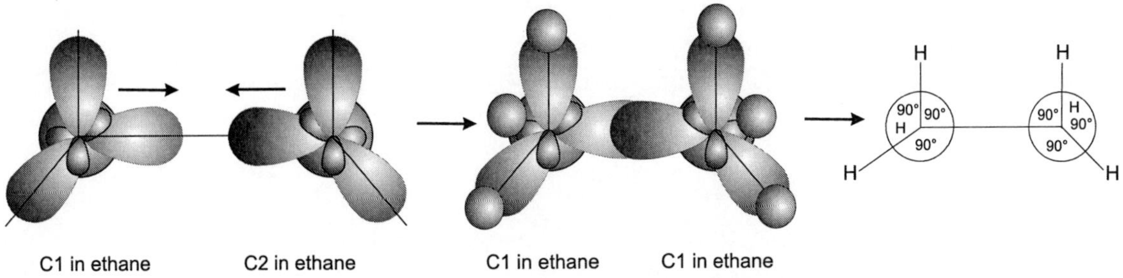

Fig 3.4 Shape of ethane molecule with unhybridised carbon atoms

It gives non-equivalent bonds in the molecules, both in energy and shape. That is not the case in real state of methane or ethane. The geometry of methane or ethane as determined by physical methods and X-ray, is tetrahedral and not as shown below. It has been determined that the four C–H bonds in methane or six C–H bonds in ethane are equivalent, both in energy and bond angle around carbon nucleus. So something is **lacking** in this type of bonding of carbon atomic orbitals. To overcome this anomaly, the concept of hybridisation was introduced.

3.4.1 The Hybridisation Concept

Linus Pauling, in 1931, gave the concept of hybridisation to explain the observed geometry of molecules. Hybridisation may be looked as a mathematical process of addition of unhybridised orbitals to form a new set of equal number and equivalent orbitals. These are **called hybrid orbitals**. Each of **new hybrid orbitals** possess properties of the original unhybridised orbitals from which these were formed.

To make the shape of methane or ethane molecule in the above case identical with natural methane or ethane, the rearrangement of atomic orbitals of carbon is essential. This change in the nature of atomic orbitals in each atom, depends upon the shape and properties of the molecules being considered. In case of methane or ethane, carbon possesses tetrahedral geometry, as determined by physical methods. For this purpose, sp^3 hybridisation is carried. There are two other types of hybridisation, called sp^2 and sp hybridisations and are considered below.

3.5 Different Ways of Hybridisation

There are three different types of hybrid orbitals for carbon compounds. These are named as sp^3, sp^2 and sp hybrid orbitals. These three types of hybrid orbitals have different orientations in space and account for the geometry of almost all type of organic compounds. sp^3 hybrid orbitals appear in saturated compounds and have tetrahedral geometry. sp^2 hybridised orbitals are present in compounds containing double bonds and have diagonal geometry. Lastly, sp hybridised orbitals are present in compounds that contain triple bond and have linear geometry. However, carbon dioxide has two double bonds and is a linear molecule. It has sp hybrid orbitals, since two double bonds are equivalent to one triple bond.

3.5.1 sp^3 Hybridisation

Four $\boldsymbol{sp^3}$ orbitals are formed from one s and three p unhybridised orbitals. The addition process takes place to form a new set of four equivalent $\boldsymbol{sp^3}$ orbitals. It forms a tetrahedral structure in conformity with that of the natural structure of saturated carbon compounds.

The simplest saturated alkane is methane, CH_4, it has four hydrogens bonded symmetrically to the carbon atom and four bonds are arranged in tetrahedral shape. This hybridisation of the orbitals of carbon in the methane molecule, is sp^3 hybridisation, *i.e.*, this involves hybridising *one s* and three *p* orbitals of carbon. It produces four single bonds symmetrically arranged around carbon atom in tetrahedral shape, as shown below:

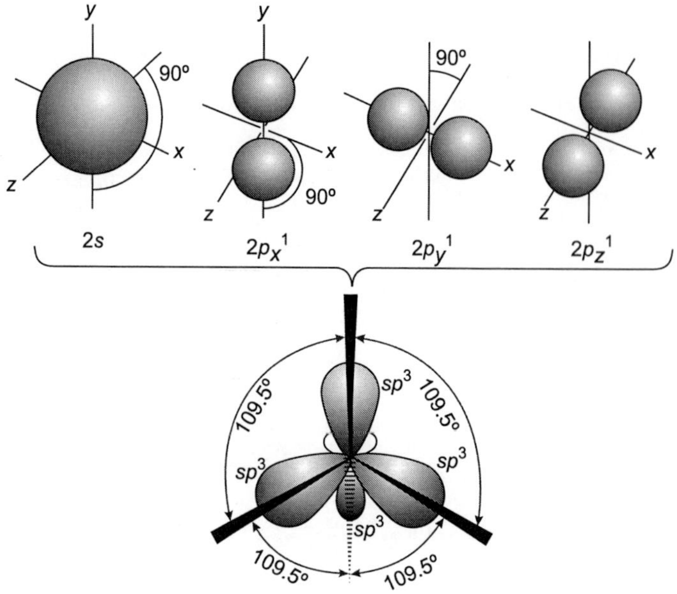

Fig. 3.5 One *s* and three *p* orbitals combine together to form a set of four sp^3 hybridised.

C–C single bond orbitals and simple representation of bonds

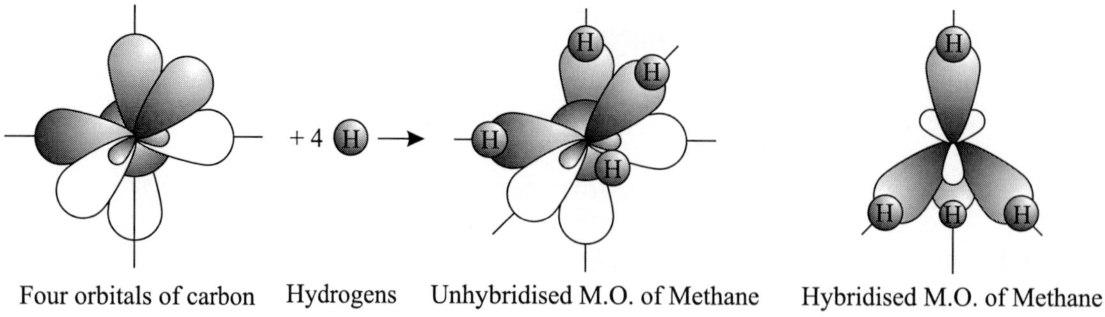

Similarly, a symmetrical compound, carbon tetrachloride has tetrahedral geometry, as shown ahead:

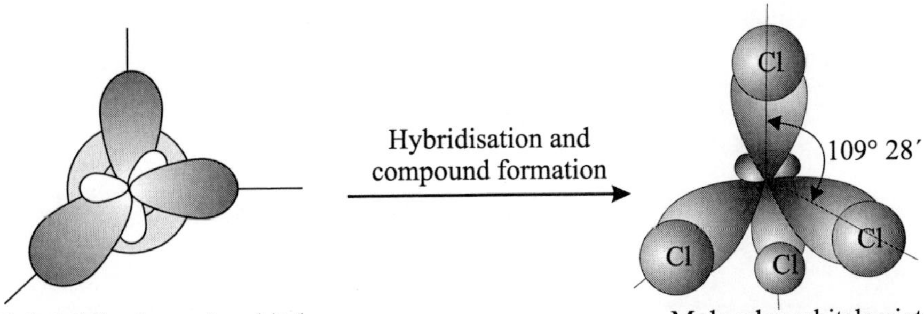

Unhybridised atomic orbitals of carbon along the cartesian co-ordinates at an angle of 90°

Molecular orbitals picture of carbon tetrachloride and is tetrahedral

Hybridisation of four orbitals of carbon in carbon tetrachloride

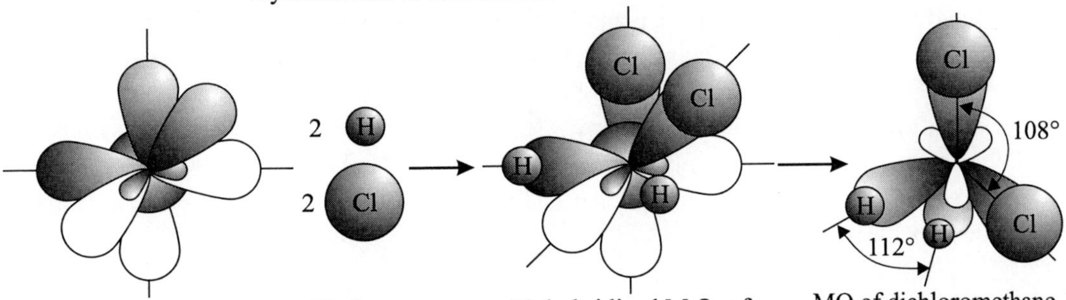

Four unhybridised Atomic orbitals of carbon

Hydrogens and Chlorine

Unhybridised M.O. of dichloromethane

MO of dichloromethane, bond angles are unexpected due to C–Cl bond electrons are at a longer distance than C–H bond electrons

Using this hybridisation, line drawings and molecular orbital picture of methane, propane and carbon tetrachloride and dichloromethane are given.

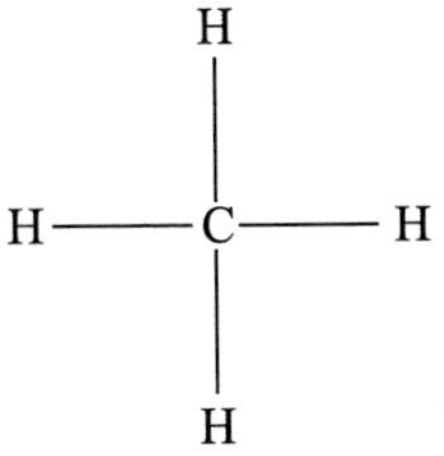

Methane

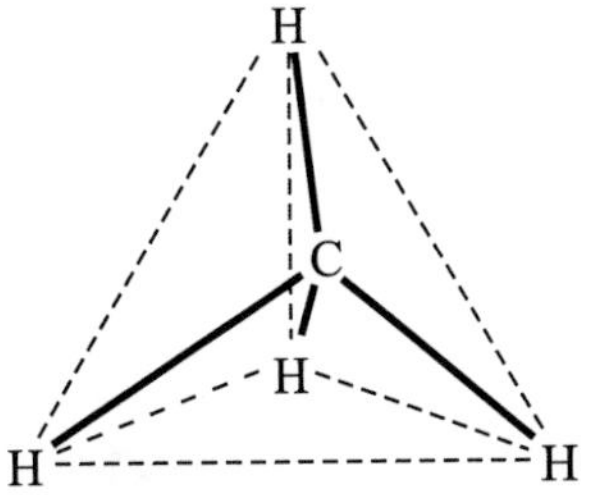

Tetrahedron methane

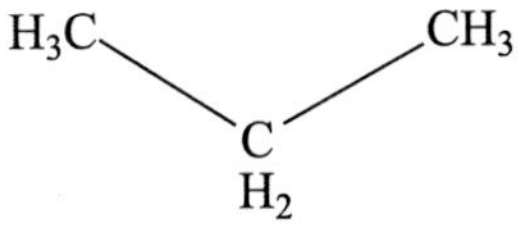

Propane

Propane orbital picture

For more complex molecules, orbital picture is difficult to make and simple line drawing are used as shown for cyclohexane.

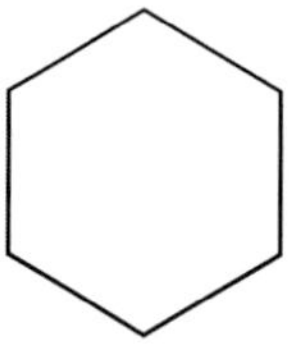

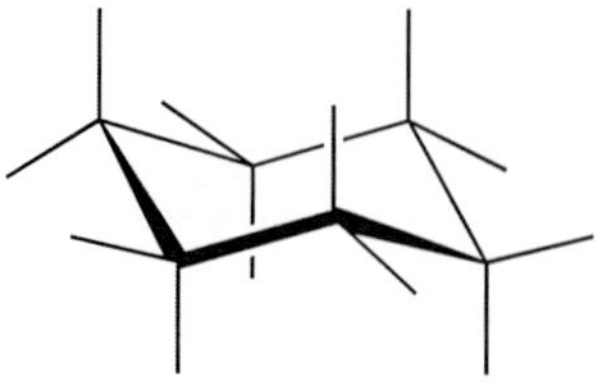

Chair form I Cyclohexane

3.5.2 *sp*² Hybridisation

The simplest unsaturated carbon compound, is ethene, C_2H_4. It has a planar structure, where the two carbon atoms and four hydrogens lie in one plane. The geometry of ethene cannot be explained by using sp^3 hybrid orbitals, so a new form of hybrid orbitals called sp^2 hybrid orbitals are used. The electronic configuration of carbon in the excited state is, $1s^2, 2s^1, 2p_x^1, 2p_y^1, 2p_z^1$. Out of these four $2s^1$ and $2p^3$ orbitals, we hybridise only three orbitals and leave one '*p*' orbital unhybridised. Thus, to obtain hybrid orbital addition of one *s* and two *p* orbitals is carried out leaving behind one *p* orbital unhybridised. The unhybridised orbital lies in a plane perpendicular to the plane of hybrid orbitals. Each of the hybrid orbitals is called sp^2 hybrid orbital.

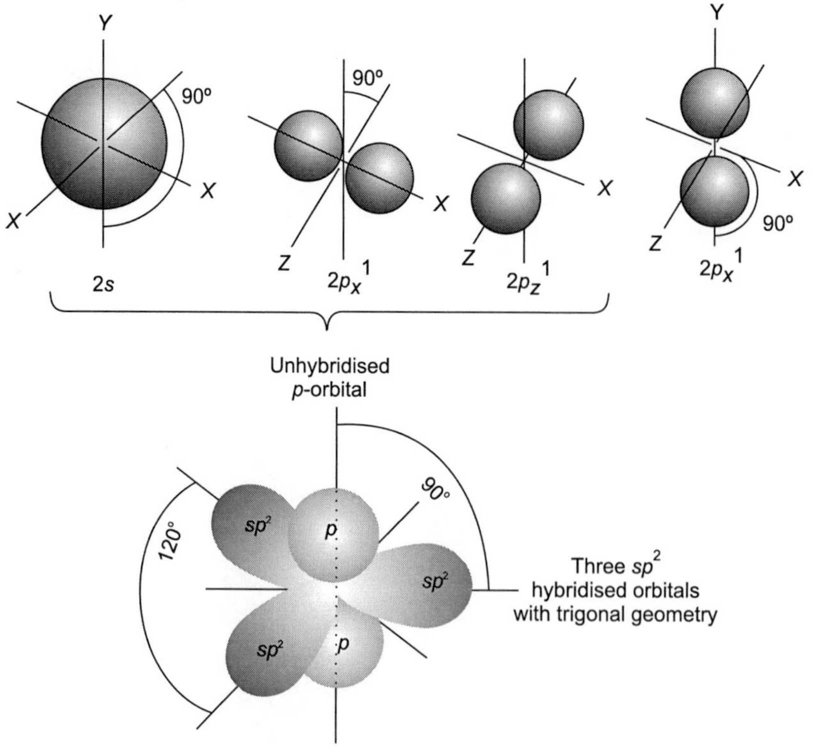

Fig. 3.6 Hybridisation of one '*s*' and two '*p*' orbitals form three '*sp*²' hybridised orbitals.

The set of three orbitals are so arranged that they lie in one plane at an angle of 120^0. It explains the geometry of molecules like ethene or other double bonded structures. The bonds between the hydrogen and carbon, that are formed by linear or end on overlap of three sp^2 hybrid **orbitals are called σ-bonds (sigma bonds). The second bond of the double bond between the carbon atoms in ethene is formed by the side-on overlap of the original unhybridised atomic orbitals (*i.e.*, the $2p_z$ (1) orbital) and are called pi (π) bond. It prevents free rotation around the carbon-carbon bond.**

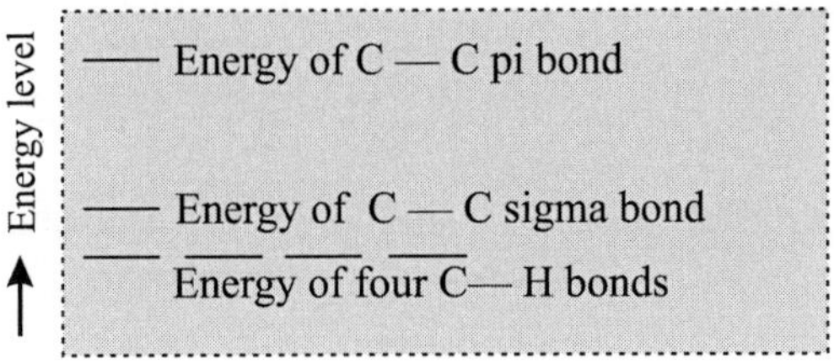

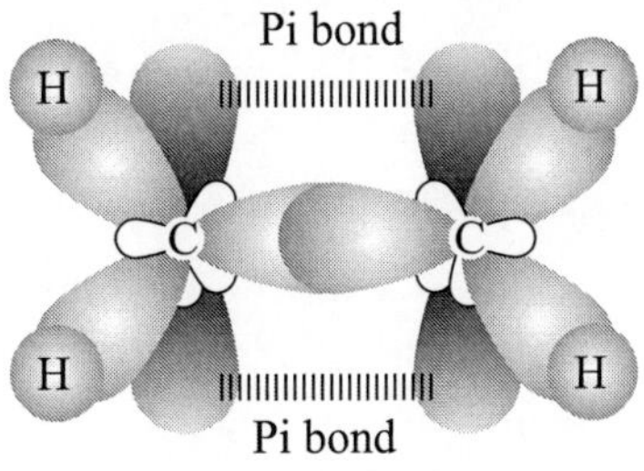

Molecular orbitals of ethene

Other examples of such bonds are acetaldehyde,

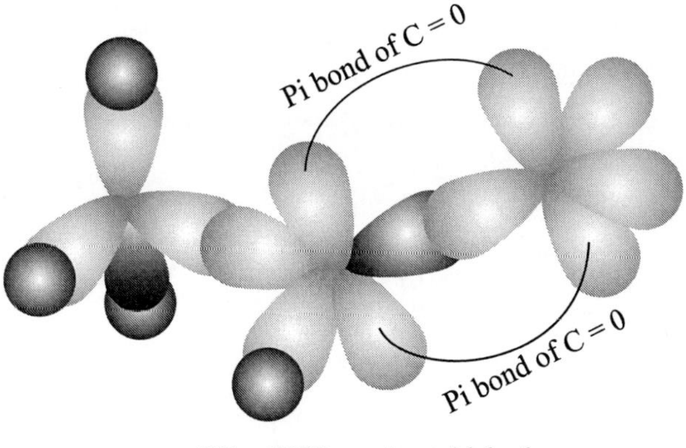

CH_3– CHO Acetaldehyde

Hybrid Orbitals of acetaldehyde

Examples of carbon-carbon double bond compounds

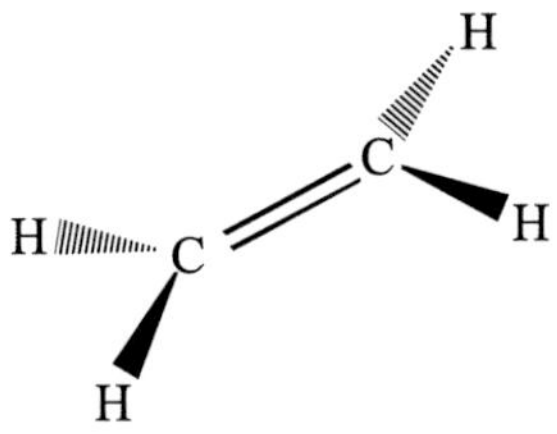

Ethene

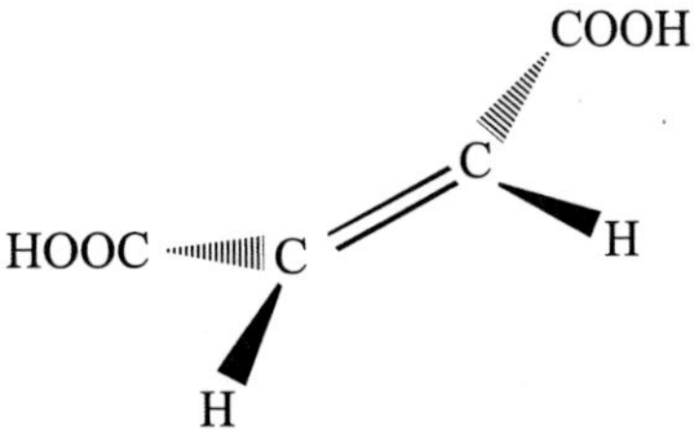

Maleic acid

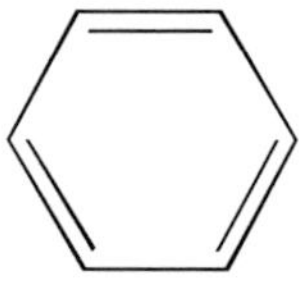

or

Benzene viewed from top of the plane

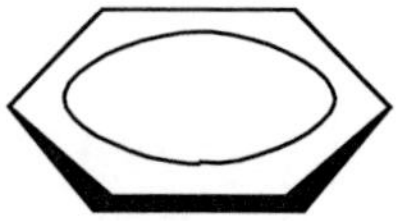

Benzene viewed from horizontal plane

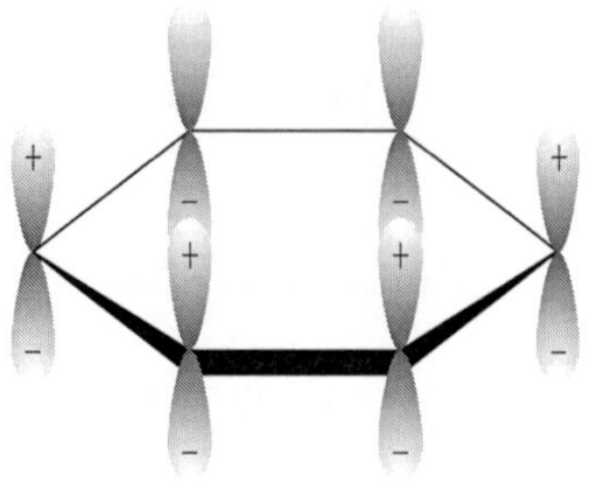

Benzene orbitals

3.5.3 *sp* Hybridisation

sp hybridisation is required for compounds containing triple bond. The simplest triple bond containing carbon compound is ethyne, C_2H_2, (*i.e.*, Acetylene). It has a linear structure, where two carbon atoms and two hydrogens are attached and lie along a straight line. The geometry of this structure cannot be explained by using the shape of the ***Atomic Orbitals*** of carbon atom in the unhybridised state or by sp^3 or sp^2 hybrid orbitals. So, a new form of hybrid orbitals called sp orbitals are used. These are used for σ-bond formation in alkynes. In the excited state of carbon, $1s^2$, $2s^1$, $2P_x^1$, $2P_y^1$, $2P_z^1$ we hybridise only two orbitals to form two equivalent orbitals called sp orbitals. Other two $2p$ orbitals (*e.g.*, the $2p_y^1$, $2p_z^1$ sub-orbitals) are left unhybridised and are used to form the second and third bond of the carbon-carbon that are the two π–bonds of triple bond. The hybridisation of the orbitals of carbon in compounds like ethyne, is called *sp* hybridisation. It explains the geometry of all alkynes.

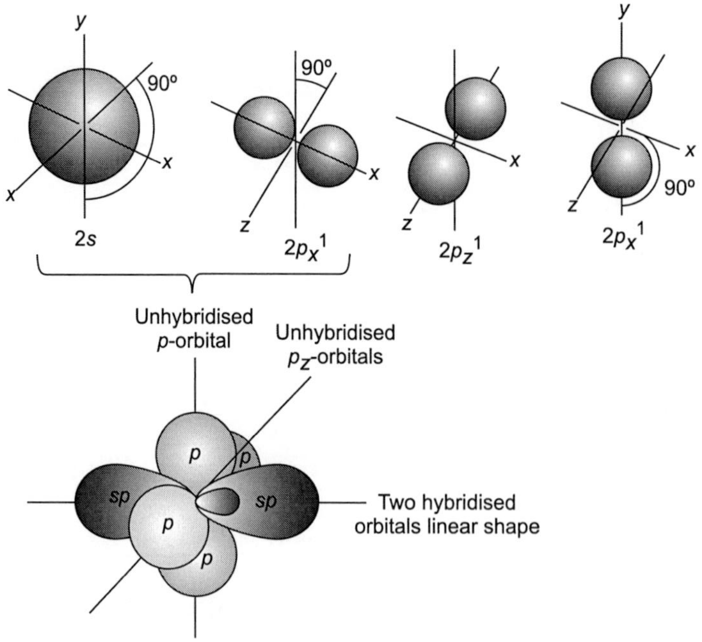

Fig. 3.7 Hybridisation of one '*s*' and one '*p*' orbital to form two '*sp*' hybridised orbitals.

The bonds between hydrogen and carbon and one bond (of the triple bond) between the carbon and carbon are formed by the end-on overlap. The σ–bond between carbon and carbon, carbon and hydrogen bonds are arranged as far apart in space as possible at 180° to each other. It accounts for the 180° bond angles observed in triple bond containing compounds. The second and third bond (of the triple bond) between the carbon atoms in ethyne are formed by the side-on overlap of the original unhybridised atomic orbitals (*i.e.*, the $2p_y^1$, $2p_z^1$ orbitals) and these are called π–bonds. The orbital picture of acetylene is shown below.

3.6 Structure of Molecules

Real form molecules are difficult to draw on paper or in book, so we have adopted representational methods by different drawings. It could be ***structural drawings*** or ***orbitals or 3-dimensional representations*** of molecules by models.

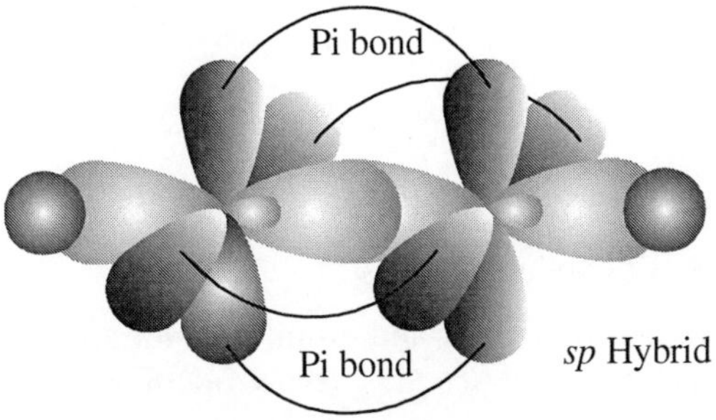

Acetylene

sp Hybrid orbitals of acetylene

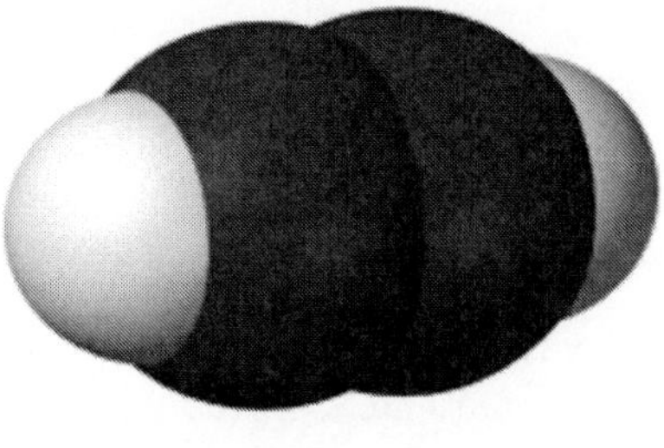

1. Structural Drawings

Kekule's Structure	Condensed Formula	Shorthand Formulae	2D-3D lines
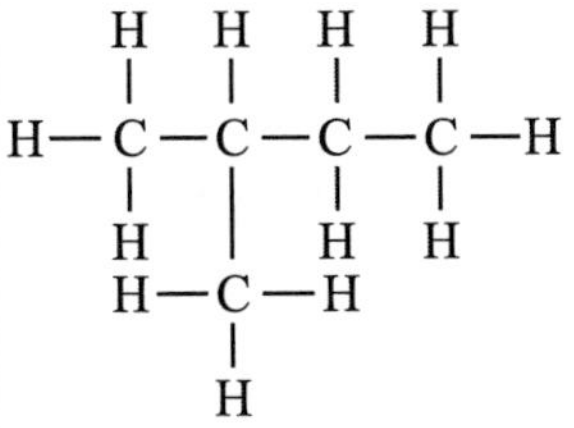	$CH_3.CH(CH_3).CH_2CH_3$	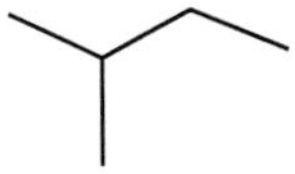	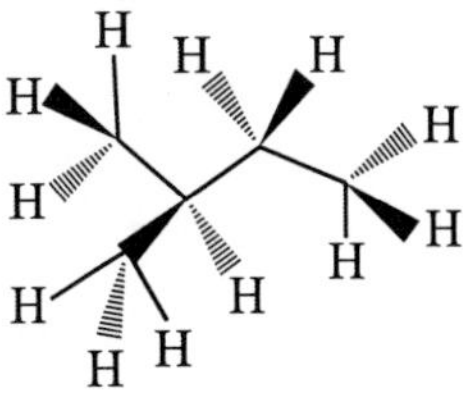
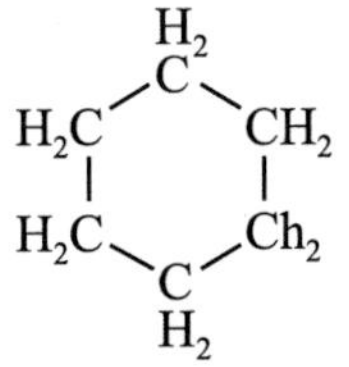	C_6H_{12}		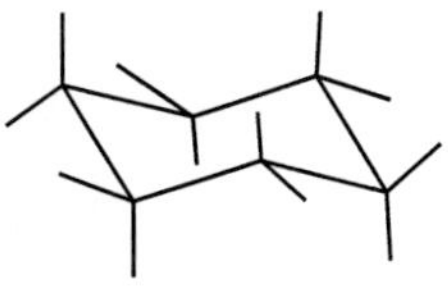

Normal bond Dashed bond Wedge bond Hatched bond

2. Orbitals Representations

A more detailed model of covalent bonding requires a consideration of valence shell atomic orbitals. For second period elements such as carbon, nitrogen and oxygen, these orbitals have been designated $2s$, $2p_x$, $2p_y$ and $2p_z$. The spatial distribution of electrons occupying each of these orbitals is shown in the diagram below.

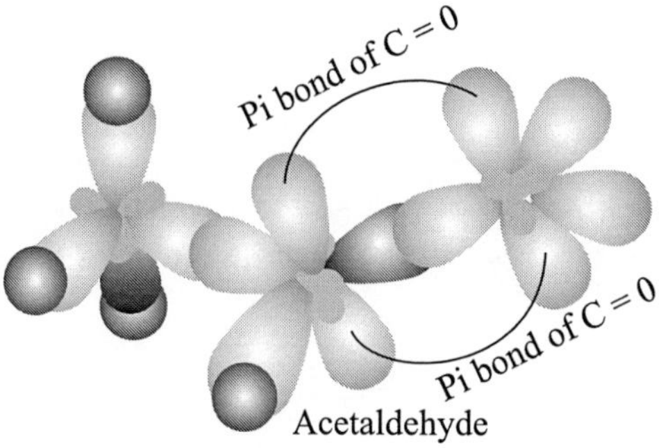

CH_3—CHO

3. Three-Dimensional Representations of Molecules by models:

There are five methods of representing molecules in **3-Dimensions.**

1. **Line** or **Stick Models:** In this type of models, we use thin pipes and atomic nuclei are not represented, atoms are implied at each vertex. The bonds are used as sticks or lines, as in 2D molecular structures.
2. **Electron Density Plot:** It shows the electron density determined either by crystallographic methods or Quantum mechanical calculations, rather than distinct atoms or bonds.
3. **Ball and Stick Models:** atomic nuclei are represented by spheres (balls) and the bonds as sticks.
4. **Spacefilling Models or CPK Models:** (Also uses an atomic coloring scheme in representations) the molecule is represented by overlapping spheres representing the atoms.
5. **Cartoon:** A representation used for proteins where loops, beta sheets, alpha helices are represented diagrammatically and no atoms or bonds are represented explicitly just the protein backbone as a smooth pipe.

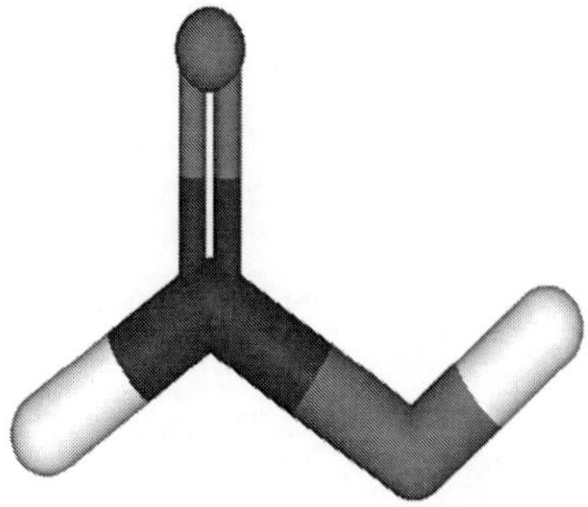

Stick Model of Formic acid

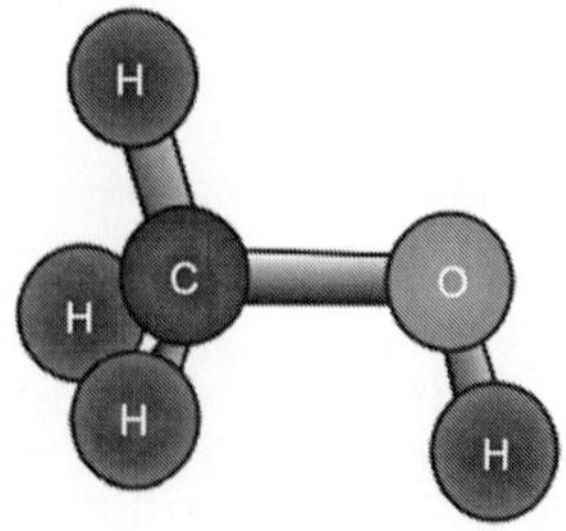

Ball and Stick model of Methanol

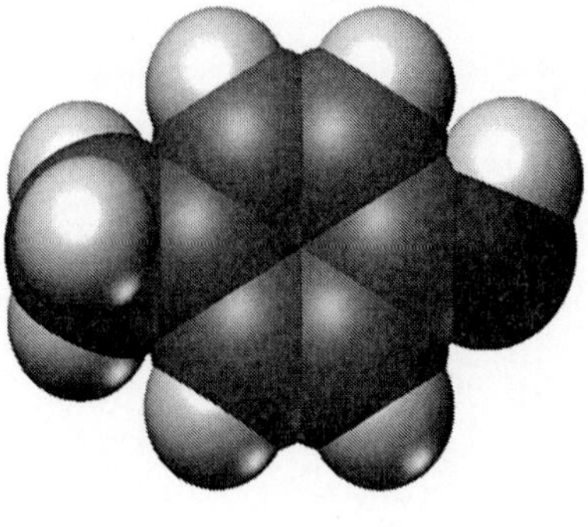

Space-filling Model of *p* cresol

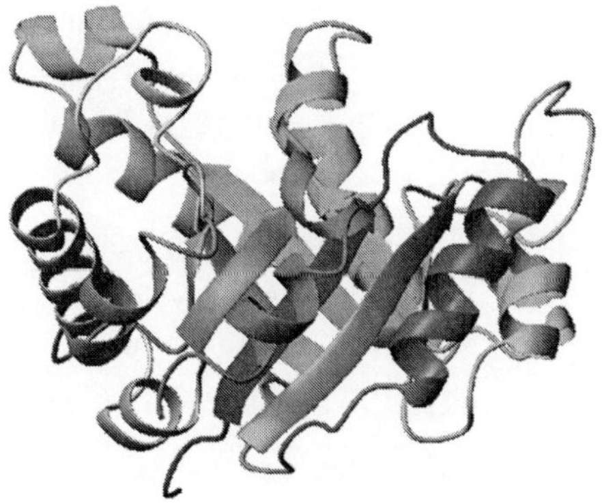

Cartoon type model for proteins

Fig. 3. 8 DIfferent kind of models.

Thus, we can represent compounds as models, orbital pictures, or wedge diagrams, stereo formulae and plane formulae. All these representations do not change the chemical nature of compound in a test tube, but help us to understand the nature and properties of molecules.

Exercises

1. Using de-Broglie relation determine the wavelength of electron moving at a velocity of 3×10^{-8} ms^{-1}. Given the value of mass of electron, $m = 9.1 \times 10^{-31}$ kg and Planck's constant, $h = 6.6 \times 10^{-34}$ kg $m^2 s^{-1}$.
2. What will be the wavelength of a foot-ball of mass 1 kg when thrown at a velocity of 50 km s^{-1}. Or will it just follow a straight track?
3. The velocity of light is 3×10^{-8} ms^{-1}, what will be the maximum velocity of an electron? Is there any correlation?
4. How do matter waves differ from the radiation waves?
5. Define Heisenberg's Uncertainty principle. Calculate the product of uncertainty of position and velocity of a micro-particle like electron, whose mass = 9.1×10^{-31} kg. Given the relation, $\Delta x \times \Delta v \geq \frac{h}{4\pi . m}$.
6. What is the idea of quantization of energy, as given by Niel Bohr? How the term *orbit* for the path of an electron was changed to *orbital* and who changed it?

7. What are the values of quantum numbers that determine the shape and orientation of 2*s* orbital and for 2*p* orbitals.

8. Give orbital notations for the following values of quantum numbers:

   (*i*) $n = 2, l = 1, m = 0$ (*ii*) $n = 2, l = 0$ (*iii*) $n = 1, l = 0$

9. Giving two examples, explain the electronic theory and quantum theory of bonding.

10. What are the Valence Bond and Molecular Orbital theories of bonding? How do they differ in their approach?

11. What is the meaning of the term hybridisation?

12. What was the necessity of using such a concept?

13. Using model of *n*-hexane, give the drawing of a corresponding cyclic compound. In hexane, all carbon atoms lie in one plane, whereas carbon atoms of the corresponding cyclic compound do not lie in one plane. Explain, why is it necessary?

14. What are the geometrical shapes of sp^3, sp^2 and *sp* orbitals?

15. What is the hybridisation status of atoms marked with asterik carbon in each of the following compounds?

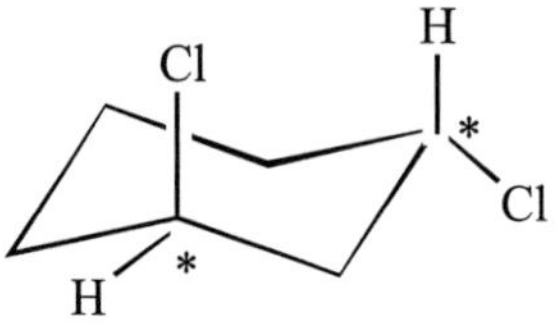

1,3- Dichlorocyclohexane

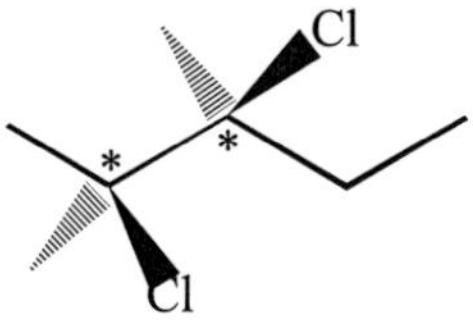

1,2- Dichloropentane

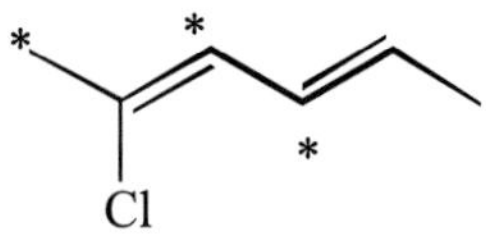

2-Chloro-hex-2,4-diene

Cyclopentadiene

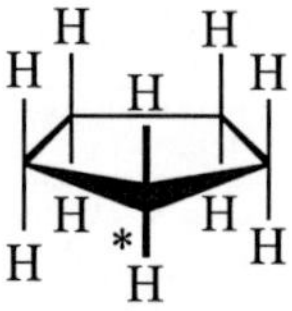

Cyclopentane

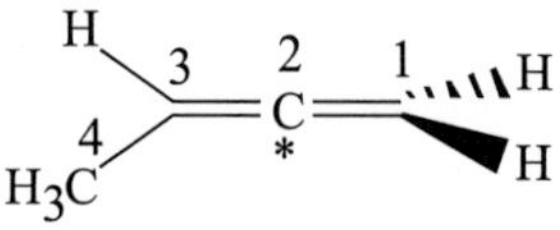

But-1,2-diene

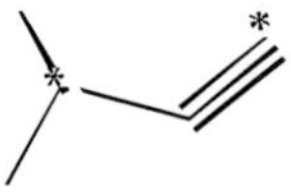

3-Methyl-But-1-yne

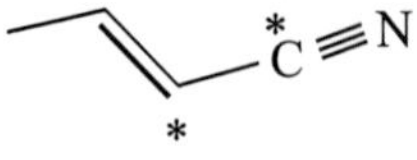

But-2-enonitrile

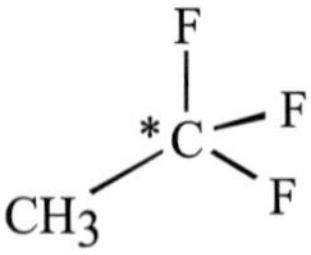

1,1,1-Trifluoroethane

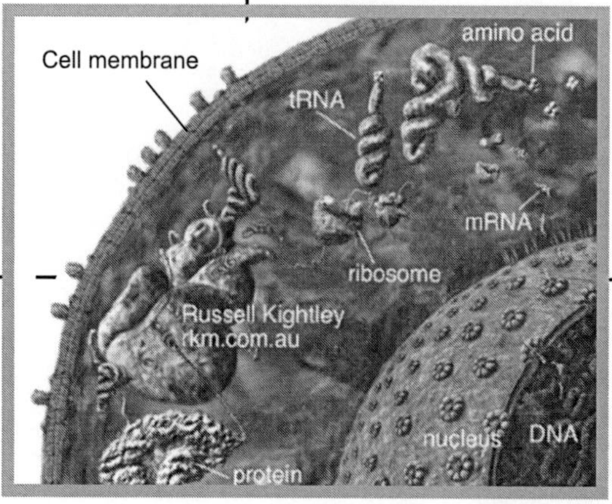

Protein Synthesis and some reactive sites

4

Electronic Displacements and Reactive Intermediates

IN THIS CHAPTER

Above is a simplified diagram showing how proteins are manufactured in a cell. At bottom right, messenger RNA (mRNA) is being produced in the nucleus. mRNA molecules are shown as yellow string like molecules. This process, which copies the information stored in the nuclear DNA into complementary code in the mRNA, is called **transcription**. The mRNA then leaves the nucleus via a nuclear pore (nuclear pores are the flower shaped structures embedded in the nuclear membrane) and moves into the cytoplasm. Ribosomes attach to the mRNA and start to **translate** the information contained in the mRNA into a sequence of **amino acids**.

Ribosomes are small (about 24nm diameter) molecular scale machines that crunch along the mRNA, reading it like a tape, and adding appropriate amino acids together to create proteins. The specific amino acids (shown as purple blobs) are recognised, captured and brought to the ribosomes by transfer RNAs (tRNAs). tRNA molecules are shown as the orange twisted rope like molecules. Inside a ribosome, the newly arrived tRNA, bearing its specific amino acid, docks next to the tRNA molecule already inside the ribosome. The polypeptide chain is then attached to the newly delivered amino acid and so comes to hang from the incoming tRNA. The first (now liberated) tRNA then exits from the ribosome. The ribosome grinds along the mRNA and a space is created to accommodate a new incoming tRNA with its attached amino acid. This process is repeated each time an amino acid is added.

This diagram is very simplified but gives the essence of the process of protein manufacture in the cell and should be very useful for illustrating articles on proteomics or as an introduction to protein synthesis. Proteomics essentially means the identification and understanding of proteins (in a cell or organism) and is comparable to the term genomics which applies to the genes (DNA). There is considerable interest in proteomics since it follows naturally from genomics (the chracterisation of the genome) and is the domain where such drug development may occur. Ref. Russell Knightley.rkm..com.au

Chapter 4

Electronic Displacements and Reactive Intermediates

Introduction

In the previous chapter, we have understood the structure of molecules and what kind of forces are operative in the bonds and in-between the molecules. Hybridisation of molecules is another concept that was introduced and it results in specific shapes of molecules. Electron distribution in bonds or around atoms of different electronegativity produces different kind of forces. In the present chapter, we shall study the effect of electron movement in molecules and the kind of species that get produced by bond breaking. It is interesting to study, how these species produced by electron movement give fairly stable or reactive entities. What is the shape of such species and how do these take part in chemical reactions?

4.1 Cleavage of a Covalent Bond

Modern concept of covalent bond formation is considered as a linear overlap of *two* atomic orbitals to give a molecular orbital. An orbital is defined as the volume in space, where the probability of finding an electron is maximum. This picture of bonding is not very convenient for rapid writing and understanding of chemical reactions. For this purpose, the formation or cleavage of a covalent bond, using electrons as a particle in between two atoms is still a very good concept. It helps us to know how bonds break and how new bonds are formed by electron movement. It gives a clearer understanding of chemical reactions and that are easy to write. There are three ways by which a bond can get cleaved or broken in a compound A-B, as shown below.

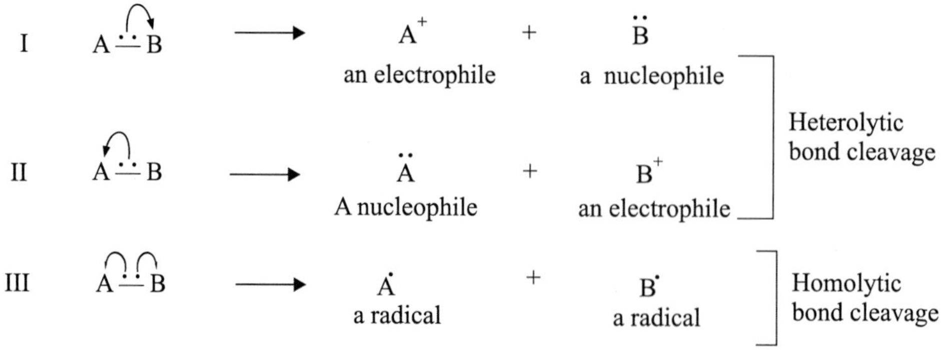

In reactions I and II, the bond breaks up by shifting two electrons to either atom. It gives positive and negatively charged species. The cation is electron deficient entity and is called an **electrophile**, while the anion is rich in electrons and is called **nucleophile**. The type I and II of bond cleavage is called *heterolytic bond cleavage*. In the III case, both the species that get formed have single electron each. These entities with single electron are called *radicals*. This bond cleavage is called *homolytic bond cleavage*.

The reverse process of bond cleavage or breaking, is bond formation. Formation of bond requires linking of species like anion with cation or cation with anion or a radical with another radical. All chemical reactions involve bond cleavage and bond formation. Thus, it will be more approriate to study such species and the factors that control their formation. Bond dissociation energy of a bond plays an important role in bond cleavage.

The factors that affect the formation of cations (electrophiles), anions (nucleophiles) and radicals in any chemical reaction, are

(*i*) structure of the substrate (the molecule undergoing chemical reaction)
(*ii*) the nature of attacking reagent,
(*iii*) temperature of reaction,
(*iv*) nature of solvent etc.

The role that these factors play in chemical reactions is being considered in subsequent discussions.

4.2 Electronic Displacements

Electronic movement in different compounds is governed by presence of electronegative atoms or double or triple bonds. For example, a compound having an electronegative atom like chlorine, shall attract the electrons towards itself from the bond it forms. It makes the neighbouring carbon atom, slightly positively charged, thus the attack by an electronegative entity from the medium shall take place only at the neighbouring carbon atom. Similarly, other factors that influence the course of chemical reactions, are disucssed in sections below.

4.2.1 Inductive Effect

Inductive effect is an experimentally observable effect that takes place by electrostatic induction. Let us consider an electronegative atom, like chlorine, attached to a carbon chain. The electrons present in C_1 and Cl bond are more shared by chlorine because of its high electronegativity. The electron pair thus remains slightly displaced towards chlorine. It results in creation of a δ^+ charge on C_1 and a δ^- charge on chlorine. The positive charge on C_1 carbon, in turn, attracts electron pair in the next C_1– C_2 bond. Thus, another δ^{+1} charge gets induced on C_2. The process of induction of positive charge on the carbon atoms continues along the carbon chain, though with diminishing magnitude of charge. The magnitude of charge is significant only up to 2 or 3 carbon atoms. Thus, inductive effect produces an electrophilic character at C_1 that easily undergoes reaction with a nucleophile.

One may ask why a stable covalent C– Cl bond in ethyl chloride undergoes reaction with 1% sodium hydroxide solution even at room temperature, though slowly. The answer is that it is due to inductive effect of chlorine that induces positive charge on carbon. The weakening of C_1– Cl bond takes place and it is easy for –OH group to displace it as shown below.

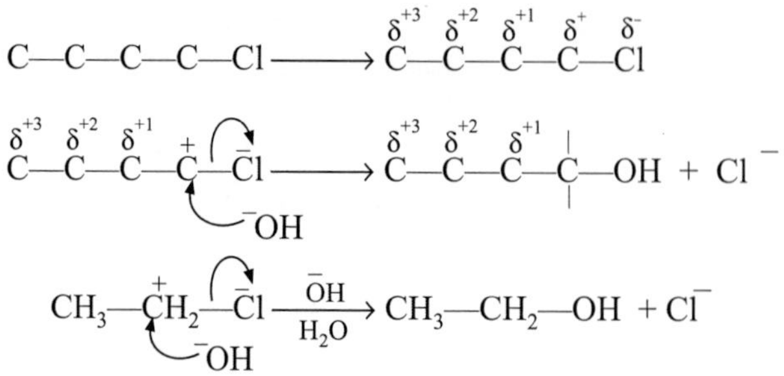

Inductive effect is of two types

1. Where electrons are pulled towards the group, it is called -I effect (electron withdrawing nature or high electronegativity) ;

   Groups with –I effect are: –F, –Cl, –Br, –I, $-NH_2$, $-NR_2$, $-NO_2$, $-CH = CH_2$, C_6H_6, –CN, –COOH,

2. Where the electrons are pushed by the group, it is called +I effect (electron donating nature or low electronegativity).

   Groups with +I effect are: –H, –D, $-CH_3$, $CH_3 - CH_2-$, other alkyl groups , (–OH, –OR, $-O^-$ in aromatic systems),

Inductive effect primarily generates a partial negative or positive charge through **partial displacement** of electrons within a **single bond**, when joined with a more electronegative atom. **These charged species assist in directing the incoming group in a chemical reaction.** It also explains the acid character of different compounds as discussed in chapter 5.

4.2.2 Electromeric Effect

Electromeric effect is another effect of electron movement within a molecule containing double or triple bond. In a double bond, pi (π) electrons are labile and get **moved** towards the more electronegative atom. It is a temporary effect, in which the pi-electrons flow to the more electronegative atom or to the attacking reagent.

For example, in a carbonyl group, the electrons in the C=O bond, move to oxygen. It makes oxygen negatively charged and carbon as positively charged. It makes carbon of carbonyl group prone to nucleophilic attack. Similar is the case with thiocarbonyl and nitrile groups in a molecule as shown below:

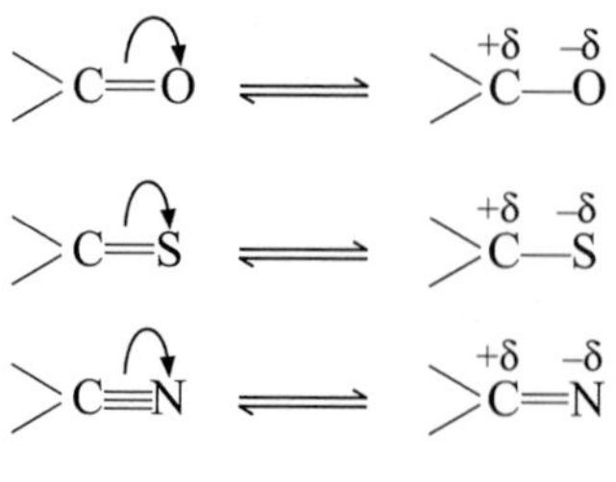

+ E effect

When the transfer of electrons takes place towards the attacking reagent, it is called +E effect (positive electromeric effect). For example addition of H^+ to double bond takes place by attack of *pi*-electron.

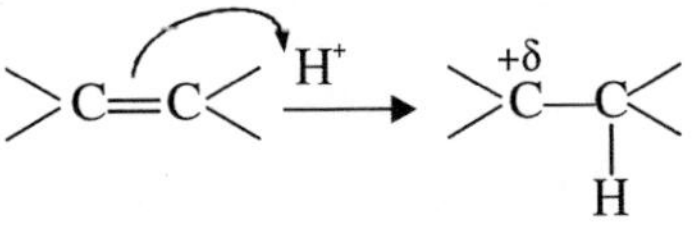

– E effect

When the transfer of π-electrons takes place toward the more electronegative atom of the double bond, the carbon becomes positively charged and the nucleophile, such as cyanide ion, from the medium attacks the carbon atom. It is called –E effect (negative electromeric effect).

Examples

Example 4.1 Find out the direction of movement of electrons between a carbon and the group, X when attached to butane. Write their final structure that get formed under the influence of the groups. What will be influence of ethyl and methyl group of butane?

Solution:

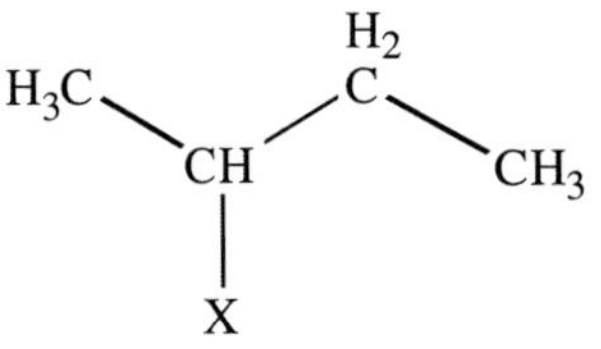

X = Cl, Br, CN, NO_2, -OH and -OCH_3

Example 4.2 What is the point of attack by a nucleophile, -CN, on butanone (ethyl methyl ketone). What will be the effect on the reaction, if the substrate is changed to 1-nitro-butan-2-one or formaldehyde.

Solution:

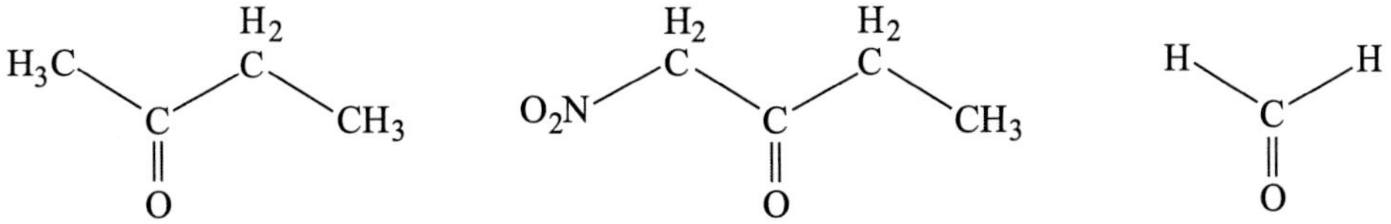

4.2.3 Resonance or Mesomeric Effect

Resonance or Mesomeric effect is very similar to electromeric effect, but the movement of electrons on a double bond is extended to the whole molecule due to conjugation. It was discovered by Ingold in 1938. In the process of movement of electrons between different atoms, it is possible to write a larger number of structures with positve and negative charges on different atoms of the chain without changing the carbon skeleton.

For example, 1,3-butadiene is represented with positive and negative charge in four different structures. Each structure has the same linkage of atoms, but differ in charges or electron position. Four structures of 1,3-butadiene are called its canonical forms and butadiene itself is a hybrid of these structures. The properties of butadiene are represented by all the structures. It explains many properties of compounds, such as, the site of chemical reactions, dipole moments, pK_a values. Resonance is influenced by the nature of groups that are attached to the conjugated chain.

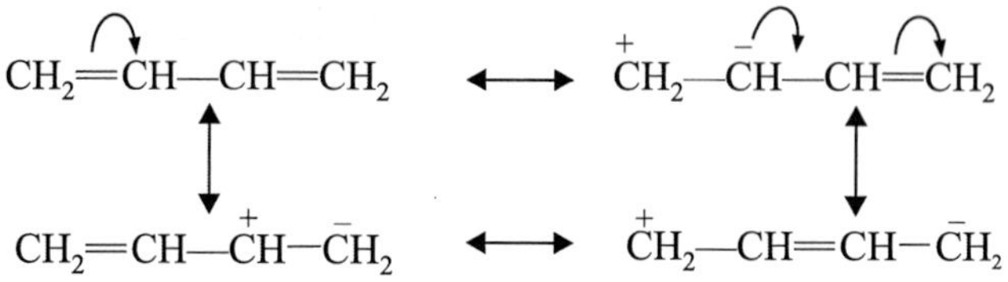

Mesomeric effects are of two types, +M and –M effect.

+ M effect

This effect is observed when the electrons get transferred to the conjugated system from the atom of group having a lone pair, such as phenol, that has electron rich oxygen in hydroxyl group. It makes the *ortho-*, *para-* positions rich in electron for attack by an electrophile. However, the acidic effect is primarily due to the inductive effect rather than the belief that it is due to mesomeric effect.

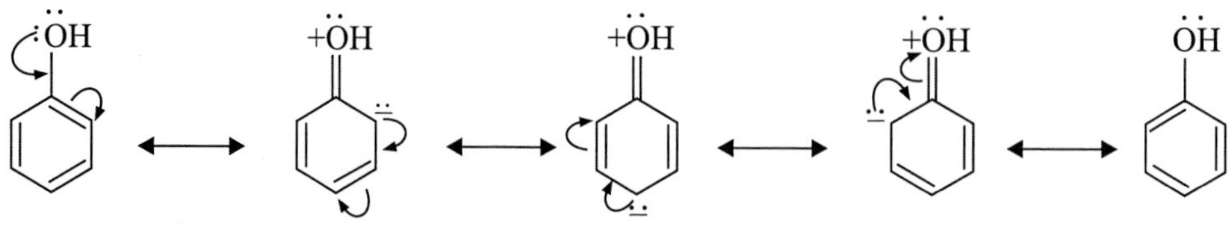

– M effect

This effect is observed when the electrons get transferred from the conjugated system to the atom or group, such as nitro group that is electron attracting and makes the meta position electron deficient. It makes the meta position susceptible to attack by a nucleophile.

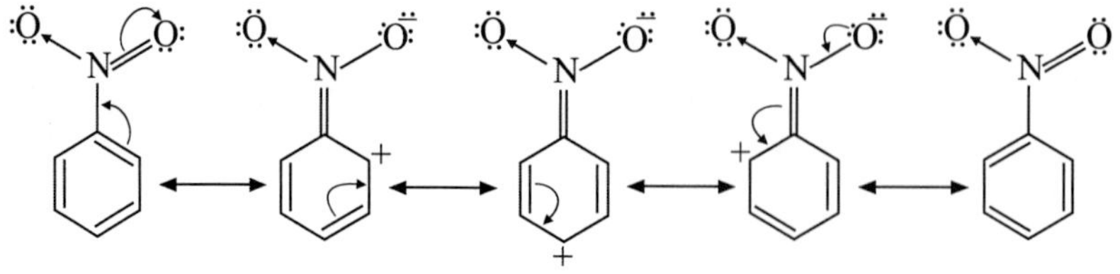

4.2.4 Hyperconjugation

This is the fourth effect involving electron movement. The term was introduced by R.S. Mulliken. We have seen above that pi (π) electrons from the double bond move towards the more electronegative atom, and makes the molecules rich and electron deficient. Hyperconjugation is the conjugation that gets extended due to sigma bond electron movement from the carbon-hydrogen bond α- to double bond. The ionisation of hydrogen from α-C, makes π-bond electrons move much faster towards the far direction as shown below. Hyperconjugation effect brings about 1000 times more activity than simple electromeric effect.

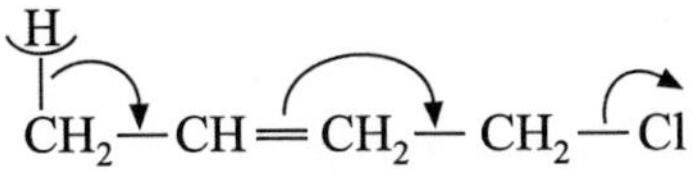

Hyperconjugation can also be looked at as the stabilising interaction that results from the interaction of the electrons in a sigma σ-bond (usually C—H) with an adjacent empty or partially filled π-orbital to give an extended molecular orbital that increases the stability of the system.

Based on the valence bond model of bonding, hyperconjugation can be described as "double bond -no bond resonance" but it is not what we would "normally" call resonance, though there is some similarity.

It is represented in terms of valence bond (I). The same thing is visualised in terms of orbital picture (II). The sigma bond electron can form bond with π-orbital, I, or a vacant π- orbital, II(C^+ cation)

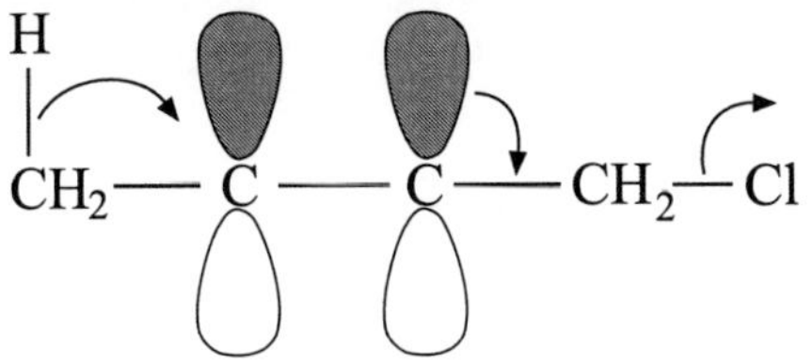

I

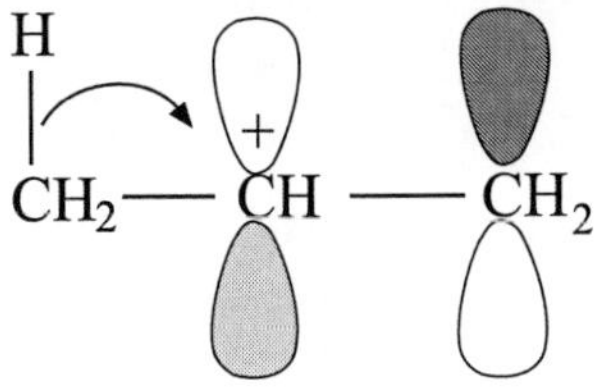

II

This effect explains

1. The stability of conjugated system.
2. Reactivity of terminal group.
3. The bond length of different carbon-carbon bonds.
4. The stability of carbocation formed by ionisation of the terminal group.
5. The dipole moment of such molecules.

4.3 Usefulness of the Concept of Electronic Displacements

1. These explain chemical reactions by showing arrows and bond breaking.
2. These explain physical properties of organic compounds.
3. These also explain the generation of polarity in a molecule, like, CH_3Cl. Thus, methyl having positive charge on carbon undergoes chemical reactions with oppositely charged species, *i.e.*, nucleophiles.

4. These explain the bond length of polar bonds. Due to generation of charges in a bond, the attractive forces between oppositely charged atoms lead to shortening of bond length. Larger the force of attraction, greater is the effect on the reduction of bond length. That is more electronegative atom at the end of a carbon chain shall have shorter bond length. For example, C–F bond is shorter as compared to C–I bond. It is important to note that the atomic size of Fluorine and Iodine are also important in determining the bond length, besides their electronegativity.

	CH_3—F,	CH_3—Cl,	CH_3—Br,	CH_3—I
Bond length	138 nm	178 nm	194 nm	214 nm

5. This also explains the generation of dipole moment. Due to the electronegativity of elements, the inductive effect becomes operative. A summation of all bond moments gives net dipole moment of molecule. From dipole mements of molecules, we can predict their geometry.

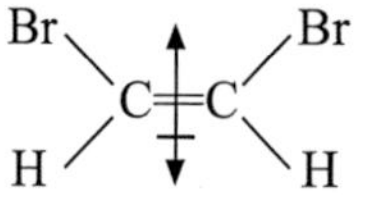

+, Dipole moment

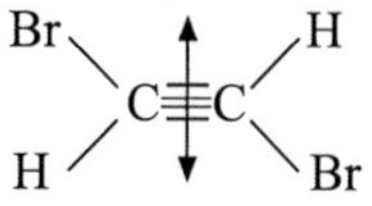

zero dipole mement

6. Electron push as in **inductive effect**, explains the strength of acids. For example, formic acid is much stronger acid than those having alkyl chain attached to -COOH. Due to electron push by alkyl group, the electrons in –O–H group of –COOH, does not allow H atom to go off. It explains the reduced acidic strength of carboxylic acids. with larger alkyl chain. More details are given in chapter 8.

7. Similarly, it also explains the strength of bases, that are strong, when electron density increases on nitrogen atom. Positive inductive effect (+I) increases the base character, especially when the alkyl groups is directly attached to nitrogen atom.

8. **Electromeric and resonance effects help in understanding reactions like nitration of phenol (an electrophilic substitution reaction).** It takes place by attack of an electrophile NO_2^+. Experiments tell us that nitration of phenol gives ortho and para nitrophenols.

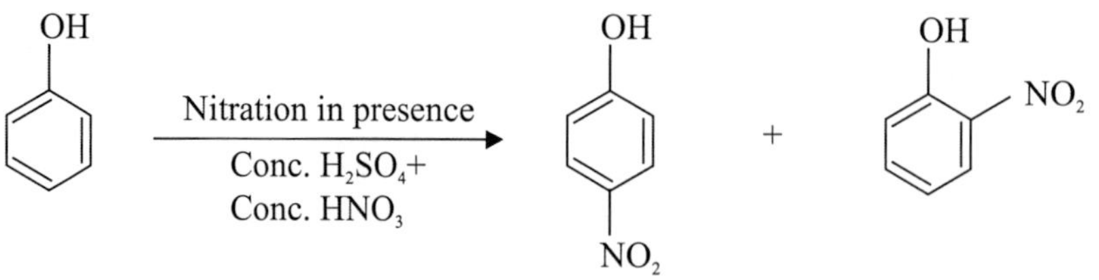

Nitration is best explained by using electromeric and resonance effects as shown below:

$$H_2SO_4 + HNO_3 \longrightarrow NO_2^+ + HSO_4^- + H_2O$$

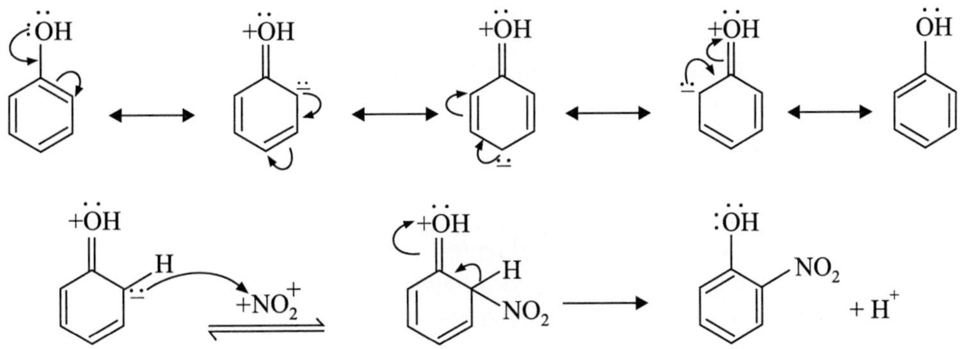

4.4 Structure and Reactivity of Carbon Species

So far, we have understood, how the electron displacements occur in molecules and thus create charges at different positions in a molecule. Also besides charged positions, many a times single electrons or radical get generated by such movement of electrons. It is important to understand further the reactivity and shape of such charged species or radicals that get created by electron movement due to different electronegativity of atoms and structure of molecules. This can help to control chemical reactions or at least understand further, how chemical reactions can be controlled at our desire.

4.4.1 Carbocations

Carbocations arc electron deficient carbon atoms in a compound. That is, out of the four electrons of carbon, only three electrons are left in the bonded state. It shall carry positive charge and is called carbocation or carbonium ion.

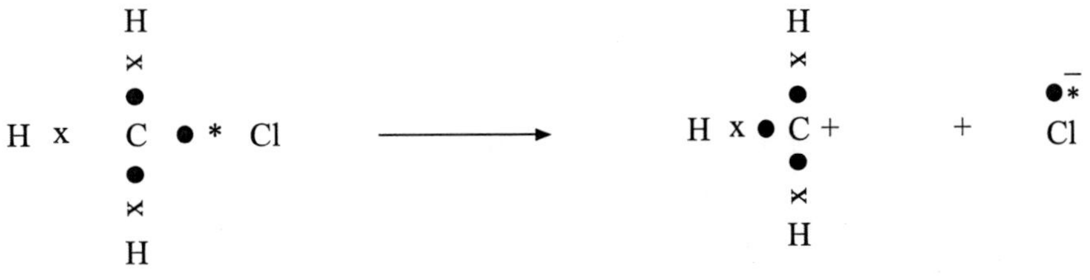

Methyl Carbocation

Electrophile is a wider term used for any electron deficient atom, an ion (positively charged) or molecule that has an affinity for an electron pair and will bond to a base or a nucleophile.

There are three types of species that can be called electrophiles.

(*i*) Those entities which have distinct plus charge formed after bond breaking, such as

$^+CH_3$, $^+NO_2$, $^+SO_3H$, Cl^+, Br^+.

$$CH_3—Cl \longrightarrow CH_3^+ + Cl^-$$

(*ii*) Those entities which have partial δ+ charge, due to high electro-negativity of one of the atoms in a bonded state, *such as alkyl halides and sulfonate esters.*

(*iii*) Those entities which are neutral and have the capacity to take a pair of electron, so as to form octet for stability, for example, $AlCl_3$, BF_3, CH_3NO_2. These are also called electron hungry entities.

Geometry of Carbonium Ions

Carbonium ion has three electrons and the orbital representing the fourth electron is vacant. This orbital is the *p*-orbital and shall hence perpendicular to the plane of other three bonds. The geometry of a carbonium ion has to be (planar with vacant *p*-orbital) as shown below:

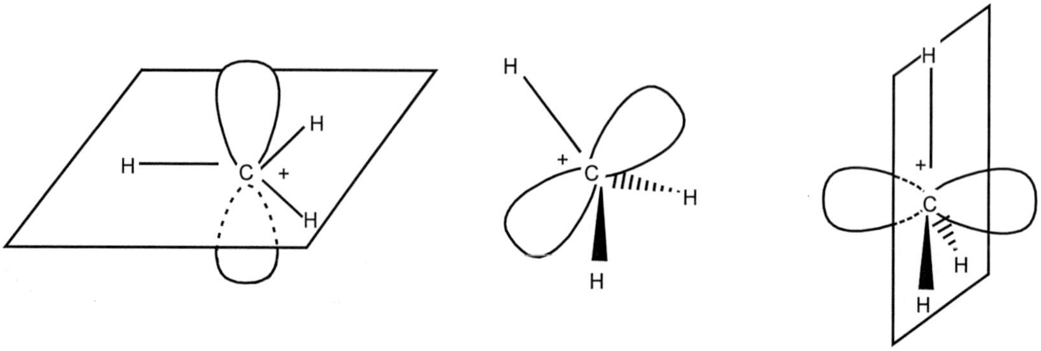

Carbocations are planar and *p*-orbital is perpendicular to the plane of other three atoms

Fig. 4.1 Geometry of carbocations

Stability of carbocations

A. Stability of carbocations due to Inductive effect

Carbonium ions carrying alkyl groups get stabilised (stability means ablity of charge species to become nearly neutral) due to +I effect (pushing electron away). Larger the number of alkyl groups on carbonium ion, greater is the stability. Thus, tert-butyl carbonium ion with three methyl groups makes it most stable, compared to ethyl carbonium ion having only one methyl group. Thus the order of stability of different carbonium ions is, tertiary > secondary > primary.

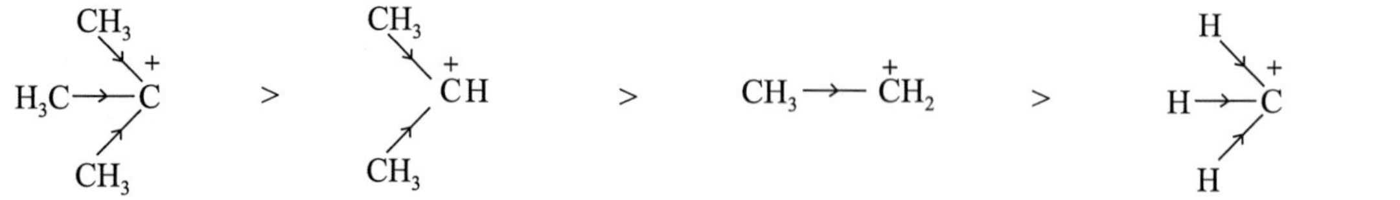

Inductive effect due to 3 methyl groups, 9 hydrogen	Inductive effect due to 2 methyl groups, total 7 hydrogen	Inductive effect due to 1 methyl groups, total 5 hydrogen	Inductive effect due to 1 methyl groups, total 3 hydrogen

B. Stability of carbocations due to hyperconjugation

Hyperconjugation is a kind of super-conjugation that gets created due to σ-bond cleavage of H- present on carbon that is α-to a double bond. The order of stability of C^+-gets reversed when hyperconjugation is operating compared to inductive effect alone as shown. In such systems, hyperconjugation effect takes priority in offering stability. It depends upon the number of ionisable hydrogens present on the carbon that is alpha to double bond. In Inductive effect order of stability is

$$(CH_3)_3\,C^+ > (CH_3)_2\,CH^+ > CH_3\,CH_2^+ > CH_3^+$$

In Hyperconjugation, a methyl group on carbon α-to double bond, has three hydrogens, compared to no hydrogen on *tert.*-butyl group. Therefore, stability order is as given below.

$$\underline{H_3C}\text{—}CH\text{=}CH\text{—}CH_2^+ > (CH_3)\,\underline{H_2C}\text{—}CH\text{=}CH\text{—}CH_2^+ >$$
$$(CH_3)_2\underline{HC}\text{—}CH\text{=}CH\text{—}CH_2^+ > (CH_3)_3\underline{C}\text{—}CH\text{=}CH\text{—}CH_2^+$$

Reason:

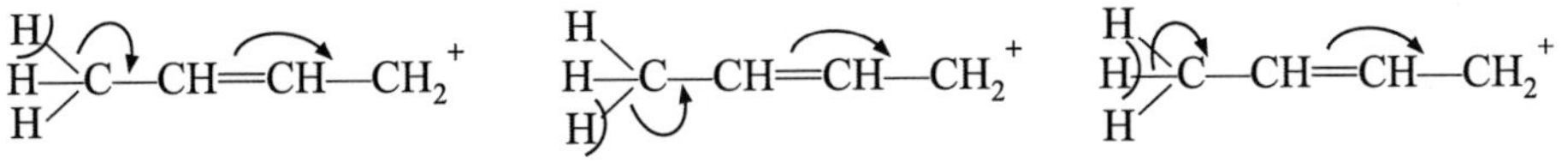

Contribution by three hydrogen ionisation to give more stability

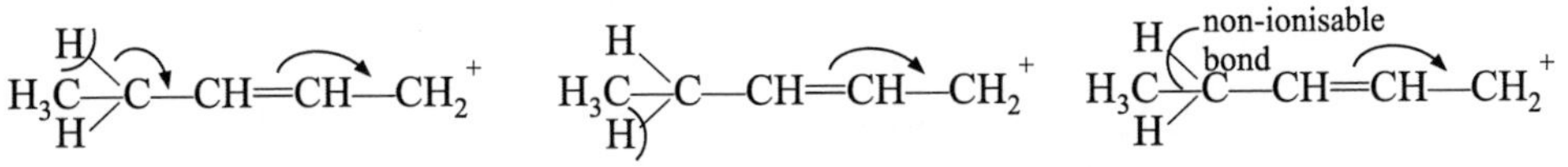

Contribution by two hydrogen ionisation to give stability

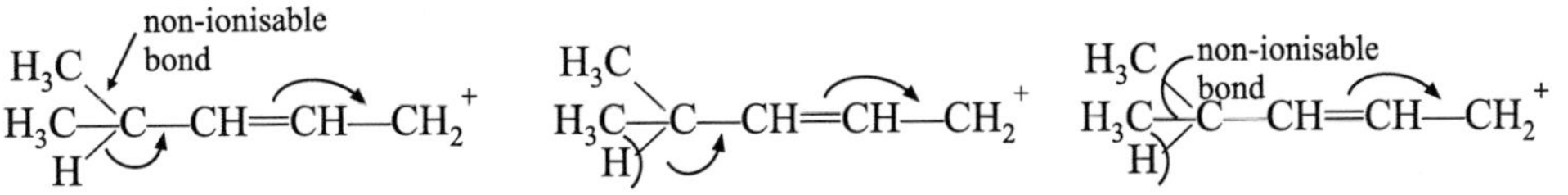

Contribution by one hydrogen ionisation to give stability

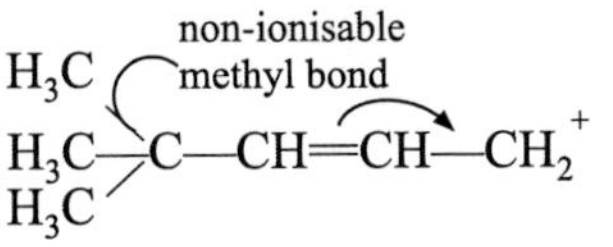

(No ionisable hydrogen on α-carbon)

Therefore, *tert.*-butyl group contribution towards carbonium ion stability in compounds with double bond as shown, is least compared to that due to methyl group, where it is higher. This order of stability is reverse of that due to inductive effect in a single bonded compound.

C. Highly substituted alkene are most stable

Highest substituted double bond are most stable. It is due to hyper-conjugation because of hydrogen carrying methyl groups provide largest number of hydrogens. For example, 2,3-dimethylbut-2-ene due to four methyl sustituents on the double bond, makes it the more stable alkene. The four methyl substituents have 12 hydrogens for ionisation and making the σ-bond electrons flow towards the carbonium ion formed by electromerisation of π-electrons in double bond. This is due to the hyperconjugation effect of 12 ionisable hydrogens.

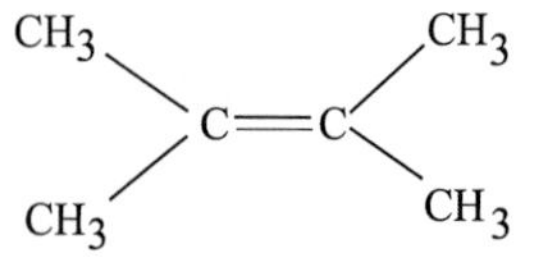

2,3-dimethylbut-2-ene

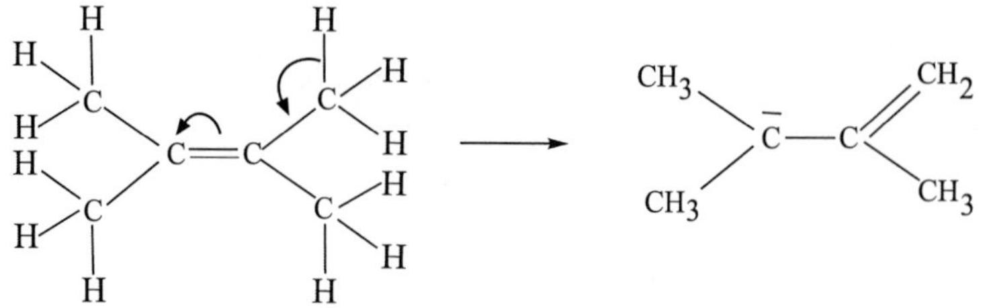

Twelve ionisable hydrogens are present on α-carbon to double bond

D. Stability of Carbocations due to Resonance

Resonance is another electron movement effect that helps to explain stability of carbonium ions. When a carbocation is linked to conjugated double bonds, resonance effect takes precedence over other effects to establish the stability of carbonium ions. For example, during reactions of benzyl chloride with aqueous sodium hydroxide, first step of ionisation of C–Cl bond gives $C_6H_5CH_2^+$, that gets stabilized due to resonance and thus, the reaction becomes favourable.

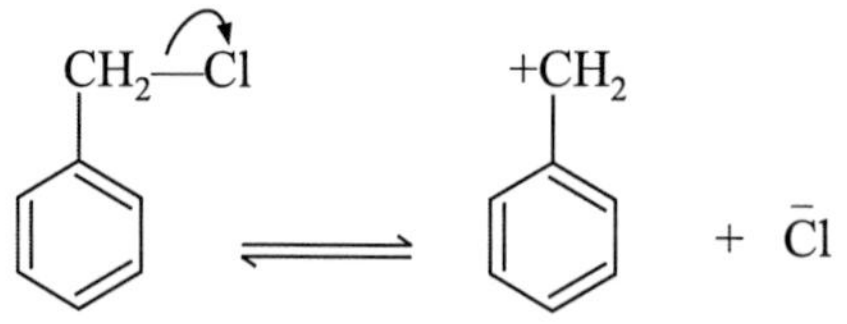

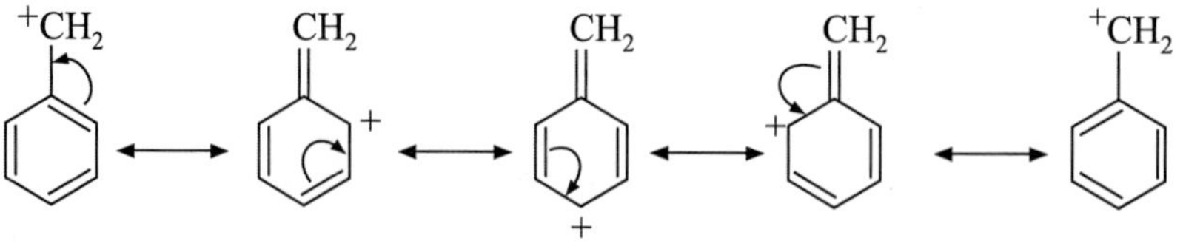

Similar reactions involving stabilization of the intermediate promotes reaction going very fast. For example, when we compare the speed of hydrolysis of propyl chloride and allyl chloride, we find that allyl chloride undergoes hydrolysis very much faster than propyl chloride. It is due to resonance stabilization of the intermediate carbocation.

$$CH_2{=}\underset{H}{C}{-}\overset{H_2}{C}{-}Cl \rightleftharpoons CH_2{=}\underset{H}{C}{-}\overset{+}{C}H_2 + \bar{Cl}$$

$$CH_2{=}\underset{H}{C}{-}\overset{+}{C}H_2 \longleftrightarrow \overset{+}{C}H_2{-}\underset{H}{C}{=}CH_2$$

Resonance stabilisation of allyl carbocation

$$CH_3{-}CH_2{-}CH_2{-}Cl \longrightarrow CH_3{-}CH_2{-}\overset{+}{C}H_2$$

Only inductive effect stabilized propyl carbocation

E. Aromaticity effect on stability

Another example of very high stability of cations is the cyclic unsaturated compound, cycloheptatriene cation. It has 6 π electrons in the cyclic ring and acccording to Huckel rule, it is aromatic in character. Thus, both aromaticity and resonance give stability to cycloheptatriene cation.

It has 6 pi electrons so is aromatic and stable

Cyclo heptatriene chloride   Cyclo heptatrienyl cation

F. Stability of carbocations due to Steric effects

Steric effect due to bulky substituents play an important role in the stability of carbocation. That is, steric hinderance at C^+ offers stability. Trimethylmethyl (*t*-butyl) and tri-isopropyl methyl carbocations show hinderance to attack by nucleophiles. Besides substitution, elimination reaction also becomes a favourable reaction if a sink for proton is available in the medium.

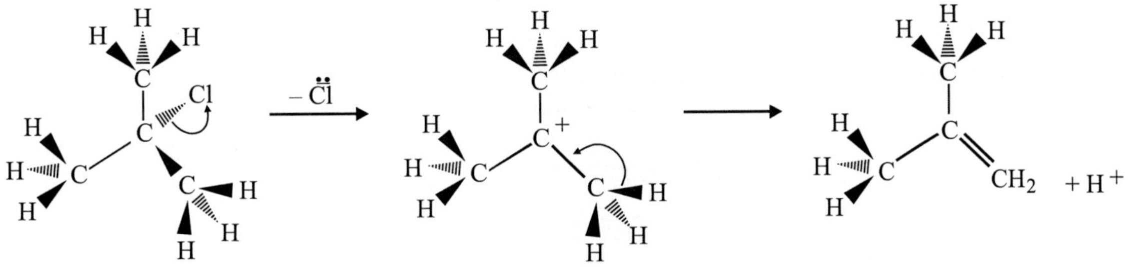

t-Butyl carbocation   Isobutene

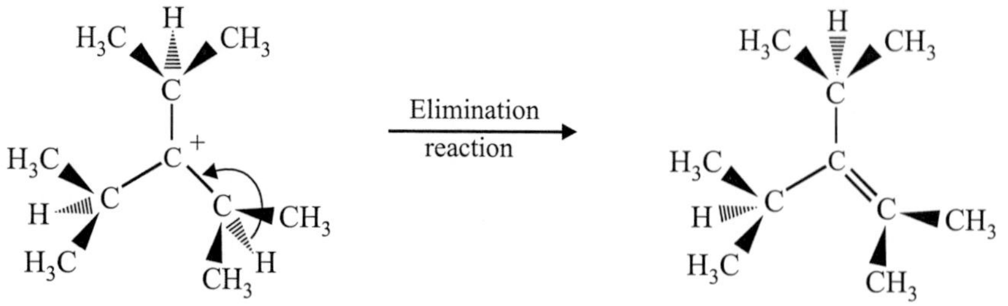

tri-isopropyl methyl carbonium ion

2,4-dimethyl-3-isopropyl pent-2-ene

Consider the substitution reaction of a halide group in tri-t-butyl methyl chloride. After ionisation of chlorine, the tri-t-butylmethyl carbocation is formed. Three bulky t-butyl groups at C^+ leave little space for the incoming nucleophile and thus offers stability to the carbocation. Triphenylmethyl carbocation is much more stable cation. It offers stability due to steric as well as resonance effects.

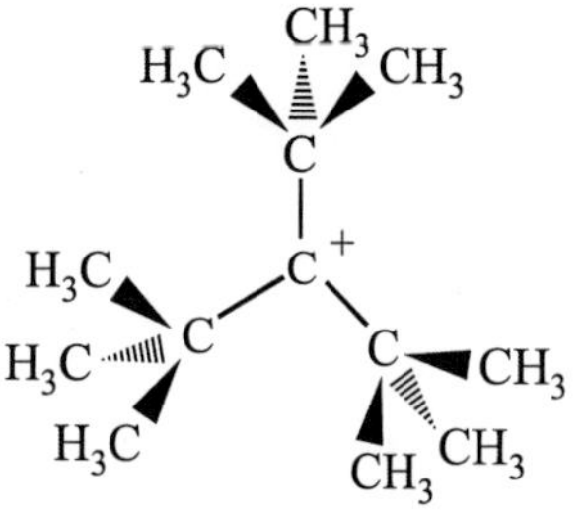

Tri-t-butylmethyl carbocation

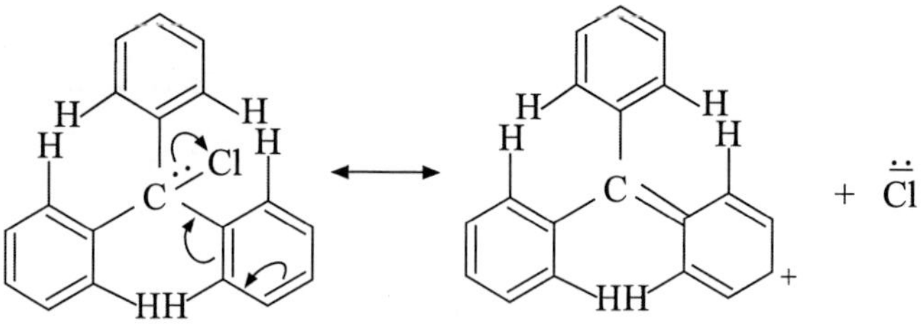

All the three phenyl rings come into resonance to form five structures as depicted earlier for nitrophenol

Triphenylmethyl chloride carbocation

Triphenylmethyl carbocation

4.4.2 Carbanions

Carbanions are negatively charged species with negative charge on carbon atom in a molecule. There is an extra electron on carbon, such as methyl carbon anion, CH_3^-, has five electrons. These are formed by heterolytic bond cleavage of a group that goes out without electrons, such as hydrogen can be taken away from C–H bond by using a strong base, *e.g.*, methyl lithium. For example, 1-chloro-1-nitroethane or isopropyl benzene, generate carbonanions by a strong base.

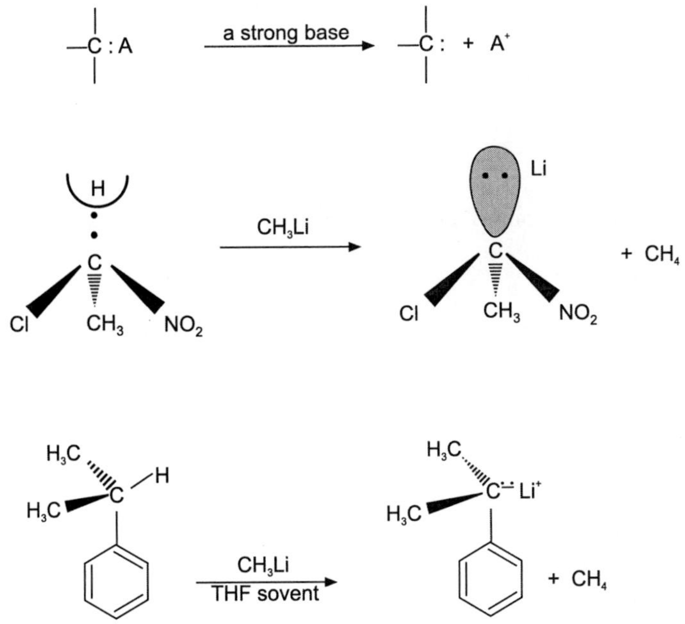

Fig. 4.2 Bond cleavage to give a carbanion.

Nucleophile is a wider term that can be used in place of carbanions. It includes the negatively charged atoms of other elements, such as nitrogen, oxygen or sulphur etc. A nucleophile by definition is an atom, an ion (negatively charged) or a molecule that has an electron pair for donation for bonding with an electrophile (or Lewis acid). Some of the examples of nucleophilic groups are

$I^{(-)}$, $Br^{(-)}$, $Cl^{(-)}$, $RO^{(-)}$, $RS^{(-)}$, $N_3^{(-)}$, $NH_2^{(-)}$, $CH_3CO_2^{(-)}$, $CN^{(-)}$, $CH_3^{(-)}$

The order of strength of nucleophiles is as shown:

$I^{(-)} < Br^{(-)} < Cl^{(-)}, RO^{(-)}, RS^{(-)}, \sim N_3^{(-)}, NH_2^{(-)} < CH_3CO_2^{(-)} < CN^{(-)} \sim CH_3S^{(-)} < CH_3O^{(-)}, CH_3^{(-)}$

(Weak nucleophile, starting point) (Strong nucleophile, at the end)

Geometry of Carbanions

Geometry of cabanions differ from that of carbonium ions. Carbanions are trigonal pyramidal. Orientation of the electron pair is along the direction of one of the bonds in a tetrahedron. It could also be looked at as the apex bond of a pyramid, as shown in Fig. 4.3.

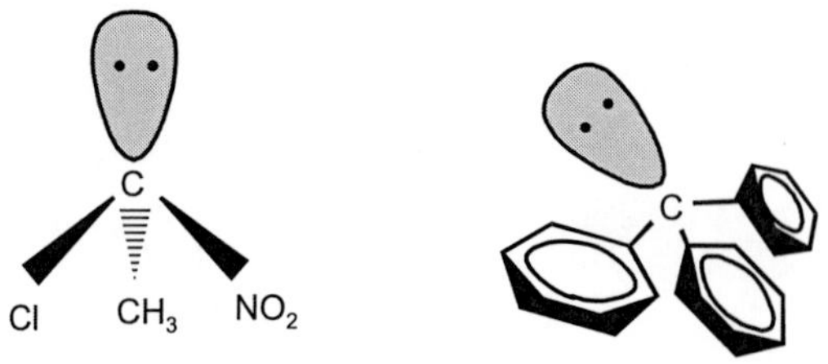

Fig. 4.3 Pyramidal or tetrahedral like structure of carbanions.

Stability of carbanions

Carbanions are nucleophiles that get stabilized due to the following factors:

(*i*) Inductive effect of groups

(*ii*) Hybridisation state

(*iii*) Extent of conjugation and resonance.

(*i*) **Inductive effect of groups:** Carbanions of saturated molecules are reactive intermediates in organic reactions, so it is difficult to talk of stability of carbanions. However, the relative stability for certain anions is worth mentioning. A molecule that is rich in electrons must be very reactive compared to carbonium ions. It has been observed that the order of stability of carbanions is opposite to that of carbonium ions.

Inductive effect of alkyl groups and stability order of carbanions :

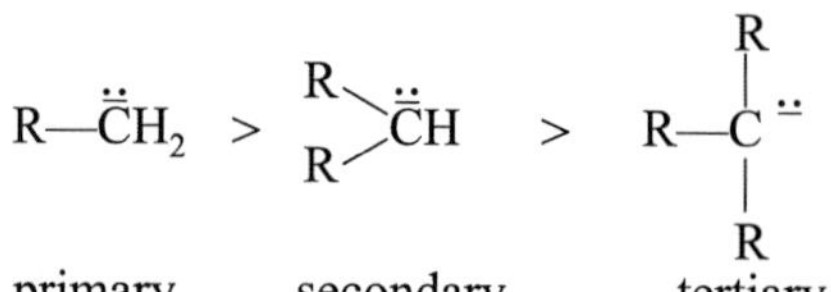

primary secondary tertiary

(*a*) When, R is the alkyl group/s with +I effects **(that repel electrons)**. It will exert its influence on the stability of these carbonanions. Stability shall decrease due to pushing of electrons to the negative charge, in other words carbanion become more reactive.

$$R= ,\quad H— > CH_3— > C_2H_5— > CH_3—CH_2—CH_2— >$$

(*b*) Similarly, if we attach an **electron withdrawing group** (–I effect), the stability factor shall increase due to decrease in electron density at the carbon anion. However, factors like, resonance and electron withdrawing groups make the negative charge spread over the whole molecule and *stability* of the carbanions shall *increase*.

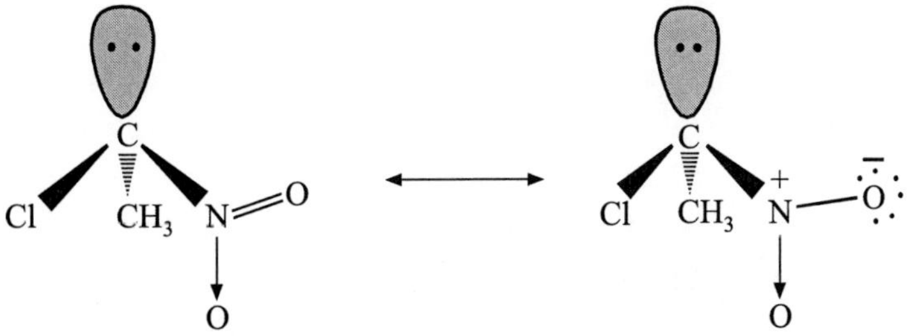

***(ii)* Stability due to hybridisation state:** It has been observed greater the '*s*'character of the charge bearing atom, greater is the stability due to larger surface area.

***(iii)* Stability due to Aromaticity, resonance:** Another example of very high stability of anions is the cyclic unsaturated compound, cyclopentadienyl anion. It has 6π-electrons in the cyclic ring and acccording to Huckel rule, it offers it a stability due to aromatic character. An other example of stability of carbon anions are benzylic anion. In benzylic anion the stabilisation occurs due to resonance.

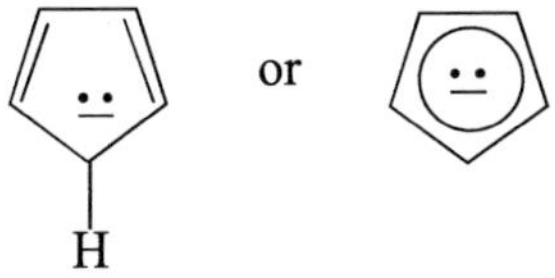

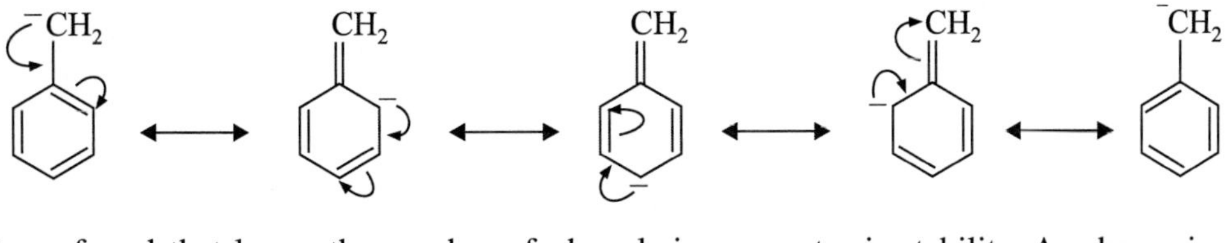

It has been found that larger the number of phenyl rings, greater is stability. As shown in case of triphenylmethane.

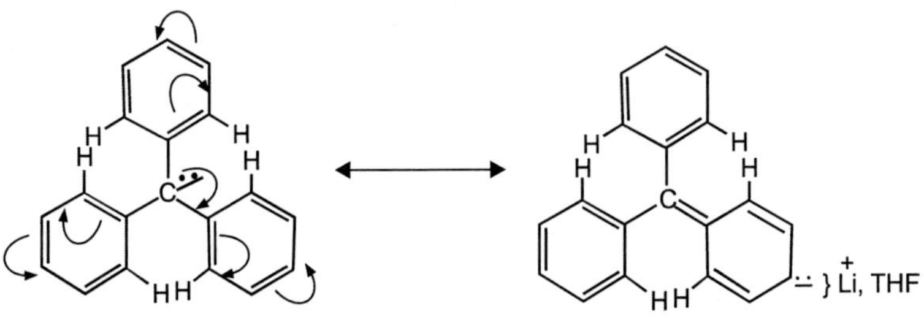

All the three phenyl rings come into resonance to form
five structures as depicted earlier for nitrophenol

Fig. 4.4 Stability of some Anions.

Molecules that lose H^+ to form a carbanion, these are also called carbon acids, with appropriate pK_a values.

4.4.3 Carbon Free Radicals

We have learnt that homolytic bond cleavage between A–B gives $\dot{A}$ and $\dot{B}$ each having one unpaired electron. In this bond cleavage, each atom retains its own share of one unpaired electron. These species are called *Radicals*. These are very reactive species.

$$A{-}B \longrightarrow \dot{A} + \dot{B}$$

Geometry of Carbon free radicals

The geometry of radicals looks very similar to that of carbonium ions. Carbonium ions have vacant *p*-orbital, whereas radicals have one electron in the *p*-orbital. In case of methyl radical, the geometry appears to be more sp^2 hybridised and the single electron occupies the *p*- orbital that is perpendicular to the plane of three bonds. Radicals are very reactive entities and dealing with it is not easy.

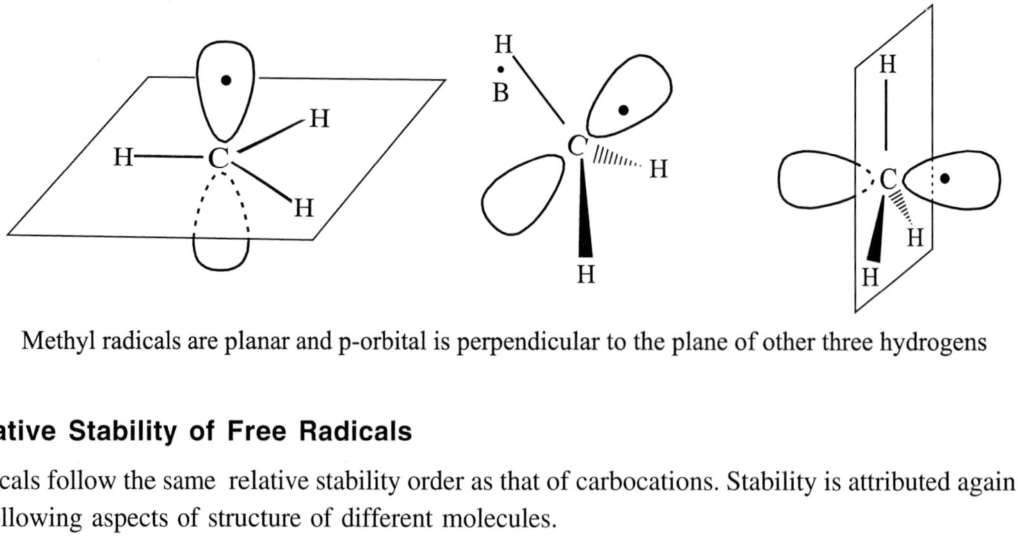

Methyl radicals are planar and p-orbital is perpendicular to the plane of other three hydrogens

Relative Stability of Free Radicals

Radicals follow the same relative stability order as that of carbocations. Stability is attributed again due to following aspects of structure of different molecules.

(*i*) Stability due to Inductive effect

Attempt of single electron in a radical is to get paired or experience the force of opposite spin of near by electrons. Thus, radicals with larger the number of alkyl groups attached to carbon (with unpaired electron) are more stable. For example, the stability order of alkyl groups is tertiary butyl radical, isopropyl radical, ethyl and methyl radical. In other words, the tertiary radical is more stable than secondary and secondary is more stable than primary radical.

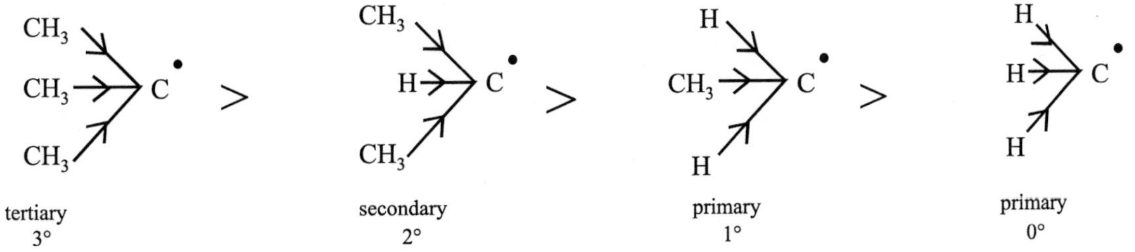

(*ii*) Stability due to limited conjugation

Allyl radicals get formed in allylic bromination and show stability due to resonance, though that is limited to one double bond. In resonance, shift of electrons takes place by electromeric effect. It is represented in case of allyl radical

$$CH_2{=}CH{-}\dot{C}H_2 \longleftrightarrow \dot{C}H_2{-}CH{=}CH_2 \longleftrightarrow [CH_2{\overset{\cdots}{=}}CH{\overset{\cdots}{=}}CH_2]$$

(*iii*) Stability due to resonance

Radicals show stability due to resonance as in case of benzyl radicals.The movement of one electron is shown by half arrow. Two electrons make one bond, so two half arrows are needed for forming a bond. Resonance stabilised aromatic radicals are more stable than aliphatic radicals.

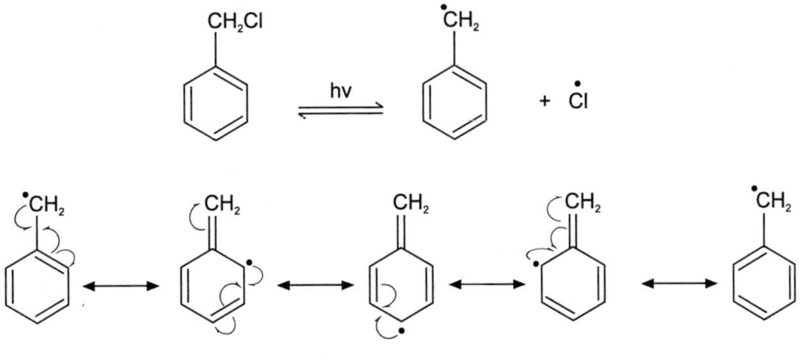

Fig. 4.5 Resonance stabilities of benzyl radical, observe half arrow movement of one electron.

(*iv*) Stability due to hyperconjugation

Hyperconjugation, in case of radical is very similar to that of carbonium ions. It requires alpha carbon to the radical to have hydrogen, that may get ionised. Alpha σ -bond cleavage leads to π - bond formation via single electron movement. It shall stabilise the radical or undergo a chemical reaction. In case of tertiary butyl radical, it is shown below. Similar, is the case of crotonyl radical. In both the cases alkenes are formed.

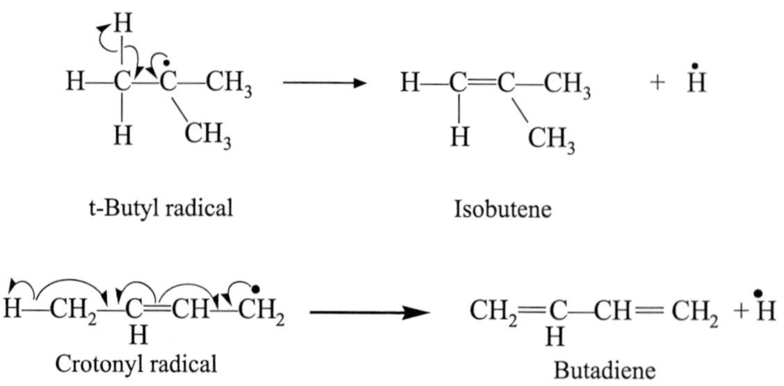

4.4.4 Carbenes

Carbenes are entities having two unpaired electrons on a carbon. These are neutral, thus, carbon is divalent in a carbene. These are also very reactive entities and undergo chemical reactions very fast. They can add on to double bond or electron rich sites in molecules. Simple example of carbene formation is the reaction of chloroform with sodium hydroxide. Reaction of chloroform with phenol to form salicylaldehyde is a well understood carbene reaction.

Let us, examine the state of two electrons in a carbene. The six electrons in carbon atom in the ground state are filled in $1s^2\ 2s^2\ 2p^2$ orbitals. In formation of methylene carbene, two hydrogens orbitals overlap with one $2s$ and one $2p$ orbital of carbon, leaving behind two electrons on carbon to represent a carbene. These two electrons with unspecified spins, under the influence of magnetic field may attain, same side or opposite spins. It can be determined by **Electron Spin Resonance** spectroscopy. These **spins splittings** are according to $(2s+1)$ formula, where s is the total spin of two electrons. For opposite spins of two electrons, $s = ½ - ½ = 0$, thus the formula becomes, $2{\times}0+1 = 1$, that is a singlet state. When these are parallel spin, then $s = ½ + ½ = 1$, thus the formula becomes, $2{\times}1+1 = 3$, that is a triplet state. In case of methylene carbene or dichlorocarbene, following positions of electrons are presented in different states.

Singlet carbenes have a pair of electrons and an sp^2 hybrid structure. Triplet carbenes have two unpaired electrons. They may be either sp^2 hybrid or linear sp hybrid. Most carbenes have a non-linear triplet ground state, except for those with nitrogen, oxygen, or sulfur atoms and dihalocarbenes.

Thus, in carbene two electrons are unhybridised and their spins become identifiable in a magnetic field, When spins of unpaired electrons are opposite to each other, it is called a **singlet state.** It is an allowed transition and is very fast. It is a lower energy state. In the other state two electrons like singlet state are unpaired but the spin of these electrons lies in the same direction. It is called **a triplet state**.

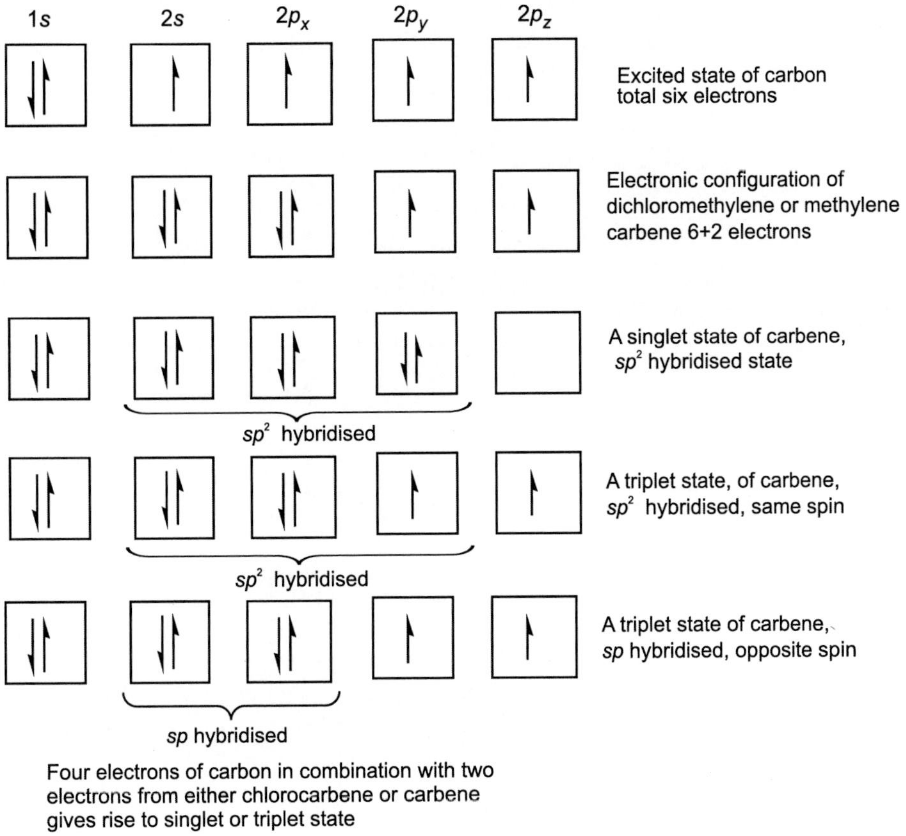

Fig. 4.6 Electron distribution in orbitals for singlet and triplet states.

Orbital Picture of Singlet carbene, sp^2 hybridised

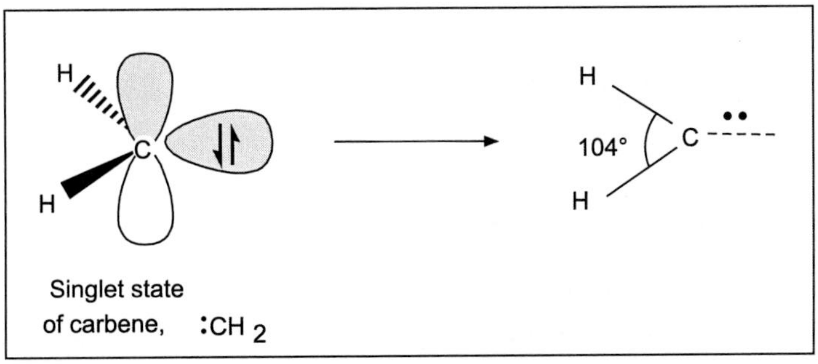

Orbital Picture of Triplet carbene, sp^2 hybridised

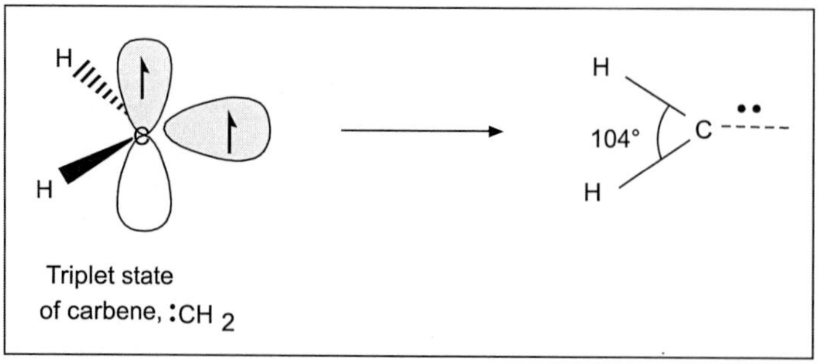

Orbital Picture of Triplet carbene, *sp* hybridised

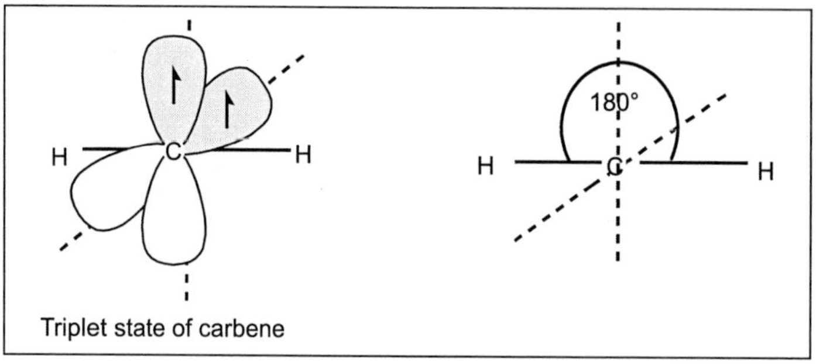

Fig. 4.7 Geometry of singlet and triplet states of carbene

Stability or reactivity of carbenes

Carbene like dichlorocarbene, become more stable due to resonance, as shown below. The resonating structures of dichlorocarbene clearly explains the extra stability of this carbene. Stability of carbenes is dependent upon the electronic and steric factors of the substituents. In general, singlet carbene having paired electrons, is more stable than the triplet state carbenes. Triplet state carbenes are considered as diradicals and thus are reactive entities. For a long time carbenes have been considered reactive and unisolable entities. More recently, some derivatives as stable carbenes have been isolated.

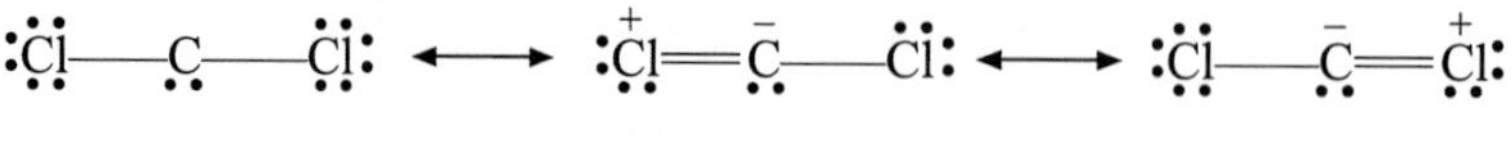

Singlet carbenes usually undergo addition reactions to double bond as an electrophiles, as well as nucleophiles. Singlet carbene can react in a single *concerted* step. Addition of singlet carbenes to olefinic double bonds is more stereoselective than that of triplet carbenes. Addition reactions with alkenes can be used to determine whether the singlet or triplet carbene is involved.

Stereochemistry of Carbene reactions

The behaviour of carbene reactions depends upon the singlet or triplet states. The concerted reactions give *cis* addition reaction products.

In Singlet state, it acts as a *p*-electron species and undergoes a concerted reaction with π-electrons of alkene

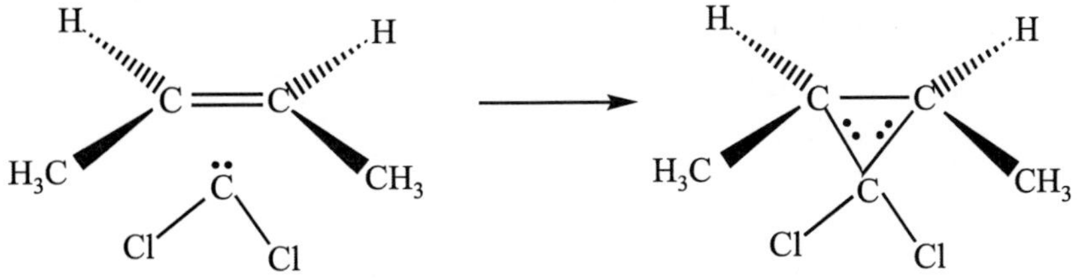

In Triplet state it acts as a diradical addition

The triplet state of carbene, acts more or less act as diradicals. Triplet carbenes have to go through an intermediate with two unpaired electrons, forming two isomeric products, because of the free rotation around single bond in the intermediate radical.

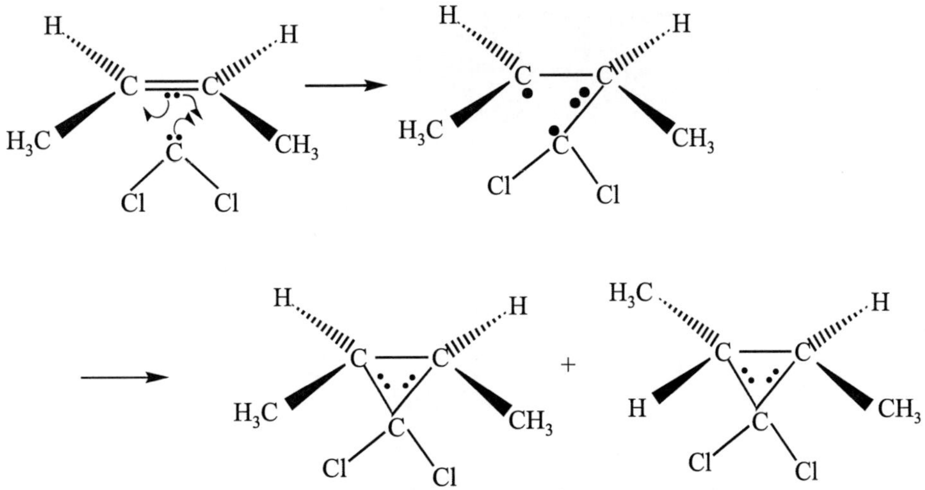

Isomeric Products, Opposite stereochemistry

Mechanism of Riemer-Tiemann Reaction, A Carbene Addition Reaction

Let us look at the stability of carbene formed from chloroform. It is considered to be in singlet state and is ordinarily more stable than other carbenes.

$$CHCl_3 + NaOH \longrightarrow :CCl_2 + HCl$$

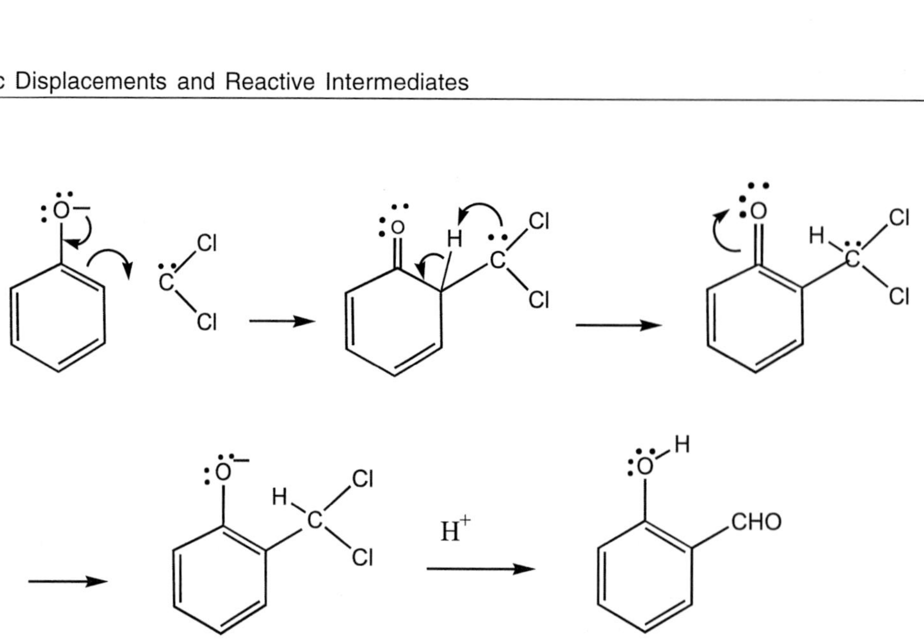

Points to Remember

1. The movement of electrons occurs in organic compounds without changing the carbon skeleton. It involves, usually π-bond, breaking and reforming at different place.
2. The electronic movements have been characterised to show the following effects:
   (*a*) **Inductive effect** is the partial movement of sigma bond electrons with in the bond. It occurs in single bonded compounds.
   (*b*) **Electromeric effect** is shown in unsaturated compounds/groups, such as C=C, C=O, C=S or CN or $CONH_2$ etc. In the process π-bond electrons move from one bond to others.
   (*c*) **Resonance** involves the movement of π-bond electrons over the whole molecule.
3. The electron movement accompanied by bond cleavage of certain atoms or groups from the molecule, leads to the formation of charged species. A positively charged ion of carbon is called carbocations. A negatively charged, anion of carbon is called carbanion.
4. The charged species mentioned above, play a very imporant role in understanding chemical reactions through their mechanism. It also helps in understanding or visualisng unknown reactions.

Exercises

1. What are the favourable factors that help in cleavage of C—H and C—C bond?
2. Which of the bonds, C—H or C—C is a stronger bond?
3. The total bond energy of methane molecule is of the order of 400 kcal /mol. What will be the average bond energy of a C—H bond?
4. Explain the concept of inductive effect as applied to isopropyl chloride and compare it with that of tert-butyl chloride and active pentyl chloride (2,2-dimethylpropyl chloride).
5. Which of the compounds in problem 4 affect the C—Cl bond polarity due to inductive effect?
6. How does inductive effect differs from electromeric effect. Do you expect a reasonable degree of electromeric effect in tert-butyl alcohol?
7. Arrange the following in order of increasing polarity, that the groups may attain due to electromeric effect.

   >C=O , >C=C<, >C=NH, —C≡N and >C=S
8. How electromeric effect differs from resonance. What is the net effect of resonance in chemical reactions?
9. 1-Chloro-but-2-ene undergoes hyperconjugation in a nucleophilic attack by a strong base. Can the flow of electrons, over all the carbon atoms be regarded as equivalent to resonance effect.
10. How hyperconjugation differs from resonance? Giving two examples to elaborate your answer.
11. What are different modes of bond breaking? To break a bond by unequal sharing of electrons between two atoms forming the bond, what is the necessary condition for such a bond breaking?
12. What are different modes of movement of electrons in organic molecules that occurs without changing the carbon skeleton?

❑❑❑

5

Strength of Acids and Bases

The colour of flowers and plant materials, such as fruits are mainly because of chlorophylls, carotenes, xanthophylls and anthocyanins. All these are in different proportion and pH gives rise to brilliant colours and hues seen in nature.

IN THIS CHAPTER

CHAPTER 5

Strength of Acids and Bases

5.1 Acid–Base Theories

There are three theories that define acid–base character of substances.

5.1.1 Arrhenius Theory: Svante Arrhenius (1857–1927)

Arrhenius (1883) stated that ionic compounds dissociate when dissolved in water and can become free ions acting as individual entities in solution. He defined acid and base as

ACID: A substance that ***produces*** H^+ in water.

BASE: A substance that ***produces*** OH^- in water.

$$HCl_{(aq)} \xrightarrow{H_2\ddot{O}:} H_3O^+ + Cl^- \quad \text{produces } H^+ \text{ in water}$$

$$NH_{3(aq)} \xrightarrow{H_2\ddot{O}:} NH_4^+ + OH^- \quad \text{produces } OH^- \text{ in water}$$

Although NH_3 does not contain ^-OH, but ^-OH ions get formed when water is added to it. Arrhenius acid and base neutralize each other to produce salt and water:

$$HCl_{(aq)} + NaOH_{(aq)} \longrightarrow NaCl_{(aq)} + H_2O_{(l)}$$

$$H^+_{(aq)} + OH^-_{(aq)} \longrightarrow H_2O_{(l)}$$

Arrenhius theory requires reactions to take place in aqueous solution and applies only to substances that produce H_3O^+ or ^-OH.

5.1.2 Brønsted–Lowry Theory

[Johannes Brønsted (1879–1947) and Thomas Lowry (1874–1936)]

In 1923, Brønsted and Lowry proposed a theory to define acid and bases. Their approach was not limited to aqueous solutions but for all proton (H^+) containing substances. They defined acid and base as donor and acceptor of protons respectively.

ACID: A substance that can ***donate*** proton, (H^+).

BASE: A substance that can ***accept*** proton, (H^+).

BrØnsted–Lowry expanded, Arrenhius theory to include any proton transfer (water is not a necessary requirement). Acids may be cations, neutral molecules, while bases may be anions or neutral molecules. A proton donor (acid) must accompany a proton acceptor (base) in a chemical reaction.

An acid transfers its proton to the medium or a base and the reaction occurs. The product of the reaction is the formation of **conjugate base (CB).** Similarly, when base accepts a proton, it becomes a **conjugate acid (CA)**. If one examines, the BrØnsted–Lowry concept, ***proton is produced*** and in the Arrenhius concept ***proton is accepted***. Both theories, look very similar. In reaction between HCl(g) and NH_3 (g), when water is not involved, the reaction can still take place. It can be explained by BrØnsted/ Lowry concept and not by Arrenhius theory.

$$\underset{\text{Acid}}{HC1(g)} + \underset{\text{Base}}{NH_3(g)} \longrightarrow \underset{\text{CA}}{NH_4^{+1}} + \underset{\text{CB}}{C1^{-1}}$$

5.1.3 Lewis Theory [Gilbert Newton Lewis (1875–1946)]

Arrhenius theory did not support acid-base behaviour in non-aqueous systems and the BrØnsted-Lowry model excluded non-protonated systems. Lewis, in 1923, suggested a theory that explained all types of acid and bases. This theory was fully developed by 1938.

ACID: *It is an electron deficient substance that can accept a pair of electrons from another substance to form a new bond.*

BASE: *It is an electron rich substance that can donate a pair of electrons to another substance to form a new bond.*

Lewis expanded acid-base reactions to include many substances without hydrogen. The electron pair acceptor and electron pair donor are the key words. For example, boron in boron trifluoride, is electron deficient and to complete its octet, accepts a pair of electrons from nitrogen atom of ammonia, as shown. This reaction is explained by Lewis theory but not by Arrenhius or BrØnsted–Lowry theories.

$$F_3B + :NH_3 \longrightarrow F_3\overset{-}{B}:\overset{+}{N}H_3$$

Problem 5.1: Which theory can explain the following reactions?

(*i*) $HI + H_2O \longrightarrow H_3O^{+1} + I^{-1}$

(*ii*) $HI + NH_3 \longrightarrow NH_4^{+1} + I^{-1}$

(*iii*) $I_2 + NH_3 \longrightarrow NH_3I^{+1} + I^{-1}$.

Solution: (*i*) Explained by all 3 theories

(*ii*) Explained by BrØnsted and Lowry and Lewis

(*iii*) Explained by Lewis.

5.2 Ionisation Constant, K_a and pK_a are Measures of Acid Strength

Carboxylic acids in solution undergo ionisation process to form associated protons in water. The extent of ionisation is determined by the concentration of ionised and unionised acid. The following equation gives the ionisation of a carboxylic acid. Let at time $t = 0$, the starting concentration of acid be c, and at that time the products shall be zero. However, at time t = t', if α is the degree of dissociation, the undissociated acid shall be $c(1 - \alpha)$. The two products shall be $c\alpha$ and $c\alpha$, as shown below.

R–COOH (aq.)	+	$H_2O(1)$	⟶	$RCOO^-$(aq.)	+	H_3O^+ (aq.)
Conc. at time, $t = 0$, c		large		0		0
at time, $t = t'$, $c(1 - \alpha)$		large		$c\alpha$		$c\alpha$

$$\text{the equilibrium constant of acid } K_a = \frac{[c\alpha][c\alpha]}{[c(1-\alpha)]} = \frac{c\alpha^2}{(1-\alpha)}$$

taking minus log of the equilibrium constant, we get –log K_a, it is equated to pK_a,

$$pK_a = -\log K_a$$

pK_a, is a constant telling us the strength of an acid, very much like pH that is = – log H^+.

Higher the value of K_a or lower the value of pK_a, stronger is the acid. Similarly lower the value of K_a or higher the value of pK_a, weaker is the acid. These are experimentally determinable constants and thus we can quantitatively determine the strength of an acid. Because of difficulty of remembering all the data, an alternative approach has been adopted. It is possible to have some rough estimate of the strength of an acid from the consideration of structures as shown next. However, the Table 8.1 gives the K_a and pK_a of different acids.

5.3 Why is–COOH, an Acidic Group?

A carboxyl group behaves like an acid because the structure permits release of proton due to electromeric effect of carbonyl group as shown in the figure below:

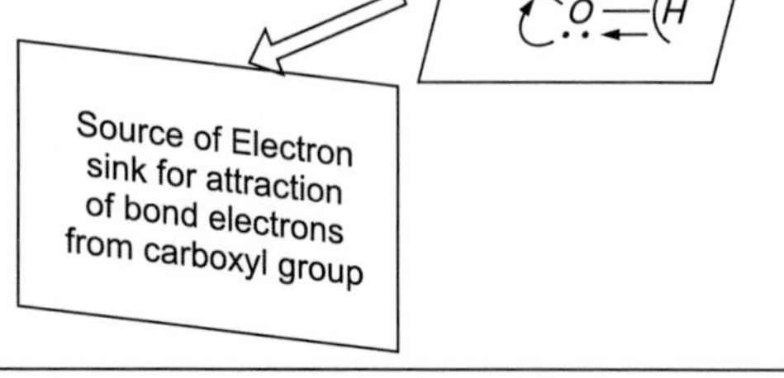

A –COOH group should have a fixed power to release a proton because of the electromeric and mesomeric effects of > C = O group. A group that is attached to it makes it a stronger or weaker acid depending upon the electron withdrawing or electron donating power respectively. Since we know the –COOH group cannot exist alone, it is attached to different groups. The simplest acid is formic acid, H–COOH. The electron donating groups like alkyl groups make it a weaker acid. The electron withdrawing groups, makes it a weaker acid. With electron withdrawing groups, like NO_2, >C=O makes it stronger acid.

Fig. 5.1 A representation of acid character of COOH group.

The acid strength due to –COO<u>H</u> depends upon the nature of groups attached to it. The electron withdrawing group increases and electron donating group decreases the proton releasing power of –COO<u>H</u>, that is the acid strength. This statement is true so long as we are talking of structure of a carboxylic acid only. We have ignored medium in which it is present, that has its own effect on the strength of acid. Let us look at effect of structure of carboxylic acids on their acid strength.

5.4 Effect of Structure on Acid Strength of Aliphatic Acids

Different kind of substituents on the carboxyl carbon can change the acid strength of a carboxylic group. Two type of effects become operative, inductive effect and electromeric or mesomeric effect. When alkyl groups are attached to a carboxyl group, such as acetic acid, propionic acid, butanoic acid, –COOH becomes a weaker acid group, because of electron pushing effect of alkyl group. It makes –OH in –COOH less ionisable. Thus, fatty acids, having a long carbon chain, are very weak acids. Formic acid with no alkyl chain is the strongest in the aliphatic series.

On the other hand, when chloro or nitro–group/s is present at the methyl group of acetic acid, the electron pulling effect of these groups makes the acetic acid a stronger acid. It is observed from the pK_a values of different acids, given in the Table 5.1.

Table 5.1 pK_a values of some acids

Basic unit		R—C(=O)—O—H	
Substituent R	**Inductive effect direction**	K_a	pK_a
H–	⟶	1.77×10^{-4}	3.75
CH_3–	⟶	1.80×10^{-5}	4.75
CH_3CH_2–	⟶	1.34×10^{-5}	4.67
$CH_3CH_2CH_2$–	⟶	1.34×10^{-5}	4.81
$(CH_3)_2CH$–	⟶	1.41×10^{-5}	4.84
$(CH_3)_2CHCH_2$–	⟶	1.7×10^{-5}	4.77
Cl–CH_2–	⟵	1.7×10^{-3}	2.86
Cl_2–CH–	⟵	1.4×10^{-2}	1.29
Cl_3C–	⟵	5.05×10^{-1}	0.65
Br–CH_2–	⟵	2.3×10^{-3}	2.86
I–CH_2–	⟵	1.3×10^{-4}	3.12
NO_2CH_2–	⟵	7.2×10^{-2}	1.68
CH_3O–CH_2–	⟵	2.2×10^{-5}	3.53

The effect of substituents on acid strength is explained by electron flow in the following examples. In acetic acid, methyl group is electron pushing and it reduces the positive charge on carbon of >C=O group. Thus, electrons of –O–H bond do not move towards carbon. A stronger bond remains intact between O– and H– . The proton is not able to get off the –COOH group easily. It makes acetic acid a weaker acid compared to formic acid,

Similarly, in case of an electron withdrawing group the electrons on –O–H group get attracted towards carbon and makes it a stronger acid. It is shown in the following structures.

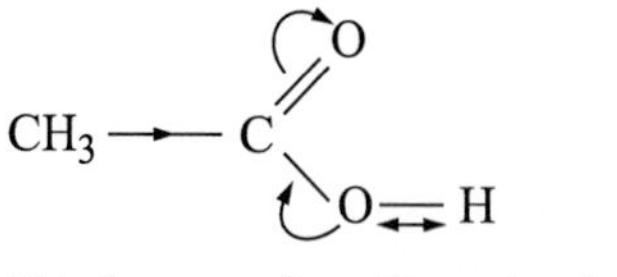

This flow towards positive carbon is inhibited, so becomes a weaker acid

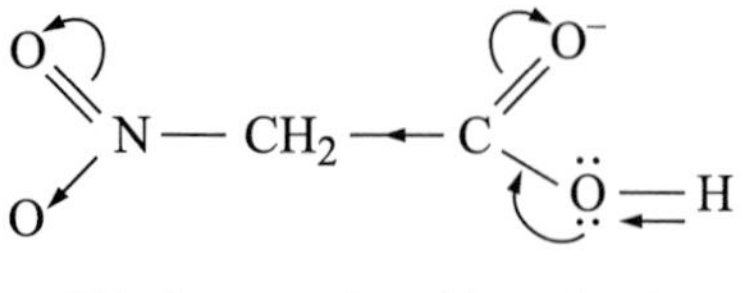

This flow towards positive carbon is facilitated, so becomes a stronger acid

5.5 Effect of Structure on Acid Strength of Aromatic Acids

The carboxyl group in benzoic acid is less acidic than in formic acid. The resonating structures of benzoic acid show that as soon as pi electrons of >C=O electromerise, the pi electrons of benzene ring come to neutralise the positive charge on carbon of $>C^+ - O^-$. The non-bonded electrons of –O–H **do not** flow towards C^+ of >C=O, that would have weakened the –O–H bond for releasing hydrogen. It is the extent of this hydrogen release that makes carboxyl group stronger or weaker acid. When different groups are attached to different positions of benzene ring, effect of electron withdrawl or electron donation becomes prominent. For example, methyl group is electron donating, this decreases the acid character. Nitro– group in nitro–benzoic acid is electron with–drawing, it enhances the pull of electrons from –O–H bond and thus, makes it a stronger acid.

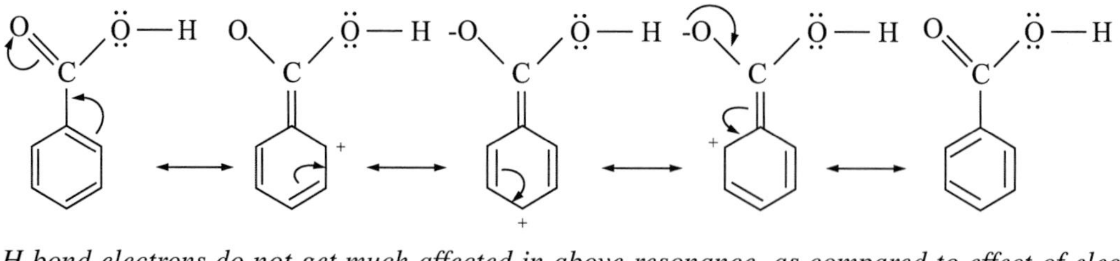

O – H bond electrons do not get much affected in above resonance, as compared to effect of electron push or pull by methyl or nitro group directly attached to –CH_2 COOH, as in acetic and nitro acetic acid. or to phenyl ring in benzoic acid.

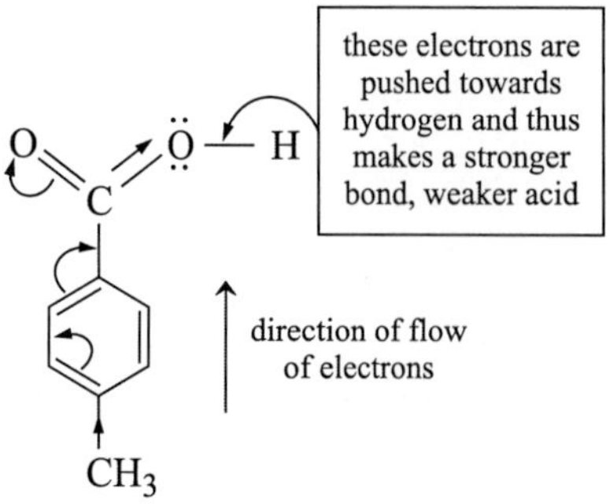

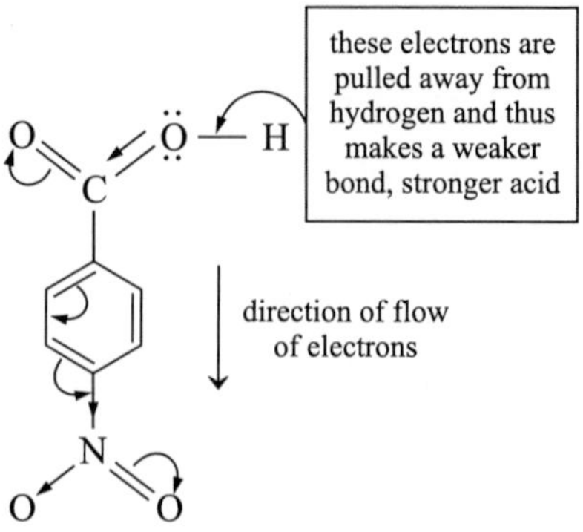

Making use of this concept, we can easily and roughly estimate the strength of an acid. In the Table 5.1, pK_a values of some of the common acids and phenol are given. We can correlate these experimentally determined pK_a values with their structure and justify our explanation.

5.6 Acid Strength of Phenols, (no COOH Group)

Phenols are another class of aromatic compounds that are acidic and may not contain any –COOH group. The ionisation of –OH proton becomes possible because of resonance in benzene ring. Let us look at ionisation of –OH group and pK_a value of some phenolic compounds. The resonating structures of phenol and its acid character is shown below:

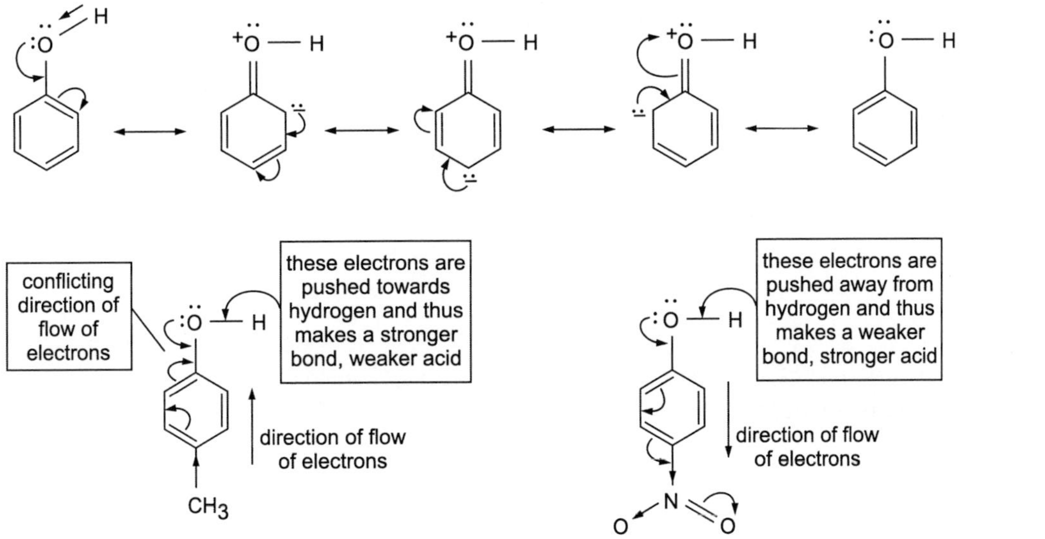

Fig. 5.2 The resonating structures of phenol and its acid character. Effect of methyl and nitro groups on acid strength is shown.

From the resonating structures of phenols, it is obvious to understand how phenols behave as acids. The effect of methyl group and that of nitro group are also similar as discussed earlier in case of carboxylic acids, *p*-methyl group makes phenol a weaker acid, whereas *p*-nitro group in phenol makes it a stronger acid.

It is clear from the Table 5.2 of pK_a value that phenols are acidic but not that strong acids like carboxylic acids. It determines experimentally that when nitro–groups are attached, phenols also become sufficiently acidic. For example, 2,4,6–trinitrophenol behaves like an acid and pK_a value is in the same range as that of a strong carboxylic acid.

Acid character can also be checked qualitatively by reaction with sodium hydroxide, sodium carbonate and sodium bicarbonate in three different test tubes. Sodium hydroxide is the stronger base compared to sodium carbonate and sodium bicarbonate. Ionisation of three takes place to form anions as shown ahead:

$$NaOH \longrightarrow Na^+ + OH^-$$
$$Na_2CO_3 \longrightarrow 2\,Na^+ + CO_3^{-2}$$
$$NaHCO_3 \longrightarrow Na^+ + HCO_3^-$$

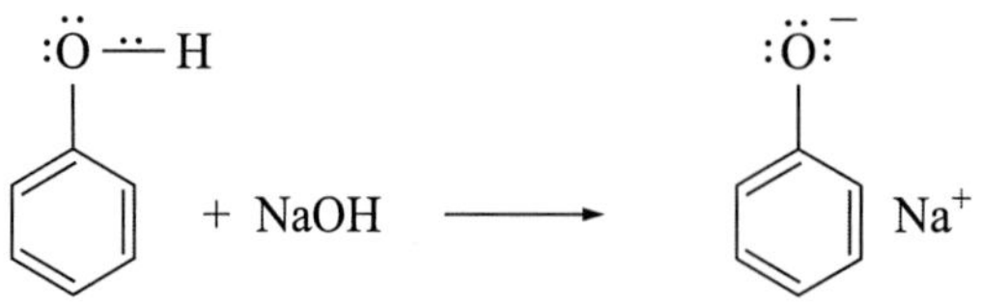

* Hydroxide a stronger base reacts even with weaker acid.
* Less basic carbonate ion shall react with a slightly stronger acid.
* Bicarbonate is slightly acid, so shall react with a strong acid.

Thus, an approximate distinction of acid strength can be made.

No reaction with Na_2CO_3 and $NaHCO_3$, but reacts with sodium hydroxide, shows it to be a weak acid).

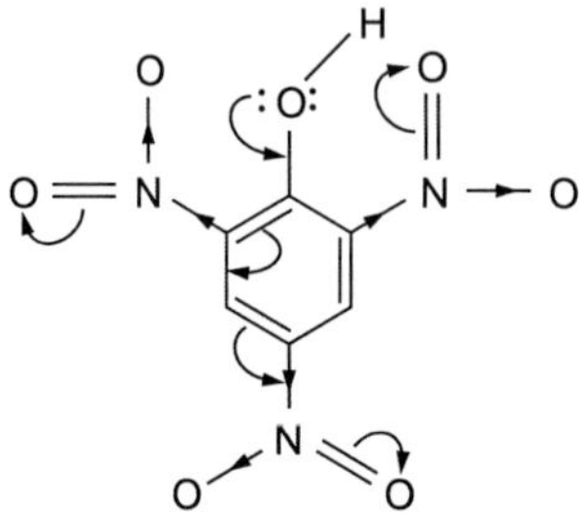

direction of flow of electrons in three direction makes it a strong acid to react with bicarbonate ion.

Fig. 5.3 Reactions of weak and strong acids with strong and a weak base.

Thus, when a sample of a compound does not react with carbonate and bicarbonate but gets dissolved rapidly in sodium hydroxide, it must be a weak acid. However, 2,4,6–trinitrophenol undergoes reaction even with sodium bicarbonate, showing it to be a strong acid. It will definitely react with sodium carbonate and sodium hydroxide.

Table 5.2 pK_a values of acids and phenols

Compound	K_a	pK_a
Benzoic acid	6.5×10^{-5}	4.20
p–Toluic acid	4.33×10^{-5}	4.36
o–Nitrobenzoic acid	6.95×10^{-3}	2.16
m–Nitrobenzoic acid	3.4×10^{-4}	3.45
p–Nitrobenzoic acid	3.9×10^{-4}	3.44
p–Hydroxy benzoic acid	3.3×10^{-4}	4.48
o–Hydroxy benzoic acid (Salicylic acid)	1.07×10^{-3}	2.97
Phenol	1.3×10^{-10}	9.89
o–Nitrophenol	6.8×10^{-8}	7.17
p–Nitrophenol	7.0×10^{-8}	7.15
2,4,6–Trinitrophenol	4.25×10^{-1}	0.38
Carbonic acid	5.6×10^{-11}	10.25
Succinic acid	(*i*) 6.9×10^{-5}	4.19
	(*ii*) 2.47×10^{-6}	5.15

5.7 Ionisation Constant is a Measure of Strength of a Base

Base strength is the capacity of a base to combine with an acid. Thus, bases are nucleophiles and make electrons or negatively charged species available for interaction with positive charged entities. In a solution, base accepts a proton and forms a conjugate acid.

$$\underset{\text{a base}}{\text{B:(aq.)}} + \underset{\text{water (acid)}}{H_2O(l)} \rightleftharpoons \underset{\text{CA}}{BH^+(aq.)} + \underset{\text{CB}}{OH^-(aq.)},$$

Equilibrium constant for this reaction, $K_b = \frac{[\text{conjugated acid}]\ [\text{conjugated base}]}{[\text{base,B}^{\cdot\cdot}]} = \frac{[\text{salt}]\ [OH^-]}{[\text{base,B}^{\cdot\cdot}]}$

$$\log K_b = \log\,[\text{salt}] + \log\,[OH^-] - \log\,[\text{base}]$$

$$-\log K_b = pK_b$$

5.8 Expression of Base Strength, pK_b, in Terms of pK_a

Base character of a base is its ability to accept a **proton**. Basic character is related to the pK_b values or pK_a of the corresponding conjugate acid, as shown below. The stronger bases have the weaker conjugate acids and *vice versa.* In case of bases also (like in case of acids) lower value of pK_b means a strong base. However, in terms of its conjugate acid, the order is reversed. Details of how pK_a and pK_b are related is given in chapter 14. At this point, only the relation is given here.

$$pH = K_a + \log\frac{[salt]}{[acid]} \qquad pOH = K_b + \log\frac{[salt]}{[base]}$$

Also remember, pH + pOH = 14 and with in the approximation limits, $pK_a + pK_b = 14$

Base	I(–)	$Cl^{(-)}$	$HO^{(-)}$	$CH_3CO_2^{(-)}$	$RS^{(-)}$	$CN^{(-)}$	$RO^{(-)}$	$NH_2^{(-)}$	$CH_3^{(-)}$
pK_b	23	21	15.7	9.2	6	4.9	–2	–19	–26

weaker base .... stronger base

Acid	HI	HCl	$H_3O^{(+)}$	CH_3CO_2H	RSH	HCN	ROH	HN_3	CH_4
pK_a	–9	–7	–1.7	4.8	8	9.1	16	33	40

stronger acid .... weaker acid

increasing order of base strength →

5.9 Effect of Structure on Base Strength of Aliphatic Bases

Organic bases are nitrogen compounds, known as amines. These are bases, because of the availability of electron pair on nitrogen. Simplest one is ammonia. Amines are alkyl derivatives of ammonia, *e.g.*, CH_3–NH_2, methyl amine. Due to inductive effect, alkyl group pushes the electrons of C–N bond towards nitrogen and thus nitrogen becomes richer in electrons. In other words, amines becomes stronger bases. When two methyl groups are attached, $(CH_3)_2NH$, dimethyl amine is still a stronger base than methyl amine. Normally, one would consider, that when three methyl groups are attached to nitrogen–trimethyl amine, would be a stronger base than dimethyl amine, but it is not true. It is because, that **not** only the inductive effect, but steric effects also play their role. It is well documented from the pK_a or pK_b values of these amines.

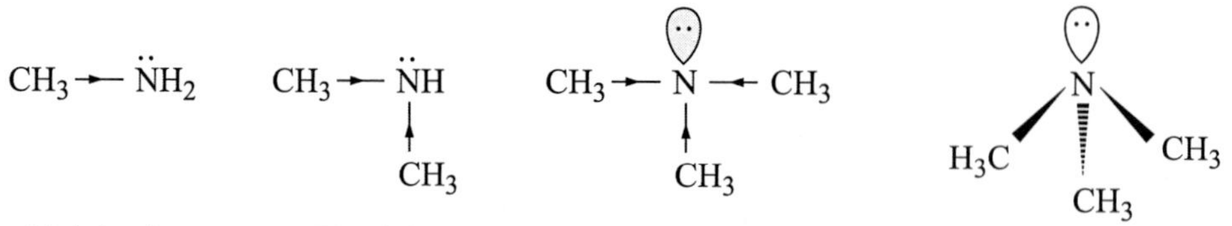

Methyl amine Dimethyl amine Trimethyl amine Lone pair in orbital

Table 5.3 Base Strength of Common Aliphatic Bases

Compound	Conjugate Acid	K_b	pK_a
NH_3	NH_4^+	1.79×10^{-5}	4.75
CH_3NH_2	$CH_3NH_3^+$	4.38×10^{4}	3.36
$(CH_3)_2NH$	$(CH_3)_2NH_2^+$	5.20×10^{-4}	3.28
$(CH_3)_3N$	$(CH_3)_3NH^+$	5.45×10^{-5}	4.26
$C_2H_5NH_2$	$C_2H_5NH_3^+$	5.60×10^{-4}	3.25
$(C_2H_5)_2NH$	$(C_2H_5)_2NH_2^+$	9.60×10^{-4}	3.02

5.10 Effect of Structure on Base Strength of Aromatic Bases

Like other amines, aniline also carries lone pair of electrons on nitrogen, thus behaves as a base. The strength of such bases shall depend upon the availability of lone pair on nitrogen. Unlike aliphatic amines, aromatic amines are weaker bases.

Reason

1. The factor that reduces the base character of $-NH_2$ group in aniline is resonance of the lone pair of nitrogen with the pi electrons of benzene ring. Benzene acts as a sink of electrons and movement of electrons away from nitrogen makes it a weaker base.

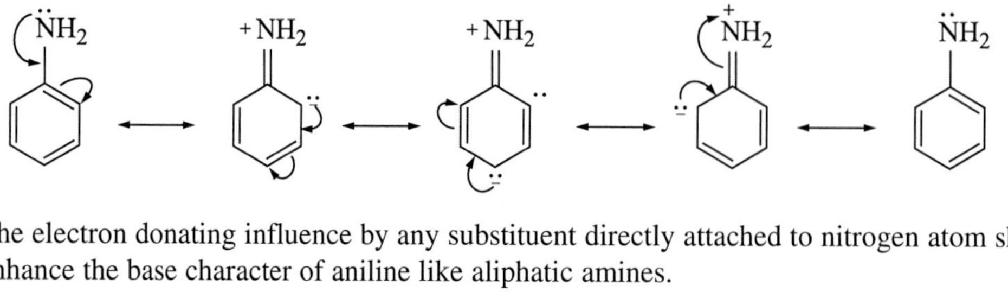

2. The electron donating influence by any substituent directly attached to nitrogen atom shall also enhance the base character of aniline like aliphatic amines.

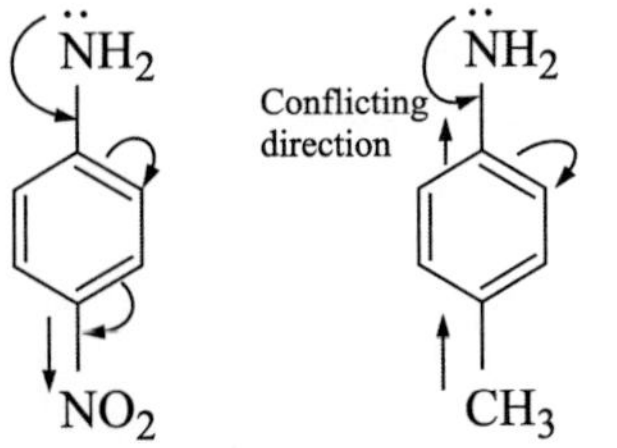

Electron withdrawing nitro group make aniline a weaker base unlike acids. Similarly, electron donating methyl group makes aniline a stronger base, unlike acids.

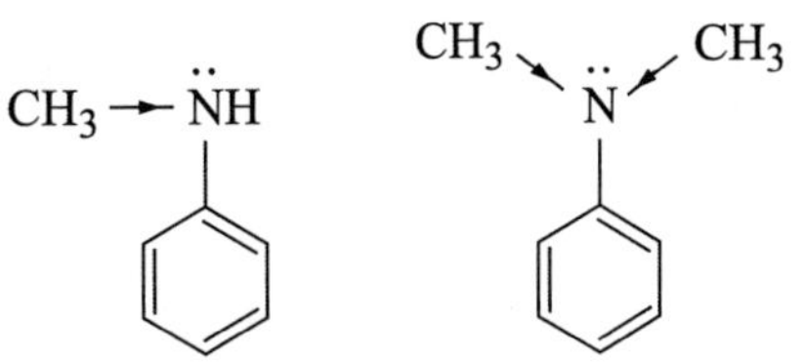

Electron donation effect of alkyl group on nitrogen makes it stronger base like aliphatic amines

3. When we attach an electron withdrawing group on the benzene ring, it further reduces the base character of anilines. When, there is an electron donating influence by any substitutent on the benzene ring, it enhances its base character, as shown by their pK_a values.

Table 5.4 Base Strength of Common Aromatic Bases

Compound	Conjugate Acid	K_a	pK_a
$C_6H_5NH_2$	$C_6H_5NH_3+$	3.82×10^{-10}	9.42
$C_6H_5NHCH_3$	$C_6H_5N^+H_2CH_3$	5.0×10^{-10}	9.30
$C_6H_5N(CH_3)_2$	$C_6H_5N^+H(CH_3)_2$	1.15×10^{-9}	8.94
$C_6H_5N(C_2H_5)_2$	$C_6H_5N^+H(C_2H_5)_2$	3.65×10^{-8}	7.44
o–$CH_3C_6H_4NH_2$	*o*–$CH_3C_6H_4N^+H_3$	2.47×10^{-10}	9.61
m–$CH_3C_6H_4NH_2$	*m*–$CH_3C_6H_4N^+H_3$	4.92×10^{-10}	9.31
p–$CH_3C_6H_4NH_2$	*p*–$CH_3C_6H_4N^+H_3$	1.18×10^{-9}	8.92
o–$NO_2C_6H_4NH_2$	*o*–$NO_2C_6H_4N^+H_3$	6.0×10^{-4}	3.22
m–$NO_2C_6H_4NH_2$	*m*–$NO_2C_6H_4N^+H_3$	2.7×10^{-5}	4.57

5.11 Effect of Structure on Base Strength of Heterocyclic Bases

Saturated heterocyclic bases, like piperidine, tetrahydropyrollidine etc, behave more or less like aliphatic amines. The effect of substituent is also very similar in both cases. In heterocyclic bases, electronic effect of substituents produces its effect within the ring. Some examples of heterocyclic bases are given ahead.

Saturated heterocyclic bases

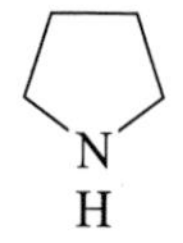

Tetrahydropyrollidine

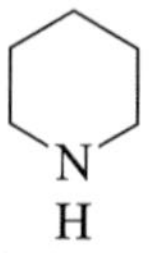

Piperidine

Hetereocyclic bases

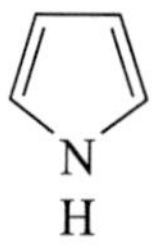

Pyrrolidine

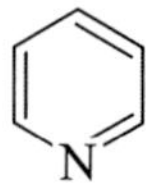

Pyridine

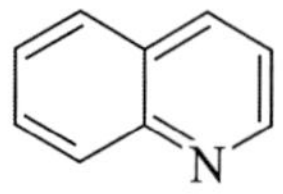

Quinoline

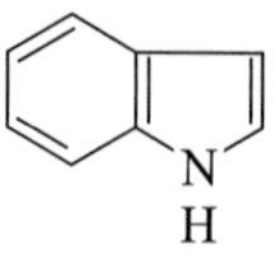

5.12 Solvent Effects

Solvation of nucleophilic anions markedly influence their reactivity. Polar, protic solvents such as water and alcohols solvate anions by hydrogen bonding. These solvated species are more stable and less reactive than the unsolvated "naked" anions. Polar, aprotic solvents such as DMSO (dimethyl sulfoxide), DMF (dimethylformamide) and acetonitrile do not solvate anions that well as methanol does.

5.13 Nucleophilicity and Basicity

Both **nucleophile or a base** work as donors of electrons to an electron deficient entity. A nucleophile attacks an electrophile, where as the base attacks a proton. Nucloephilicity is the capacity of a nucleophile to combine with an electrophile and basicity is capacity of a base to combine with porton. Thus the base character of a negatively charged entity is considered, when it combines with a **proton** and **nucleophilicity** is the term used when it combines with **any electron deficient entity, such as carbonium ion.**

In a nucleophilic substitution reaction of halide such as, CH_3–Br, the rate depends upon the nucleophilicity of the nucleophile. It shall attack on the carbonium ion that gets formed in the course of reaction by slow ionisation of C-Br bond or via a one step reaction.

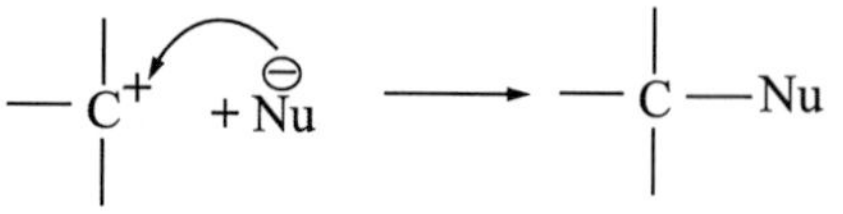

strength of this nucleophile is referred to as nucleophilicity

$$CH_3 - Br + \ddot{Nu} \xrightarrow{CH_3OH} CH_3 - Nu + \ddot{Br}$$

In nucleophilic reactions, it is the kinetic control of the electron pair donation to as electrophile *e.g.*, carbonium ion as given above, that takes its precedence over thermodynanic control.

Basicity is the capacity of a negatively charged group of a neutral entity having high electron density to combine with a proton to form a neutral molecule or plus charged enity.

$$\overset{\ominus}{B} + H{-}A \rightleftharpoons B-H + \overset{\ominus}{A}$$

Base character in this case is referred to as basicity

In base reactions, the equilibrium (thermodynamic control) of electron donation to a proton to form BH, takes precedence over the kinetic control.

A negatively charged species is always a stronger nucleophile than its conjugate acid, provided it also has an unshared pair of electrons. For example, ^{-}OH is a stronger nucleophile than water and so is $^{-}NH_2$ a stronger nucleophile than NH_3.

5.13.1 Nucleophilicity Order

Nucleophilicity of nucleophile, Nu:$^{(-)}$ varies depending upon its capacity to give or share electrons. A highly electronegative atom with a negative charge has low nucleophilicity, but a stronger base character. The order of nucleophilicity of some negatively charged species is :

$$R_3C^{(-)} > CH_3^{(-)} > {}^{(-)}NR_2 > {}^{(-)}NH_2 > RO^{(-)} > {}^{(-)}OH > ArO^{(-)} > NH_3 > F^{(-)} > ClO_4^{(-)}$$

In the halide series, the nucleophilicity has the following order:

$$F^{-} < Cl^{(-)} < Br^{(-)} < I^{(-)}$$

The further discussion of the topic is beyond the scope of this chapter. More of this discussion , you can find under Ionisation chapter in physical chemistry section.

Points to Remember

- There are three theories for defining an acid and a base, the most versatile is the **Lewis Theory**. Other two, namely, **Arrhenius and Brønsted theories** were defined earlier, but could not completely or universally define the acid and a base. Thus, according to **Lewis** theory, an acid is any molecule or substance that can accept a pair of electrons.
- The strength of an acid or a base is quantitatively defined by ionisation constant, K_a or its minus log, that is pK_a values. Similarly, we define the strength of a base by K_b or pK_b constants.
- Instead of remembering the pK_a data for all the acids and bases, alternatively approach of analysis of the structure in relation to the acid character is equally convenient.

- **–COOH** group is acidic group, because of the fact that carbonyl group in carboxyl serve as a sink of electron and –O–H bond breaking becomes easy. Similarly, the phenolic –OH is acidic because the benzene ring attached to –OH withdraws electrons, thus makes –O–H breaking easy.
- Attaching an additional **electron withdrawing group**, like chloro-, bromo- or nitro or a carbonyl, makes –COOH **more acidic** and attaching electron donating groups, like alkyl-, amino- or hydroxyl or alkoxyl, makes these groups less acidic, as explained by different structures in the text. Similar is the case with benzoic acid series.
- **–NH_2** group or its substituted group in open chain or cyclic structure act as bases, because the electron pair available on nitrogen makes it basic or attracts an electron deficient entity.
- Attaching an additional electron withdrawing group, like chloro-, bromo- or nitro or a carbonyl, to the carbon chain makes nitrogen base, **less basic** and attaching electron donating groups, like alkyl-, amino- or hydroxyl or alkoxyl, makes these groups more basic, as explained by different structures in the text. The attachement of alkyl group, (one or two) to nitrogen makes it more basic. Similar is the case when such groups get attached to benzene ring.
- Effect of structure and solvent have also been discussed in the text.

Exercises

1. Identify the Arrhenius, Brønsted–Lowry acids and bases in the following reactions. Explain the conjugate acids and conjugate bases.

(*a*) $CH_3—C(=O)—CH_2—COOC_2H_5 + {}^{-}\!:\!\ddot{O}H \longrightarrow CH_3—C(—{}^{-}O—Na^+) = CH—COOC_2H_5 + H_2O$

(*b*) $CH_3—C(=O)—CH_3 + :\!\ddot{O}H \longrightarrow CH_3—C(—{}^{-}\!:\!\ddot{O}—Na^+) = CH_2 + H_2O$

(*c*) $H—Cl + H_2\ddot{O}: \longrightarrow H_3O:^+ + Cl^+$

(*d*) $CH_3—COOH + H_2\ddot{O}: \longrightarrow CH_3—COO^- + H_3O:^+$

(e) 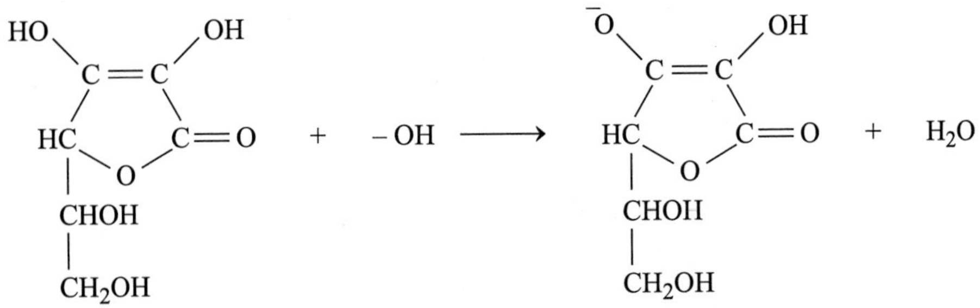

(f) $\ddot{N}H_3 + H_2\ddot{O}: \longrightarrow NH_4^+ + -:\ddot{O}H$

2. Identify the Lewis Acids and Lewis bases in the following reactions

(a) $AlCl_3 + 3\,H_2O \longrightarrow Al(OH)_3 + 3\,HCl$

(b) $CH_3^+ + Cl^- \longrightarrow CH_3 + Cl$

(c) $CH_2=CH_2 + Cl-Cl + NaBr \longrightarrow Cl-CH_2-CH_2^- + Br- \longrightarrow Cl-CH_2-CH_2-Br$

(d) $CH_3-CH = CH_2-H^+ \longrightarrow CH_3-{}^+CH-CH_3$

3. Arrange the following acids in increasing order of their acid strength.

CH_3COOH, $HCOOH$, $O_2N\text{-}CH_2COOH$, $ClCH_2COOH$

4. Explain, which one of the following isomers is a stronger acid at its first ionisation level.

Maleic acid Fumaric acid

5. Arrange the following acids in the decreasing order of their acid strength. Give reasons.

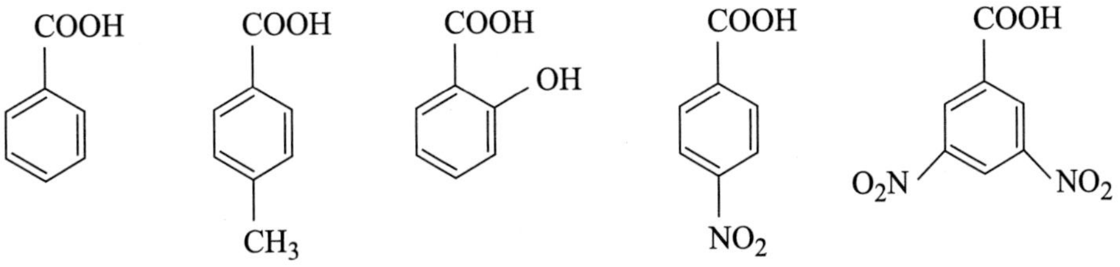

6. Will Cl–CH_2 CH_2 COOH acid be more acidic than CH_3CH COOH ?

7. Arrange the following phenols in order of their decreasing acidic strength. Justify your answer.

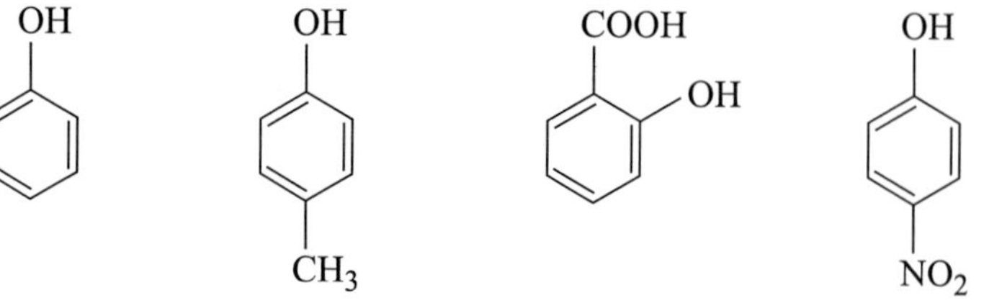

8. Given the pK_a value of acetic acid as, 1.8×10^{-5}, calculate the pH of the solution, if the concentration of acetic acid in water solution is 0.5 mol L^{-1} .

9. Given the pK_a value of Phenol to be10, calculate the degree of ionisation of solution of phenol in water, with concentration equal to 0.1 mol L^{-1}.

10. The pK_a value of ethyl alcohol is 17, its conjugate base shall be a strong base. Do you expect alpha hydrogen in a ketone like ethyl methyl ketone or ethyl acetate to undergo proton loss with this base. What should be the pK_a value requirement for the acid proton to undergo this type of reaction.

11. In acetoacetic ester, there are three types of hydrogens marked 1, 2 and 3. Write these three types of hydrogens in order of their increasing K_a value.

$$C\underline{H}_3^1 - \underset{\underset{O}{\|}}{C} - C\underline{H}_2^2 - COO\underline{H}^3$$

12. Write the following bases in order of their increasing base strength.

$$CH_3NH_2, (CH_3)_2NH, (CH_3)_3N \quad \text{and} \quad CH_3CH_2CH_2CH_2NH_2$$

13. How do you account for the fact that aliphatic amines like methyl amine is a stronger base than the aromatic amine, like aniline?

$$CH_3NH_2 \quad \text{and} \quad C_6H_5NH_2$$

14. Benzoic acid gets ionised in accordance to the following equation. If the starting concentration of benzoic acid is 0.10 mol L^{-1}. After one hour, analysis shows that 50% of benzoic acid has got ionised. Calculate the value pK_a. Also show that under such conditions, pK_a = –log(H_3O^+) = pH.

$$C_6H_5\text{—}COOH + H_2O \longrightarrow C_6H_5\text{—}COO^- + H_3O^+$$

15. What is the effect of nitro group on the acid character of p–nitrobenzoic acid and base character of *p*–nitroaniline? Compare both with the corresponding ortho isomers.

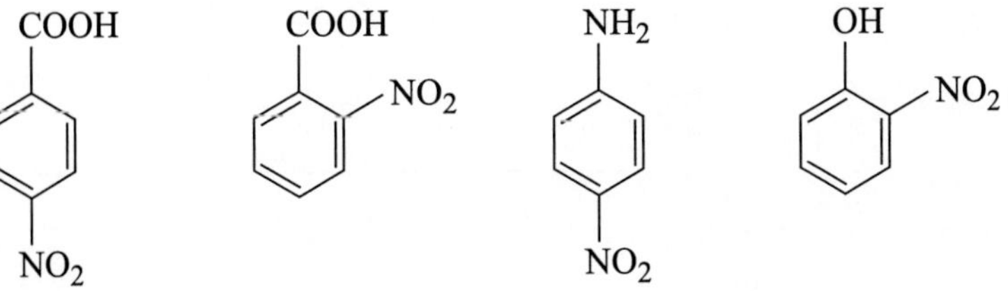

❑❑❑

6

Aromaticity

Aroma means, a characteristic flavour of organic compounds obtainable from plants or flowers or spices as natural sources. Aromatherapy is a form of alternative medicine that uses volatile plant materials, known as essential oils, and other aromatic compounds for the purpose of altering a person's mood, cognitive function or health. The effectiveness of aromatherapy is yet to be scientifically proven, however some evidence exists that essential oils may have therapeutic potential.

IN THIS CHAPTER

The concept of aromatherapy was first mooted by a small number of European scientists and doctors, in about 1907. In 1937, the word first appeared in print Huiles Essentielles, Hormones Bégétales by René-Maurice Gattefossé, chemist. An English version was published in 1993. In 1910, Gattefossé burned a hand very badly in a laboratory explosion. The hand developed gas gangrene, which he successfully, and intentionally, treated with lavender oil. Essential oils have been in use in India, both for bringing happiness from flavours and for therapy purposes.

CHAPTER 6

Aromaticity

6.1 Definition of Aromaticity

The word aromatic has entirely different meaning to common man than to a chemist. It is the aroma or pleasant odour that these aromatic compounds possess. For example, in south India, there are lot of plants containing aromatic compounds that are planted and irrigated for their fruits (spices) and oils for their fragrance and medicinal properties. To mention a few, we have the botanical and common names as below.

1. *Cinnamomum verum* (Cinnamom)
2. *Ginger Elettaria* (Cardamom)
3. *Syzygium aromaticum* (Clove)
4. *Eucalyptus melliodora* (Eucalyptus)
5. *Jasminum nudiflorum* (Jasmine)
6. *Lavandula angustifolia* (Lavender)

For a chemist, aromaticity is a special characteristic of cyclic fully unsaturated compounds containing benzene or heterocyclic ring. Due to aromatic character the double bonds of cyclic system do not behave like double bonds of aliphatic compounds. The double bonds of benzene are more inert (that is not reactive) compared to those of ethylene in respect of usual addition reactions of double bond. For example, addition of chlorine to ethylene takes place at room temperature, but benzene requires high energy like irradiation with high power bulb or sun rays. There are many reasons to account for unreactivity of double bonds in aromatic system that have been proposed by chemists over a long period and are discussed here.

6.2 Usual Properties of Benzene

For a chemist the aromaticity is associated with the unusual properties that aromatic cyclic compounds possess. Unusual reactions that benzene shows are:

1. **Carbon, hydrogen analysis** and molecular weight determination shows **molecular formula of benzene to be C_6H_6** (not C_6H_{12}, C_6H_{10} or C_6H_8 like aliphatic compounds).

2. Benzene **adds on three hydrogen molecules**, showing the presence of three double bonds. The molecular formula of the alkatriene would be C_6H_6. That is a saturated alkane, C_6H_{12} minus 6H for three double bonds, $C_6H_{12} - 6H = C_6H_6$.
3. Benzene **undergoes ozonolysis with three ozone** molecules to form a triozonide, again showing the presence of three double bonds.
4. Benzene undergoes substitution reaction with halogens in presence of halogen carrier like iron under milder condition of temperature, *e.g.*, ~ 100°C. Substitution reaction is not a reaction of alkenes at the double bond.

To account for six carbon atoms and three double bonds, it took many years for chemists to write the structure of benzene. We know today that benzene as suggested by Kekule, is written in two structures corresponding to resonance hybrids or canonical forms as given below.

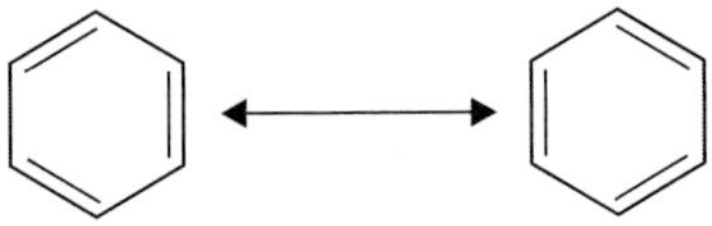

Kekule structure of benzene

Above two electronic structures for benzene that are structurally and energetically equivalent are written. No single structure provides an accurate or even an adequate representation of the true benzene molecule. Even these two structures do not account for ***unusual properties*** that the three double bonds of benzene possess.

6.3 Unusual Properties of Benzene

1. Chemical Stability of Double Bond Towards

A. Potassium Permanganate Reaction: When we treat ethylene with $KMnO_4 / H_2SO_4$ at room temperature, it gets oxidised to ethylene glycol. Similar treatment with benzene has no reaction. Even on boiling no reaction is observed. Only strong oxidising agents like vanadium pentoxide at high temperature can oxidise benzene with rupture of benzene ring in low yield and to maleic anhydride.

B. Addition Reaction with Halogens: Halogens like chlorine adds on to ethylene double bond at room temperature or any moderate temperature to give addition product. However, in case of benzene, exposure to sun light or a strong bulb light is necessary for addition reactions to takes place to form benzene hexachloride (BHC). When chlorine and benzene are treated at a moderate temperature in presence of halogen carriers like iron, instead of addition reaction to double bond, substitution of hydrogen takes place to form chlorobenzene.

All the **usual and unusual reactions** are summarised in the chart below:

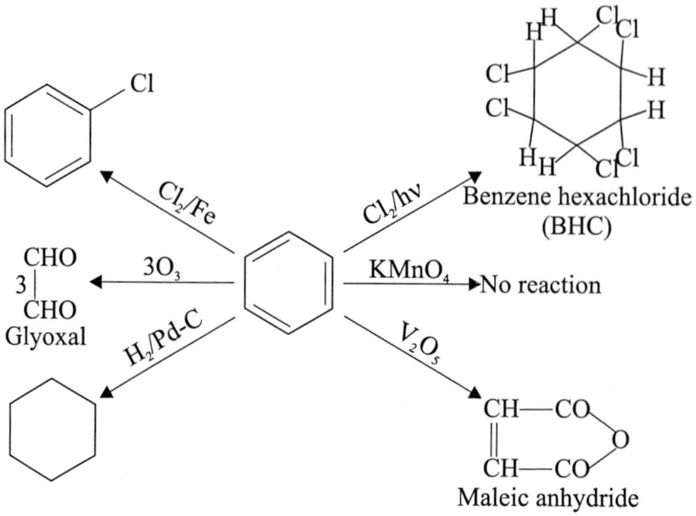

The usual reactions of ethylene double bond are given in the following chart

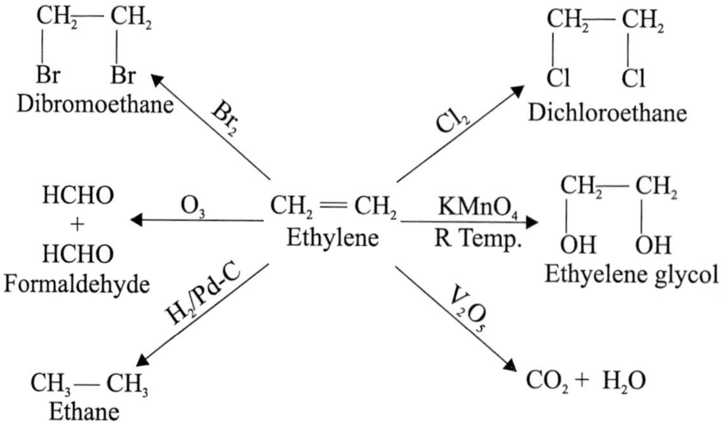

2. Unusual Physical Properties

A. C–C and C=C bond lengths of benzene are same. In benzene, the six-membered ring is a perfect hexagon. All carbon-carbon bonds have an identical bond length of 140 pm, (1.40 Å). The cyclohexatriene structures would be expected to show double bond length as 134 pm (1.34 Å) and that of single bond as 155 pm (1.55 Å). The shortening of C–C bond length is unusual for benzene. It is considered to be because of jumping of double bonds in a benzene ring, as shown by two ring structures. An alternative representation for benzene (circle within a hexagon) emphasizes the pi-electron delocalization in this molecule and it has the advantage of being a single diagram. The electron delocalization described by resonance enhances the stability of the molecule.

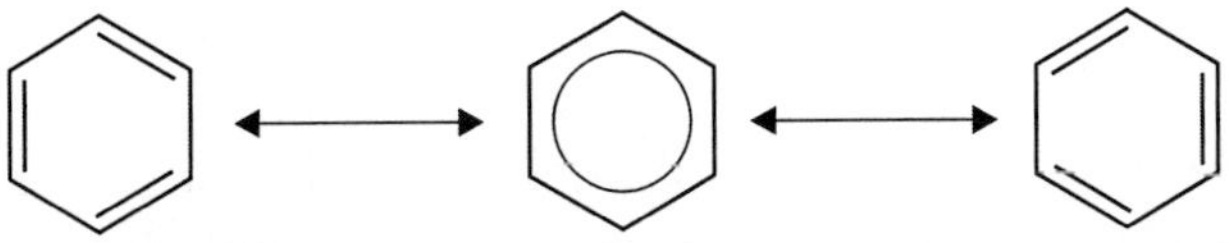

Expected single bond length = 155nm | Each bond length 140nm | Expected double bond length = 134nm

3. Unusual Thermodynamical Values

A. Heat of Hydrogenation: It has been observed experimentally that when cyclohexene is hydrogenated, it releases 28.6 kcal/mol. When cyclohexadiene is hydrogenated catalytically to give cyclohexane, it releases 55.6 kcal/ mol as compared to the expected value of $28.6 \times 2 = 56.2$ kcal/mol. Similarly, 1,3,5-cyclohexatriene (so called benzene) on hydrogenation should give out $28.6 \times 3 = 85.8$ kcal/mol. However, experimentally in case of benzene, we get only 49.8 kcal/mol. That is 36 kcal/mol less heat is obtained. This **heats of hydrogenation** reflect the relative thermodynamic stability of benzene compared to 1,3,5-cyclohexatriene. **Benzene is an extra-ordinary more stable by 36 kcal/mol than expected**. This sort of stability enhancement is now accepted as a general characteristic of all aromatic compounds.

Compound	Experimental value	Expected value
Cyclohexene	28.6 kcal/mol	28.6 kcal/mol
Cyclohex1,3-diene	55.6 kcal/mol	55.6 kcal/mol
Cyclohex-1,3,5-triene	....	85.8 kcal/mol
Benzene	49.8 kcal/mol	85.8 kcal/mol

B. Heat of Formation: Heat of formation of a compound is the amount of heat evolved or absorbed in its formation from elements in their naturally occurring state. Naturally occurring elements have been *considered* to have zero internal energy. For example, formation of methane takes place from one atom of carbon and four atoms of hydrogen. The total amount of heat absorbed will be contained in the four bonds of methane molecule. It can also be called internal energy of the methane molecule. It can also be calculated from the summation of bond energy of all the bonds of a molecule.

For benzene, that has three C–C bonds, six C–H bonds, and three C=C bonds.

The summation of bond energies of each as shown shall give heat of formation of benzene.

$\Delta_f H$ benzene = $3\,\Delta H\ C—C + 3\,\Delta H\ C{=}C + 6\,\Delta H\ C—H$

$\Delta_f H$ benzene = $3 \times 347 + 3 \times 614 + 6 \times 413 = -5361$ kJ/mol

$\Delta_f H$ benzene = -5361 kJ/mol

Experimentally $\Delta_f H$ benzene value can also be determined from acetylene values.

$$2C + H_2 \longrightarrow C_2H_2 \text{ (acetylene)} \quad \Delta H = 1A \text{ kJ/mol}$$

$$3\ C_2H_2 \longrightarrow C_6H_6 \text{ (benzene)} \quad \Delta H = 3A \text{ kJ/mol}$$

$$\Delta_f H_{benzene} = -4995 \text{ kJ/mol}$$

C. Resonance Energy: When a given structure of a compound is not able to show all its properties, we write down different possible structure for the compound with out changing the linkage of atoms but differing in position of electrons. All the possible structure so formed are called canonical structures. All the canonical structures together is the structure of a compound. Benzene has five canonical structures as shown below. All these structures together give the properties of benzene.

Resonance energy appears to have been lost in interconversion of different canonical forms. The total energy expected from a single structure of a molecule is more by the amount spent in forming different canonical structures. The difference between the expected and experimental value of heat of hydrogenation or of heat of formation is equal to 36 kcal/mol and is called **resonance energy of benzene**.

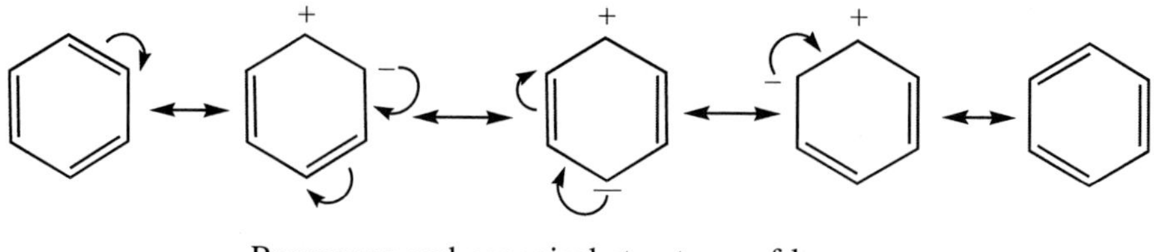

Resonance and canonical structures of benzene

Resonating energy = 36 kcal./mol

Summary of Unusual Reactions/Properties of Benzene

Unusual Chemical Reactions are shown by the following

1. Stability of double bond towards:
   A. Oxidation of double bond with $KMnO_4$
   B. Halogen addition reactions
2. Unusual physical properties
   A. C–C and C=C bond lengths are same in benzene as compared to two values in ethylene single and double bond
3. Unusual thermodynamical values
   A. Heat of hydrogenation
   B. Heat of formation
   C. Resonance energy

6.4 Cause of Aromaticity

To answer why the pi bond electrons of benzenoid and aromatic compounds are not reactive, one need go in to examinatin of the state of these pi electrons. On examination we notice that all six carbons have one each p-orbital electron unhybridised that are parallel to each other. When they form molecular orbitals, a new picture emerges as shown. It helps us to understand the stability of aromatic compounds.

1. Molecular Orbital description

Molecular Orbital of benzene provide us a more satisfying and more general treatment of "aromaticity". Benzene has a planar hexagonal structure in which all the carbon atoms are sp^2 hybridized. All the carbon-carbon bonds are of equal length. The cyclic array of six *p*-orbitals (one on each carbon) overlap to generate six molecular orbitals, three bonding and three antibonding. The plus and minus signs shown in the diagram do not represent electrostatic charge, but refer to phase signs in the equations that describe these orbitals (in the diagram the phases are also color coded). When the phases correspond, the orbitals overlap to generate a common region of like phase, with those orbitals having the greatest overlap (*e.g.*, π_1) being lowest in energy. The remaining carbon valence electrons then occupy these molecular orbitals in pairs, resulting in a fully occupied (6 electrons) set of bonding molecular orbitals. It is this completely filled set of bonding orbitals, or **closed shell**, that gives the benzene ring its thermodynamic and chemical stability, just as a filled valence shell octet confers stability on the inert gases.

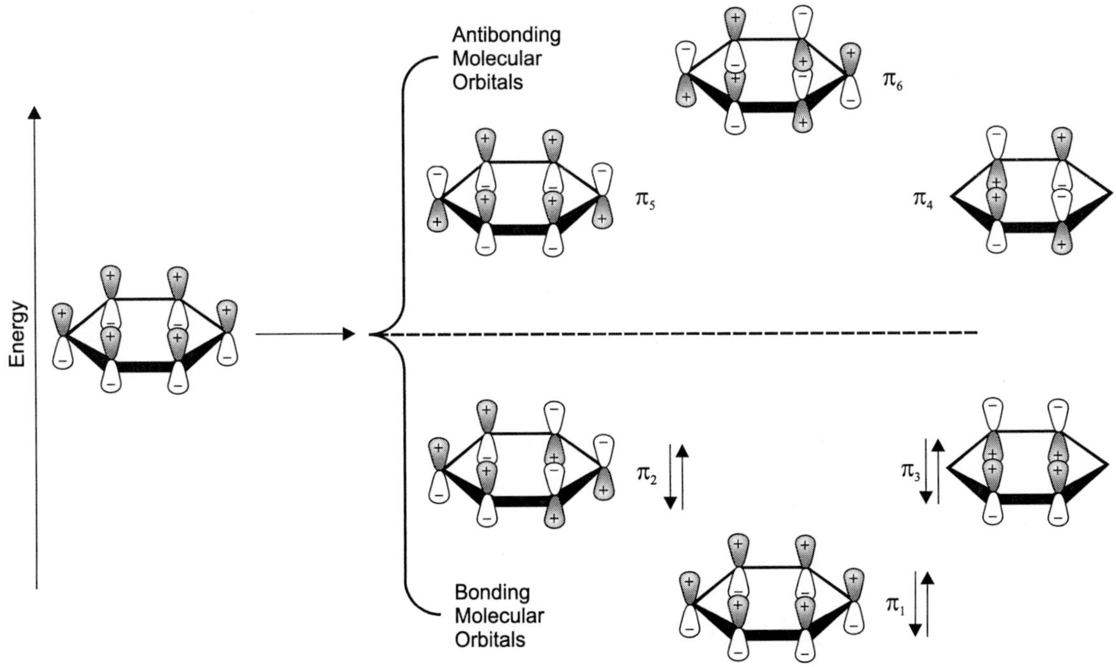

Fig. 6.1 The Molecular Orbitals of Benzene.

2. Identical Six Carbons and Six Hydrogens of Benzene

It is shown by PMR and C^{13} magnetic resonance studies. It has been found from the NMR that six protons in benzene show a single peak at δ 7.22 and in C^{13} magnetic resonance studies, it is found that all six carbons give a single peak at δ 128.7. These spectroscopic studies do not reveal any difference in the C–C single bond or C=C double bond.

6.5 How do we Determine Whether a given Compound is Aromatic or Not?

Criteria of aromaticity

1. **One way is to determine** aromaticity of a compound is to determine the values of physical **or** thermal constants **or** perform the chemical reactions to show stability, as discussed above. These parameters shall determine the aromaticity of a given compound.
2. **Another way to find out aromaticity of** any given cyclic compound is to apply Huckel's Rule. **According to Huckel's rule**, any cyclic unsaturated compound that has $(4n + 2)$ π-electrons (where n is an integer or zero) will be aromatic. Benzene, the aromatic compound has 6π -electrons ($n = 1$). It is aromatic. It shows all the properties, such as stability, physical constants etc.as discussed above. Likewise for other values of n, see the table given.

Table 6.1

Value of n	No. of p electrons	Examples
$n = 1$	6π	Benzene, thiophene, pyridine, pyrrole cyclopentadiene anion,
$n = 2$	10π	Naphthlene,
$n = 3$	14π	Anthracene, Phenanthrene
$n = 4$	18π	Annulenes

3. The third way is to build the Dreiding Model of the molecule and look for planarity of the ring system. **A planar (or near planar) cycle of *sp*² hybridized atoms,** the *p*-orbitals of which are oriented parallel to each other shall overlap to generate an array of *pi*-molecular orbitals.

6.6 Structural Types of Aromatic Compounds

Aromatic compounds are of three types, the benzenoids and heterocyclic. The third category obtains aromaticity after becoming an ion. The third category is cyclic compounds which show aromaticity are formed from nonaromatic compounds by losing or gaining one or two electrons to form cyclic ionic structures. These then observe Huckel's Rule and thus become aromatic compounds. To understand the validity of Huckel's Rule in these types of compounds, let us examine some of the examples of aromatic systems.

(*a*) **Homonuclear aromatic compounds,** those cyclic compounds which have only one type of element in the ring. For example, in benzene ring there are carbon atoms only in the ring.

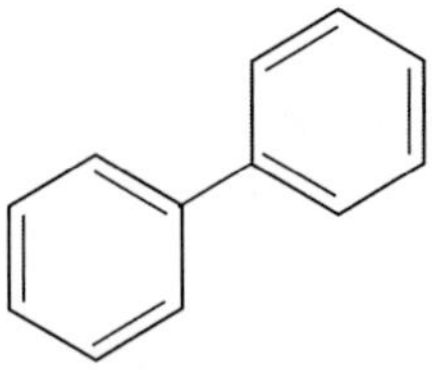

6 *pi* + 6 *pi*-electron
Diphenyl

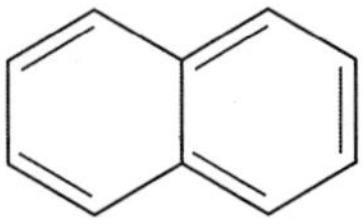

10 *pi*-electrons
Naphthalene

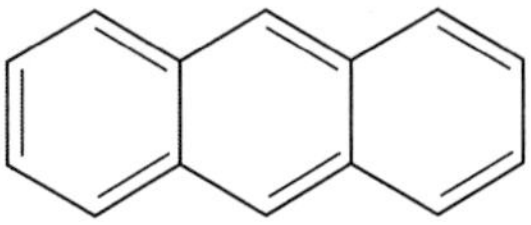

14 *pi*-electrons
Anthracene

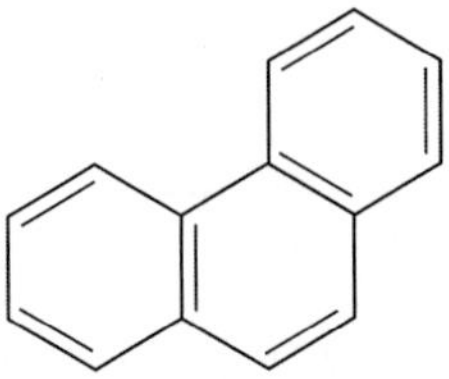

14 *pi*-electrons
Phenanthrene

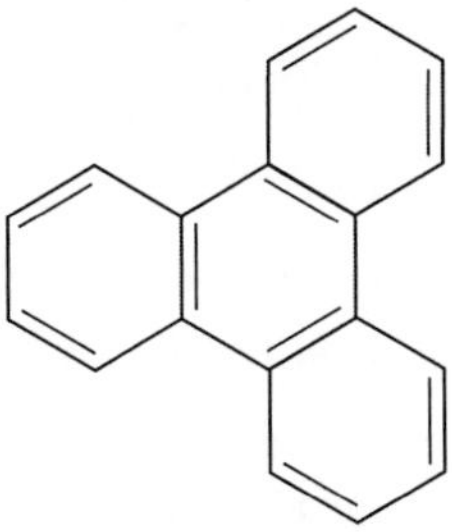

18 *pi*-electrons
Benzophenanthrene

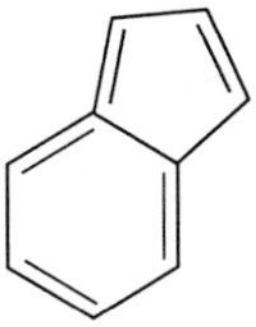

Biclo–[0,3,5]–
pentadecene

(*b*) **Heteronuclear aromatic compounds**, those compounds which have other hetero atom in the ring, besides carbon atoms. For example, furan, thiophene, pyrrole, pyridine etc.

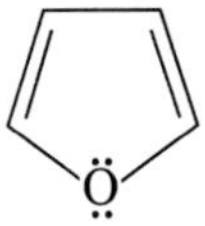

Furan

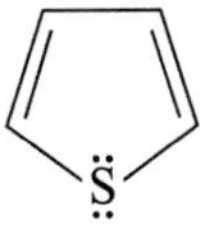

Pyridine

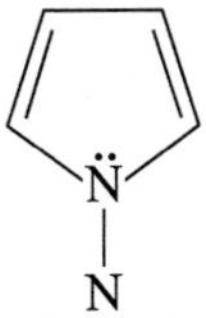

Pyrrole

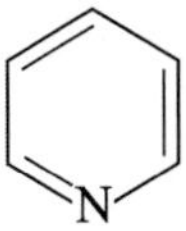

Pyridine

All are 6 *pi*-electron system

(*c*) **Aromatic Ions,** there are certain heterocyclic compounds that do not correspond to the Huckel's rule, However, after losing or gaining some electrons,that is cyclic carbanions and carbocations may also show aromatic stabilization.

The three-membered ring cation has 2 *pi*-electrons and is surprisingly stable, considering its ring strain. Cyclopentadiene is as acidic as ethanol, reflecting the stability of its 6 *pi*-electron conjugate base. Salts of cycloheptatrienyl cation (tropylium ion) are stable in water solution, again reflecting the stability of this 6 *pi*-electron cation.

Some examples are:

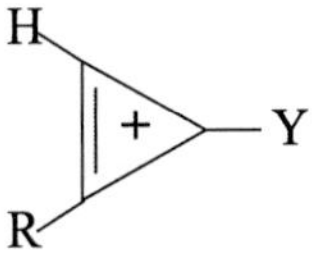

Cyclopropenyl
cation,

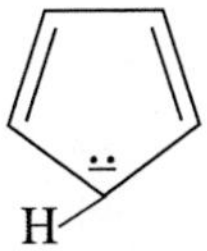

Cycclopentadienyl
anion,

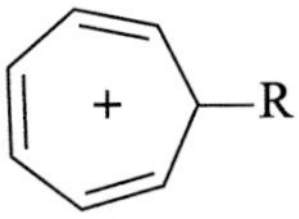

Cycloheptatriene
cation,

B. Cyclic Non-aromatic Compounds

The following compounds do not possess $(4n + 2)\pi$ electrons, so are not aromatic. It is more appropriate to define another class called anti-aromatic compounds

Examples:

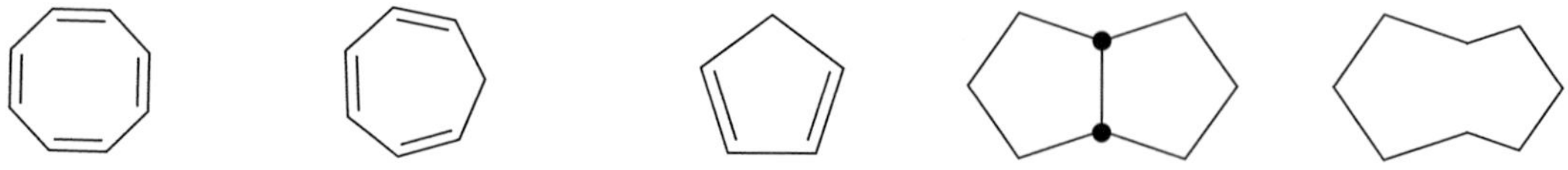

Cyclotetraoctene Cycloheptatriene Cyclopentadiene Bicyclo-octane Cyclononane

C. Anti-aromatic Compounds

There are compound that possess only $4n$ *pi*-electrons, they are termed as antiaromatic compounds. The unexpected instability of $4n$ *pi*-electron annulenes has been termed "antiaromaticity".

Conjugated ring systems having $4n$ *pi*-electrons (*e.g.*, 4, 8, 12 etc. electrons) not only fail to show any aromatic properties, but appear to be less stable and more reactive than expected.

As noted above, 1,3,5,7-cyclooctatetraene is non-planar and adopts a tub-shaped conformation.

(*a*) The compound is readily prepared, and undergoes addition reactions typical of alkenes.

(*b*) Catalytic hydrogenation of this tetraene produces cyclooctane.

(*c*) Planar bridged annulenes having $4n$ *pi*-electrons have proven to be relatively unstable.

Examples of 8 and 12 *pi*-electron systems are shown below, together with a similar 10 *pi*-electron aromatic compound.

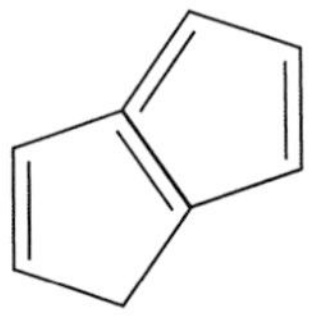

Pentalene, planar
very unstable
8 *pi*-electrons

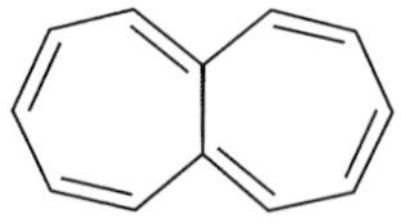

Heptalene, planar
very unstable
12 *pi*-electrons

D. Allotropes of Carbon

Element carbon has three allotropic modifications. Everyone is aware of **diamond** and **graphite**. The third form is called Buckminster fullerene. If we extend the structure of corannulene by adding similar cycles of five benzene rings, the curvature of the resulting molecule should increase, and eventually close into a sphere of carbon atoms. The archtypical compound of this kind (C_{60}) has been named buckminster fullerene because of its resemblance to the geodesic structures created by Buckminster Fuller. It is a member of a family of similar carbon structures that are called **fullerenes**. These materials represent a third class of carbon allotropes. Alternating views of the C_{60} fullerene structure are shown on the right, together with a soccer ball-like representation of the 12 five and 20 six-membered rings composing its surface.

Furthermore, fullerenes are very important in pharmaceutical applications. Wudl synthesised a derivative of a fullerene which is soluble in water. It seems to inhibit the activity of HIV and could be used in a medicine for the treatment of AIDS. Nanotechnology is thus a domain of central importance, at the forefront of chemistry, physics, engineering and materials science. Nanotechnology find ways of manipulating atoms and molecules in an effort to construct new materials and different molecular stratagems.

In conclusion, carbon is truly important for chemistry and for life. It is essential that we continue to study the chemistry of carbon to increase understanding of ourselves and our environment and to improve the quality of life.

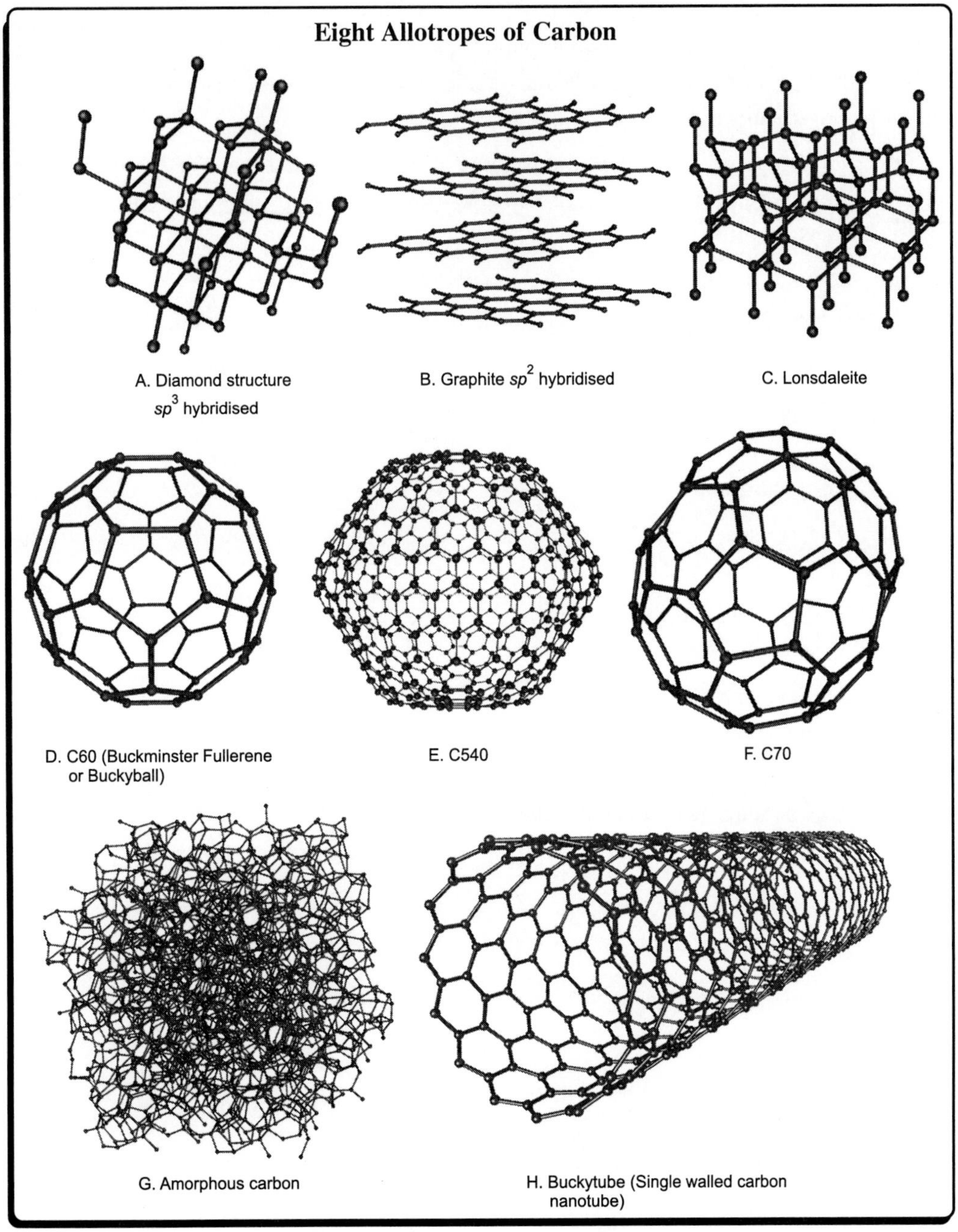

Fig. 6.2 Allotropic forms of carbon

Points to Remember

- Aromaticity is characteristic property of certain organic compounds that possess characteristic odour as well as special structural features.
- Structurally, aromaticity of compounds has been attributed to the presence of cyclic structure with alternate single and double bonds. It constitutes a 6π -electron system in six-membered ring system.
- In a general way, all cyclic unsaturated systems are aromatic if these obey the Huckel's rule. It states that presence of $(4n + 2)\pi$ electrons in a cyclic structures makes it aromatic.
- Aromatic compounds possess characteristic properties, such as,
  (*i*) Double bonds are not reactive double bonds like those of alkenes
  (*ii*) Instead of addition reactions, characteristic of double bond, aromatic compounds undergo substitution reactions more easily.
  (*iii*) For all addition reactions, aromatic compounds require strong reaction conditions.
  (*iv*) Presence of number of double bonds can be demonstrated by ozonolysis, as in case of any other olefin.
- Physical as well as thermodynamic properties of aromatic compounds are unusual. For example,
  (*i*) Bond length of all carbon-carbon bonds, irrespective of single or double bond, are same.
  (*ii*) Experimental values of heats of hydrogenation of aromatic compounds differ markedly from the calculated values.
  (*iii*) All aromatic compounds show resonance energy as a characteristic of stability of such compounds.
- Besides benzenoids and heterocyclics, certain fully unsaturated alicyclic compounds can also show aromaticity, provided they possess $(4n + 2)\pi$ electron system.
- Compounds that possess $4n$ *pi*-electrons are known as anti-aromatic systems.
- Carbon exists in many allotropic forms. The most common ones are graphite, diamond and fullerenes. Recently more allotropic forms of carbon have been discovered. They differ at their molecular level structures.

Exercises

1. How do you define aromaticity of a compound?
2. What are the characteristic reactions of an aromatic double bond and that of ethylene double bond?
3. What is the difference in the reaction of bromine with toluene and propene under different set of reaction conditions.
4. How do you experimentally account for two structrures for *o*-xylene?

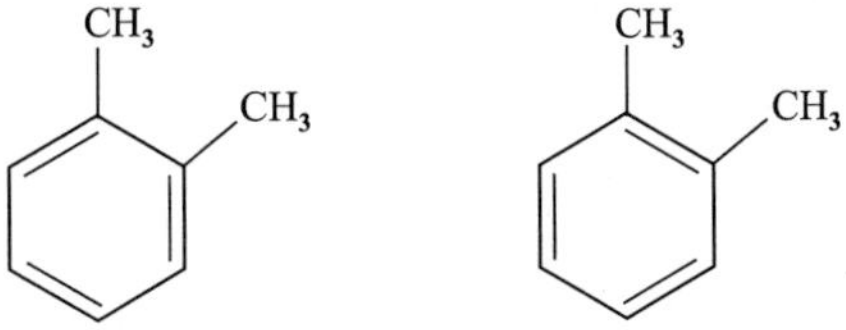

(Hint: Use ozonolysis reaction and analyse the products)

5. Comment on 'aromatic compounds possess flavour due to easy volatility of vapours, still many compounds have high boiling points'.
6. Is borazine, an aromatic compound? Draw the electronic structure for each bond and justify your answer.

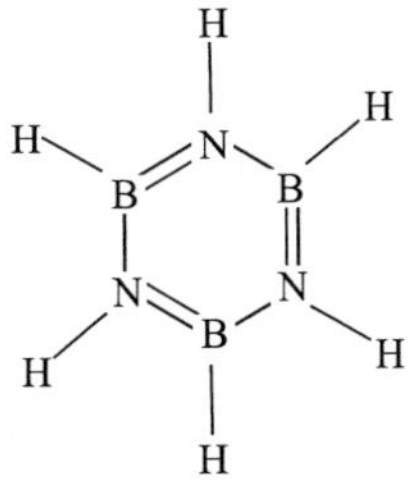

7. Which of the following compounds would you regard as aromatic? Apply Huckel's rule and justify the answers.

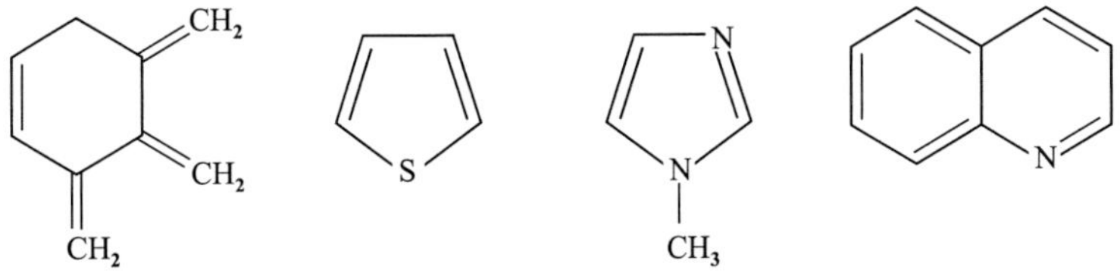

Emil Fischer

7

Stereochemistry

Herman Emil Fischer, (9 October 1852 – 15 July 1919) was a German chemist and recipient of the Nobel Prize for Chemistry in 1902, for his work on carbohydrates and purines. He was considered the greatest organic chemist of his time. In 1875, von Baeyer was asked to succeed Liebig at the University of Munich and Fischer went there with him to become an assistant in organic chemistry. In 1885, he was asked to become Professor of Chemistry at the University of Würzburg and here he remained until 1892, when he was asked to succeed A. W. Hofmann in the Chair of Chemistry at the University of Berlin. Here he remained until his death in 1919.

Research: His discovery that phenylhydrazine reacts with carbohydrates to form osazones enabled him to elucidate the stereochemistry of sugars. In 1890, by epimerization between gluconic and mannonic acids, he established the stereochemical and isomeric nature of the sugars, and between 1891 and 1894 he established the stereochemical configuration of all the known sugars and exactly foretold the possible isomers, by an ingenious application of the theory of the asymmetrical carbon atom of van't Hoff and Le Bel, published in 1874. Fischer studied the active principles of tea, coffee and cocoa, namely, caffeine and theobromine, and established the constitution of a series of compounds in this field, eventually synthesizing them. First to

IN THIS CHAPTER

synthesize adenine and guanine, xanthine, uric acid, he made the first synthetic nucleotides. This monumental work on the sugars, carried out between 1884 and 1894, was extended by other work, the most important being his studies of the glucosides. Fischer also studied the enzymes and the chemical substances in the lichens which he found during his frequent holidays in the Black Forest, and also substances used in tanning and, during the final years of his life, the fats. In 1890, he also proposed a "Lock and Key Model" to visualize the substrate and enzyme interaction. Though, later studies did not support this model in all enzymatic reactions.

Fischer is noted for his work on sugars among other work, the organic synthesis of (+) glucose and purines. Among many other contributions was his work on protein structure and synthesis, the "lock-and-key" model for enzyme reactions, structure of *tri*-arylmethane dyes, and the design of laboratory hoods and safety equipment.

Between 1899 and 1908, Fischer made his great contributions to knowledge of the proteins. He sought effective analytical methods of separating and identifying the individual amino acids, discovering a new type, the cyclic amino acids: proline and oxyproline. He also studied the synthesis of proteins by obtaining the various amino acids in an optically active form in order to unite them. He was able to establish the type of bond that would connect them together in chains, namely, the peptide bond, and by means of this he obtained the dipeptides and later the tripeptides and polypeptides.

CHAPTER 7

Stereochemistry

Introduction

Stereochemistry deals with molecular architecture in three dimension and the behaviour of molecules. Stereochemistry has been existing since the dawn of life on this planet, but became to be understood only after fourth quarter of 19th century. Biot in 1815, discovered that certain organic compounds, like turpentine, sugar and camphor possessed the special property of rotating the plane of plane polarised light. The property existed, whether the compound was in solution or in gas form. Thus, this special property of rotation of plane of polarised light was attributed to certain characteristics of molecular structure. This property of rotation of plane of plane polarised is known as optical activity. L. Pasteur (1848) by crystallisation of tartaric acid, obtained two forms of tartaric acid by hand picking method, these two forms behaved differently towards plane polarised light, but had identical chemical properties. These two crystalline forms, possessed crystalline faces, that were mirror images to each other. The existence of two structures for a single compound became obvious, but it could not be explained till, **van't Hoff and LeBel** in 1874, proposed the idea of tetrahedral valency of carbon that is three dimensional bonds of carbon. A newer isomerism called stereoisomerism emerged.

Stereochemistry and Its Importance

Research work over a century, has revealed a number of facts about three dimensional molecular architecture and its impact on numerous aspects of chemical and biological reactions. Stereochemistry has gained importance because of its relation with specificity of chemical reactions in laboratory as well as in life processes, such as, enzyme reactions, specific protein synthesis involving 3D structure of DNA, RNA etc. Effect of drugs on human beings is another area, that is specific in respect of the stereochemistry of drug molecules. Drugs, like, ***penicillin***, ***amoxycillin***, ***tetracyclines*** etc., have been used for a long time, but the relation of their stereospecificity and drug effect have been revealed only recently.

Cetrizine, a drug molecule has been used as a mixture of two stereoisomers. It has been discovered that only leavo-rotatory (rotation of plane of polarised light towards the reducing value) form of cetrizine is active as *antihistamine* and other is toxic. It has changed the drug marketing strategy and a new form of drug called ***leavocet*** has come in to market. ***Pseudoephedrine hydrochloride***, is another drug molecule, taken by a large number of persons suffering from *allergic rhinitis.*

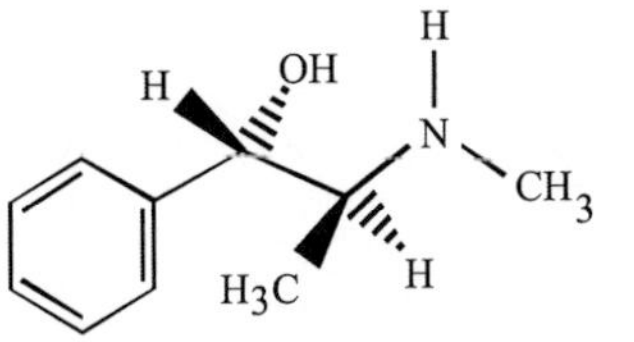

Pseudo-ephedrine

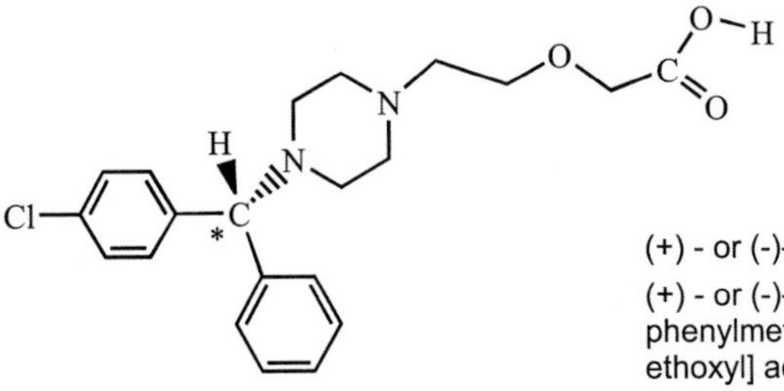

(+) - or (-)- Cetrizine

(+) - or (-)- [2-[(4-chlorophenyl) phenylmethyl]- 1 piperazinyl] ethoxyl] acetic acid

Only one isomer with specific stereochemistry is used as a decongestant for the relief of nasal congestion due to *allergic rhinitis*. Thus, research and knowledge of stereochemistry of molecules has become very useful in all aspects of human life.

7.1 Architecture and Projection of Molecules

Consider, for example, the optically active lactic acid sample placed in a flask or a bottle and learn different steps to represent its architecture. The first step needed is to know molecular formula of the compound. It is done by C, H and O percentage analysis and molecular weight determinations and subsequent steps as given here.

1. The **molecular formula** of lactic acid is $\mathbf{C_3H_6O_3}$.
2. The **constitutional structure** of lactic acid is

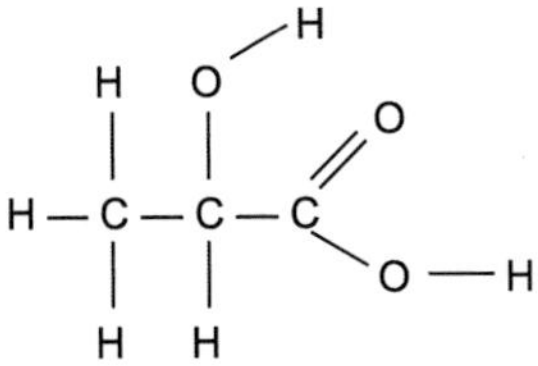

3. The **bond structure** of lactic acid is

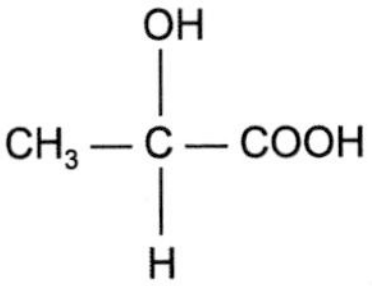

4. Following are the ball and stick models of isomeric forms of of lactic acid.

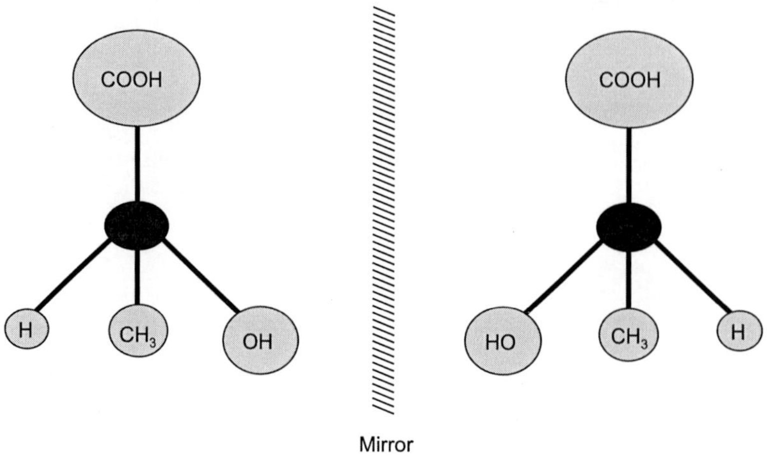

5. Wedge diagrams of two forms of lactic acid are:

6. Fischer's Projections of lactic acid are

Fig.7.1 Different presentation of structures and models.

A projection is a kind of shadow of a molecule in three dimensional (3D) on a plane. Different methods for drawing projectons are in use and the most commonly used is the Fischer's projection. It is a crossed projection, as shown above. However, every crossed representation is not a Fischer Projection. Certain prescribed rules are used in drawing Fischer's projection as well as for other projections in organic chemistry.

7.1.1 Fischer's Projection

Fischer's projection is a cross representation of the four bonds of carbon. Certain conditions are followed for holding the model and casting the projection. These are given as under the Fig.7.2.

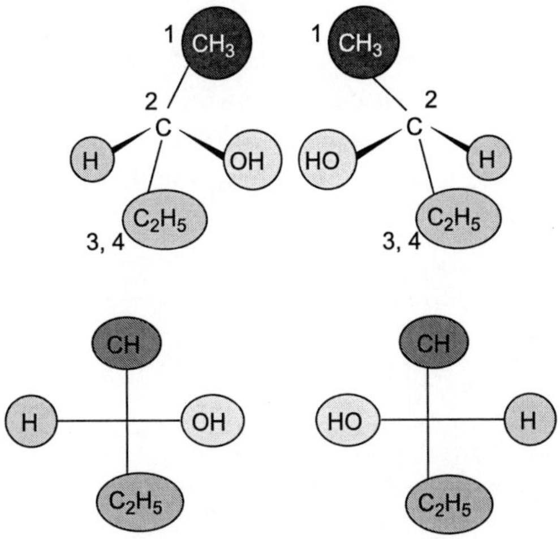

Fig. 7.2 Fischer's Projections of two isomers of butan-2-ol.

1. Hold the molecule in such a way that the central carbon lies in the plane of projection.
2. The groups on carbon held at the **top and bottom** must be inclined **behind the plane**. (These are represented by dotted lines in the wedge diagram.)
3. The groups on carbon at the **right and left** must be inclined **above the plane**. (These are represented by thick lines in the wedge diagram.)
4. The most oxidised carbon (or number 1 group as per the IUPAC nomenclature) be kept at the top.
5. The shadow of the model of molecule, then gives Fischer's projection. It is like pressing the model against the plane of projection.

7.1.2 Newman Projection

Newman's Projections are obtained by looking at the molecule/model exactly along the bond joining the two carbon atoms. The three bonds at the front carbon (from the viewer's side) are represented by full lines enclosed in a circle at an angle of 120^0. The rear carbon bonds are represented by half lines outside the circle. Rotation of groups at the rear carbon with reference to the front carbon atom and *vice-versa*, gives different isomers as represented in Fig. 7.3.

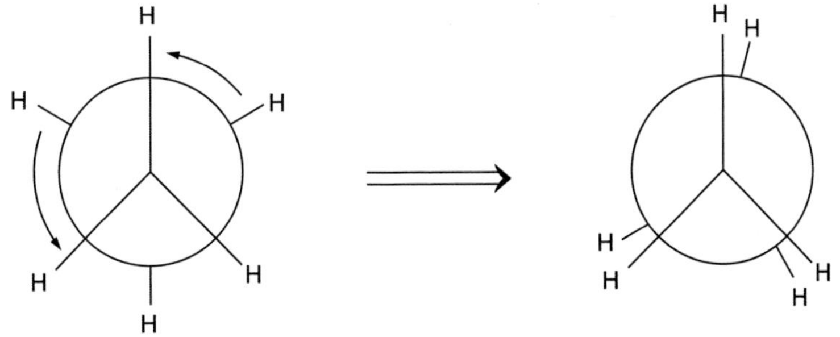

Fig.7.3 Newman's projection for two Conformations of ethane.

7.1.3 Sawhorse Projection

Sawhorse projections are obtained by looking at the molecule, at a small angle to the bond joining the two carbon atoms. Three bonds at each carbon are drawn in the shape 'Y or inverted Y'. The three groups at one carbon can be rotated with reference to the three groups at the other carbon. Such projections for butane are shown in Fig.7.4. In this projection the rear groups are rotated through 180° to get the right hand side projection

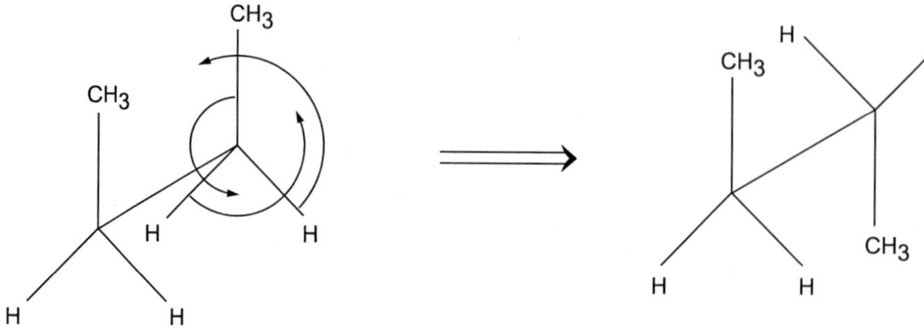

Fig. 7.4 Four Sawhorse projections of butane conformations.

7.1.4 Flying Wedge Projection

It is the three dimensional (3D) projection of a model or the molecule on plane of the paper. The bonds are represented as dotted lines for bonds behind the plane, thick lines are for the bonds above the plane and regular line are for bonds in the plane. Thick or dotted lines can be narrow at one end depicting the distance from the viewer. Just as when we stand in between the railway track, two tracks appear to merge at distance.

Fig. 7.5 Wedge diagram of Lactic Acid and butan-3-ol.

7.1.5 Conversion of Wedge into Fischer's Projection

Wedge diagram is a three dimensional representation using strong, dotted and thin lines as bonds. In a tetrahedral model, either three of the four bonds are out of plane and one bond in plane or we can also have two bonds out of plane and two bonds in plane. To convert these into Fischer's projection, we need to bring the horizontal out of plane bonds in front and automatically, the top and bottoms bonds shall go behind the plane.

The Fig.7.6 gives the method used for changing the wedge into Fischer projection . It is followed by four examples to practice in writing the Fischer's projection.

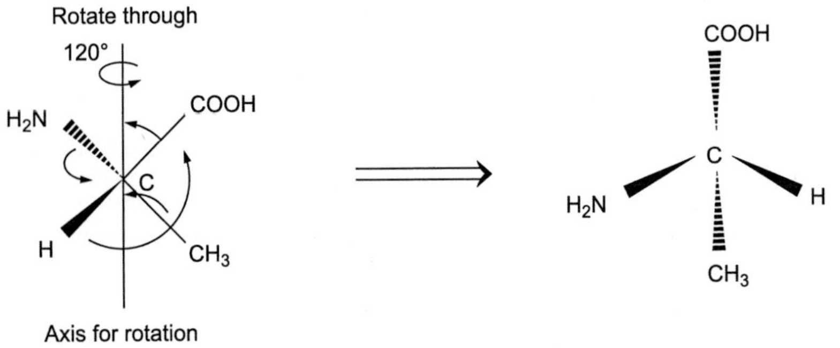

Fig. 7.6 Process of rotation of wedge diagram around the axis to get Fischer's Projection.

Example 7.1: Given below are some examples of wedge projections, namely: (I) 2-Methyl butan-1-ol, (II) 2-Amino propanoic acid (Alanine), (III) Glyceraldehyde and (IV) Bromochlorofluoromethane. On the extreme right are the Fischer's Projections of these molecules.

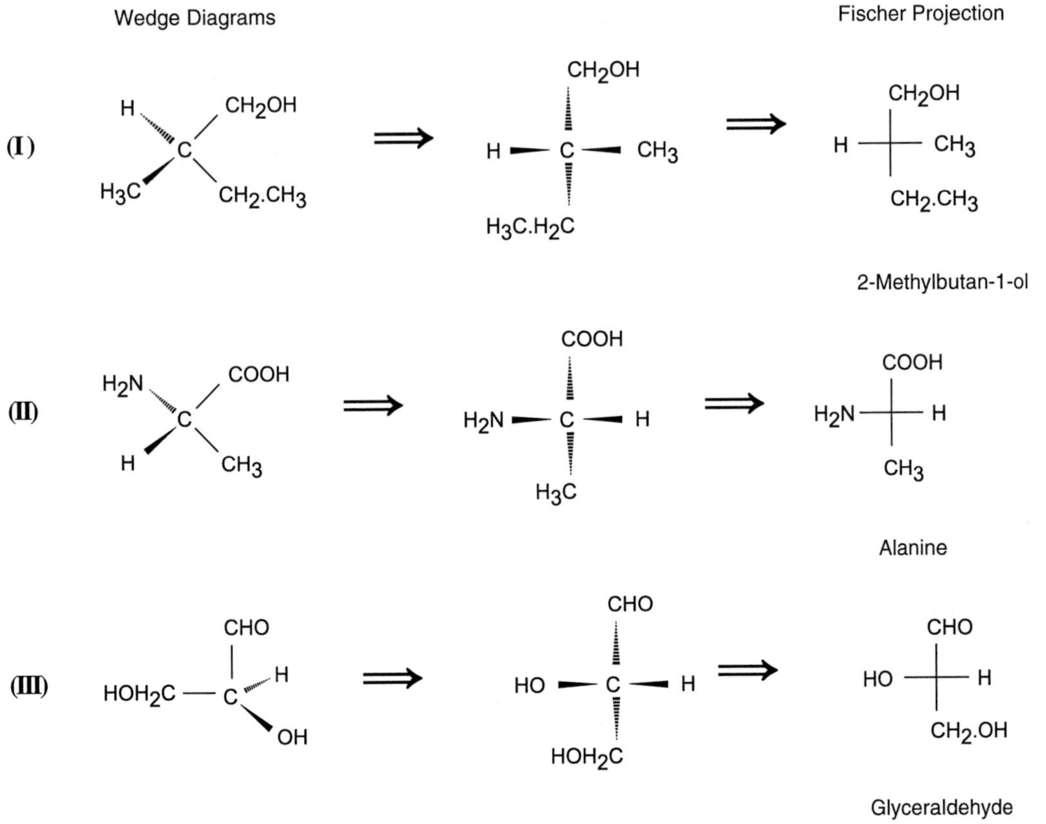

(IV)

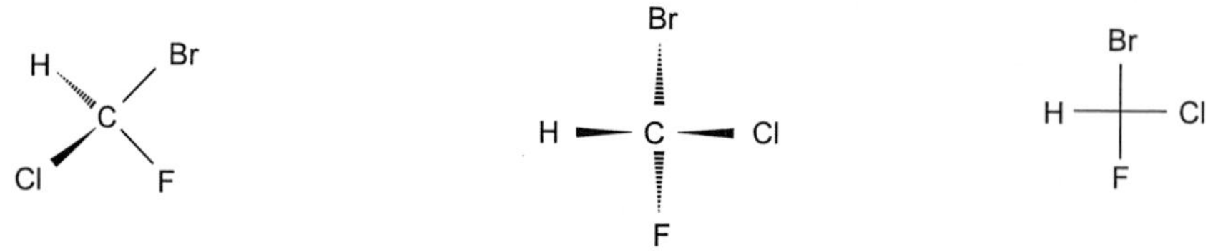

Fig. 7.7 A conversion of Wedge into Fischer projections of various molecules.

Sawhorse and Newman's projections can also be converted into Fischer's projection. You have to be very sure of the angle of vision of the molecule. Given below is one example of 2-bromo-2-nitro propanoic acid to represent Sawhorse, Newman and Fischer's projections.

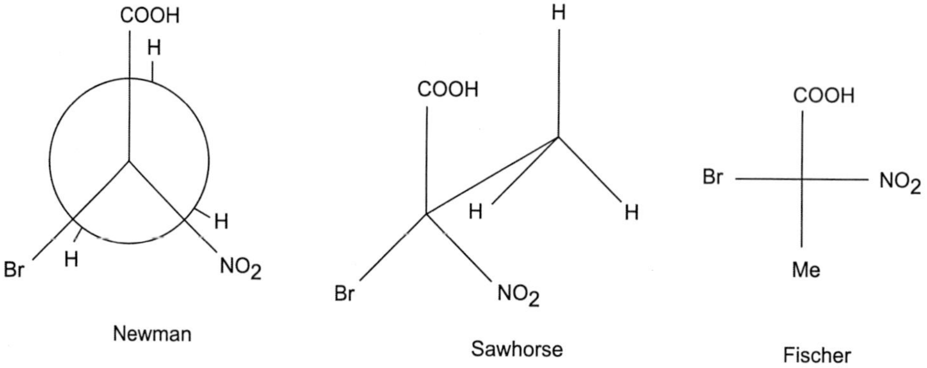

Fig. 7.8 Three types of projections of 2-bromo-2-nitro propanoic acid.

7.2 Rotational or Conformational Isomerism

A conformation is one of the infinite number of momentary arrangements that are obtained by rotation of groups at a carbon, reference to groups at the adjacent carbon joined through a single bond. Thus, rotational or conformational isomerism is the property of compounds due to which they exhibit the formation of different conformers.

Another definition: A conformation is one or more different arrangements of atoms or groups of a molecule in three dimensions that are interconvertible merely by rotation around single bond. These different arrangements are called conformations of the molecule.

Characteristics of Conformers

1. Different conformers are not separable or isolable easily, because they rapidly interconvert with slight change of temperature.
2. The conformers can be separated by lowering temperature to near freezing point of a compound. This lowering of temperature is to stop or freeze molecular movements or rotation of groups around single bond.

3. Only those conformers shall be abundant at a given temperature that have lower energy and are more stable. Conformations are visualised by rotation around C – C single bond and is explained by different structures of ethane.

Sir Derek H R Barton, (1918-1998) *recevied Nobel Prize for his work on Conformational Analysis of Steroids, 1969. He was elected Fellow of the Royal Society. He was interested in Basic Organic Chemistry. Barton Reaction, Synthesis of tetracycline, penicillins and cephlosporins etc. are important contributions. He had a great insight of the Molecular Architecture in relation to the chemical reactions. He worked at Harvard, USA; Imperial College, UK and many other places. One of the authors of this book, S. C. Datta carried out the synthesis of tetracycline in his laboratories at Imperial College of Science, Technology and Medicine, London.*

7.2.1 Conformations of Ethane, Their Energy Diagram and Stability

Consider ethane molecule in the Sawhorse projection as shown in Fig.7.9. Examine the position obtained by rotation of rear methyl group hydrogens relative to the front methyl group hydrogens around C – C bond in clockwise direction through 60° angle variation.

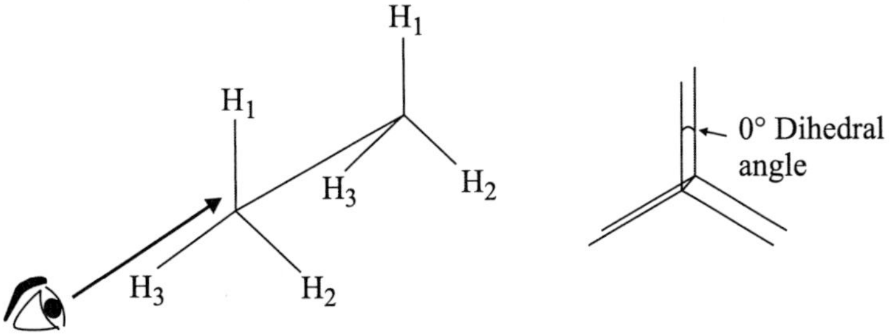

at 0° dihedral angle,
Eclipsed

The angle between the front and rear bonds on carbon-carbon bond is called dihedral angle. Six different positions are possible. Analysis of the figures shows that there are infact only two different positions and others are identical. The two differentiated conformations are called ***eclipsed*** at dihedral angles of 0°, 120° and 240° and ***staggered*** at dihedral angles of 60°, 180° and 300°.

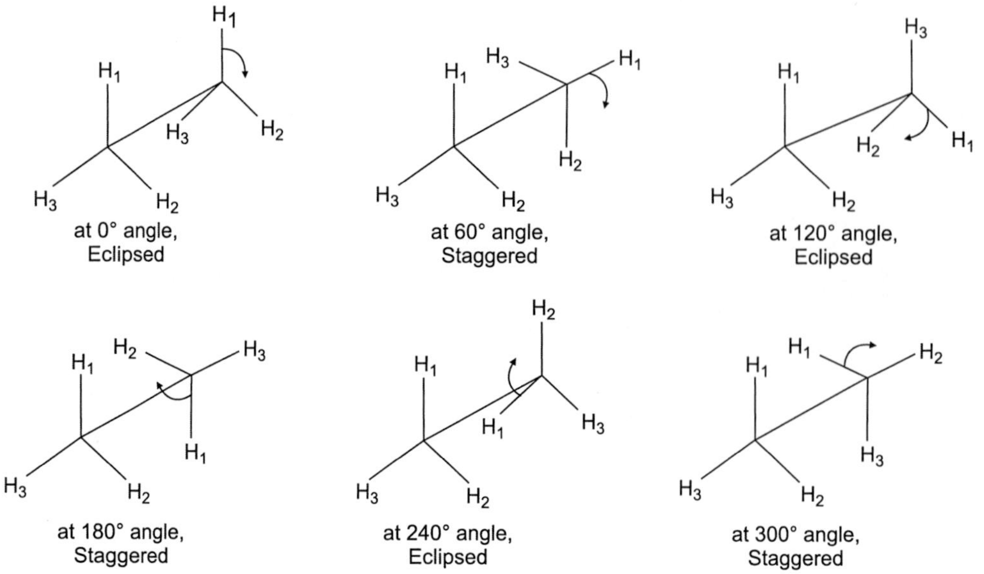

Fig. 7.9 Eclipsed and staggered conformation of ethane at different dihedral angles shown as Sawhorse projections.

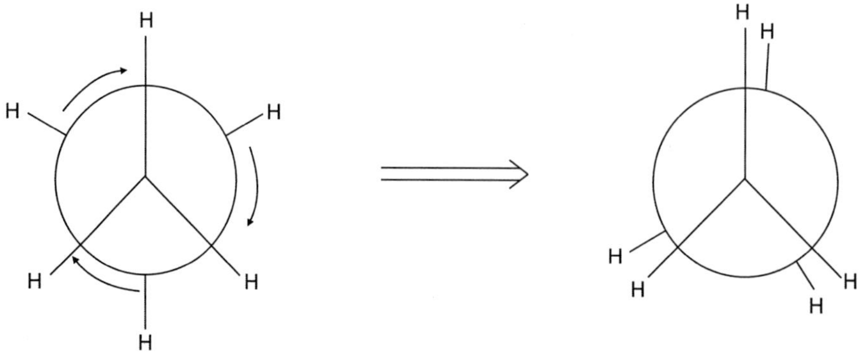

Fig. 7.10 Staggered and Eclipsed Newman's Projections of ethane.

Let us examine energy diagram and analyse the stability of all these structures of ethane. What we find that at zero angle between front methyl hydrogens and rear methyl hydrogens, the distance between the hydrogens is smallest, thus repulsive forces are maximum; same is the case at 120°, 240° angle. However at angles of 60°, 180° and 300° the distance between front and rear hydrogens is maximum. Thus, they have least interaction, consequently the energy of the molecule is minimum. That is what we observed as the maxima and minima in the energy curve as shown in Fig. 7.11.

Draw the conformation of propane like ethane and shows that it also has two conformations. However, the energy barrier is different.

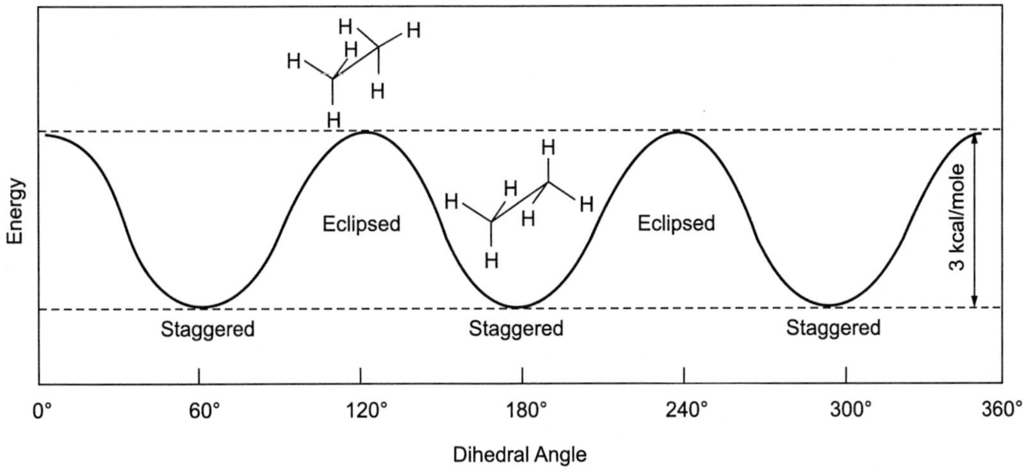

Fig. 7.11 Energy diagram of ethane conformations.

7.2.2 Conformations of Butane, Their Energy Diagram and Stability

Consider the butane $(\overset{1}{CH_3}-\overset{2}{CH_2}-\overset{3}{CH_2}-\overset{4}{CH_3})$ molecule. It has two methyl groups in place of two hydrogens in ethane molecule. At 0° dihedral angle, methyl groups on two carbons are very close to each other, interactions are maximum. Two hydrogens on C2 also are close to two hydrogens on C3 carbon. It gives a high energy state to the molecule. It is called the **eclipsed conformation**.

When the rear methyl and two hydrogens rotate around C2–C3 single bond to dihedral 60°, methyl group lies in between two hydrogens and so are two hydrogens. It is called **gauche form**. van der Waals forces play their role in imparting energy to the molecule. Because of larger distance between groups, the molecule has lower energy. Further rotation of rear groups relative to the groups on the front carbon atom, gives a state at 120° rotation with higher energy. It is less stable than the gauche form by 3.5 kcal/mole. It is also called skew or gauche conformation.

At 180° rotation of rear groups, two methyl groups are at the maximum distance possible between them. It is called the **staggered conformation**. It is the most stable state of butane molecule. This variation of energy is repeated at 240° and 300° angle of rotation. Butane thus, has four conformers compared to two conformers in case of ethane and propane. The total energy barrier of an eclipsed and staggered conformation is 4.5 to 6.1 kcal per mole. A graph depicting energy changes with change in dihedral angle is given in Fig. 7.12.

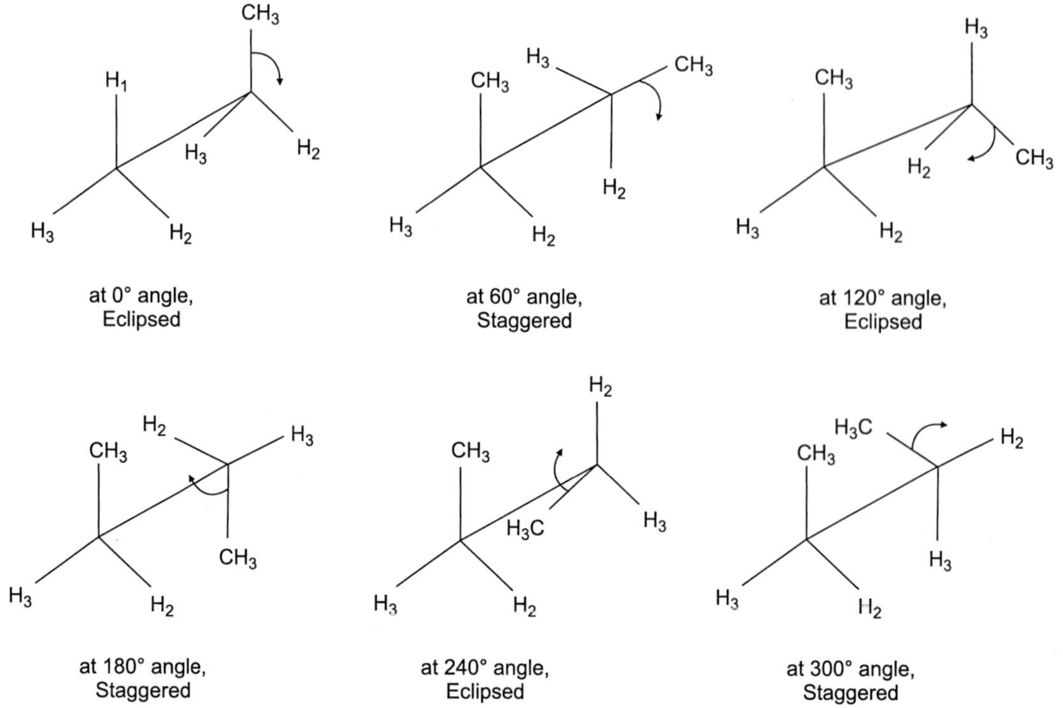

Fig 7.12 Conformations obtained by rotation through different angles.

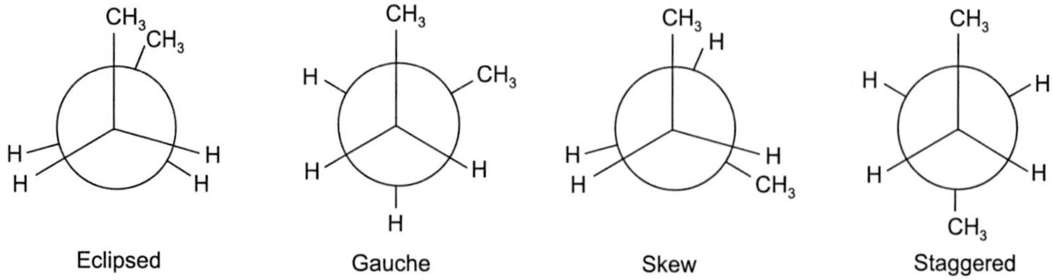

Fig 7.13 Different Conformations of butane.

Butane, has two'gauche' conformation, which are obtained at dihedral angle of 60° and 120°. Thermodynamic calculations have shown that these conformations differ from each other in very small energy of the order of 3 kcal/mol. The barrier of this energy is required to be crossed to change the staggered conformation to eclipsed conformation. In the Fig. 7.14 Newman's representation on the curve has been used to show structures of four conformations as these exist with reference to their dihedral angle. Thus, a staggered conformation is the most stable conformation in butane as in energy curve, Fig. 7.14.

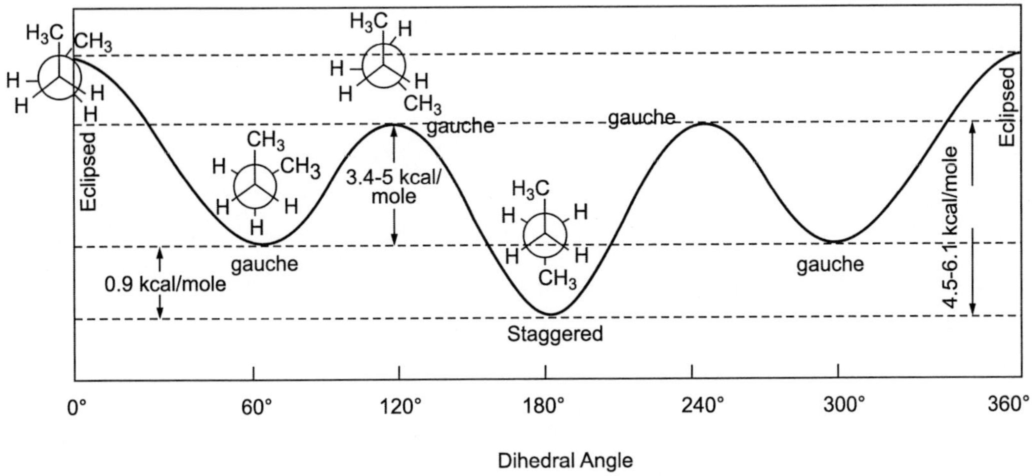

Fig. 7.14 Energy diagram of butane for different conformations

7.3 Relative Stability of Cycloalkanes

Conformations are momentary arrangements obtained by rotation around single bond in a compound. However, in case of cyclic compounds this free rotation is not there, but it is flip over and change of orientation of groups that takes place. The relative stability of conformations cycloalkanes is important to study and know more about the nature, chemical reactions of cyclic compounds. Different theories for their stability were proposed such as Baeyer's Strain theory and Sachse theory.

7.3.1 Baeyer's Strain Theory

A large number of cyclic compounds that occur in nature are six membered ring structures. These are the most common size of rings. Five membered ring compounds are less and in comparison, the four and three carbon ring compounds are still less prevalent. The stability of ring systems and their natural occcurrence appears to bear a direct relation. Baeyer studied a large number of ring compounds and based on bending of valence bond to form rings, he proposed a theory called "Baeyer's Strain Theory". According to this theory, small and large sized ring compounds are not stable because they involve lot of angle strain or tension in the carbon–carbon bonds of the ring. He calculated the tension of different sized ring compounds from distortion in bond angle from the normal value of 109° 28′.

For example, In cyclopropane, the C–C–C bond angle (as of a triangle) is 60°. That is each bond has to come closer from their normal angle to form a 60° angle between two bonds. Hence distortion is

$$= \frac{1}{2}(109^\circ 28' - 60^\circ) = 24^\circ 44'$$

Similarly, the distortion in four, five and six-membered ring structures, are as shown below:

4 member ring distortion = ½ (109° 28′ – 90°) = 9° 44′

5 member ring distortion = ½ (109° 28′ – 108°) = 0° 44′

6 member ring distortion = ½ (109° 28′–120°) = –5° 16′

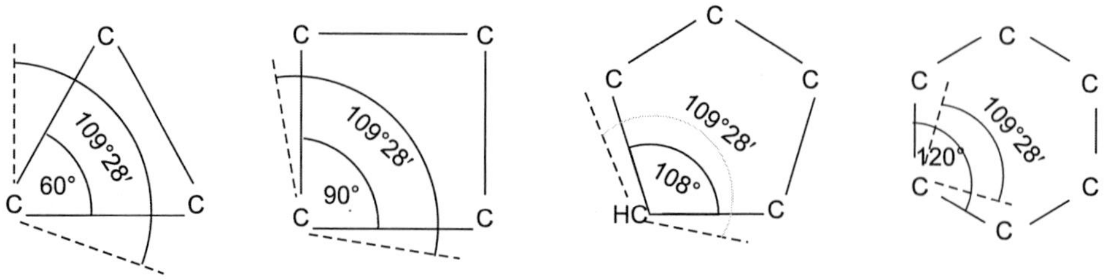

Fig. 7.15 Showing different cyclic structures and bond angle deviation.

Thus, Baeyer was able to explain the instability of three and four membered ring structures. According to his theory, five membered rings have the least strain and so should be most stable ring system. They should also occur widely in nature compared to six membered ring system. It is not the reality of nature. Six-membered ring system either alone or in combined state with other rings are the most abundant. Hence, there is some thing wrong with his theory. Later, Sachse explained the stablity of cyclic compounds based on strainless rings and that is more acceptable.

There are two kind of strain in compounds, namely, the Ring strain and Steric Strain.

1. **The ring strain consists of two kind of strains, the Bond Angle Strain and Torsional Strain.**

   (*a*) **Bond Angle Strain** arises when the C-C-C bonds of the ring depart (because of geometric necessity) from the ideal tetrahedral angle 109°28′ as in sp^3 hybridised carbon.

   (*b*) **Torsional Strain**, arises when bonds are not ideally staggered. In most ring systems, it is impossible to perfectly stagger all bonds. In cyclohexane, fortuitously, perfect bond staggering occurs in the chair conformation, but no conformation in any other ring allows for such perfect staggering.

2. **Steric Strain:** It is due to approach of atoms near to each other and because of their volume and repulsive forces, steric strain becomes operative.

7.3.2 Ring Strain in Cyclopropane

Cyclopropane, the smallest cycloalkane, is the most reactive amongst ring compounds. Instead of undergoing substitution reaction, it undergoes addition type reaction with halogens. Its cyclic ring comes from three points on a plane, so is most strained ring. The ring strain has been estimated, from heats of combustion, to be 28 kcal/mol. It thus weaken the C–C bond. The other kind of strain is the angle strain (49.5°), where angle changes from 109.5° to 60°. It is best explained on the basis of overlap of atomic orbitals as shown in the Fig. 7.15. There is a very little overlap, so the bonds are bent and weak. Also these are not pure *sigma* bonds, there is a partial *pi* bond character, These are sometime called *banana bonds*. That is why experimentally, instead of substitution, addition reaction takes place as mentioned above.

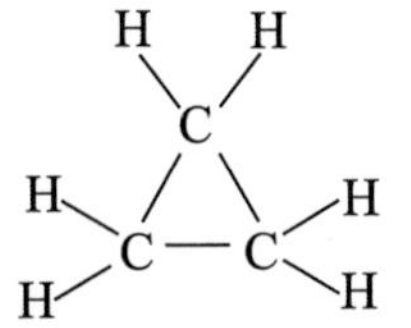

$$+ \ Br_2 \longrightarrow Br-CH_2-CH_2-CH_2-Br$$

7.3.3 Ring Strain in Cyclobutane

The stability of cyclobutane is slightly more compared to cyclopropane. Ring strain per $-CH_2-$ of cyclobutane has been estimated to be 26 kcal/mole (110 kJ/mol). The four points in a plane form cyclobutane, that is a planar structure, it involves an angle strain (29.5°). It weakens the C–C bond, but is not that much, as in cyclopropane. It can be explained on the basis of overlap of atomic orbitals. The overlap is not linear, but angular. Thus cyclobutane also does not have pure sigma bond, but a partial pi bond. It also undergoes addition reaction, but at higher temperature. It undergoes substitution reactions also, that are carried out at –78°C *via* carbanion generation. Substituted cyclobutanes are well-known compounds.

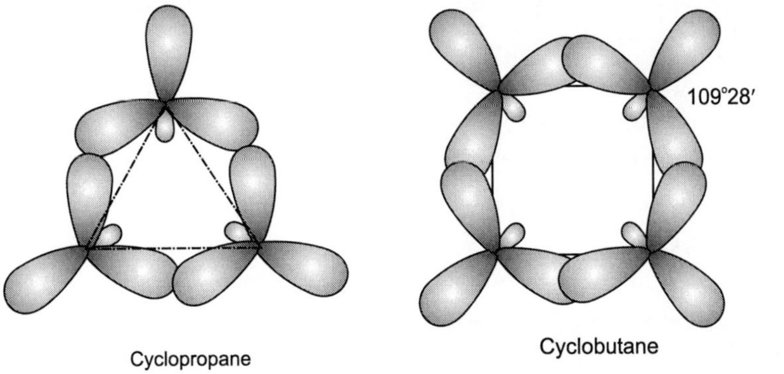

Fig.7.16 Cyclic structures formed by overlap of orbitals.

7.3.4 Sachse Explanation of Stability of Cyclic Compounds: Strainless Rings

Baeyer had considered all rings as planar and ended up in wrong conclusion about the stability of cycloalkanes. **Sachse (1890) proposed that structure of cyclic compounds are not planar**. He proposed **puckered rings structures** of cyclohexane. Puckering means non-planar rings and the carbon-carbon bond angles remain at $109^0\ 28'$ (tetrahedaral). **It makes the rings - strainless** (for carbon-carbon bond in the ring, thus there is no change) and hence it offers stability to the cyclic compounds.

7.3.5 Conformations of Cyclohexane and Their Stability

Sachse had proposed the puckered ring structure for cyclohexane, so cyclohexane ring acquires a completely strainless structure. When we make models of cyclohexane, that are not planar, we find that there can be more than one way the model of cyclohexane can be made. There are many models that are available. Two pictures of cyclohexane made from a model called flexible model are shown below.

Fig. 7.17 Models of chair and boat forms of cyclohexane using flexible models.

Analysis of models show that the left hand side structure has a shape of a chair and the right hand structure has a shape of a boat. So, these structures are referred to a chair form and boat form structures of cyclohexane.

In chair cyclohexane, this angle just happens to be virtually identical to the tetrahedral angle, so that neither angle nor torsional strain occurs in cyclohexane in the chair form. *Therefore, the total ring strain is nil for cyclohexane.* It exists strainless, because bonds are ideally staggered,thus has least steric strain as well. In most ring systems, it is impossible to perfectly stagger all bonds.

Other models that are used in chemistry are known as space filling model, Dreiding models, ball and stick models etc.. Some of these are given below:

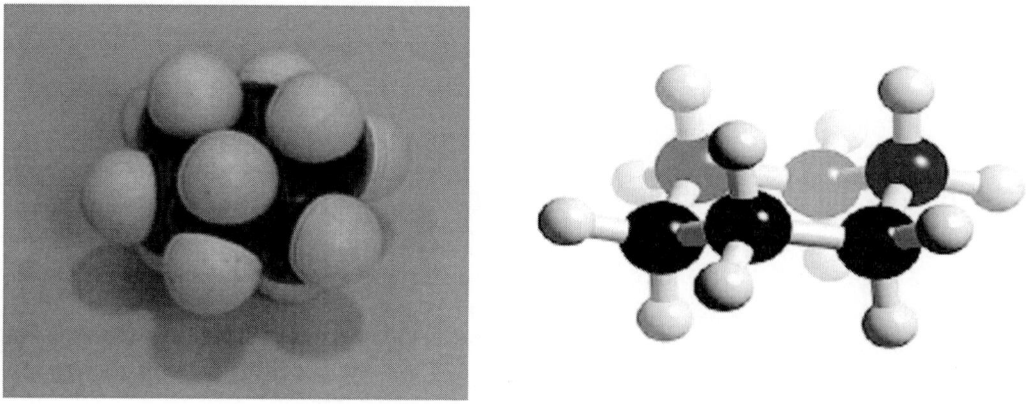

Fig. 7.18 Space filling and ball and stick model.

Geometry of Cyclohexane

There is another way of understanding the geometry of these models by making a chair and boat and letting the six atoms sit on chair or four atoms of boat as shown below.

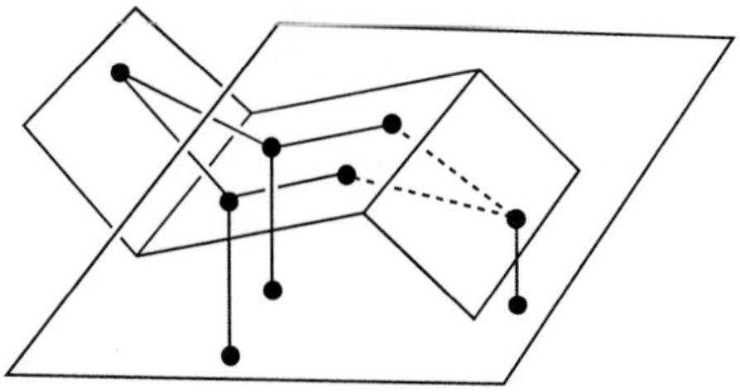

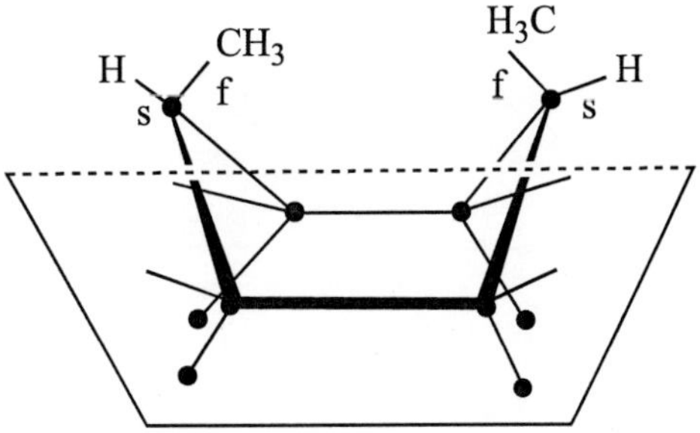

From the cyclohexane models, one can determine the orientation of stubstituents on the chair and boat conformations. These are represented further below.

On analysis we find that chairform of cyclohexane has two kind of bonds, namely, axial and equatorial. The axial bonds are parallel to the virtual axis that passes through the centre of cyclohexane ring. The equatorial bonds are oriented away from the ring and make an angle of 109 28' with other bonds of carbon. All this results in strainless ring. It is also flexible ring structure and change over to different forms.

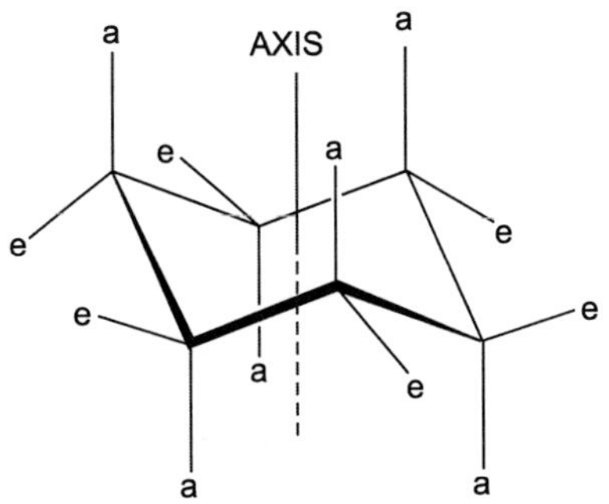

Chair form of cyclohexane

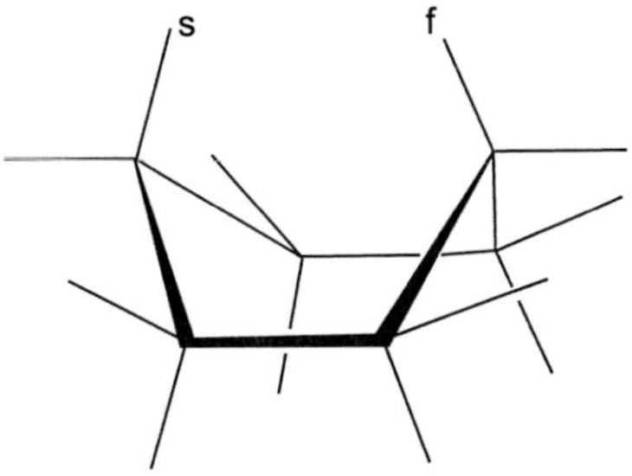

Boat form of cyclohexane

's' stands for blowsprit and 'f' for flagpole equatorial bonds. Axial bonds are oriented parallel to the axis of ring, and equatorial bonds are oriented away from ring.

'a' stand for axial bonds and 'e' stand for atoms in the boat form of cyclohexane. They exert good interactions because of their nearness.

Fig. 7.19 Chair and boat structures of cyclohexane.

Totally four different forms have been recognised that are chair, twisted, boat and skew-boat and/or their mirror images etc.. Look at the model of **chair** form, if one lifts the lowest leg up, it gets converted into **boat** form. Different structures of cyclohexane are represented ahead.

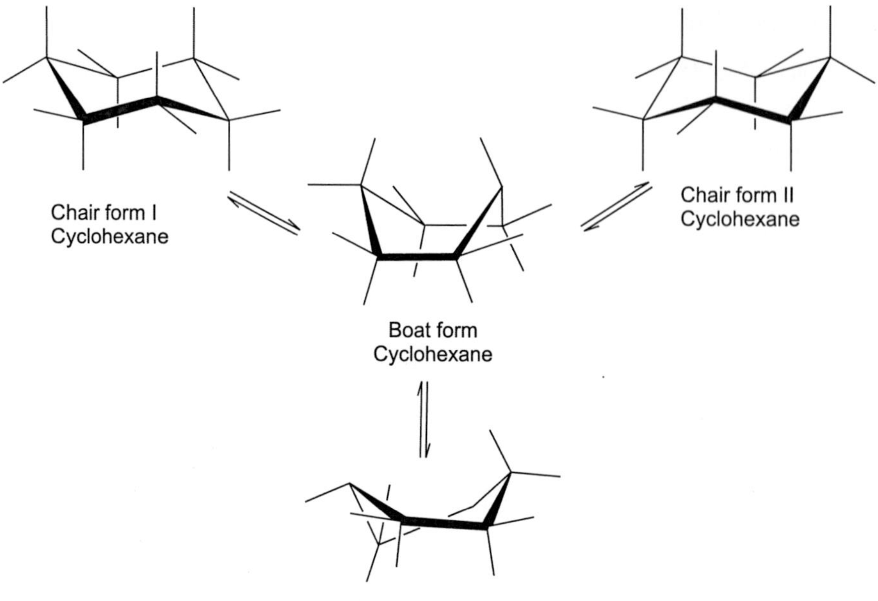

Fig. 7.20 Interconversion of different cyclohexanes.

The interconversion of chair and boat forms is shown using conformations of cyclohexane, Fig. 7.20. During the process, energy is spent and so is the state in real molecules. Interconversion of chair form and boat form takes place via another conformation called skew or twisted conformation.

A graph of free energy of chair, boat and skew boat forms, is given in Fig.7.21. The interconversion is accompanied with a transition state of high energy. The energy of highest peak representing transition state is of the order of 10–11 kcal/mole (41.8–45.6 kJ/mol). This range of variation of energy is because of different possible substitution in cyclohexane.

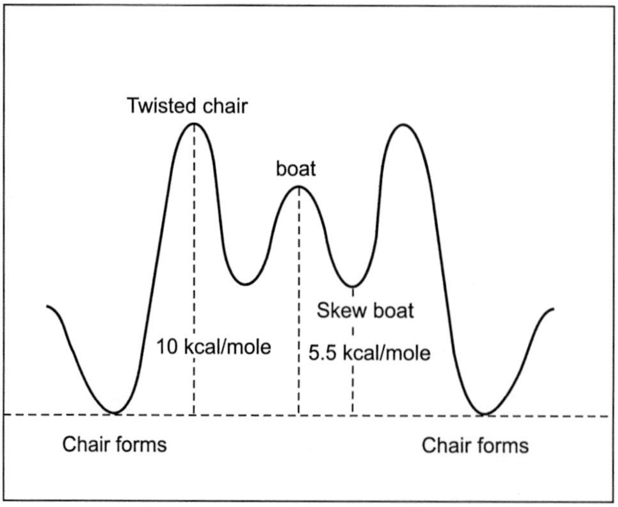

Fig. 7.21 Energy diagram of cyclohexane in chair, boat and skew forms.

From the graph we find that chair form is the most stable and passes through the energy barrier to get converted into skew and boat forms. The energy calculations are based on heat of combustion and/ or equilibrium constant determination. It has been estimated by energy calculations that the population of chair form of cyclohexane is maximum.

7.4 Chirality

Chirality is a property of certain objects, alphabets and of molecules due to which they become non-superimposable with their respective mirror images. It is also called handedness of objects or molecules. The term handedness has been derived from hands of a person. Two hands, left and right hand are not superimposable with each other, so are **chiral**. Certain objects or molecules are superimposable with their mirror images, these are called '**Achiral**'. Chirality or achirality term can be applied to house hold objects as well. However, molecules in particular, when chiral, become optically active, a property of molecules that has been discussed earlier. Given below are the examples of each category of chiral and achiral objects, alphabets and molecules.

1. Hands are Chiral:

Two hands are not superimposable, so are chiral in nature

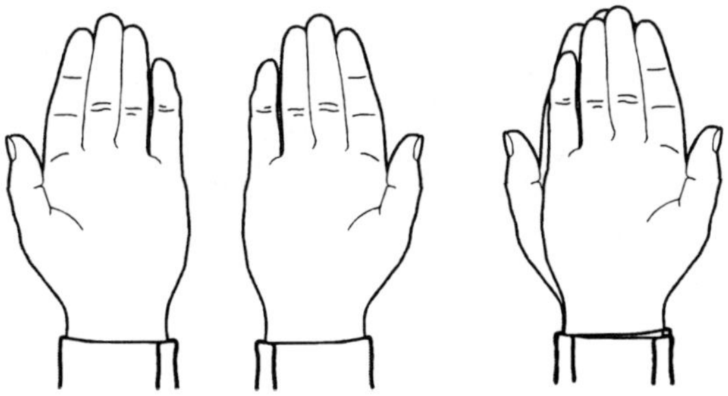

2. Chiral and Achiral Alphabets:

Certain alphabets of English language are chiral, while others are achiral. There are two sets of alphabets given below, one set is of chiral alphabets and other is of achiral alphabets. You practice yourself by sliding or rotating the alphabets in plane and try to find out if they overlap. We notice that group A alphabets do not overlap either by sliding the mirror image or by rotating, thus, are chiral. Whereas the group B alphabets do overlap, thus, are achiral.

Group A: Chiral Alphabets

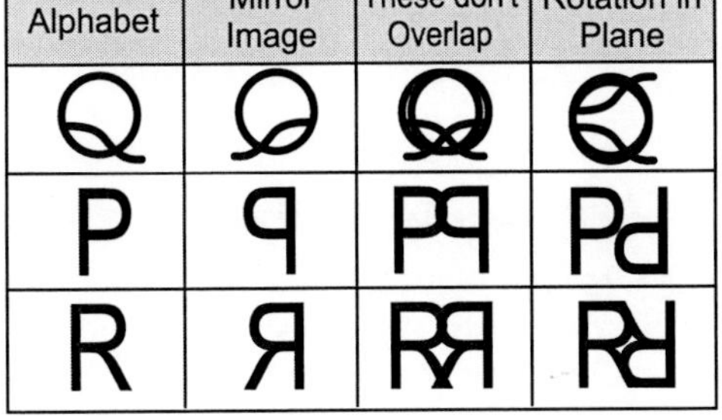

Group B: Achiral Alphabets

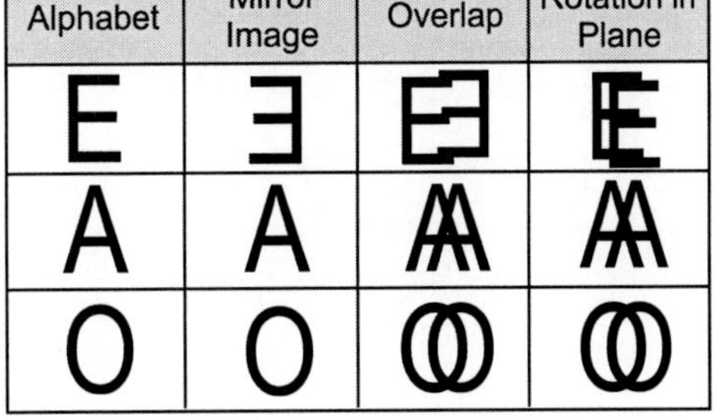

3. Chiral Molecules:

Model of chiral molecule, butan-2-ol, two are non-superimposable, so are chiral

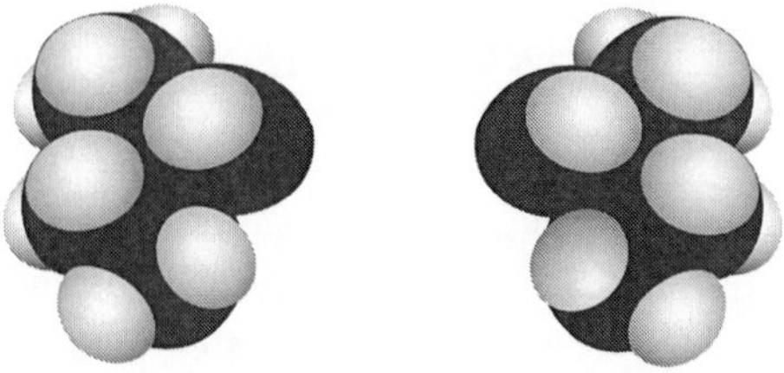

Wedge diagram of the model, butan-2-ol, chiral molecule.

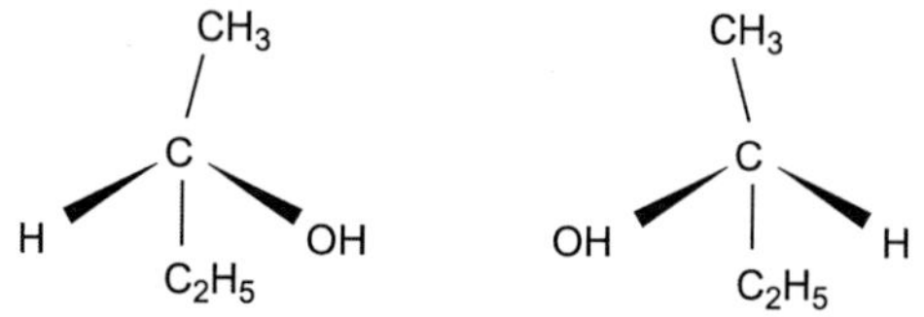

Fig. 7.22 A pair of enantiomers of butan-2-ol

Example 7.2: Match the following structures in pairs of object and mirror image. Which of the these structures represent optically active compounds.

1. 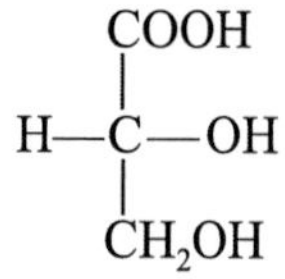

2. 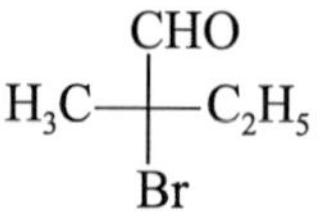

3. 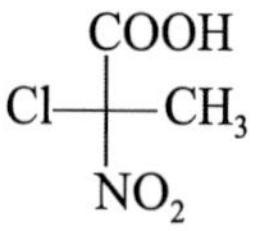

4. 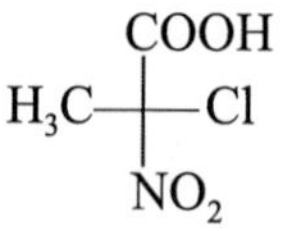

5. 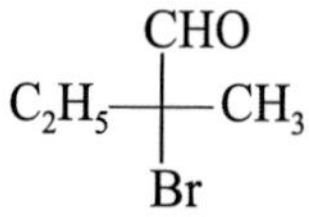

6. 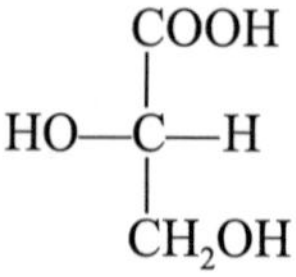

7. 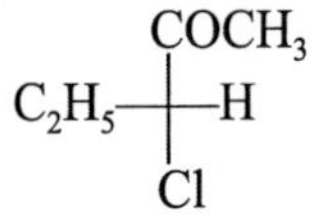

8. 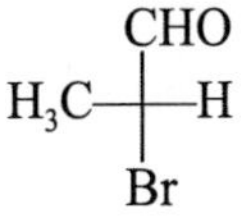

9. 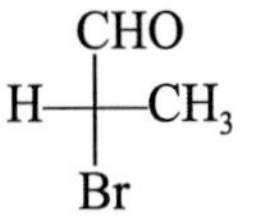

10. 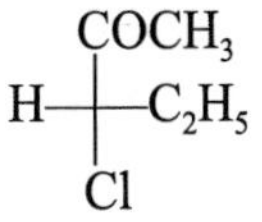

Solution: All are chiral molecules, thus optically active. The structures that exist as object-mirror pair are.

1 and 6; 2 and 5; 3 and 4; 7 and 10 ; 8 and 9.

7.5 Optical Activity

7.5.1 Plane Polarised Light

A ray of light having uniform wavelength is called monochromatic light. It consists of waves having vibrations in all directions but perpendicular to the direction of propagation. When such a ray of light is passed through a 'Nicol prism', the multi-directional vibrations get restricted to uniplanar or unidirectional vibrations. *The resultant monochromatic ray of light with uniplanar vibrations in a plane perpendicular to the direction of propagation, is called plane polarised light.*

7.5.2 Optical Activity

When an organic compound in solution or in gas phase is placed in the path of plane polarised light and if the plane of polarisation gets rotated through a certain angle, the compound is said to be optically active. If rotation of the plane of plane polarised light is to right hand side, the compound is called a dextro-rotatory. When it gets rotated to the left hand side, the compound is called leavo-rotatory. ***Dextro** and **leavo**-rotation of compounds are represented by* (+) *and* (–) *signs respectively.*

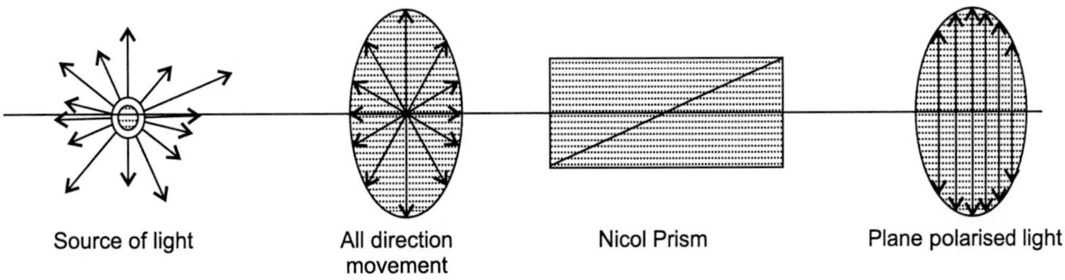

Fig. 7.23 Plane polarised light.

The instrument used for determining optical rotation is called ***polarimeter***. Right hand side rotation of plane polarised light in the polarimeter actually means, the positive value of (final reading - initial reading) and that is called dextro-rotatory. If the (final reading -initial reading) in the polarimeter is negative, it is called leavo-rotation, corresponding to left hand rotation of the plane of plane polarised light.

Dextro-rotation = +ve value of (final reading – initial reading) in the polarimeter

Leavo-rotation = –ve value of (final reading – initial reading) in the polarimeter

7.5.3 Polarimeter

The schematic diagram of a polarimeter is given in Fig. 7.24. Different parts are labelled. It consists of a source of sodium light (light of fixed wavelength). This light after getting narrowed at the slit, passes through a 'NICOL' prism to get plane polarised light. This light is then passed through a solution of an organic compound placed in the tube. If the compound is optical active, the plane of polarisation gets rotated to right or left direction. The angle 'θ' is the degree of rotation of the plane of plane polarised light and it is determined by analyser prism. In fact, it is the angle through which the analyser is rotated to get back the original position of the plane of polarised light. The angle 'θ' gives the extent of optical activity of the compound. Specific activity is a standard value of a compound and is related to angel 'θ' as given next.

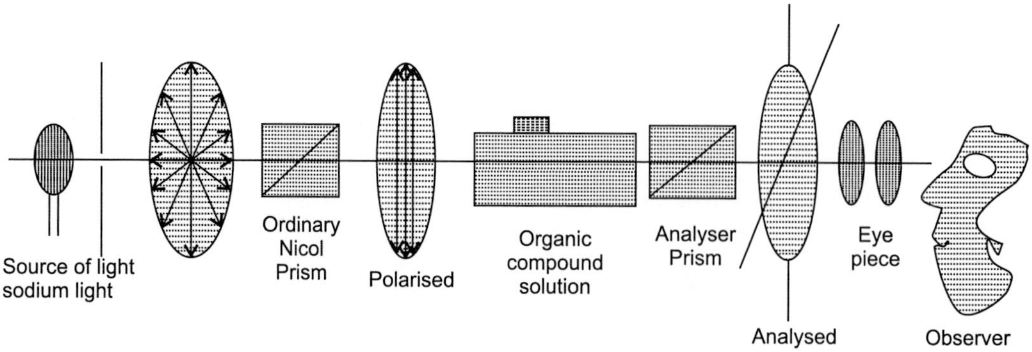

Fig. 7.24 Polarisation of monochromatic light and a schematic representation of a polarimeter.

7.5.4 Specific Rotation

Specific Rotation is defined as the optical rotation 'θ', due to a solution of unit concentration (g/ mL), that is, 1g in 1 mL in a polarimeter, having 1 decimeter (l dm) long tube. The definition is obtained from the equation based on the fact that the optical rotation, 'θ', of a compound in polarimeter is found to be directly proportional to concentration, c and length of the tube (path of light through the compound), l.

$\theta \propto c$, the concentration in terms of g/mL and

$\theta \propto l$, the length of the polarimter tube for solution in decimeter (dm)

therefore, $\theta \propto c.l$, $\theta = [\alpha]_D.c.l$

the **constant of proportionality $[\alpha]_D$, is called *Specific Rotation*.**

Rewriting the above equation, we have

$$[\alpha]_D^{t^oC} = \frac{\theta\,(\text{degree})}{c(g/mL) \times l(dm)}$$

Many a times, it is difficult to dissolve the compound is 1 mL, so we dissolve the compound in 100 mL and change the above formula for specific rotation by multiplying the numerator by 100.

Thus, we get,

$$[\alpha]_D^{t^oC} = \frac{\theta\,(\text{degree}) \times 100}{c(g/100mL) \times l(dm)}$$

7.5.5 Representation of the Value of Specific Rotation

While representing value of specific rotation, it is customary to include the temperature and solvent used. Say for example, for lactic acid, if the specific rotation is +2.9°, it is represented as below.

$$[\alpha]_D^t = +\,2.9^o \text{ (Ethanol, } c = 0.095 \text{ g/mL)}$$

Example 7.3: *Calculate the specific rotation of a solution of an optically active compound, that has 6.75 g of compound dissolved in 250 mL of ethanol. The polarimeter used has a sodium lamp (λ =589 nanometers) and a tube length of 20 cm, at a temperature of 35 degree centigrade. It showed final reading of optical rotation as 190°.*

Solution: The equation for optical rotation is

$$[\alpha]_D^{35^oC} = \frac{\theta\,(\text{degree}) \times 100}{c(g/100mL) \times l(dm)}$$

filling the values of parameters, we get

$$[\alpha]_D^{35^oC} = \frac{(190 - 180)\text{ degree } . \times 250\ mL}{6.75\ \dfrac{\text{g}}{250\ \text{mL}} \times \dfrac{20\ cm}{10\ \dfrac{cm}{dm}}}$$

$$= \frac{10 \times 250}{6.75 \times 2} = \frac{2500}{12.5} = 200 \text{ degree}$$

$$[\alpha]_D^{35^oC} = +200\ ^o(c = 6.75\ g/250\ mL, ethanol\ )$$

7.5.6 Molar Rotation

Besides specific rotation another type of rotation values that involve molecular mass is called molar rotation. It is defined as the specific rotation, $[\alpha]_D^t$ multiplied by molecular mass,M of the compound.

Thus, Molar Rotation = $[\alpha]_D^t \times M$

OR

$$Molar\ Rotation,\ [\alpha]^t_{M,D} = [\alpha]^t_D \times \frac{M}{100}$$

$$= \frac{\alpha \times 100 \times M}{c(mol/100mL) \times l(dm) \times 100}$$

The division by 100 is made arbitrarily to keep the value of molar rotation, reasonably low.

7.6 Stereoisomerism

Broadly, isomerism is of two types, namely, **structural isomerism and stereoisomerism**. Structural isomerism we have studied in senior secondary classes. Let us now understand different types and different aspects of stereo-isomerism.

Stereoisomerism is an isomerism that is exhibited by certain organic compounds that have different possible spatial arrangements of four different groups around carbon (considered as the rigid part of a molecule). There are **three types of stereo-isomerism**

(*i*) **Optical isomerism:** It is an isomerism exhibited by compounds having molecular asymmetry. Asymmetry arises due to different stereo-arrangement of four different groups around carbon in isomeric compounds. Certain molecules ***may not*** have four different groups around carbon atom, still can exhibit optical isomerism. This type of isomerism is called **Atropisomerism**. It produces asymmetry due to **restricted rotation around single bond**.

Optical isomers have identical chemical and physical properties. They differ from each other only in their behaviour towards plane polarised light and bear an object and mirror image relationship to each other and are non-superimposable. This property of non-superimposability of object-mirror image, is called **chirality.** It is the basic requirement for a compound to be optically active. Optically active molecules are called **chiral molecules.**

(*ii*) **Geometrical isomerism** is exhibited by compounds that have **restricted rotation due to double bond**. These isomers ***do not*** exhibit optical activity. However, optical activity may arise due to the presence of certain asymmetric units in other parts of the molecule.

(*iii*) **Conformational isomerism** is the third kind of isomerism arising in molecules due to free rotation around single bond. Usually these isomers differ from each other in energy by the a small amount and are interconvertible at room temperature. At extreme lower temperature, conformational isomers can be isolated. It is important to tell at this stage that conformational isomerism is of great significance in case of cyclic compounds, where free rotation is not possible, but molecules can still opt for different orientations or conformation in space. Given below is the flow sheet diagram showing relation of each type of isomerism.

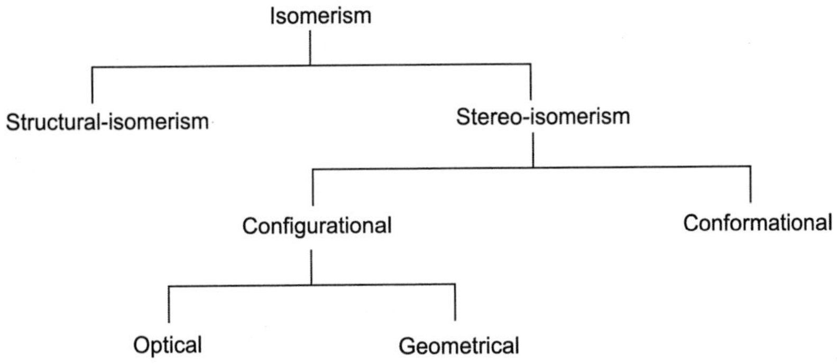

Fig. 7.25 Types of Stereo-isomerism.

7.6.1 Stereoisomerism in Compounds Containing One Chiral Centre

Each of the molecules like, butan-2-ol and lactic acid have been isolated in two optically active forms, differing from each other only in the sign of optical rotation.The only possible way, the single structure of such molecules, can be written in two different forms, is to use the three dimensional structure representations. When we make two three dimensional models, each bearing an object- mirror image relationship, for butan-2-ol or lactic acid, we find that object and mirror image model of each of these compounds do not overlap each. Thus, these are different, hence exhibit isomerism, called **stereoisomerism**. These are optically active, so are called optical isomers. Such non-superimposable object-mirror image are called 'enantiomers', the isomerism is called **enantiomerism**.

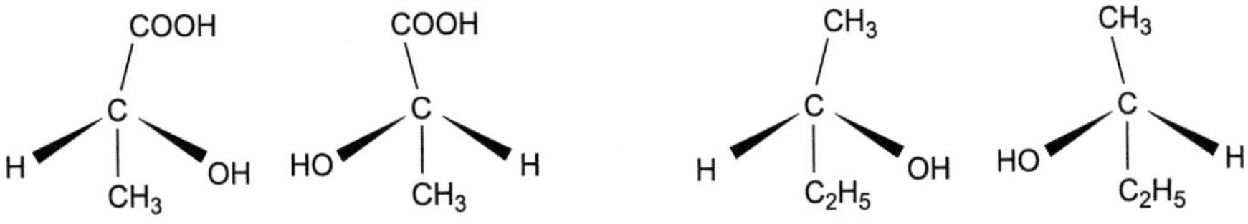

Lactic acid stereo-isomers Butan-2-ol stereo-isomers

7.6.2 Stereoisomerism in Compounds Containing Two Chiral Centres

Let us now consider compounds containing two chiral centres, such as given below in Fig.7.26 and marked I to IV. It has two asymmetric center, therefore, it has $2^2 = 4$ isomers. All these are isomeric and have molecular formula, $C_4H_8O_4$. There are two sets enantiomers I and II, that are called erythroses and other two, III and IV are called threoses. Erythroses are related to eachother as non-superimposable mirror images, so are **enantiomeric**. The structures III and IV called threoses, are also non-superimposable mirror images, so are also enantiomeric to eachother. However, the relationship of I and III or II and IV are such that they do not form mirror images, but are stereoisomers, and are called **diastereomers**.

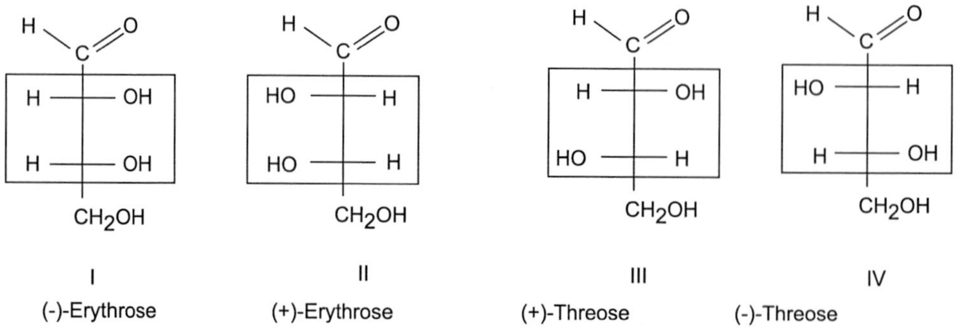

Fig. 7.26 Distinction between enantiomers and diastereomers.

The inter-relationship of these four structure is given in Fig. 7.27.

There is **another nomenclature** for stereo-orientation of substituents at two chiral centres as presented in structures I to IV, that uses the prefixe, **Threo- and Erythro-**. The same side orientation of substituents at two adjacent chiral centres, as in structure I and II, is called **erythro**- orientation. Same is the case with its mirror image, II, that is called **erythro**- orientation. Other isomeric compounds (III, IV) having orientation of two substituents on adjacent carbons on the opposite side, are called **threo-orientation**.

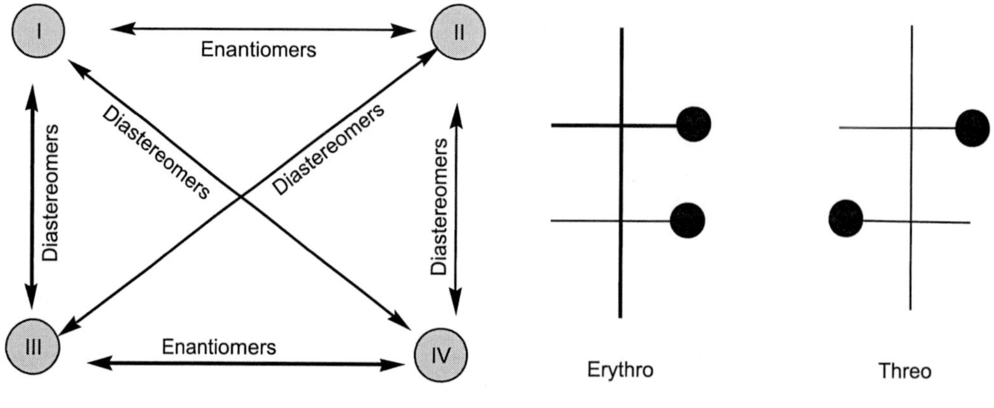

Fig. 7.27 Interlink of Threo and erythro isomers.

Some other examples of erythro and threo isomers and their nomenclature are given ahead:

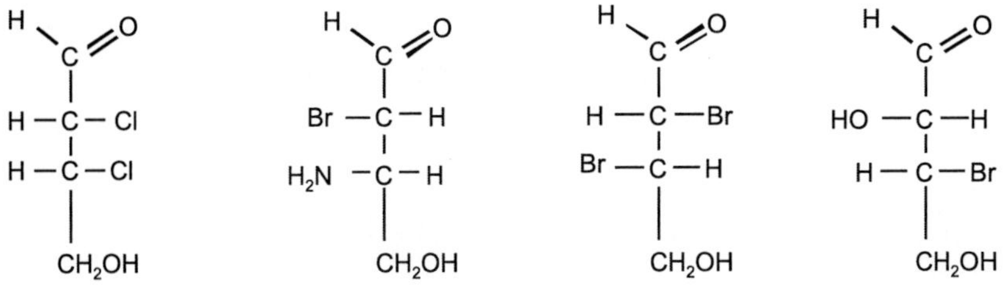

V VI VII VIII

Erythro Erythro Threo Threo

Fig. 7.28 Erythro, Threo configuration of compounds.

Structures V and VI separately are called **erythro structures**. Basically, it means that the substitutents on two adjacent chiral carbon atoms are on the same side. Structures VII and VIII are called **threo structures**, meaning thereby that the substituents on the adjacent chiral carbons are on opposite side.

7.6.2.1 Tartaric Acid Isomerism

Tartaric is one of the most important optically active compound that was first to be studied exhaustively. Its molecular formula is $C_4H_6O_6$. It is a dicarboxylic acid, named as 2,3-dihydroxy butanedioic acid. It has two asymmetric centres and thus has four isomers. The structure of these four isomers are given below. First one, IX is **dextro-rotatory** and its mirror image, X is **leavo-rotatory**. The other two isomers XI and XII are not optically active. These are called **meso-structures**. These isomers are optically inactive, but inactivity comes by **internal compensation**. A mixture of (IX+X) having 50% of each of the optically active tartaric acids shall produce optically inactive mixture called, **racemic modification**. This mixture is optically inactive by **external compenstion**. This mixture can be separated in to its components IX and X by suitable procedure discussed later.

As late as, 1951, Bijvoet, determined the structure of dextro-rotatory form of tartaric acid by X-ray crystallography of its potassium rubidium salt and arrived at its structure as IX. Thus, it becomes the absolute structure of (+)-tartaric acid.

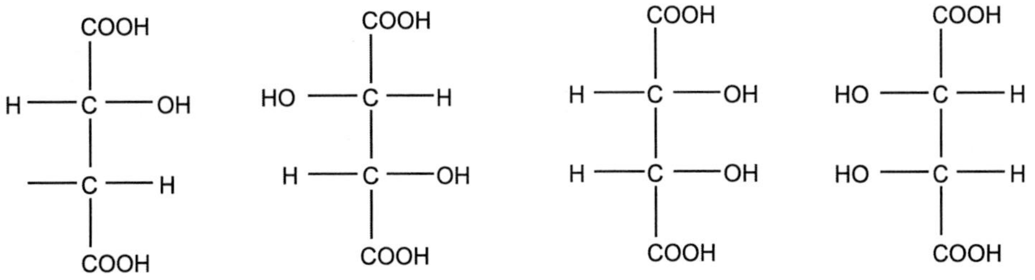

(+)-Tartaric acid IX (–)- Tartaric acid X Meso tartaric acid XI Meso-tartaric acid XII

Fig. 7.29 Different forms of tartaric acids.

Meso form: The **third and fourth forms** of configuration, are called ***meso forms (XI and XII)***. These are optically inactive that is the zero optical rotation. It is due to internal compensation of optical rotations. It arises only in case of two or more asymmetric carbon containing compounds. It is a single component and cannot be separated in to any other component, as was possible in case of racemic modification.

7.6.3 Properties of Enantiomers and Diastereomers

Enantiomers have identical chemical and physical properties, except their optical rotation. In comparison, the diastereomers have different physical properties and sometimes different chemical properties also. The Table 7.1 given below shows the similarity and differences of properties of enantiomers and diastereomers.

Table 7.1 Comparison of the Properties of Enantiomers and Diastereomers

S.No.	Property	Enantiomers	Diastereomers
1.	M.P.	Identical M.P.	Different M.P.
2.	B.P.	Idenitcal B.P.	Different B.P.
3.	Solubility	Same	Different
4.	Optical Rotation	Same, but opposite sign	Different values may have same or opposite sign
4.	Chemical Properties	Similar	May be different
5.	Free Energy	Same	Different

Properties of four isomers of tartaric acid

Example:	Isomer	$[\alpha]_D$	M.P.
	(+)-tartaric acid:	+13°	172 °C
	(–)-tartaric acid:	–13°	172 °C
	racemic-tartaric acid:	0°	206 °C
	meso-tartaric acid:	0°	140 °C

7.7 Resolution of Racemic Modification

Resolution of racemic modification means separation of optically active components of the mixture. There are four methods that are used for resolution of racemic mixtures.

7.7.1 Crystallisation

Crystallisation of mixture and mechanical separation of the crystals based on their crystal structure is a difficult process. When we crystallise slowly, the crystals get formed with very clear crystal structure and can be distinguished under a microscope. It is a tedious method of separation but works well in certain cases. Infact, in case of tartaric acid, the ammonium potassium tartarate crystallised very well to distinguish two forms of crystals under microsocpe. This pioneering work was done by Louis Pasteur.

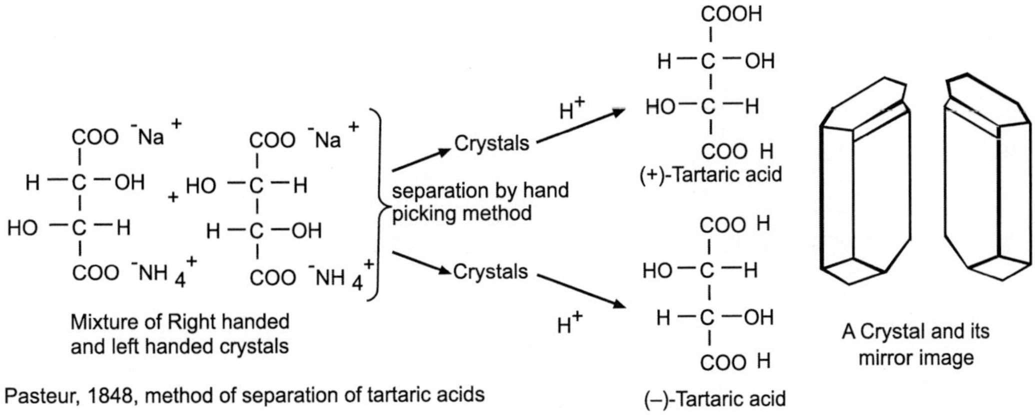

Fig. 7.30 Resolution by salt formation, crystallisation and hand picking method.

7.7.2 Chemical Method

Chemical method is based on the preparation of salt, say, we have (±) -1-phenylethylamine , a basic compound or (±) -phenyl glycine, also contains a basic group. It can react with a chiral acid, like (+)-tartaric acid or (+)-camphor-sulphonic acid. Such a reaction forms a mixture of two diastereomers.The mixture is fractionally crystallised. Separation of one form of diastereomer crystals can be preferentially done. Thus, after separation, salts are hydrolysed and pure enantiomers can be isolated. However, the method involves loss of compound, so other methods are adopted.

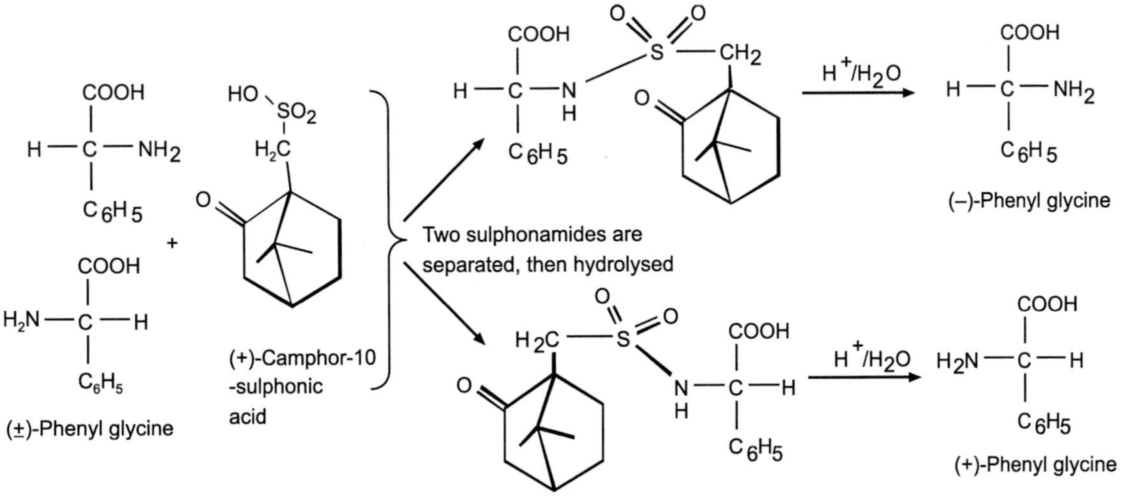

Fig. 7.31 Resolution of racemic modification by salt formation, crystallisation and separation based on solubility or other characteristics of crystals of diasteromers.

7.7.3 Biological Methods of Separation

Biological methods are based on the use or growth of certain moulds, microbes and insects on a medium containing a racemic compound. These grow by using one of the enantiomer, leaving behind the other isomer. The choice of insect or microbe is highly specific for different compounds. For example, *Penicillium glaucum* is used for preparation of (–)-tartaric acid. This mould attacks (+)-tartaric acid and leaves behind the leavo-rotatory compound.

This method has its limitations:

1. The availability and search of such moulds that are compound specific is a difficult process.
2. More so, we lose one enantiomer in the process. So is not a convenient process to be adopted for resolution of racemic modification.
3. It is a slow process and only dilute solution can be used for mould or bacteria to grow.

7.7.4 High Pressure Liquid Chromatography Using Chiral Support

Chromatography was invented by Twestt in 1916, but High Pressure Liquid Chromatography is one of the newer chromatographic technique of separation of a mixture of organic compounds. It has been developed only after 1970. In the case of enantiomers, conventional chromatography or other separation processes are incapable of separating them. To enable chiral separations to take place effecively, either the mobile phase or the stationary phase must itself be made chiral. It gives different affinities between the analytes. When the adsorbent is also using chiral support, the formation of π-complexes with chiral molecules becomes distinctive. It is because of the force of attraction between two isomers and the chiral adsorbent is different. It may be looked as if the polar functional groups of two enantiomeric molecules have different attraction due to different orientation reference to the active site of the support. May be one enantiomer has orientation of interactive group towards the solid support in the column and other has orientation away from support in the column. Thus, producing different binding that become operative and help in separation of chiral molecules during chromatography .

S.A. Matlin etal, in the year 2000, found that there is a great influence of support surface chemistry on the enantioselectivity and column performance. One of the support system like, cellulose coated with tris(3,5-dimethylphenyl carbamate)- or underivatised, aminopropylated and octadecylated silica have been found to be excellent for carrying out chiral HPLC. For separation of many racemates, underivatised, silica at a 20% (w/w) loading was found to be the most efficient phase.

7.8 Problem Faced by Chemists

Optical Activity of two enantiomers of a compound, as dextro- or leavo-rotatory, can be determined by using a polarimeter. By using the tetrahedral valency concept of the carbon, two stereo structures of enantiomers of a compound can also be written. An example of two enantiomers butan-2-ol is given in Fig.7.33. It remains a problem to correlate the two structures of enantiomers and their optical activity. In other words, which of the two stereoisomers is dextro-rotatory or which one is leavo-rotatotry.

7.8.1 Relationship of Optical Activity and Structure

Optical Activity of compounds, contained in bottles I (say, (+)-butan-2-ol) and II (say, (-)- butan-2-ol) is determined experimentally, by using a polarimeter (refer to Fig.7.32). From the **determined** optical rotation, 'θ'. their specific rotation can be calculated by using the formula, given earlier. The results are as shown below.

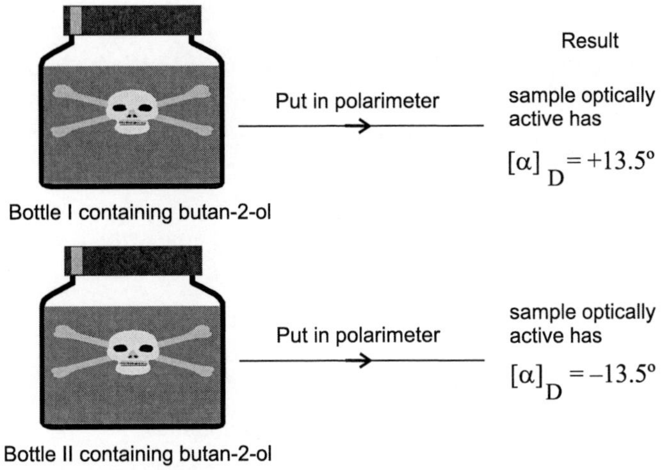

Fig. 7.32 Two optical isomers of butanol, whose optical rotations have been determined experimentally.

Based on the tetrahedral valency concept given by van't Hoff and LeBel,the wedge diagram of two optically active isomers of butan-2-ol is represented in Fig.7.33. Both are optially active and their specific rotations are as given above.

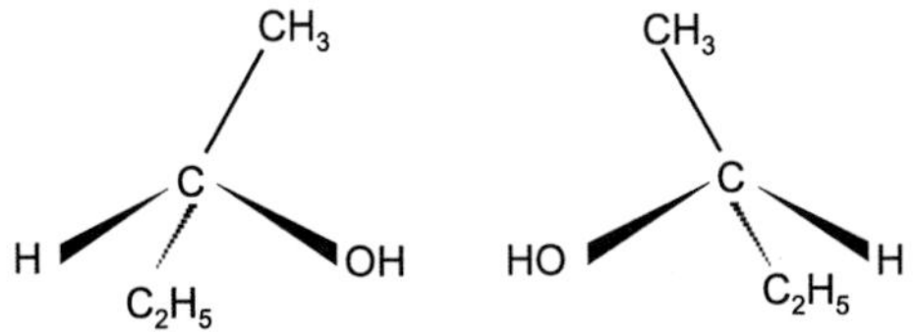

Fig. 7.33 A pair of enantiomers of butan-2-ol.

One structure represents (+), the dextro-rotatory butan-2-ol and other must be (–), the leavo-rotatory butan-2-ol. It is difficult to find out, which structure represents which isomer, unless we do a detailed study. These structures are also termed as **'configuration'** , the spatial arrangement of substituents or groups around the carbon atom.

Similarly, the structure of lactic acid (CH_3 CHOH COOH), based on the tetrahedral valency concept given by van't Hoff and LeBel could be written in two different Wedge diagrams or stereo- structures. These represent two optically active isomers of lactic acid as given below, Fig.7.34.

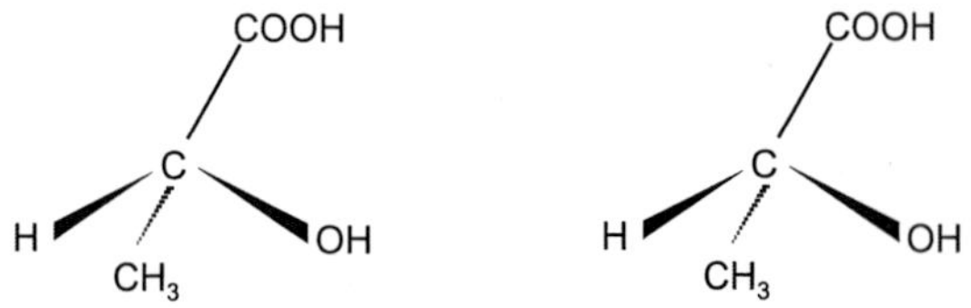

Fig. 7.34 A pair of stereoisomers of lactic acid. One structure represents (+), the dextro-rotatory lactic acid and other must be (–) , the leavo-rotatory lactic acid.

"Now the main problem for chemists was to correlate the liquid samples in bottles I and II, whose optical rotations have been determined, to the structures given in Fig. 7.33. Similarly, if bottles contained lactic acid, then correlation is needed to the structure of lactic acids given in Fig. 7.34.

There are two ways that can be adopted,

1. One way, is to correlate these compounds by chemical reactions, with a compound whose structure has been determined unequivocally.
2. **To be absolutely sure, alternative procedure** is the X-ray crystallographic study of the compound. This, no doubt, gives a sure solution to the structure of the compound using a sample from a labelled bottle and whose optical activity, (+)- or (-)- has been determined. However, the fact remains that it is not easy to determine X-ray of all compounds.

To resolve this baffling problem of correlation, chemists side lined the correlation problem and adopted an arbitrary method of assigning configurational notations to the molecules for establishing the identity of the compounds. Just from the written structure, it is possible to assign the configurational notations, that are arbitrary assignments but still based on certain rules. The beauty of the arbitrary configurational assignments made during 1891-1920, is that these were proved to be correct in 1951, by Bijvoet, who used X-ray crystallographic study on potassium rubidium salt of tartaric acid. Thus, the books did not change the configurational notation, after the X-ray crystallographic result were declared in 1951.

It may be borne in mind, that, there is no direct relation of the optical activity or (+)- and (–)- signs with the configurational notations. These are separate entities that are discussed below.

7.9 Configuration

A Configuration of a compound is the spatial arrangement of groups around rigid part of the molecule. In organic molecules, carbon atom is the rigid part that joins other atoms or groups in a spatially arranged fashion. In case, there are four different groups, there can be more than one configuration of the compound. For example, two models of butan-2-ol that are related as object and its mirror image to each other are shown in Fig. 7.35. On examining two models, we find, that these are not super-imposable to each other and thus, these represent two different compounds. They constitute a pair of stereoisomers called enantiomers. One of these represents (+)-or dextro-rotatory form and the other is (–)- or leavo-rotatory form. The molecule is represented by Wedge diagrams or Fishcer's projections also.

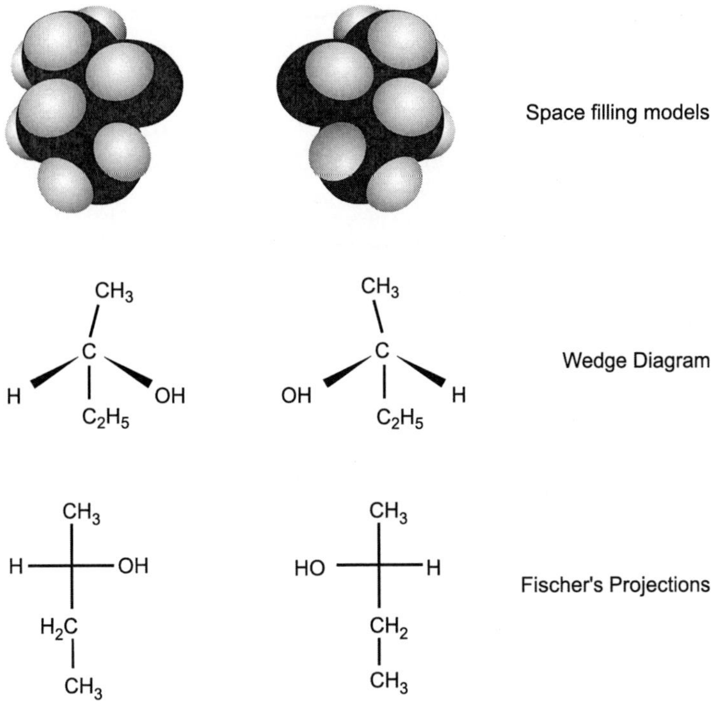

Fig. 7.35 Two Configurations of butan-2-ol, representing one dextro-or (+)- and other leavo or (–)- forms.

To differentiate these two models, we use the configurational notation, first proposed by Emil Fischer, while using a reference compound. **We have seen above, that it is very easy** to write the structures, but very difficult to remember their optical activity data for the large number of organic compounds. We only need to memorise the structrure and (+) or (–)-signs relation.

To avoid the memorisation of configuration for every compound, obtained by a long and tedious method of structure and configuration determination, chemists have adopted a simple method of assigning configurational notation of molecules, ***relative*** to another molecule, whose configuration has been determined by rigorous methods. In the beginning, the reference compound was only arbitrarily chosen, relative to which, configuration of other compounds could be assigned. It is called **relative configuration** assignment method.

7.9.1 Relative Configuration

Amongst the two isomeric structures, representing its dextro- or leavo- rotatory isomers, it is difficult to know which one is which. The identity of the isomers is carried out by arbitrarily assigning their configuration reference to another compound of known configuration. That is why, it is called **relative configuration** of a given molecule. It is just a kind of nomenclature that is used for a given structure and has not direct relation with optical rotation, (+)- or (–)-signs.

Herman Emil Fischer *(1852-1919) was a German Chemist and received Nobel Prize for Chemistry in 1902. His contributions to the field of sugars and protein chemistry have been commendable. His name is associated with configuration of (+)-glucose and projection of stereoisomers, Indole synthesis, Osazone reaction, peptide synthesis , caffeine (tea) , theobromine.*

7.9.2 *d,l*-Notations (Fischer, 1891)

Emil Fischer, the father of carbohydrate chemistry, arrived at the structure of naturally occurring dextro-rotatory glucose. This is a six carbon sugar, belonging to aldohexose series. Using the orientation of C_5-hydroxyl and hydrogen atom, he introduced a method of assigning the configuration of all other compounds. Fischer assumed that **C_5-hydroxyl group** orientated on the right hand side, as having '*d*' configuration and when orientated on the left hand side it represents '*l*' configuration. These **terms, *d* and *l* assigned by Fischer are arbitrary assignment** of nomenclature to the right or left side orientation of hydroxyl groups respectively**.** Relative to glucose notations, many of the then known compounds, were assigned configurational notations.

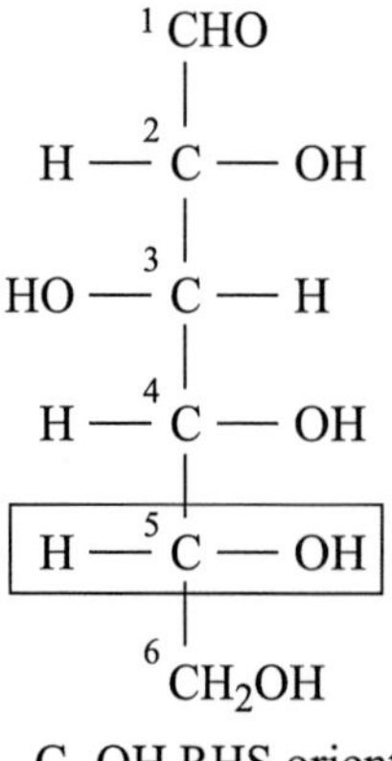

C_5-OH RHS oriented
d-configuration
(+)-*d*-glucose

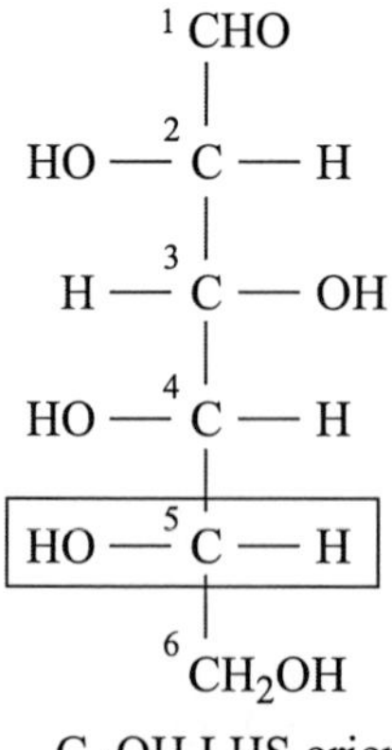

C_5-OH LHS oriented
l-configuration
(–)-*l*-glucose

Following compounds have been assigned the appropriate configurational notation with reference to glucose notation.

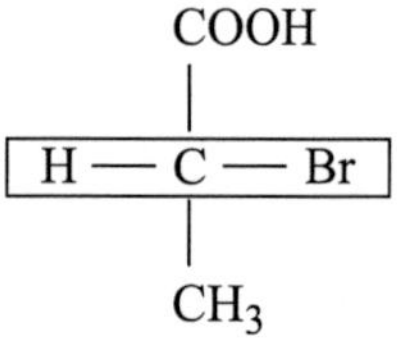

d-2-Bromo propanoic acid

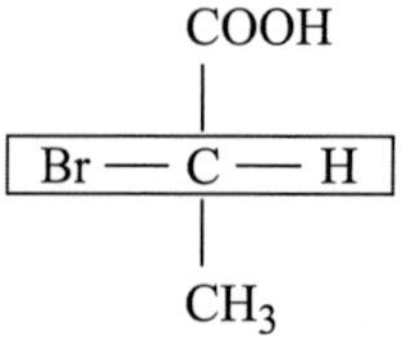

l-2-Bromo propanoic acid

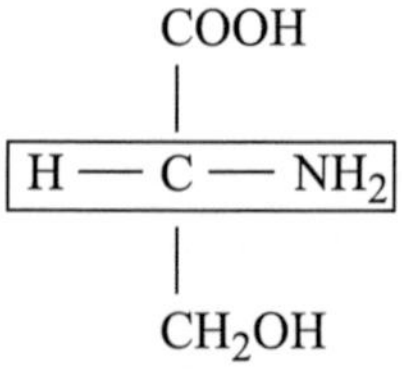

d-Serine

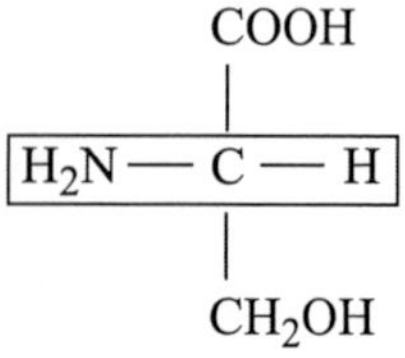

l-Serine

Remember :

> ***"The configurational notations used here, have nothing to do with optical rotation of a compound. It is just a method of writing the substituents at the chiral centre of a molecule; hydroxyl on the right hand is marked 'd' configuration and hydroxyl on the left hand side is marked 'l' configuration."***

The method has its own limitations and was revised by Rosanoff.

7.9.3 *d,l* Changed to D,L Notations (Rosanoff, 1906)

He argued that

(*i*) When there are four asymmetric centres in glucose, why only the fifth carbon was considered as a reference.

(*ii*) Also the use of symbols '*d*' and '*l*' created confusion with the terms dextro and leavo- .

To over come this ambiguity, he used a compound having only one asymmetric centre, namely, glyceraldehyde, as the reference compound. However, Rosanoff retained the direction of orientation of –OH and –H as was used by Fischer. He designated the right hand side orientation of hydroxyl group as D-configuration and left hand side orientation of hydroxyl group as L-configuration. Therefore, the name of two glyceraldehydes are

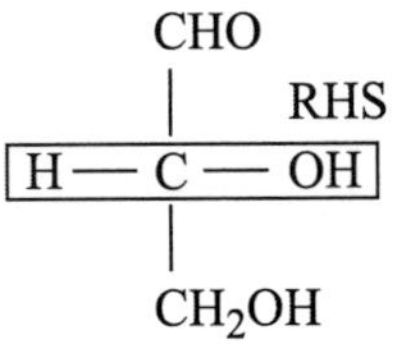

D-Glyceraldehyde

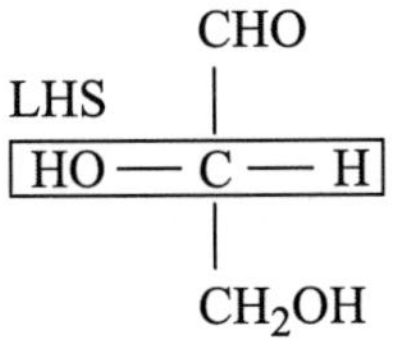

L-Glyceraldehyde

Naming of stereoisomers having only one asymmetric carbon atom, can be easily carried out in accordance to the Rosanoff nomenclature of configuration, that is, D and L notations. It is explained by the following examples.

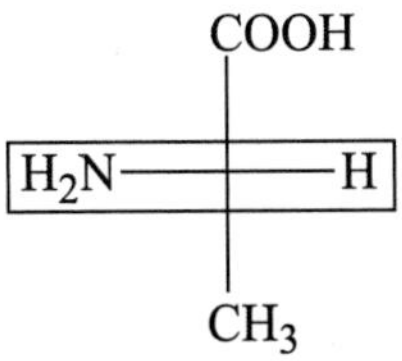

L-A lanine

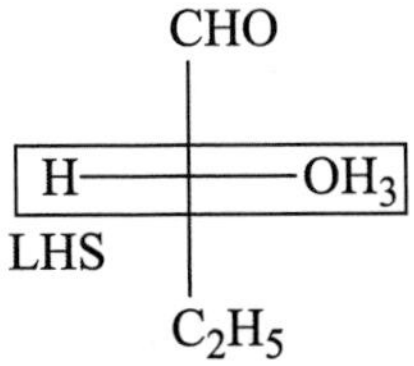

2-L-Bromo-2-D-methyl butanal

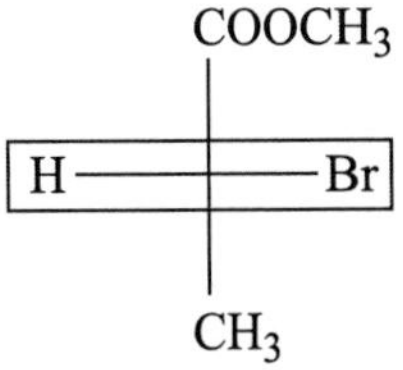

Methyl-2D-Bromo-propanoate

7.10 Configuration Notations Assignment

Likewise, we can write the configuration of in accordance to this nomenclature system. We only look for the last but one carbon substituent orientation. It may be observed here that the (+)- or (–)- sign of optical rotation of sugars, we have to memorise, then write the complete name as shown like, D-(–)- . You will find later, this method has been revised.

7.10.1 D and L Notations for Configuration of Amino Acids

Proteins are build upon α- amino acids units, of L-amino acids series. All amino acids have substitution of amino group at the α-carbon, oriented towards the left hand side. Its mirror image has right hand side orientation of amino group. The D-series amino acids are not found in natural proteins.

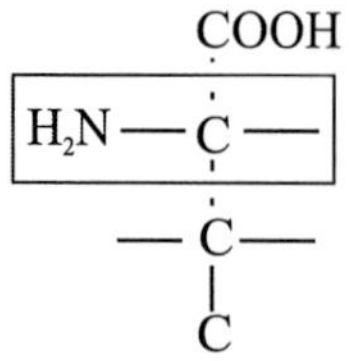

L-Series, Amino acids
LHS orientation of NH_2

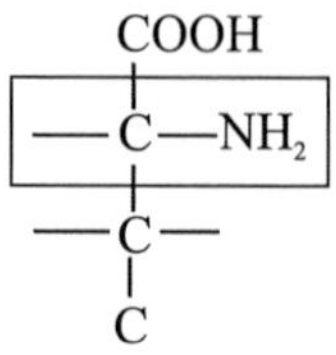

D-Series, Amino Acids
RHS orientation of NH_2

Some examples of amino acids are given below:

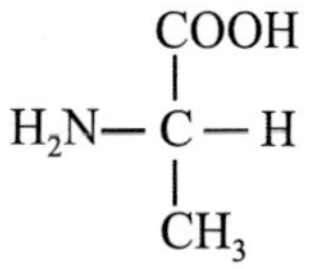

L-Alanine
It is (+)-

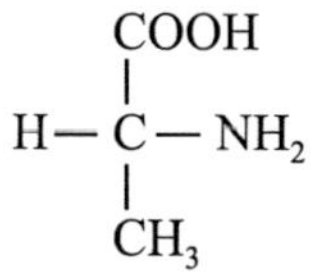

D-Alanine
It is (-)-

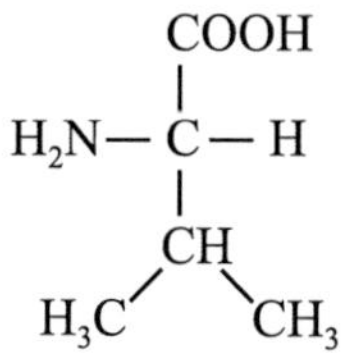

L-Valine
It is (+)-

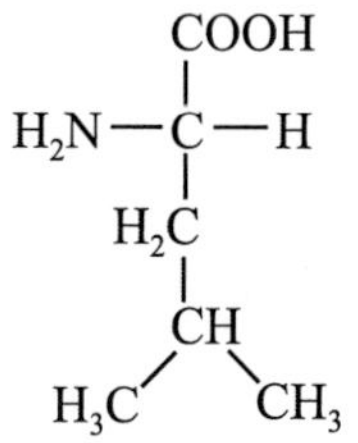

L-Leucine
It is (-)-

7.10.2 D and L Notations for Configuration of Tartaric Acid

However, there was a problem in assigning configurational notation to compounds with more than one asymmetric carbon. It is difficult to assign configuration, especially in symmetrical compounds, such as tartaric acid. Different forms tartaric acid are, enanmeric, meso and racemic form as given below.

Enantiomers

Tartaric acid, has two asymmetric centres. There are four isomeric structures. Structures I and II have non-superimposable object and mirror image relationship and are optically active isomers. These are called enantiomers.

7.10.3 Mesomers

Structures III and IV are mirror images but are superimposable, so are same compounds. The III and IV forms of tartaric acid are optically inactive, where cancellation of optical rotation takes place by internal compensation. It is called the **Meso forms** of tartaric acid.

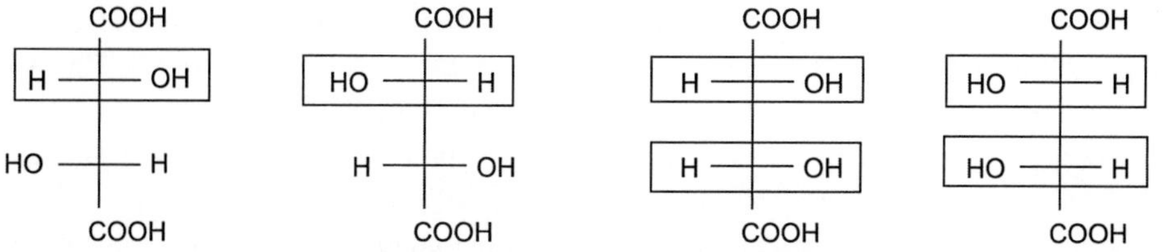

D-tartaric acid I | L-tartaric acid II | meso-tartaric acid III | meso-tartaric acid IV

Fig. 7.36 Four forms of tartaric acid.

Racemic Form

When 50% of structure I, having (+)- rotation is mixed in equal proportion with 50% of structure II, having (-) rotation, the mixture so produced shall have zero optical rotation, the mixture is called racemic mixture.

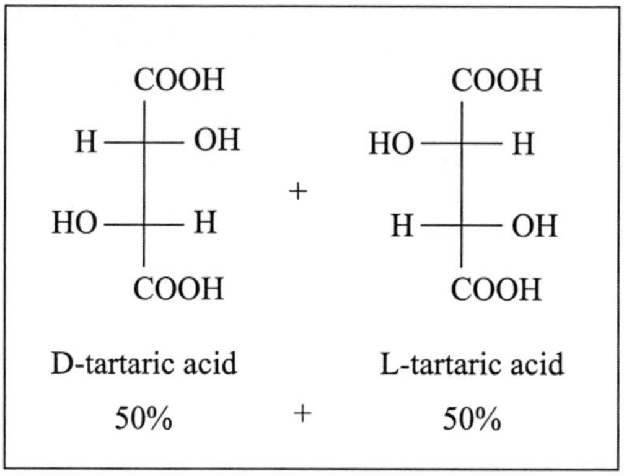

= Racemic Mixture

Also known as *dl*-pair

= 100% racemic mixture to two compounds

7.10.4 Problem of Tartaric Acid Configuration

According to Rosanoff's configuration notation, the right hand side orientation of hydroxyl shall mean D-configuration and Left hand side orientation of hydroxyl means L-configuration.

You will notice that the structure I has C3 carbon hydroxyl on the left hand and number C2 carbon hydroxyl has orientation towards right hand side, thus accordingly it **could** mean C2 as D-configuration and C-3 as L-configuration. **But it is not true. Experimentally, it has been found that structure I is dextro-rotatory, (+)- and II is leavo-rotatory, (-)-. So there may be something wrong in presentation of structures or method of notations .** A detailed examination solves that problem.

Solution of Tartaric Acid Configuration Problem

It is clarified by first rotating the molecule-I through 180° angle and then recounting the carbon chain from the top. (we notice that rotation of I through 180° gives back structure I) The original C3 hydroxyl thus also becomes D- Configuration. It is explained in Fig.7.37. This process gives C2 as well as C3 configuration in tartaric acid as same, so both the carbon atoms are of D and D configuration. This explains non-cancellation of optical rotation due to C2 and C3 -OH group orientations in structure-I of tartaric acid. Similarly, we consider its enantiomer, that shall become L and L configurations. If Structure I is (+)- tartaric acid, then structure II becomes (–)- tartaric acid.

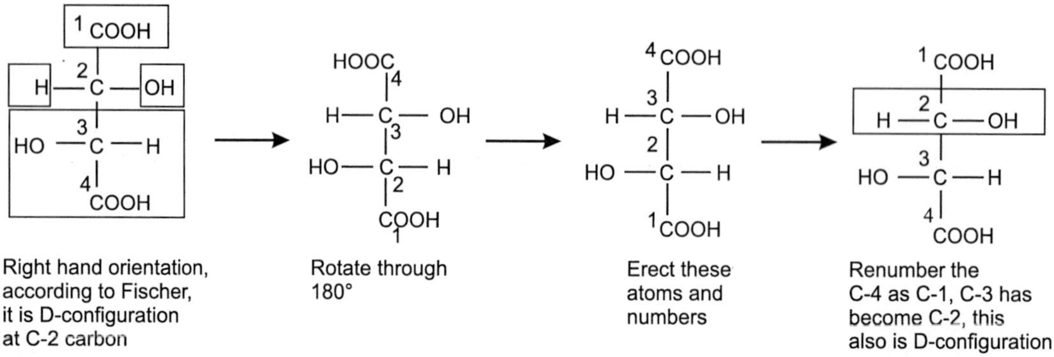

Fig. 7.37 Understanding that the structure I, both C2 and C3 are having D-configuration.

7.11 Configuration of Unsymmetrical Compounds having more than one Asymmetric Chiral Centre

More so, **in sugar series,** all the below given compounds have been named as D-series compounds.

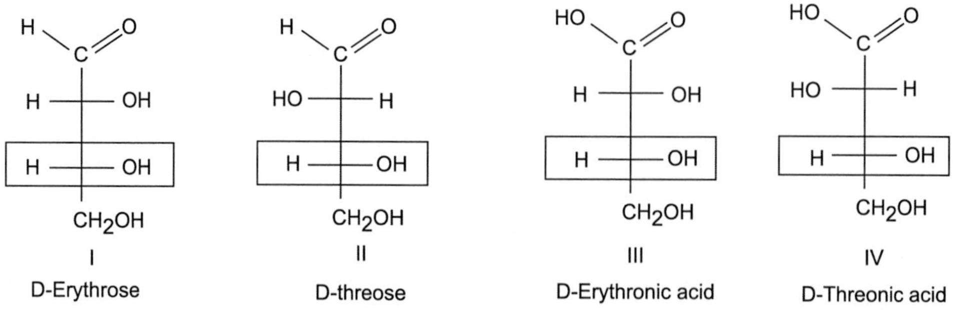

Fig. 7.38 D-notations mean RHS Orientation of –OH group.

Thus, the D and L notation procedure has it its own limitations.

1. The method of projection of three dimentional structure in to Fischer projection or other projection, need be stated all the times. It must be known, how the projection of the three molecule has been drawn.

2. Reference compound requirement for configuration assignment is a must, in this case is glyceraldehyde.

Thus, there is a need to use some alternative method of nomenclature of optically active compounds.

To overcome this problem, a newer system of nomenclature of configuration, where we **do not** use the projections, but the three dimensional (3D) model or the structure itself. It was introduced by Cahn-Ingold and Prolog (CIP method). It is called **Absolute Configuration** and is discussed below.

7.12 Cahn Ingold and Prelog Rules for Absolute Configuration (C I P Rules)- R and S Notations

1. List four groups in the given molecule in decreasing order of the **atomic number** of the first atom directly attached to the rigid part of molecule, that is the asymmetric carbon atom.
2. Observe the molecule from the side away from the atom or group of least priority.
3. If the priority of other three groups is clock-wise, the molecule is said to have 'R' configuration.
4. If the priority of other three groups is anti-clock-wise, the molecule is said to have 'S' configuration.
5. When the first atom of two or more groups are the same; look for the second atom in line in each group to decide their priority, again in decreasing order of their atomic number.

For example, two groups like –COOH and –$COOCH_3$, have same first atom joined to asymmetric atom. Its priorities are decided as below.

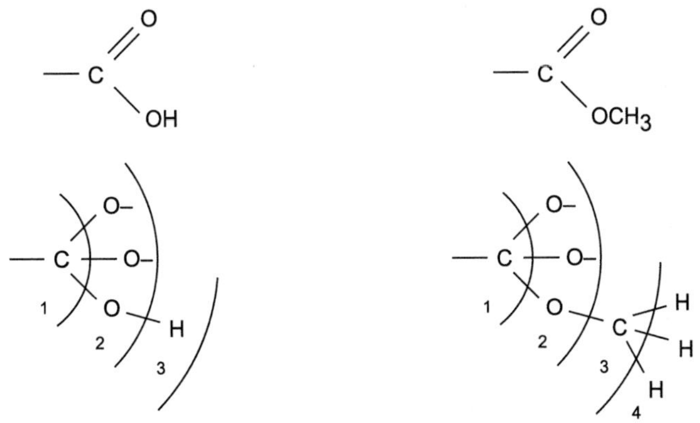

Numbers 1, 2, 3, 4 represent the first, second, third and fourth atom from the atom directly attached to the asymmetric centre

In these two groups first and second atoms are identical so we look for the third atom. Thus, the group $COOCH_3$ has priority over the COOH group

Fig. 7.39 Priority of substituents in nearly identical groups.

7.12.1 Examples for Configuration Assignment

Example 7.4 Assign the R,S,configuration of 2-bromopropanoic acid.

Solution: Given below is the structure of 2-bromopropanoic acid along with its mirror image. The priority of groups is Br > COOH > CH_3 > H. We look at the molecule from the side away from the atom of least priority and then count other three groups in order of their priorities as shown. The clockwise order in structure I, means 'R' configuration and anticlockwise order in structure II, means 'S' configuration.

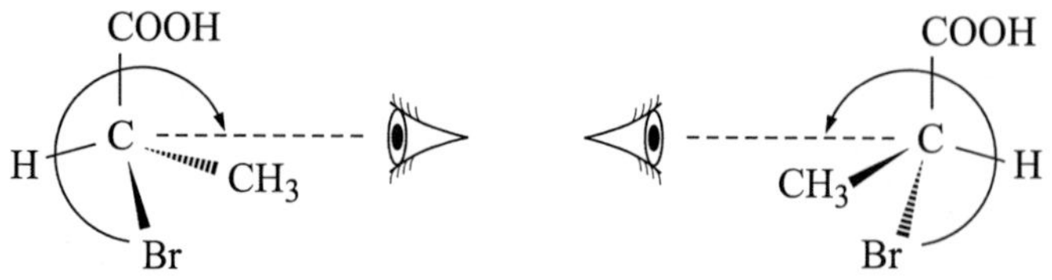

I, 'R'   II, 'S'

Example 7.5 Assign the R,S, configuration of 2-methylbutanoic acid.

Solution: Given below on LHS is one isomer of 2-methylbutanoic acid, andon the RHS is its mirror image. The priority of groups is COOH > CH_2CH_3 > CH_3 > H. We look at the molecule from the side away from the atom of least priority, that is H; then count other three groups in order of their priorities as shown. The clockwise rotation in structure II, means 'R' configuration and anticlockwise rotation in structure I, means 'S' configuration.

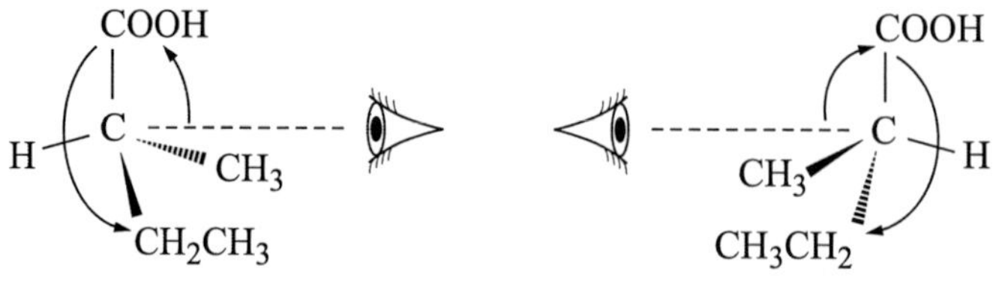

I, 'S'   II, 'R'

Priority of groups COOH > CH_2CH_3 > CH_3 > H,

7.12.2 An Alternate Method of Configuration Assignment from Fischer Projection

A method of mutual transfer of groups

Convert the 3D picture of 2-bromopropanoic acid as given in example 1 above in to Fischer projection and determine its configuration. Let us prove that the configuration of the molecule is same as given earlier.

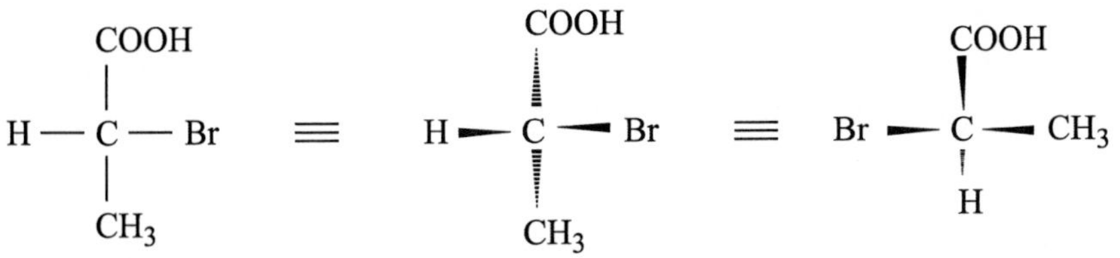

In this example, the priority of groups is Br > COOH > CH_3 > H

One of the requirement for CIP conventions, is view the molecule from the side away from group of least priority; in this case, it is H group. **A method of mutual transfer of groups** is used to bring the group of lowest priority in to the lower position of the cross representation.

To understand the interchange of groups and configuration assignment, we need to know the number of interchanges of groups. The rule followed is **"Odd number of interchange means opposite configuration, even number changes means same configuration as that of the original compound."**

The priority of groups is Br > COOH > CH_3 > H. This order of groups traces clockwise movement, therefore, the given compound is of **'R'** configuration.

Problem 7.1: Use CIP rules and determine the R or S configuration of the following compounds in their Fischer's projections.

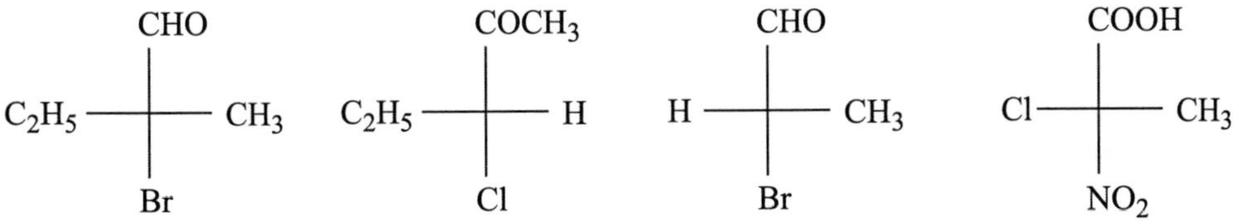

I II III IV

Problem 7.2: Given below are the Sawhorse projections of certain compounds, convert them in to Fischer's projection and determine their R or S configuration.

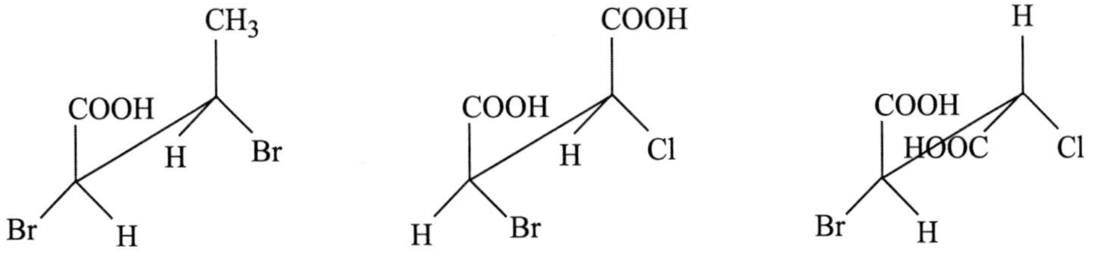

I II III

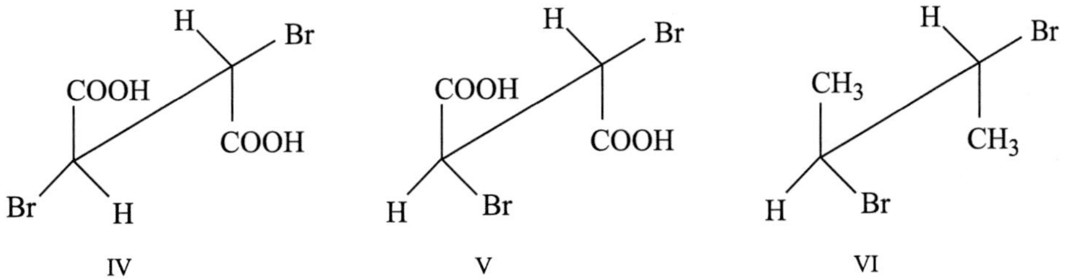

IV V VI

Problem 7.3: Given below are the Newman' projections of compounds having two asymmetric centres. Determine, which of these are identical, or different. Write the procedure of converting these structures into Fischer's projections and determine their configurations.

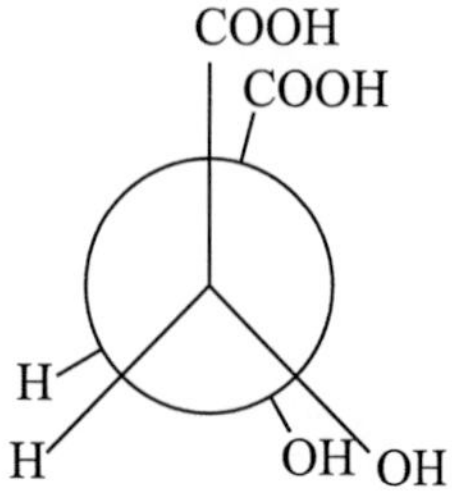

I

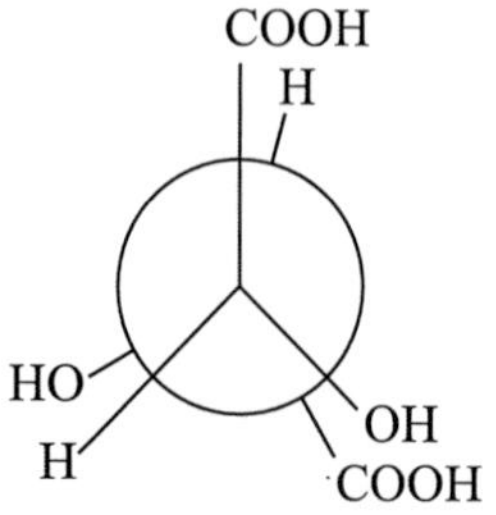

II

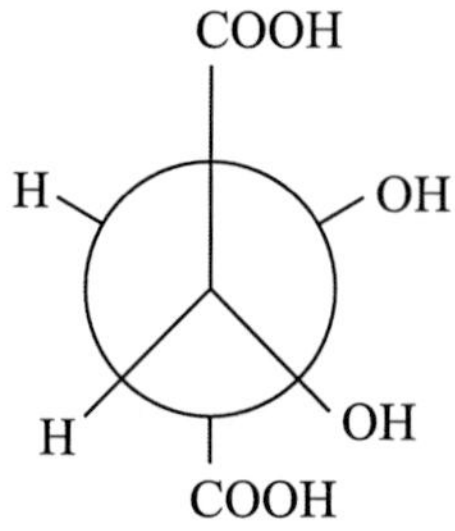

III

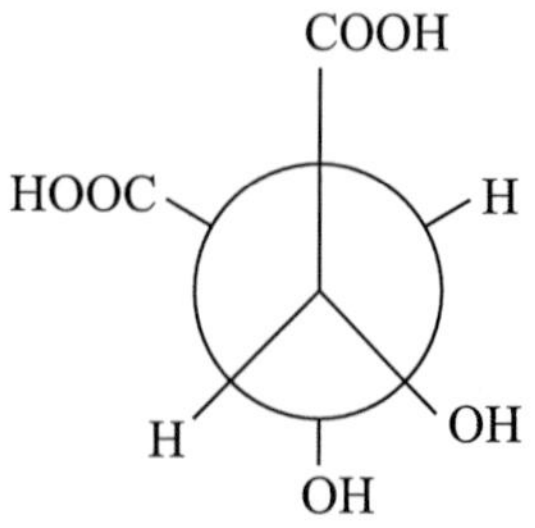

IV

Problem 7.4: Write the IUPAC names of the following compounds and determine the 'R' and 'S' configuration of all the asymmetric centres in the following Fischer's projections.

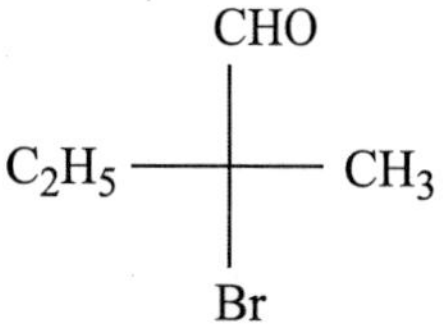

I

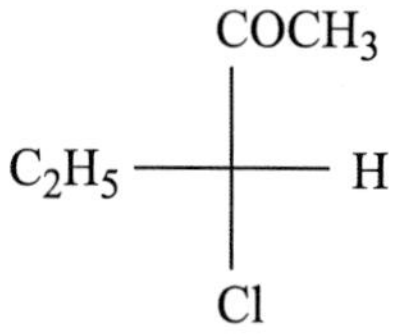

II

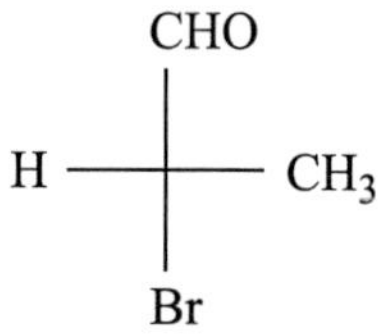

III

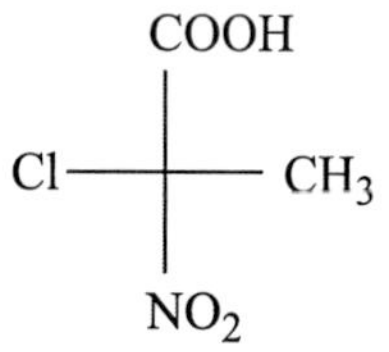

IV

Answers to Problems:

Problem 7.1. I = R, II = R, III = S and IV = R

Problem 7.2. I, C2 = R and C3 = S; II, C2= R and C3 = S; III, C2 = S and C3 = R; IV, C2 = S and C3 = R; V, C2= S and C3 = S; VI, C2 = S and C3 = S

Problem 7.3. I, C2 = S and C3 = R, it is identical with II,
III, C2 = S and C3 = S, it is identical with IV

Problem 7.4. I, C2 = R, C3 = R; II, C2 = R, C3 = S; III, C2 = S, C3 = S; IV, C2 = R, C3 = S

7.13 Geometrical Isomerism

It is an isomerism caused by restricted rotation due to double bond between a carbon and carbon.

7.13.1 Isomerism Caused by Double Bond

Goemetrical isomerism is exhibited by compounds containing double bond. When two carbon atoms are linked through a double bond, the free rotation between them is restricted, as shown in the molecular orbital picture of ethylene. This creates a specific configuration of the molecule at the time of formation of double bond. Ethylene does not exhibit geometrical isomerism, but but-2-ene forms two geometrical isomers.

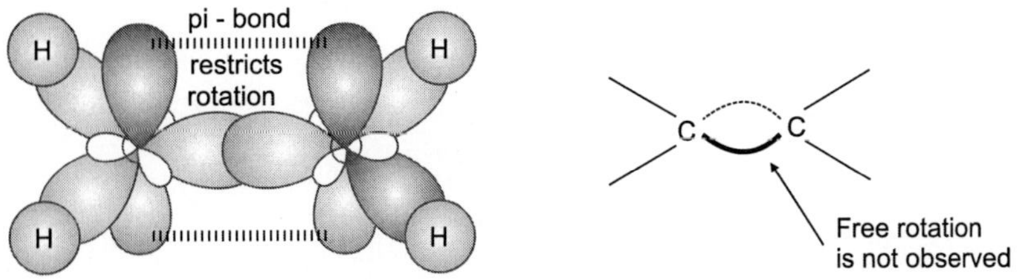

Fig. 7.40 Molecular Orbital picture of ethylene and geometrical isomers of but-2-ene

Two geometrical isomers get formed due to different spatial orientation of groups when elimination takes place to form a double bond. It is because of the free rotation of substituents in a single bonded compound.

Consider CH_3–CH = CH – CH_3, but-2-ene, two groups on C_2 are CH_3 and H and same are the substitutents at C3. Two possible orientations of groups around C = C are as given below.

cis - But-2-ene *trans* - But-2-ene

When two identical groups,say **two methyl groups in but-2-ene are placed on the same side of the double bond, the isomer is called *cis*- isomer** and when these are on **opposite side of double bond, the isomer is called *trans*-isomer.**

Kurl Alder Born in Königshütte, Germany (1902-58). Alder took his Ph.D. degree with Diels at the University of Kiel in 1926.Alder is best known for his discovery with Otto Diels of the cycloaddition reaction that bears their names, the Diels-Alder (DA) reaction. For this discovery they received the 1950 Nobel Prize in Chemistry. The Diels-Alder reaction, involves 1,4-addition of diene and dienophile. their examples included cyclopentadiene, acyclic 1,3-dienes, furans and anthracene. It is a great reaction of building up the carbon skeletons with great ease. In 1955, Alder signed, with 17 other Nobel laureates, a declaration calling on nations to renounce war as a means of settling disputes

7.13.2 Necessary Condition of Substitution to Show Geometrical Isomerism

In case of but-2-ene, alignment of two methyl groups on the same or opposite side of double bond, helped us to know whether it is *cis*- or *trans*- isomer. When groups at two carbon of double bond are not identical, newer condition is introduced.

Situations of substitutions like given below can arise.

1. Two methyl groups of 2-methyl prop-1-ene lie on the same carbon, so does not form geometrical isomers.
2. Similarly, but-1-ene, has two hydrogens on the same carbon, so does not form geometrical isomers.
3. Also when four groups are same , it shall not form geometrical isomers.

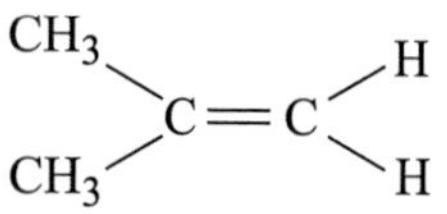

2-Methyl prop-1-ene

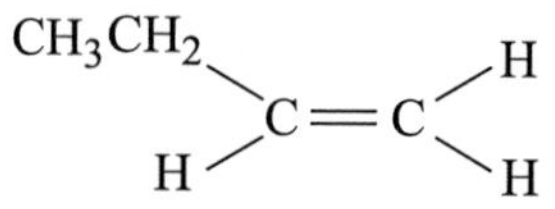

But-1-ene

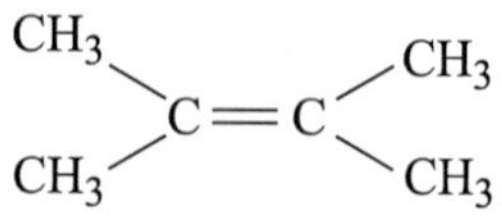

2,3-Dimethyl but-2-ene

4. More so, when all the four groups are different, naming process as above is ambiguous.

Some examples of **alkenes that do not show geometrical isomerism**.

1. Pent-1-ene, $CH_3 - CH_2 - CH_2 - CH = CH_2$
2. 2-Methylbut-2-ene $CH_3 - CH = C - CH_3$ (with CH_3 on the C)
3. But-1-ene, $CH_3 - CH_2 - CH = CH_2$
4. Propene, $CH_3CH = CH_2$
5. 1,1-Dichloroethene

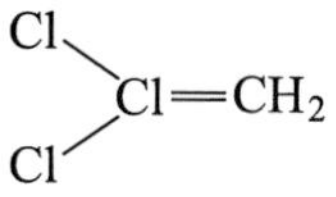

6. 2-Chloropropene

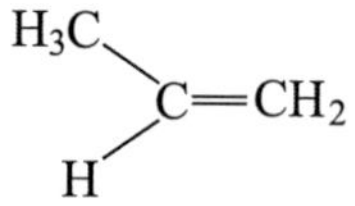

A different method of nomenclature is adopted, where a condition necessary for a molecule to show geometrical isomerism is as given below.

For an alkene of the type *abC = Cde*, the necessary condition for it to show geometrical isomerism is that *a*, *b* and *d*, *e* are not identical.

The necessary condition for an alkene show geometrical isomerism in abC=Cde, is $a \neq b$ **and** $d \neq e$

7.13.3 E, Z-Notations for Geometrical Isomers

Most common nomenclature of geometrical isomers is *cis* and *trans* . It is quite easy in case of alkenes having symmetrical substitution pattern, *e.g.*, but-2-ene. In case of unsymmmetrical substitution, *e.g.*, 3-methyl-hex-2-ene, this nomenclature is not used that conveniently.

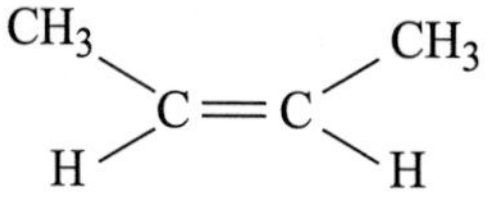

cis - But-2-ene

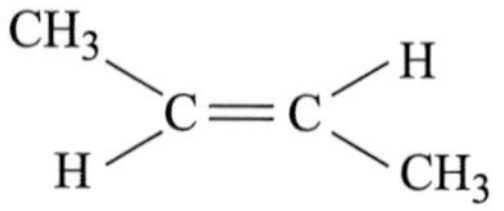

trans - But-2-ene

When the alkene is unsymmetrical, then we can't talk of similar groups in an alkene lying on the same or opposite side of double bond. We can still name these compounds as cis and trans reference to particular groups. But is not a better method of naming of compounds. More covenient method of naming of geometrical isomers of unsymmetrical alkenes, is to use **E and Z notations**. These notations have been derived from german words. E- standing for '*entgegan*' which means opposite side and Z- standing for '*zusammen*', which means together. It uses the set of rules for priority of groups for designating a compound as E or Z.

According to **these rules,**

1. **If two groups of higher priority on two carbon atom forming the double bond lie on the same side, the isomer is said to have Z-configuration (zusammen-together).**
2. **If the two groups of higher priority on two carbon atoms of double bond lie on the opposite side, they are said to have E-configuration (entgegan, opposite).**

Summary of Rules for assigning the Priority of groups on carbon atoms of double bond

The priority rules used for giving priority of groups in geometrical isomerism are also same, as were used for R/S notations by Cahn-Ingold and Prelong. For revision, these are given here again.

1. Arrange the groups on one carbon atom forming the double bond, in **decreasing order of atomic number** of the first atom directly attached to the carbon atom. The priority of common groups is given in order as below.

$$-I > -Br > -Cl > -SH > -SO_3H > -NH_2 > -COOCH_3 > -COOH > -CH_3$$

2. If first atom is the same, look for the second atom on the substituents. Thus, priority for amines is

$$-N(CH_3)_2 > -NHCH_3 > -NH_2$$

3. If second atom is also same, look for the third atom on the stubstituents.

$$-COOCH_3 > COOH$$

4. A doubly bonded atom in a group, such as oxygen in C=O, the oxygen atom is taken as equivalent to two oxygen atoms bonded to the same carbon atom.

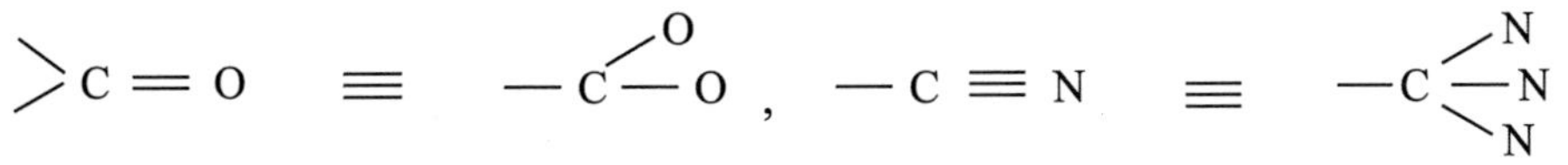

7.13.4 Examples of Geometrical Isomers

1. Let us take an example of 2-chloro-pent-2-ene, (CH_3– CH_2– CH = C(Cl)CH_3). It is possible to write two structures for this compound, I and II. In structure I, we have methyl and chorine on C2, out of these because of higher atomic number of first atom, that is, chlorine (atomic number,17), it gets priority over methyl, whose first atom is carbon (atomic number, 6). On C3 two groups are ethyl and hydrogen, out of these, ethyl has higher priority. The higher priority groups on C2 and C3 lie on the opposite side, therefore, its configuration is E. In structure II, the higher priority groups are on the same side, therefore, its configuration is Z. In both structures, the higher priority groups are encircled.

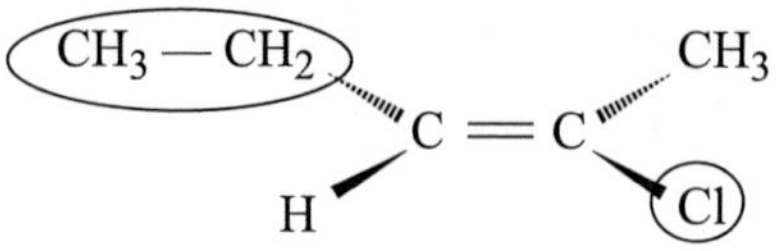

I

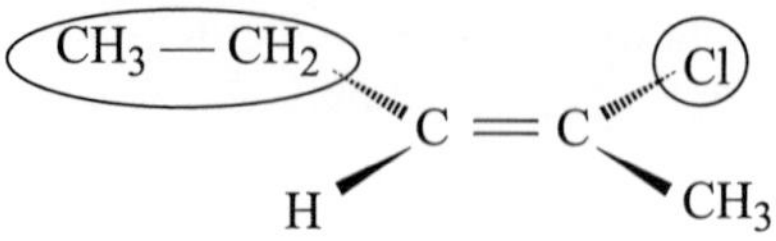

II

E - Isomer

Higher priority groups, Cl and Ethyl on opposite side

Z - Isomer

Higher priority groups, Cl and Ethyl are on same side

2. Other examples are, 1,2-dichloropropene, but-2-endioic acid and 1,2-dibromopropene, etc. The

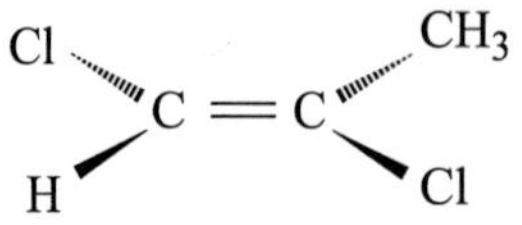

E-1, 2-Dichloropropene
trans-isomer

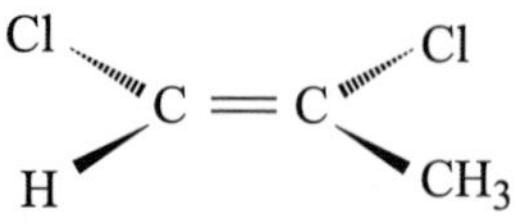

Z-1, 2-Dichloropropene
cis-isomer

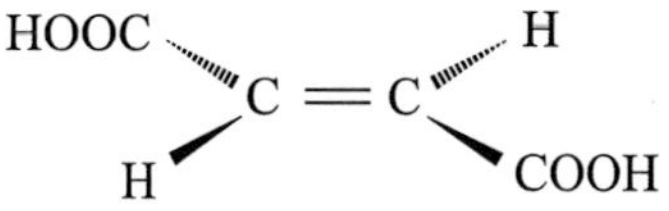

E-But-2-en-dioic acid
Fumaric acid
trans-isomer

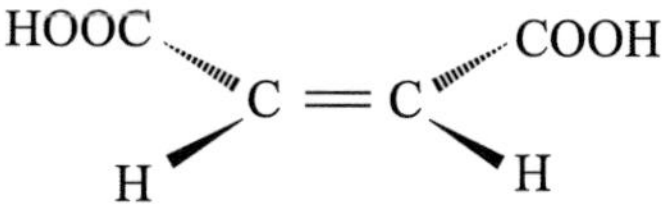

Z-But-2-en-dioic acid
Maleicacid
cis-isomer

Problem 7.5: Assign configuration of the following alkenes.

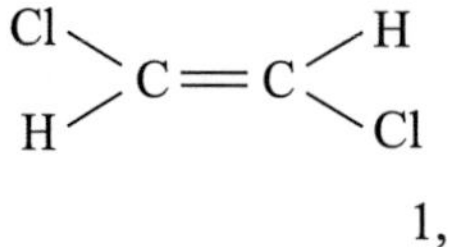

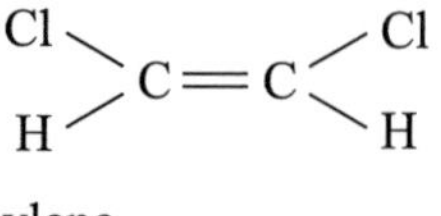

1, 2-Dichloroethylene

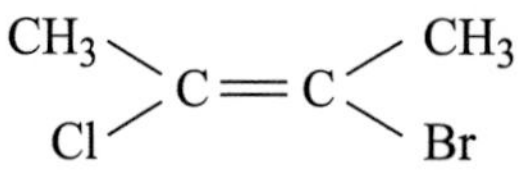

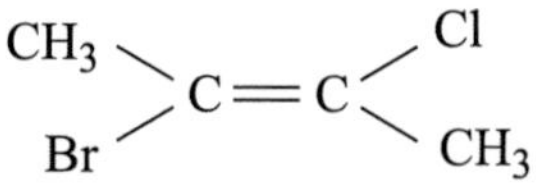

2,3-Dichlorobut-2-ene

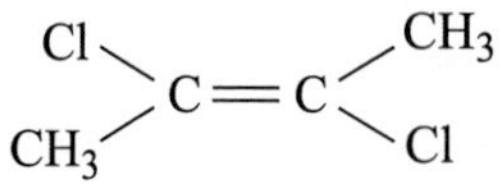

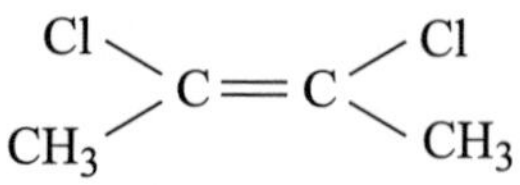

Pent-2-ene

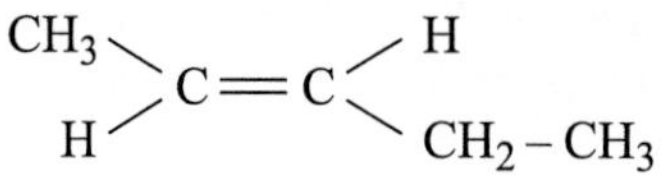

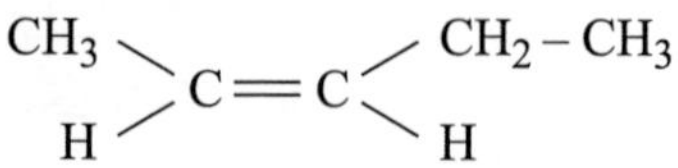

2-Bromo-3-chloro-but-2-ene

7.14 Absolute Configuration

The configuration determined by an unambiguous method, such as X-ray analysis is termed as absolute configuration. As late as 1951, Bijvoet, analysed the X-ray crystallographs and arrived at the structure of potassium rubidium salt of tartaric acid. Thus, he showed unambiguously, the configuration of dextro-rotatory, (the naturally occurring tartaric acid) as IX in Fig 7.15 and I in Fig. 7.22. The term absolute relates to the fact that configurational assignment have been carried out with out any reference compound. However, it is a tedious method and we cannot determine X-ray of every compound. A simpler and alternate method was then adopted by Cahn, Ingold and Prelog. It uses no reference compound and in that sense is also called absolute method of configuration determination. It uses model or three dimensional structure or their Fischer projection, and based on the following rules, the R and S notations have been adopted.

7.15 Revisiting Some Terms Used in Stereochemistry

Let us now learn some distinguishing details about the **terms used in stereochemistry**.

(*i*) Constitution

The term constitution describes the linkages of the atoms involved in the structure of a given molecular formula. For example, the constitution of ethanaol, C_2H_5OH , can be written as follows:

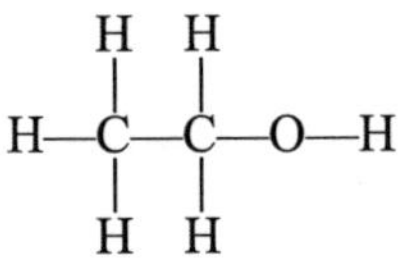

(*ii*) Configuration

The spatial arrangement of groups or atoms around a stereocentre (the rigid part of a molecule or carbon atom) is termed as its configuration. Lactic acid has two different configurations as shown below. That represents its pair of two configurations called enantiomers (defined below).

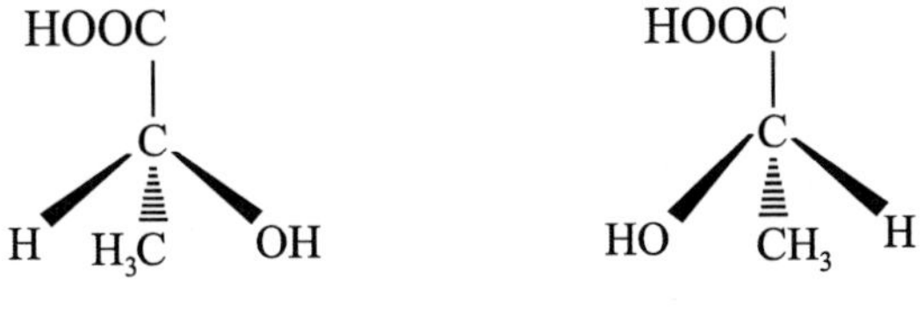

Two different configurations representing two enatiomers

(*iii*) Optical Rotation

It is the property of optical isomers expressing rotation of plane polarised light, due to a compound placed in its path in solution or gas form.Such a compound is said to be an optically active compound. The rotation is distinctly observed through the eye piece of the polarimeter.

(*iv*) Enantiomers

Chiral molecules may have one or more chiral centres. Considering one chiral centre in molecules, like lactic acid, butan-2-ol and alanine etc., we find that these form a pair of an object and its mirror image that are not superimposable to each other. A pair of molecule and its mirror image that are non-superimposable, are called **Enantiomers.** Two enantiomers of lactic acid are shown above and both are optically active, showing dextro and leavo rotation.

(*v*) Racemic Modification

A mixture of two enantiomers (optically active isomers) in equal amounts (each is 50%), it is called racemic mixture or **Racemic Modification**. Such mixtures possess zero optical rotation, that is, mixture is optically inactive mixtures due to **external compensation**. Such mixtures can be separated into its components, the dextro-rotatory and leavo-rotatory compounds. This process is called resolution.

(*vi*) Mesomers

When a compound has two chiral centres, it is the configuration of these chiral centres, that makes the molecule optically active or optically inactive (zero rotation). When it is optically inactive, it has zero rotation due to **internal compensation**. In such a compound, cancellation of contribution of rotation due to two chiral centres takes place in opposite direction. Compared to racemic modification, a mesomer cannot be separated in to any optically active compound, since it is a single compound.

(*vii*) Diastereomers

When there are more than one chiral centre, it has 2^n isomers. For example, the four carbon containing sugars, (molecular forumula, $C_4H_8O_4$) called , tetroses have two chiral centres Thus, there are four isomers. Tetrose isomers are represented in Fig.7.26. Non-superimposable mirror images are the enantiomers and other **stereoisomers that are not enantiomers are called diastereomers.** Enantiomers and diastereomers of a compound, have same molecular formula, same linkage of atoms and thus same structutre but the stereostructures are different. Diastereomers do not form superimposable mirror images. Diastereomers may have many identical chemical properties, but the physical properties are different.

(*viii*) Chiral Molecules

Chiral molecules are those molecules that possess handedness. It is the handedness of molecules due to which they become non-superimposable on their mirror image and thus optically active.The chiral word is derived form Greek word (meaning hand). The word chiral can be used for an object as well as for molecules, that has right handed or left handed disposition. The property of handedness of molecules is termed as **'Chirality'** in stereochemistry. It has been clarified in section 7.4.

(*ix*) Erythro

Erythro is the term given to substituents situated on same side of a chiral carbon atom.

(*x*) Threo

Threo is the term given to substitutents situated on opposite side of a chiral carbon chain.

(*xi*) Conformation

A conformation is one of the infinite number of possible arrangements that can be obtained by free rotation of substituents relative to those on adjacent carbon atoms joined through a single bond.

Points to Remember

- Conformation of a molecule is one of the infinite number of possible arrangements that can be obtained by rotation of groups at one carbon reference to other joined through single bond.
- Saturated compounds like ethane, propane and butane or cyclic compounds show conformations.
- Ethane and propane have two prominent conformations, called eclipsed and staggered conformations.
- The energy difference between these two conformations is very small. These can only be isolated at low temperature near to its freezing point to stop molecular rotation of groups.
- In case of butane, four different conformations are possible. Two are the same as ecipsed and staggered and other two are known as gauche conformations.
- The most stable conformation in each of the molecules are the staggered conformations.
- In case of cyclic compounds, cyclopropane and cyclobutane are relatively unstable, but cyclopentane and cyclohexane are relatively stable.
- Cyclohexane exists in chair and boat forms. Chair is more stable conformation. In case of substituted cyclohexanes, configurations are also possible and some of the compounds can be optically active.
- Attempts were made to explain the stability of these ring systems by Baeyer. The Baeyer ring strain theory survived for many years, but later was rejected due to wrong explanations in case of five and six membered ring systems.
- Sachse later introduced the idea of puckered ring systems. It very well explained the formation of strainless ring system and thereby offered explanation for their stability.

Exercises

1. Draw the Wedge diagram and Fischer projection for all chiral aliphatic mono alcohols ontaining five carbons, with formula, $C_5H_{12}O$.
2. Draw all the possible cyclic structures of chiral mono alcohol containing six carbon atoms, $C_6H_{14}O$.

3. Draw the pictures of five chiral objects in your study room.
4. Menthol and camphor are sweet smelling compounds, search a website and write their structures, and put astrisk mark on the chiral centre.
5. Proteins are made up of α -amino acids, draw the structure of any three chiral α -amino acids. Draw the structure of an amino acid without chiral centre.
6. Sugars are also optically active compounds, but not all. Write the structure of glucose and another hexose that is not optically active.
7. Given below are the structures of (-)- Lactic acid and (+)- lactic acid, assign their R- and S- configurations. Extend the knowledge to write the structures of R- and S- glyceraldehyde. Is it possible from the given structure, to determine which one has dextro- and which one has leavo- rotation.

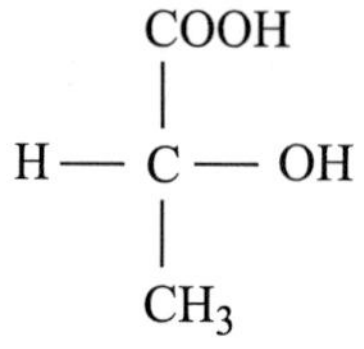

(–) – Lactic Acid

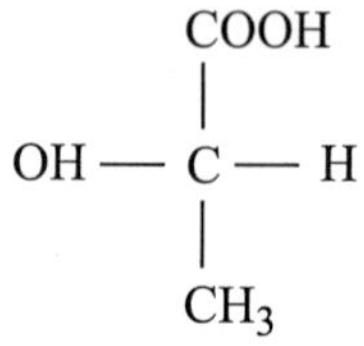

(+) – Lactic Acid

8. Draw the tetrahedral representation of (R)- 2-chloropentane. What configuration shall you assign if the given name of an isomer is just 3-chloropentane.
9. Write the IUPAC names of the following compounds and determine the R or S configuration of all the asymmetric centres in the following Fischer's projections.

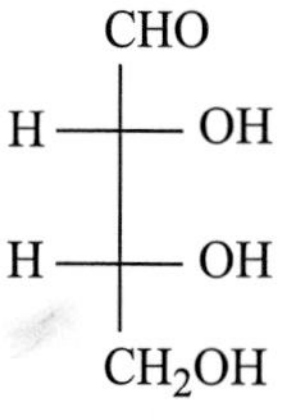

I

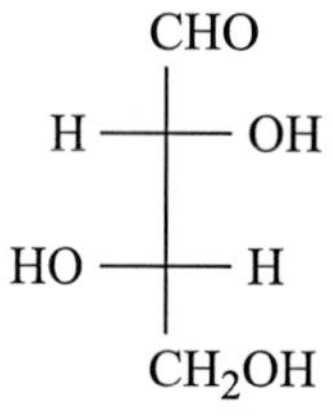

II

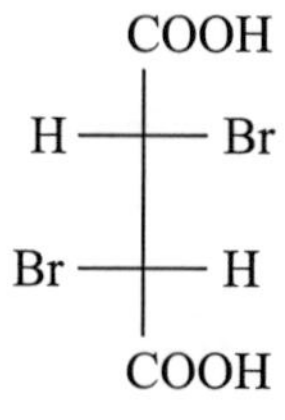

III

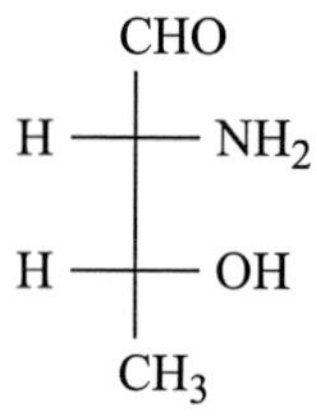

IV

10. In a given molecule, if following are substitutents at the chiral centre, assign their priorities according to Cahn-Ingold Prelog conventions.

    (*a*) CH_3, H, COOH and $COOCH_3$ (*b*) CHO, H, CH_2OH and CH_3

    (*c*) Cl, Br, H, CH_3 (*d*) Br, Cl, CH_2Cl, CH_2Br

    (*e*) CH_3, H, OCH_3, COOH (*f*) CH_2OH, NH_2, H, COOH

11. Draw the Fischer Projection of RR- tartaric acid and S-malic acid. How many stereo-isomers of tartaric acid and malic acid are possible.

12. Assign the R and S configuration of each of the stereo-centre of the following molecules:

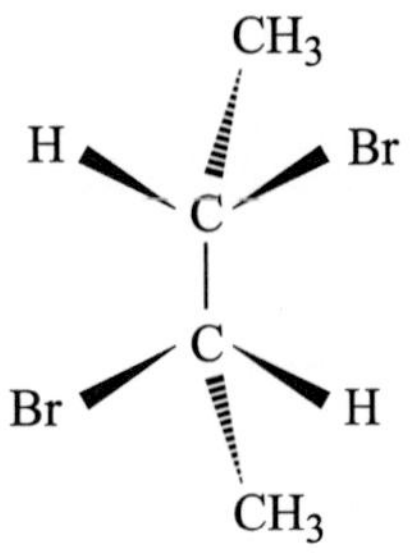

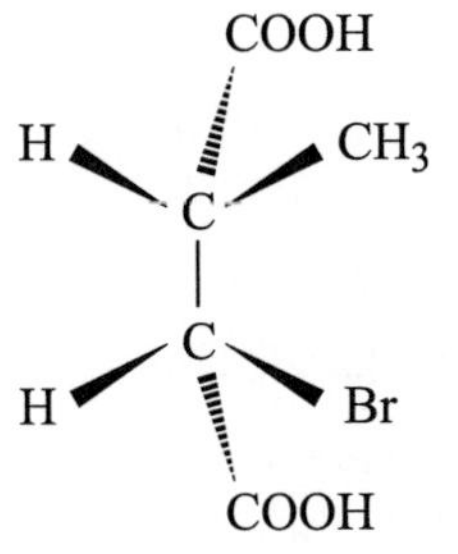

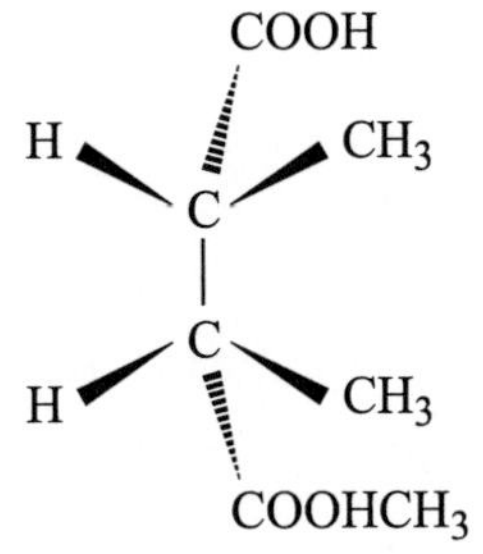

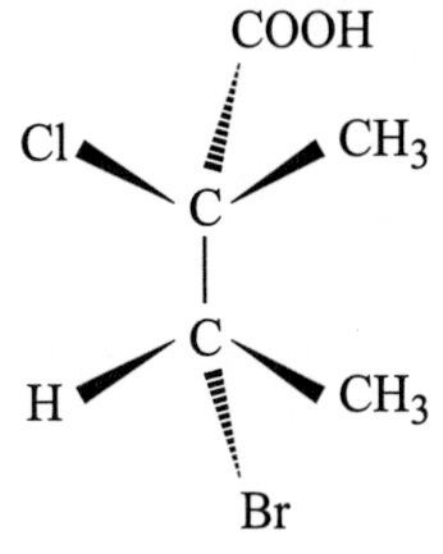

I II III IV

13. Given the molecular formula, $C_5H_{12}O$, find out how many optically active structures including enantiomeric structures can be written.

14. Menthol is a terpene, its structure is given below. Find out how many isomers it can have. Write the structures of all possible stereo-isomers.

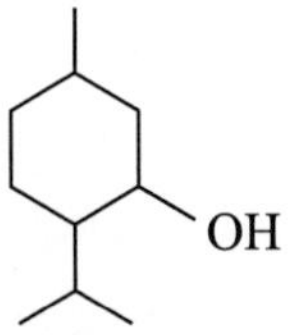

15. Penicillin is an antibiotic isolated from the mould of *penicillium* species. It has three asymmetric centres marked in the structure given below. If the stereo-chemistry of any one of the centre is changed, the biologically penicillin becomes inactive. Explain, why it happens so? Also label the chiral centres.

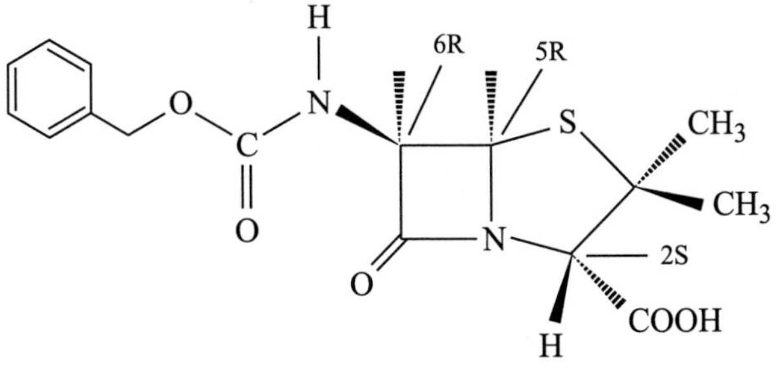

Penicillin G, 2S, 5R, 6R configuration

16. Threonine is an amino acid (2-amino-3-hydroxy butanoic acid) having two asymmetric centres. Write all the possible stereo-isomers and label the stereo-centres with R and S nomenclature.

17. Tartaric acid, (2,3- dihydroxy butanoic acid) has two asymmetric centres. Draw the stereostructures of all possible isomers and label the stereochemistry with R and S notation of nomenclature.

18. Ribose is an aldopentose sugar, (CH_2OH–*CHOH–*CHOH–*CHOH–CHO), having three asymmetric centres. How many stereoisomers are possible? Write the structure of all possible stereoisomers.

19. In the above problem, distinguish the enantiomers and diastereoisomers

20. How many of the isomers are optically active and what is the relationship among them self of optical activity of different isomers written by you?

21. Naturally occurring 2R,3R- tartaric acid has optical activity, $[\alpha]_d = +12°$. What shall be the optical activity of 2S, 3S- tartaric acid?

22. What do you understand by the term conformation? How many conformations of chloropropane are possible when present in a cylinder at room temperature?

23. What is the cause of restricted rotation around single bond? How many conformations for diphenic acid ($2,2'$ -Dinitro-diphenyl- $6,6'$ -dicarboxylic acid) molecule are possible?

24. Write the structures of geometrical isomers of the following compounds.

1. Pent-1-ene, $CH_3-CH_2-CH_2-CH=CH_2$
2. Pent-2-ene, $CH_3-CH_2-CH=CH-CH_3$
3. Isoprene, 2-Methylbut-2-ene, $CH_3-CH(CH_3)=CH-CH_3$
4. Chloroprene, 2-Chlorobut-2-ene, $CH_3-CH(Cl)=CH-CH_3$
5. But-1-ene, $CH_3CH_2CH=CH_{2.}$

* While attempting this, we are likely to face some anamolies for deciding the *cis* and *trans* isomers of these compounds. Some of these do not exhibit *cis - trans* isomerism.

25. How many geometrical isomers exist for the following compounds? Draw their structures.

(*a*) 3-Bromo-but-1-ene

(*b*) 1,3-Dichlorobut-1-ene

(*c*) 1,4-Dichlorobut-1,3-diene.

26. Which of the following compounds shall have higher dipole moment ?

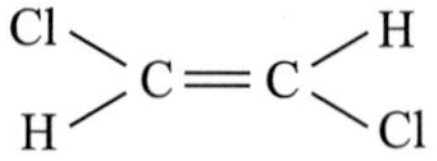

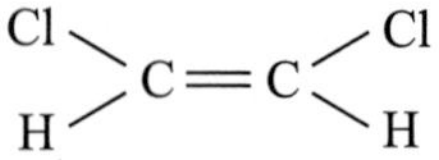

27. Which compound in the following pairs has higher stability or lower ΔH ?

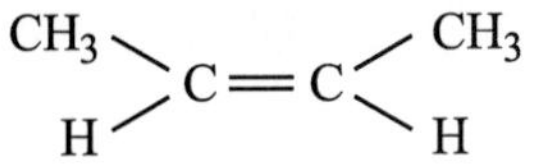

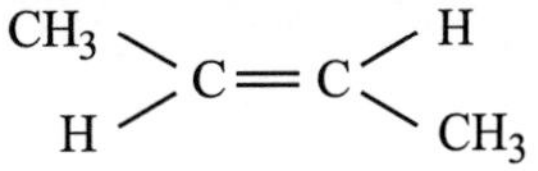

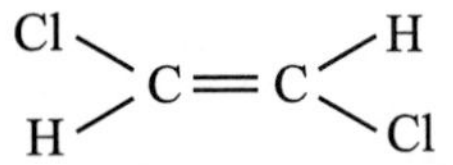

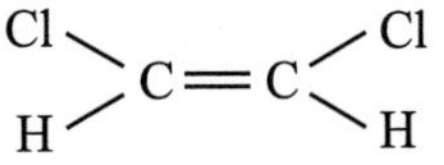

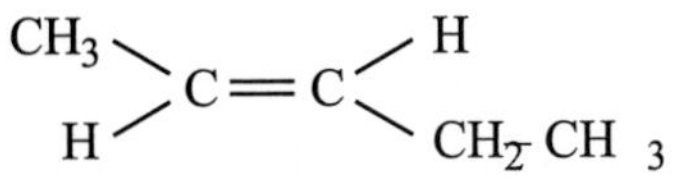

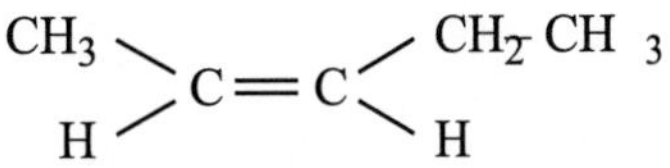

28. What will be the product of reaction in the following isomers

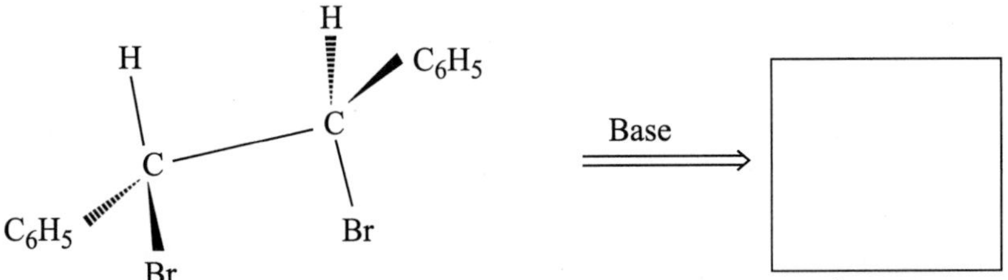

cis or *trans*

(1R, 2R)-1,2-dibromo-1, 2-diphenyl ethane

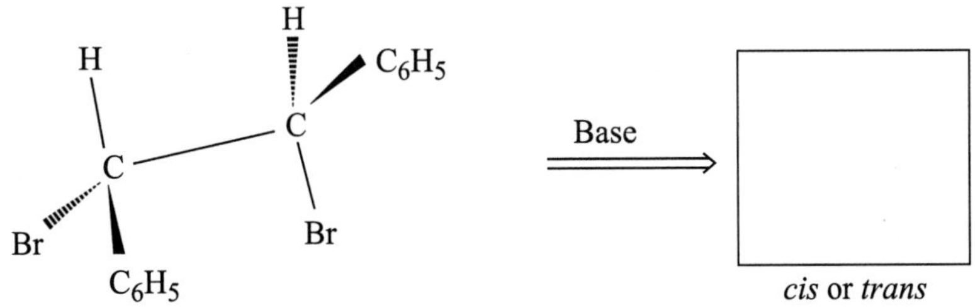

cis or *trans*

meso- 1, 2-dibromo- 1, 2-diphenyl ethane

29. Which compound of the following pairs shall be more stable and which of the structure can show geometrical isomerism ?

(*i*) 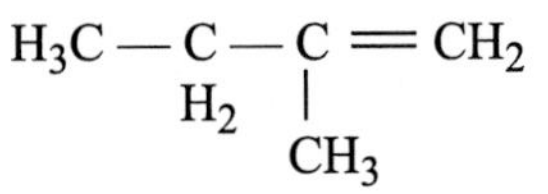 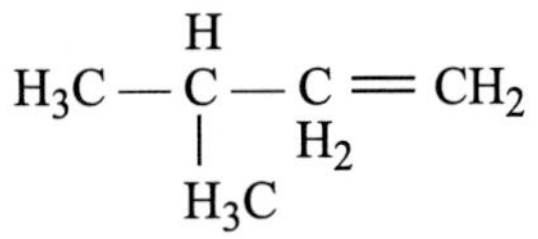

(ii) 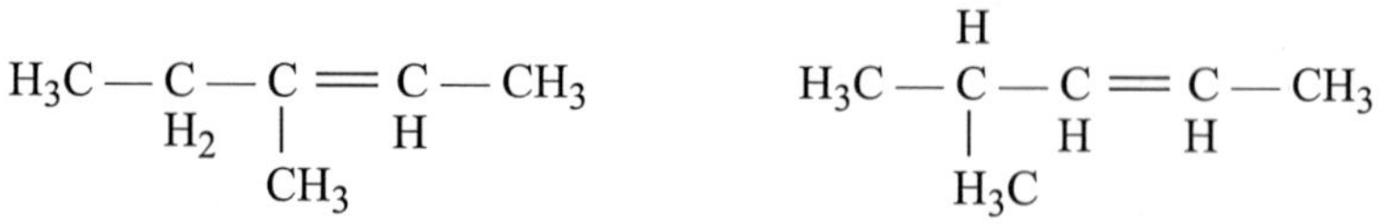

(iii) 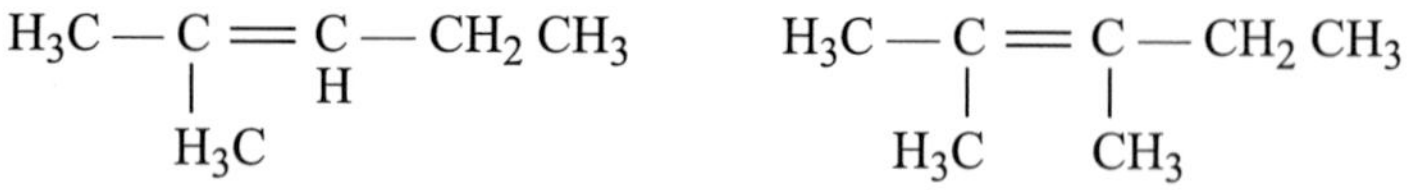

(iv) 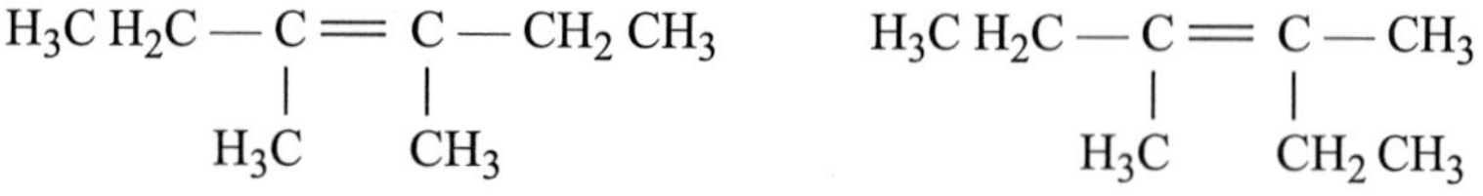

30. How do you account for the boiling point of the following alkenes as per value given against each alkene?

(i) 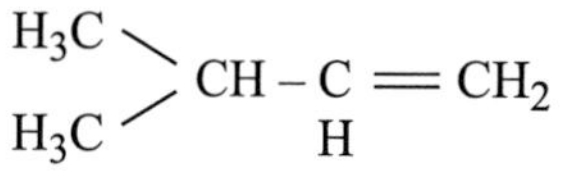 25°C

(ii) 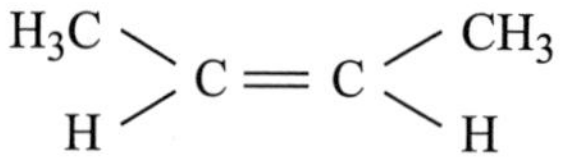 4°C

(iii) 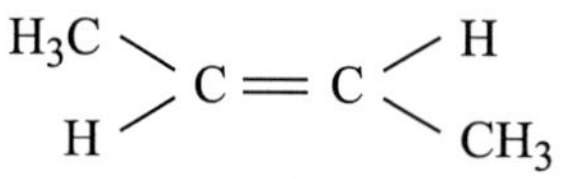 1°C

(iv) 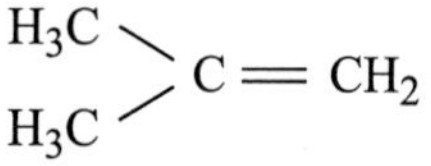 −6°C

31. Which of the following pairs shall have higher value of dipole moment?

(i) *Cis*-1,2-dichloroethene and *trans*-1,2-dichloroethene

(ii) *Cis*-1,2-dichloroethene and *trans*-1,2-dibromoethene

(iii) *Cis*-1,2-dichlorocyclohexene and *trans*-1,2-dichlorocyclohexene

32. Carry out the addition of bromine to *cis*- and trans-pent-2-enes and give mechanism of the reactions. Find out how many stereo-isomers shall be formed in each case.

33. Which of the isomers, 1e,2e-dimethyl cyclohexane or 1a,2a-dimethyl hexane shall have greater heat of combustion?

35. Draw the mirror images of 1a,2e- dichlorocyclohexane and 1a,4a- dichlorocyclohexane and determine, which of the compounds shall show optical activity.

❑❑❑

Adolphe Wurtz (1817–1884)

8

Aliphatic Hydrocarbons –Alkanes

Adolphe Wurtz (26 November, 1817–10 May, 1884) was an Alsatian French chemist. He is well known by organic chemists for thc Wurtz reaction, to form carbon-carbon bonds by reacting alkyl halides with sodium, and for his discoveries of ethylamine, ethylene glycol, and the aldol reaction. When he left the Protestant gymnasium at Strasbourg in 1834, his father allowed him to study medicine as next best to theology. He devoted himself specially to the chemical side of his profession with such success that in 1839 he was appointed Chef des travaux chimiques at the Strasbourg faculty of medicine. For the summer semester of 1842 he studied under Justus von Liebig at the University of Giessen. After graduating from Strasbourg as M.D. in 1843, with a thesis on albumin and fibrin, he went to Paris, where he worked in Jean.

The new chemistry of the 1850s took the idea of chemical atoms seriously, adopted atomic weights for the elements that strongly resemble the modern ones, and proposed a unitary schematic plan that opposed the dualistic theory derived from the work of Jons Jacob Berzelius. Soon thereafter, Wurtz also adopted the new structural theory that was developing from the work of younger chemists such as August Kekule. Influenced by such leading figures as Liebig and Dumas, by 1856 Wurtz became a powerful advocate of a reform in chemical theory. Wurtz's worked on hypophosphorous acid and discovered sulfophos-phoric acid and phosphorus oxychloride, as well as of copper hydride, cyanic ethers, methylamine, ammonia, ureas, glycerin, glycols and ethylene oxide, homologues of lactic acid. He died in 1884.

CHAPTER 8

Aliphatic Hydrocarbons –Alkanes

Introduction

Aliphatic hydrocarbons are the open chain compound, when saturated, these are called alkanes. The general formula of alkanes is C_nH_{2n+2}. The number of isomers that get formed are many, depending upon the number of carbon atoms the hydrocarbon contains. For example, C_4H_{10}, butane has two isomeric structures and C_5H_{12}, pentane has three isomeric structures.The number of structures increase enormously, even if one of the hydrogen is replaced by a chlorine atom or a hydroxyl group. $C_5H_{11}Cl$ or $C_5H_{11}OH$ each have eight isomers.

Butane

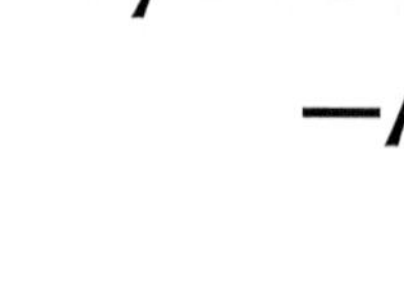

Isobutane

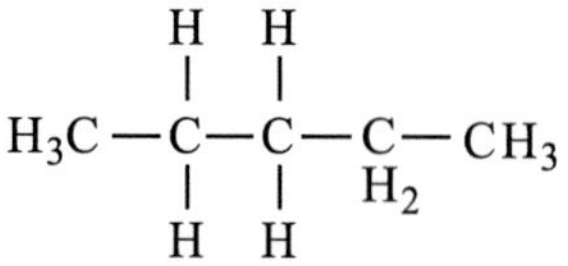

Pentane

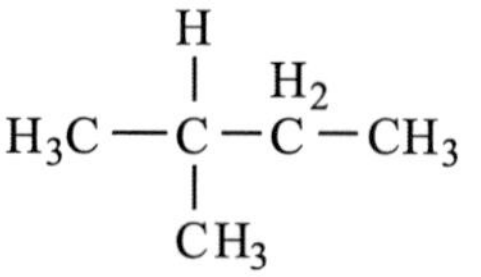

Isopentane

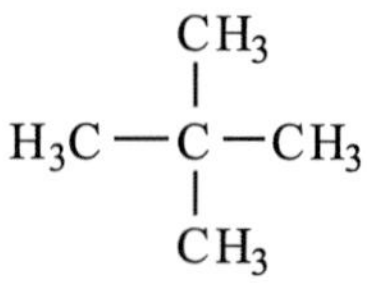

Neopentane

We have already seen the conformation of alkanes and know their architecture. To recapitulate, given below are structure of some long chain hydrocarbons. The shape of these hydrocarbons is due to the tetrahedral valency of each carbon. It is the relative position of substituents at each carbon that we look for in the conformation of a molecule. The relative position of substituents varies due to rotation around single bond.

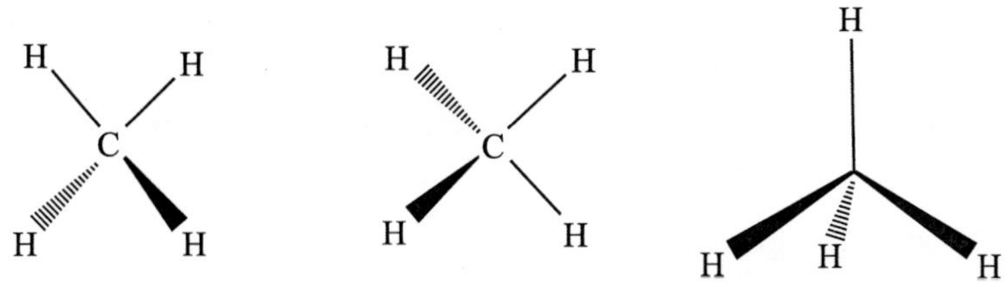

Methane structure as looked from different directions

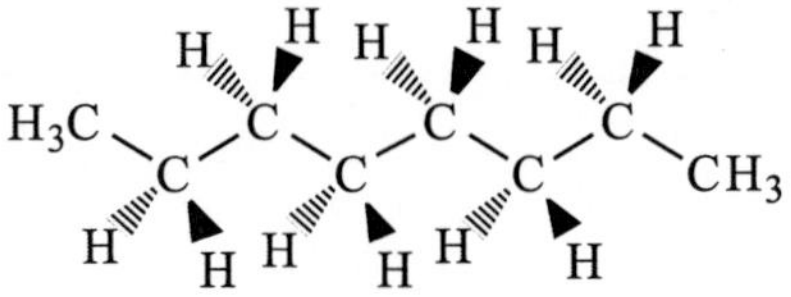

octane

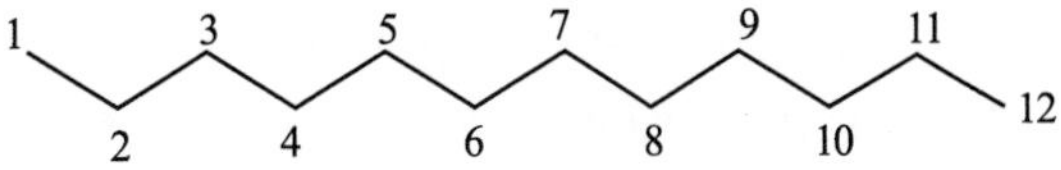

Simplified line drawing of dodecane

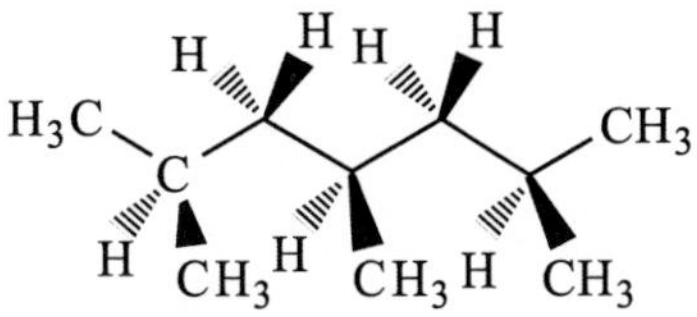

2,4,6-trimethylheptane

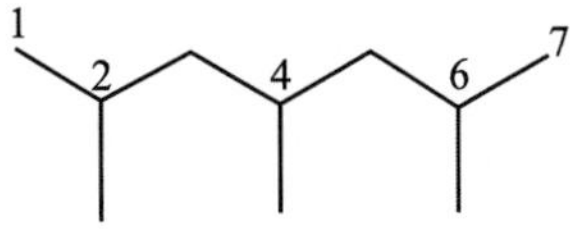

Simplified line drawing of 2,4,6-trimethylheptane

Sawhorse projections of butane

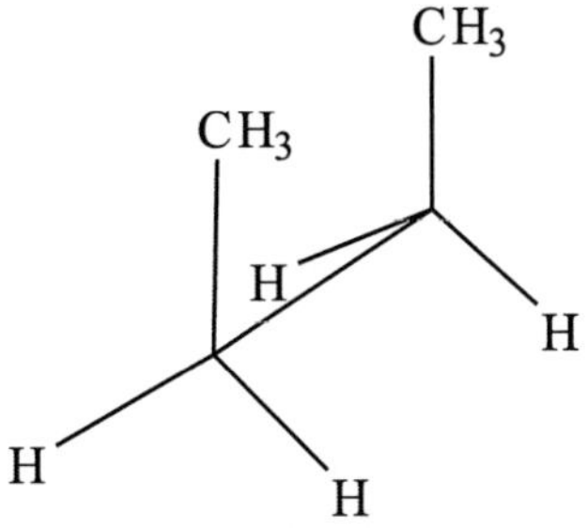

Eclipsed conformation of butane

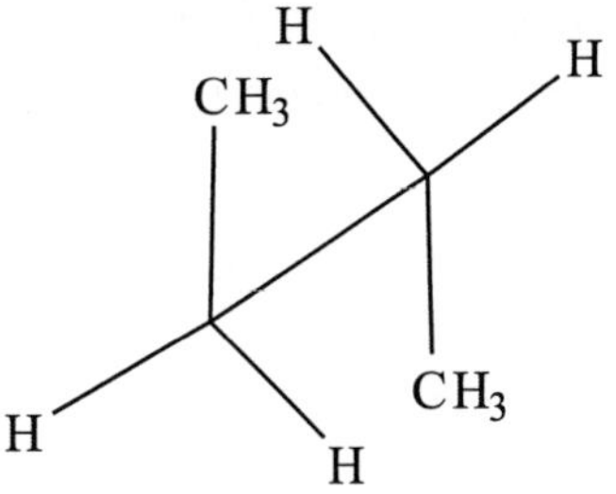

Staggered conformation of butane

Newman projections of butane

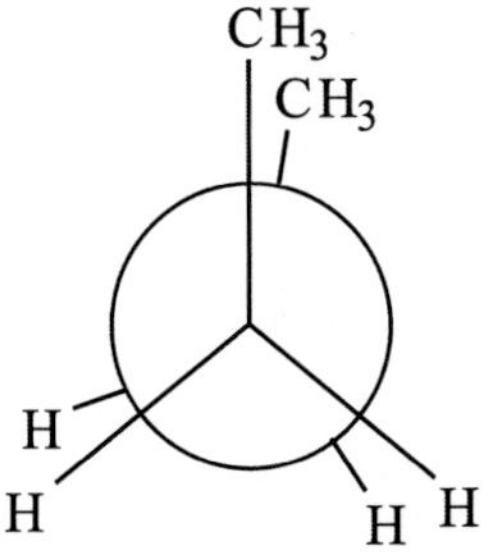

Eclipsed conformation of butane

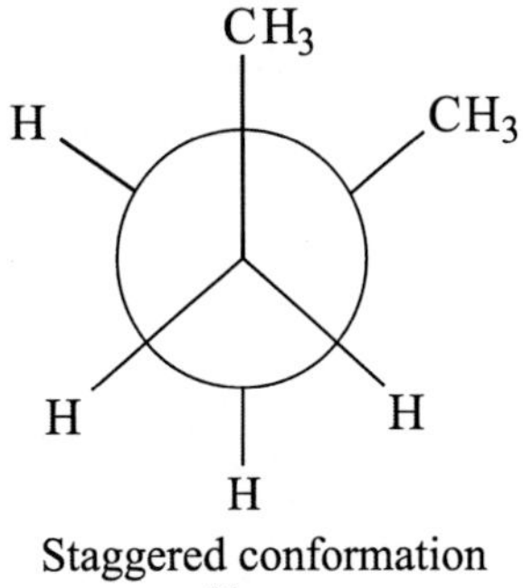

Staggered conformation of butane

8.1 Nomenclature

IUPAC Nomenclature of Aliphatic Hydrocarbons (International Union of Pure and Applied Chemistry)

A. Parts of a name of an Organic Compound

The total name of an organic compound consists of the following parts in sequence as given below:

1. The number of position of primary prefix.
2. The name of primary prefix.
3. **The word root.**
4. The number of position of primary suffix.
5. The name of primary suffix.

Primary Prefix – Word Root – Primary Suffix

Word root of a compound is taken from names of hydrocabons. The names of hydrocarbons are derived on the basis of the number of carbon atoms that they contain, except the first four members of the series. It forms the basis of nomenclature of all compounds. Given below is the list of various saturated hydrocarbons *along with the word root, which are used in the names of all other type of compounds.*

For example, the name of a five carbon saturated hydrocarbon is pentane. It has pent- as the word-root and -ane as the suffix to form pentane as the complete name.

Name of a saturated hydrocarbon

C_5H_{12} =	———	Pent +	ane =	Pentane
Formula	No prefix	word-root	suffix	name

Based on this method of naming, the names of various saturated hydrocarbons, carrying suffix -ane are given in the Table 8.1.

Table 8.1 Names of Saturated Hydrocarbons–Alkanes

C_nH_{2n+2} is the general formula for saturated hydrocarbons, called alkanes

Number of carbon atoms	Word root	Primary suffix	Formula	Name of Alkane
C_1	Meth-	-ane	CH_4	Methane
C_2	Eth-	-ane	C_2H_6	Ethane
C_3	Prop-	-ane	C_3H_8	Propane
C_4	But-	-ane	C_4H_{10}	Butane
C_5	Pent-	-ane	C_5H_{12}	Pentane
C_6	Hex-	-ane	C_6H_{14}	Hexane
C_7	Hept-	-ane	C_7H_{16}	Heptane
C_8	Oct-	-ane	C_8H_{18}	Octane
C_9	Non-	-ane	C_9H_{20}	Nonane
C_{10}	Dec-	-ane	$C_{10}H_{22}$	Decane

B. The Alkyl Groups

An alkyl group is derived from the parent hydrocarbon, alkane, of general formula, C_nH_{2n+2}. When one hydrogen is detached from an alkane, its general formula becomes, C_nH_{2n+1}. It is called alkyl group that has one valency still unsatisfied. Thus, methane with molecular formula, CH_4 less one hydrogen, that is, CH_3—, it is called methyl. The general method of writing the name of an alkyl group is as shown below.

Alk*ane* – 1 H = Alk*yl* group

remove *-ane* and add *-yl*

$$C_nH_{2n+2} - 1\ H = C_nH_{2n+1}$$

Following this procedure, we write the examples of alkyl groups as follows.

C_nH_{2n+2}	C_nH_{2n+1}	Formula	C_nH_{2n+2}	C_nH_{2n+1}	Structure
Methane	Methyl-	CH_3-	2-Methyl propane	2-Methyl propyl	CH_3—CH(CH_3)—CH_2—
Ethane	Ethyl-	C_2H_5-			
Propane	Propyl-	C_3H_7-			
Butane	Butyl-	C_4H_9-			
Pentane	Pentyl-	C_5H_{11}-			
Hexane	Hexyl-	C_6H_{13}-			
Octane	Octyl	C_8H_{17}-			
Nonane	Nonyl	C_9H_{19}-	2,2-Dimethyl propane	2,2-Dimethyl propyl	CH_3—CH(CH_3)(CH_3)—
Decane	Decyl	$C_{10}H_{21}$-			

Problem : Write the name of the following alkyl groups.

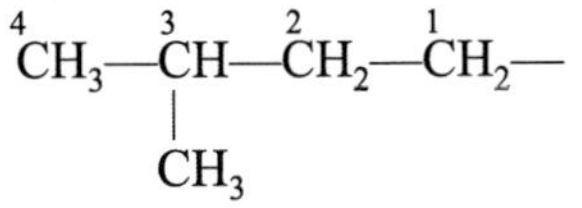

I

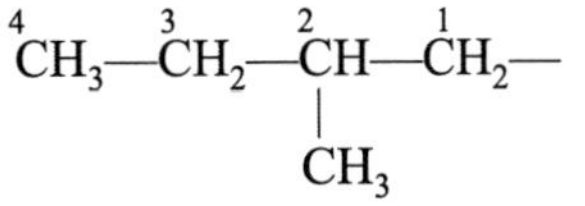

II

Solution: Both groups contain four carbon chain. In one case, the methyl group is at 3-position and other case the methyl group is at 2-position. So their names are

3-Methylbutyl   2-Methylbutyl

I   II

This method of writing numbers and position of substituents is given in details as below.

8.1.1 IUPAC Rules of Nomenclature

All classes of compounds are derived from hydrocarbons; the nomenclature of hydrocarbons is the basis for all others. The IUPAC rules applicable are given ahead.

(i) IUPAC Nomenclature of Alkanes

Rule 1. Longest Carbon Chain Rule: Select the longest carbon chain in the given compound. The word-root name of the compound is derived from the maximum number of carbon atoms present in a chain.

The following examples shall clarify the rule.

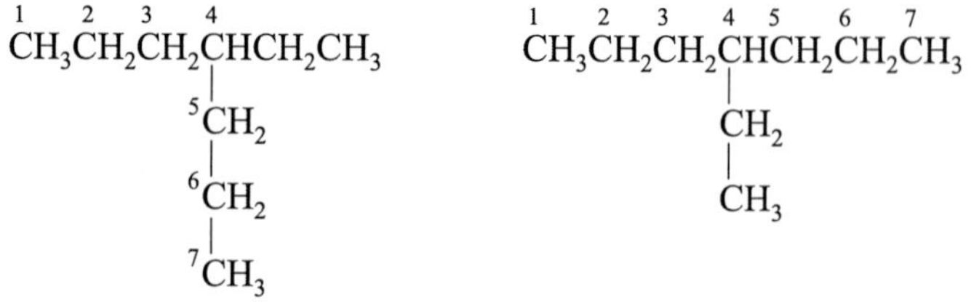

Thus, this compound having seven carbons is heptane

The above given two structures are identical, but the carbon chain is drawn differently. It is immaterial which way the carbon chain is drawn, so long the longest chain is selected. Given below is another example.

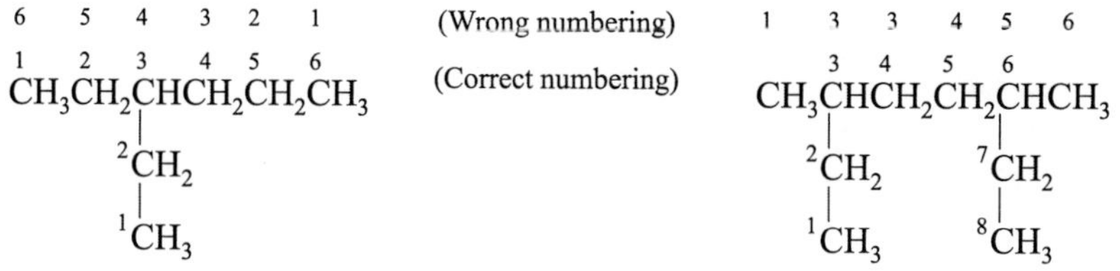

Six carbon compound is hexane
3-Ethyl-hexane

Eight carbon compound is octane.
3:6-Dimethyloctane

The correct numbering of carbon atoms is marked correct, whatever way we count the carbon atoms. The chain has to contain maximum number of carbon atoms.

Rule 2. Lowest Number for Side Chain: Counting of carbon atoms in a chain is done from the side, where branching starts first. That is, the lowest number is given to the side chain that occurs first.

Many a times, more than one alternate numbering is possible. That numbering is correct, which gives the maximum number of branching or least sub-branching in a chain.

Following examples shall illustrate the use of these two rules in naming organic compounds.

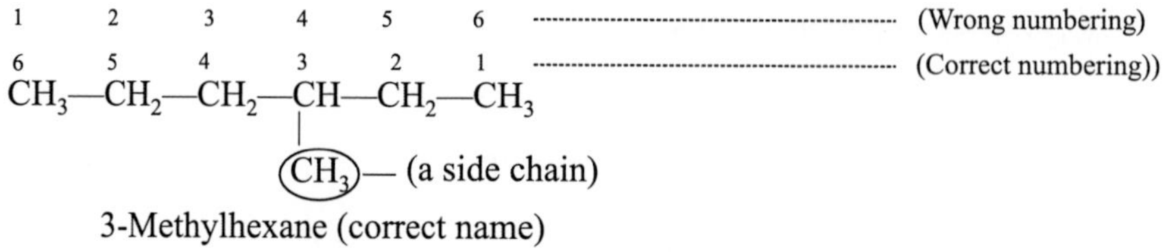

3-Methylhexane (correct name)

Encircled methyl group is a side chain, since it does not appear in the main chain. That numbering which gives lower number to this methyl group is correct, as shown above.

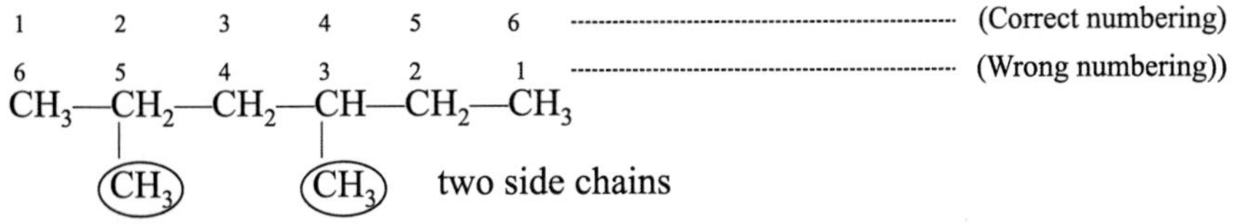

2, 4-Dimethylhexane (correct name)

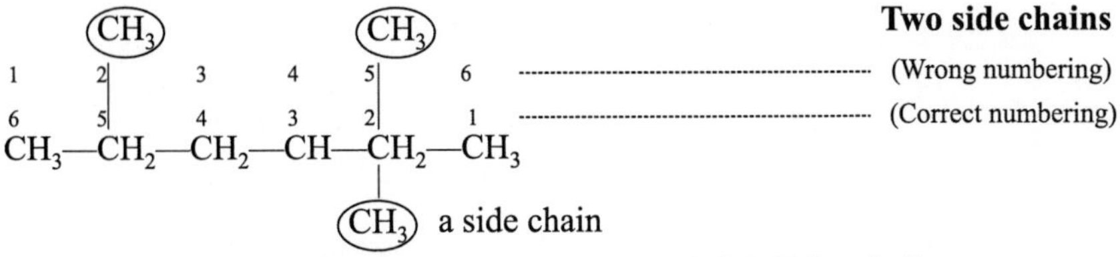

2,2,5 -Trimethylhexane (correct name)

2,5,5 -Trimethylhexane (wrong name)

In this structure, both the numbering start from equal numbered carbon atom. To correctly name this, we follow the next rule.

Rule 3. Minimum Summation Rule: In case the branching of carbon chain *starts at equal numbered carbon atoms*, counting of carbon atoms must be carried from the side, where *the sum of the positions of all the substituents, is lowest.*

In the above example, there are three side chains, which are encircled. These are at 2, 2, 5 or 2, 5, 5 numbered carbon atoms. The sum of one alternative is (2 + 2 + 5 = 9) that is lower than the other

(2 + 5 + 5 = 12). So 2, 2, 5 – numbering is followed.

That is the **lowest summation rule is followed. The other examples are:**

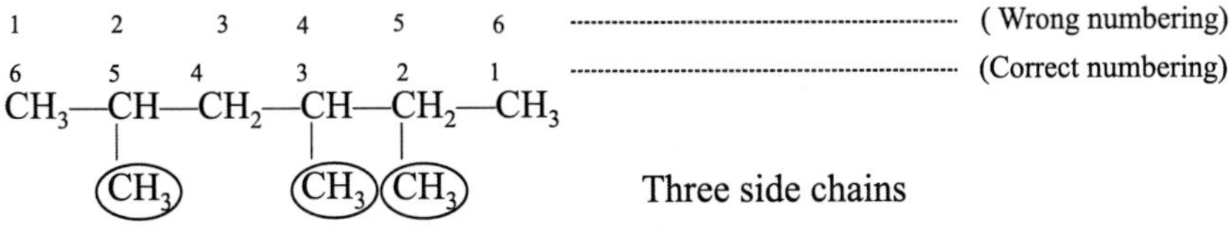

2,3,5-Trimethylhexane

Three side chains (correct name)

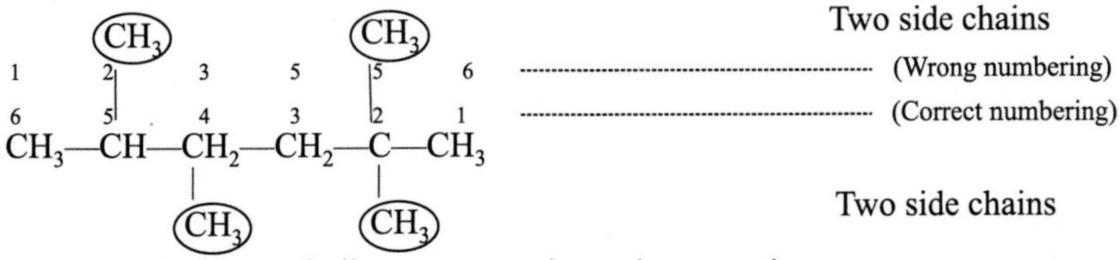

2, 2,4,5-Tetramethylhexane (correct name)

Alternative name is wrong 2,3,5, 5-Tetramethylhexane (wrong name)

Following are the structures with **sub-side chain.** Let us decide their correct name.

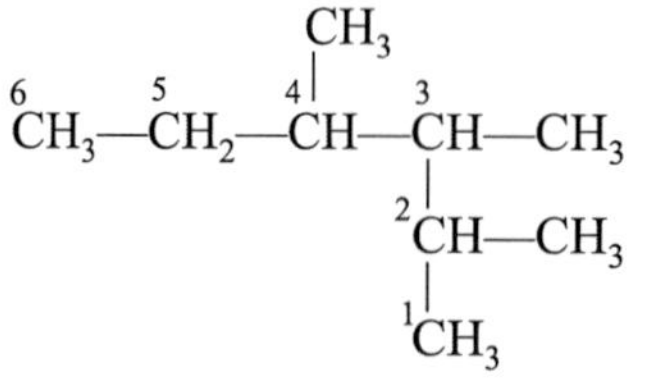

2,3,4-Trimethylhexane
(correct name)

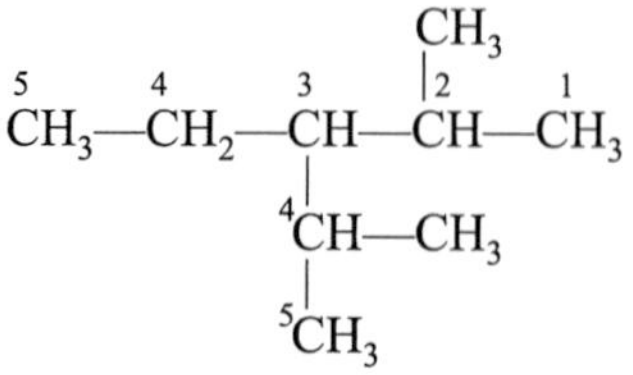

3-Ethyl-2,4-dimethylpentane
(correct name)
2-methyl-3-Isopropylpentane
(correct name)

Both the alternatives contain same number of carbon atoms, so the maximum number of side chain rule is being followed.

When the number of carbon atoms in the main chain increase further than the above example, the rule of maximum number of carbon atoms in a chain, leaving the substitutents with subside chain is adopted. The sub-side chain is named as methylethyl on carbon 3 or as isopropyl in the name as shown below.

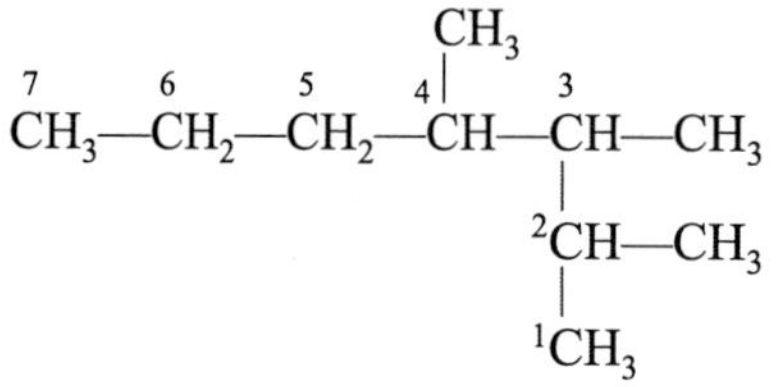

2,3,4-Trimethylheptane

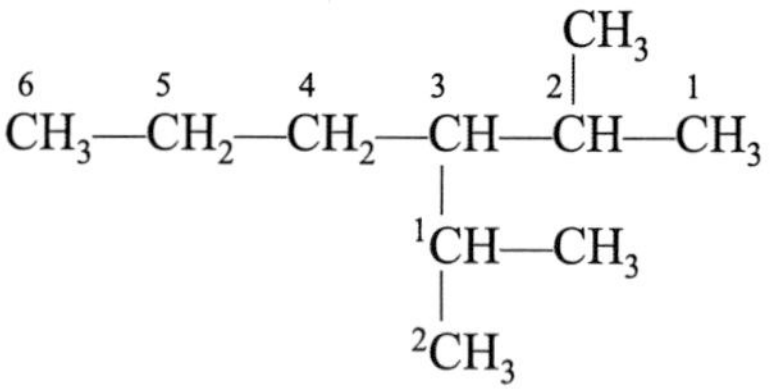

2-Methyl-3-(methylethyl) hexane
or 2-methyl-3-Isopropylhexane

(correct names)

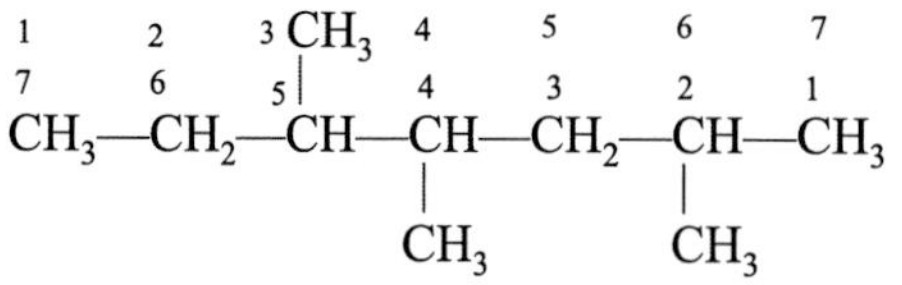

(wrong numbering)
(correct numbering)
2, 4,5- Trimethylheptane (correct name)
3,4,6-Trimethylheptane (wrong name)

The terms like, iso-, neo, sec-, tert- were not used earlier in IUPAC nomenclature of compounds. However according to 1993 IUPAC nomenclature, side chain groups can be named as t-butyl, iso-propyl, neo-pentyl etc., provided the longest carbon rule is applicable to the main chain.

Rule 4: Alphabetical Order of the Substituents: The name of side chains or substituents in an organic compound is written in alphabetical order. Examine this rule in the following example,

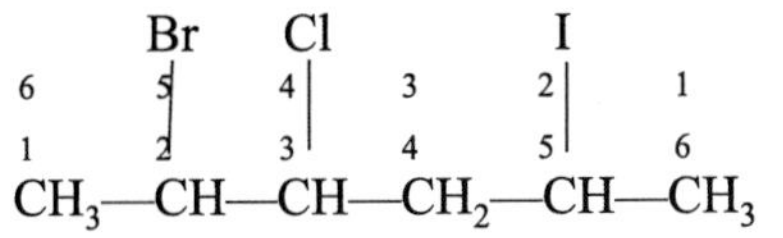

(wrong numbering)
(correct numbering)
2-Bromo-3-chloro-5-iodohexane

The above formula has two possible numbering. The one starting from left hand side is correct. It is based on the lowest summation rule, as well as the alphabetical order of the substituents. The sum of positions, in two alternatives are 2+4+5=11 and 2+3+5=10.The substituent, bromo should be given preference in writing the name.

Consider another example of molecule containing chlorine and bromine atoms.

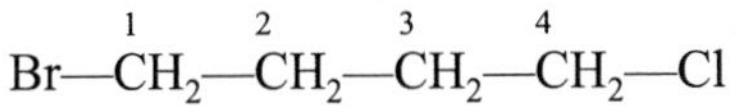

1-Bromo-4-chlorobutane
(correct name)

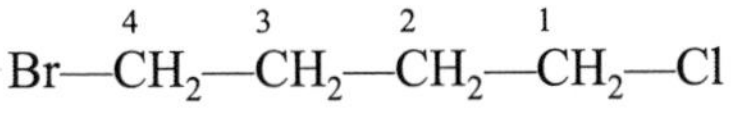

1-Chloro-4-bromobutane (wrong names)
or 4-Bromo-1-chlorobutane

That is when bromine and chlorine are attached at equal numbered carbon atoms from either side, preference is given to alphabetical order i.e., bromine numbering comes first.

Attach one more carbon atom at one end of the carbon chain, the numbering of carbon atoms must start from the side where the substituent occurs first, irrespective of the alphabetical order.

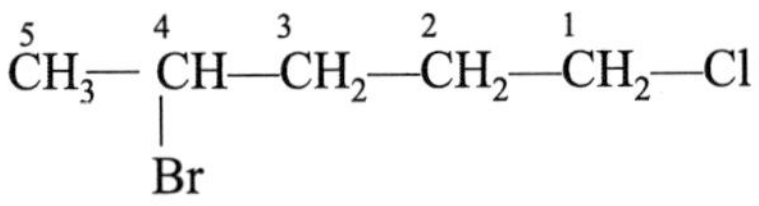

4-Bromo-1-chloropentane
(Correct name)

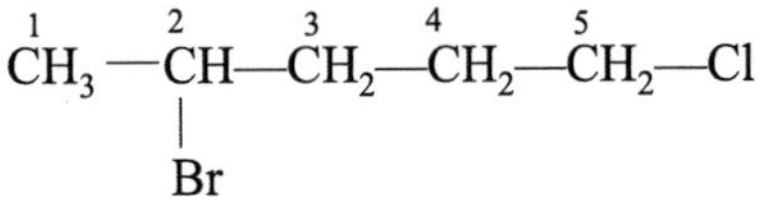

2-Bromo-5-chloropentane
(Wrong name)

Rule 5: Lowest Number for a Functional Group: Whenever a functional group is attached to a carbon chain, it must be assigned the lowest number. We follow this rule for functional group, even if the rule for counting of carbon atoms **from the side where branching starts first** need be violated.

Replacement of two hydrogens from two adjacent carbon atoms in an alkane, leads to creation of a double bond. Similarly, replacement of four hydrogens from two adjacent carbon atoms in an alkane gives a triple bond between two carbon atoms. For the purpose of nomenclature **double bond** or **a triple bond** are considered as functional groups. The rule used for their nomenclature is, while numbering carbon change double or triple bonds take preference over side chain.

The following examples shall illustrate the use of this rule.

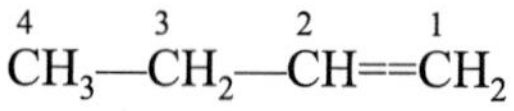

1-Butene

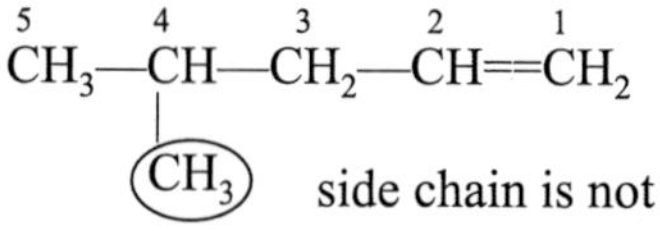

side chain is not preferred, but double bond

4-Methyl-1-pentene or 4-Methylpent-1-ene

The double bond is given the numbering 1 and 2, but the substituent is given the postion 4. The double bond between carbon atoms 1 and 2 is numbered in the name as at 1-position.

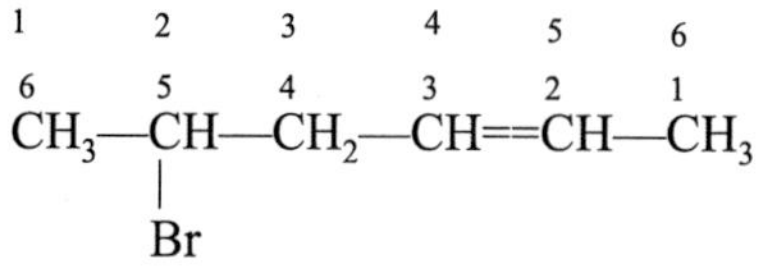

5-Bromo-hex-2-ene
(correct name)
2-Bromo-hex-3-ene
(wrong name)

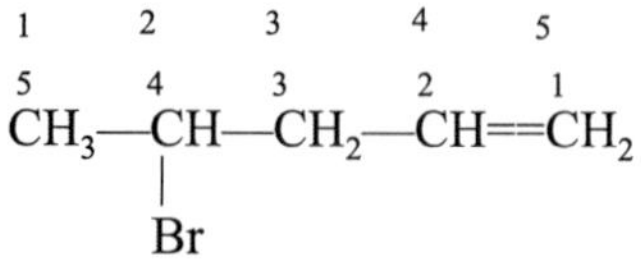

4-Bromo-pent-1-ene
(correct name)
2-Bromo-pent-4-ene
(wrong name)

The following example shall further illustrate the use of the above rule

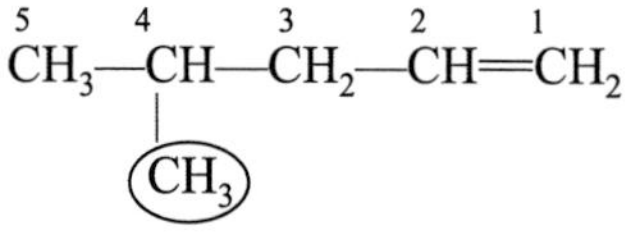

4-Methyl-pent-1-ene
(correct name)

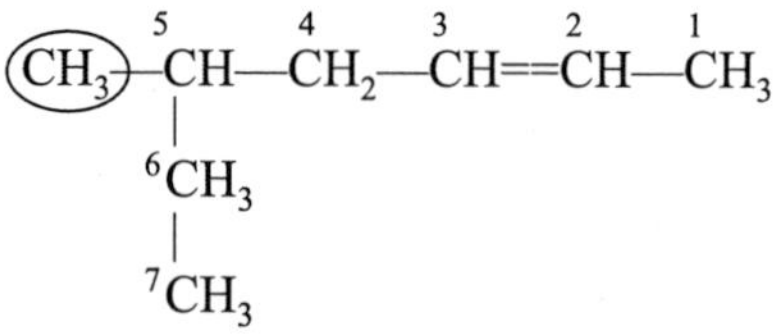

5-Methyl-hept-2-ene
(correct name)

Before 1993 IUPAC recommendations, it was written as 4-Methyl-1-pentene and 5-methyl-2-heptene.

"Rule now followed is that the numerals representing the position of a substituent or functional group, should be written immediately before the group" as given above.

8.2 Preparation of Alkanes

The methods that are used for the preparation of alkanes involves.

8.2.1 Linkage of Carbon with Another Carbon

The method of linkage of carbon and carbon helps in the formation of carbon chain of any desired length.

(*i*) Wurtz Method

It is a method that uses reaction of alkyl halide in ethereal solution with sodium metal to link carbon with another carbon. For two halides, RX and R′X the reaction is shown as below.

$$RX + R'X + 2Na \longrightarrow R - R' + 2\,NaX + R - R + R' - R'$$

Thus, best yields are obtained when the two alkyl halides are same. The mechanism that has been suggested for Wurtz reactions are as follows:

(*i*) **Free radical mechanism**, *e.g.*,

$$\underset{\text{bromomethane}}{H_3C-Br} + Na^{\bullet} \longrightarrow \underset{\text{methyl radical}}{CH_3^{\bullet}} + Na-Br$$

$$CH_3^{\bullet} + CH_3^{\bullet} \longrightarrow \underset{\text{ethane}}{H_3C-CH_3}$$

(*ii*) The **other mechanism i**nvolves intermediate organometallic compound, e.g.,

$$H_3C-Br + 2\,Na^{\bullet} \longrightarrow CH_3^{-}\text{-}Na^{+} + Na-Br$$

$$CH_3^{-}\text{-}Na^{+} + H_3C-Br \longrightarrow \underset{\text{ethane}}{H_3C-CH_3} + Na-Br$$

Besides using a single alkyl halide, still there is a possibility of disproportionation of radicals, to form in this case ethane and ethene, usually the quantity of alkene is small. This problem for yield of the reaction can be overcome, by reduction of the mixture to yield one product.

$$CH_3^{\bullet} + CH_3^{\bullet} \longrightarrow \underset{\text{ethane}}{H_3C-CH_3} + \underset{\text{ethene}}{H_2C=CH_2}$$

$$\underset{\text{ethane}}{H_3C-CH_3} + \underset{\text{ethene}}{H_2C=CH_2} \xrightarrow{H_2/Pd\text{-}C} 2\,\underset{\text{ethane}}{H_3C-CH_3}$$

(*ii*) Kolbe's Electrolytic Method

Kolbe's used sodium or potassium salt of a carboxylic acid for electrolysis. In general, the electrolysis of a carboxylic acid links two alkyl groups involving decarboxylation of carboxylic acid.

$$R\text{-}COO^{-}K^{+} + R'\text{-}COO^{-}K^{+} \longrightarrow R-R' + 2\,CO_2 + H_2 + 2KOH$$

For example, the electrolysis of potassium propionate gives butane.

$$\underset{\text{Pot. propionate}}{C_2H_5\text{-}COO^{-}K^{+}} + \underset{\text{Pot. propionate}}{C_2H_5\text{-}COO^{-}K^{+}} \longrightarrow \underset{\text{butane}}{C_2H_5-C_2H_5} + 2\,CO_2 + H_2 + 2KOH$$

Mechanism of the reaction proceeds as follows:

$$C_2H_5\text{-}COO^{\ddot{-}} \longrightarrow C_2H_5\text{-}COO^{\bullet} + e^{\bullet}$$

Pot. propionate

$$C_2H_5\text{-}COO^{\bullet} \longrightarrow C_2H_5^{\bullet} + CO_2$$

$$C_2H_5^{\bullet} + C_2H_5^{\bullet} \longrightarrow C_4H_{10}$$

Other reactions that can occur are either two radicals can combine to give ester or disproportionation to form an alkane and alkene.

$$C_2H_5\text{-}COO^{\bullet} + C_2H_5^{\bullet} \longrightarrow C_2H_5\text{-}COO{:}C_2H_5$$

Two radicals combine — Ethyl propionate

$$C_2H_5^{\bullet} + C_2H_5^{\bullet} \longrightarrow C_2H_6 + C_2H_4$$

Ethane — Ethene

(*iii*) Coupling of alkyl borane

Borane hydride adds on to alkene to form trialkylborane. This on reaction with alkali in presence of silver nitrate forms hydrocarbon by coupling of two alkyl groups. The reaction takes place in very good yield. This reaction can also be used for preparation of alcohols, when a reducing agent is used for breaking the complex.

For example, reaction of 2-methylbut-1-ene with diborane forms alkyl borane. The reaction was discovered by H. C. Brown *et al.* 1961.

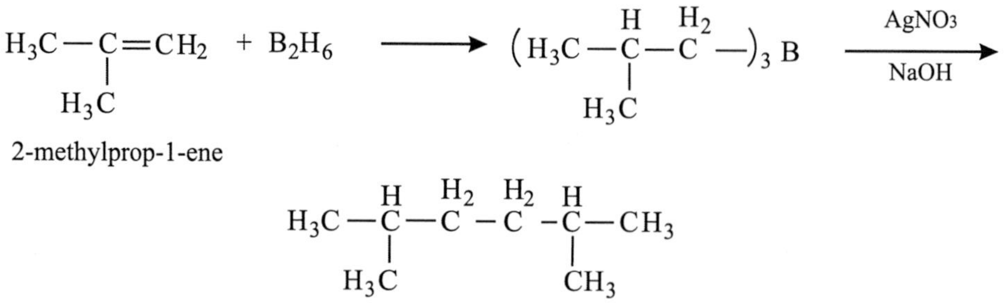

2,5-dimethylhexane

This reaction helps to form any length of carbon chain. More so, the coupling can also be affected between unequal length of carbon chains.

8.2.2 Modification of Functional Groups already present in a starting material.

(*i*) Hydrogenation of double or triple bond

Full reduction of double or triple bond gives a completely saturated hydrocarbon. The best method to carry out double or triple bond reduction is to use Raney-Ni/H_2 catalyst.

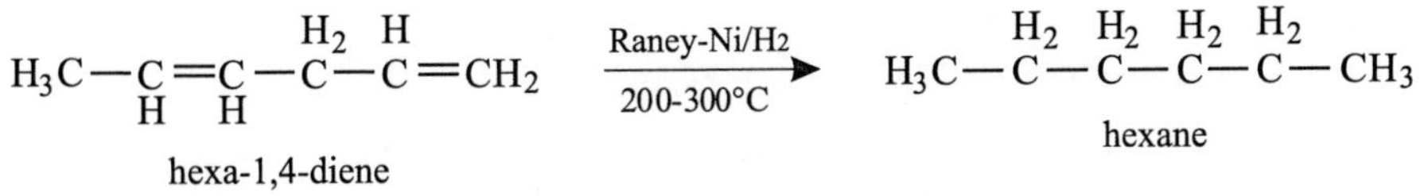

Catalytic reduction is *cis-* addition reaction, since both the hydrogen atoms are linked to the catalyst and so become available to double bond from the same side, thus, are *cis*-to double bond. However, it can add from below or above the double bond. In symmetric or unsubstitued alkenes, it really does not matter, but unsymmetric alkene give rise to optical isomers depending upon, whether the starting material is *cis-* or *trans-* isomer. For example, *cis*-3-methylhexa-1,3-diene on catalytic reduction gives stereoisomeric product than that obtained from the catalytic reduction of *trans-* 3-methylhexa-1,3-diene .

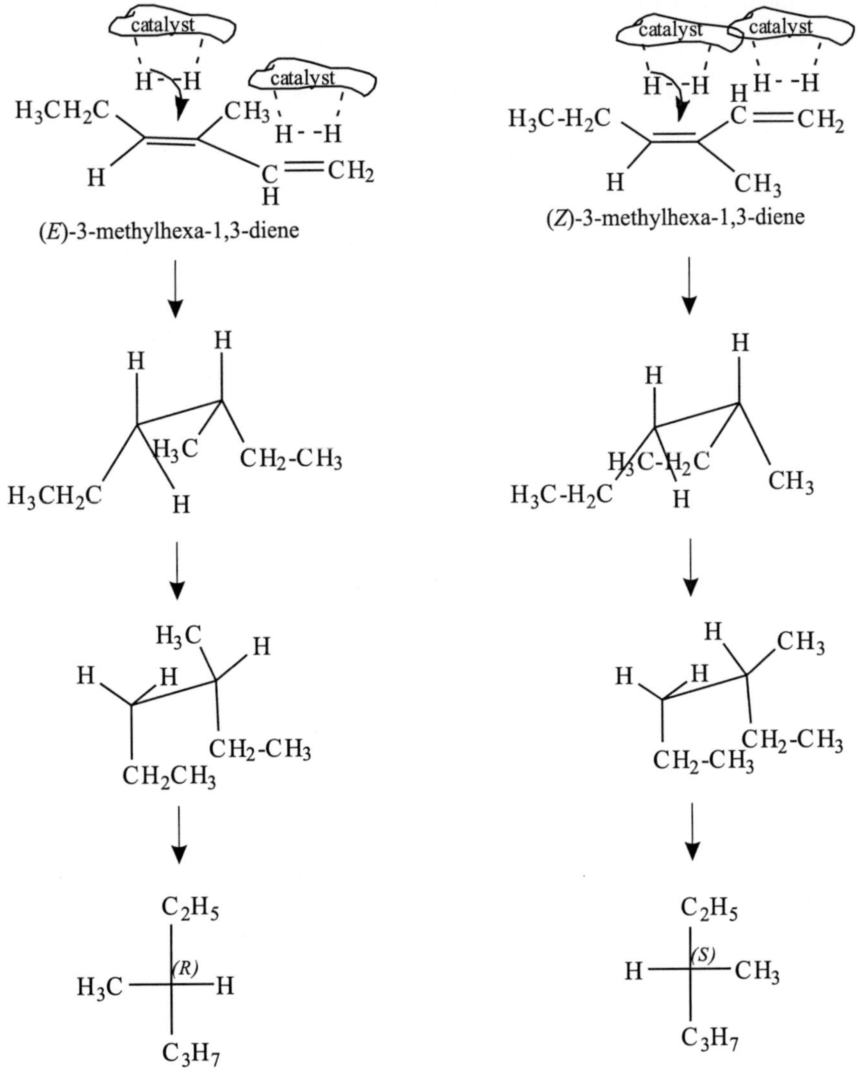

(*b*) **Reduction of acetylene** and its derivatives also form alkanes. If the supply of hydrogen is less, acetylenes may form alkenes. The reaction conditions and catalyst is same as in case of alkene reduction.

$$R_1{-}C{\equiv}C{-}R_2 \xrightarrow{H_2/\text{Raney Ni}} R_1{-}\overset{H_2}{C}{-}\overset{H_2}{C}{-}R_2$$

This can also give the isomeric products depending upon the nature of the starting material.

(*ii*) Decarboxylation of a carboxylic acid

Removal of carbon dioxide from -COOH group in a molecule is called decarboxylation. It leaves a hydrogen atom in place of -COOH and thus, give rise to hydrocarbons, provided no other functional group is present. It can be affected by heating with sodalime. For example, the sodium salt of acetic acid on heating with sodalime, gives methane, as shown.

$$R{-}COO^- Na^+ \;+\; NaOH(CaO) \longrightarrow RH + Na_2CO_3 \quad (R{=}CH_3)$$

Most of the higher carboxylic acids decompose to give mixture of products, for example, sodium propanoate gives, ethane, ethene and hydrogen.

$$C_2H_5{-}COO^- Na^+ \;+\; NaOH(CaO) \longrightarrow C_2H_6 + CH_4 + H_2 + Na_2CO_3$$

(*iii*) Reduction of alkyl halides

Reduction of alkyl halides can be carrried out in number of ways.

(*a*) **Metal/acid reaction:** metals like zinc or zinc-copper couples are often used for reduction reactions. Zinc-acetic acid or hydrochloric acid reduce the C-Cl or C-Br bond very conveniently. The reaction takes place by transfer of an electron or electrons from the metal.

$$Zn \longrightarrow Zn^{2+} + 2e$$

$$RX + e \longrightarrow \bar{X} + R^{\cdot} \xrightarrow{e} R\bar{:}$$

$$R\bar{:} + C_2H_5OH \longrightarrow R{-}H + \bar{O}C_2H_5$$

(*b*) The **metal hydrides** are very convenient and excellent reducing agents for alkyl halides. Thus, reaction of $LiAlH_4$, reduces primary as well as secondary halides, where as $NaBH_4$ reduces secondary and tertiary halides. Triphenyl tin hydride reduces all the three kind of alkyl halides to form hydrocarbons.

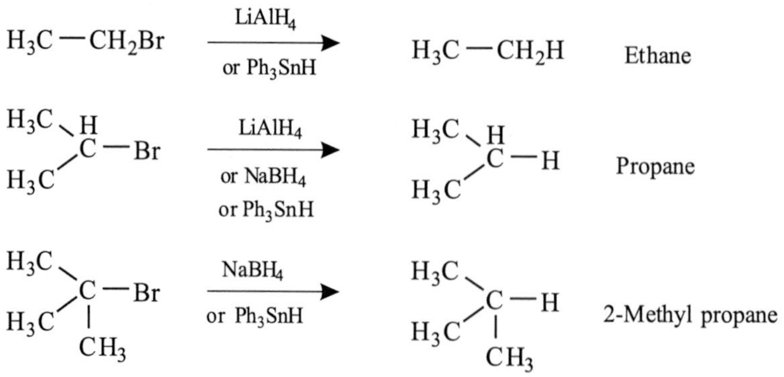

(*c*) Reduction reaction using **hydroioidic acid and red phosphorus,** reduces alkyl halides to alkanes. Alkyl iodides in particular are reduced to alkanes.

$$H_3C-CH_2I \xrightarrow{\text{HI/Red P}} H_3C-CH_2H \quad \text{Ethane}$$

(*iv*) Reduction reactions of compounds containing carbonyl functional groups

Compounds containing aldehyde or ketone and carboxylic acid can be reduced with hydroiodic acid, to form alkanes. A more specific method of reducing aldehydes, and ketones is by Clemmenson's, or Wolff-Kishner's reductions. The reduction of carboxylic acid or acid chlorides can be carried out by Rosenmund reduction reaction. These are exemplified as below.

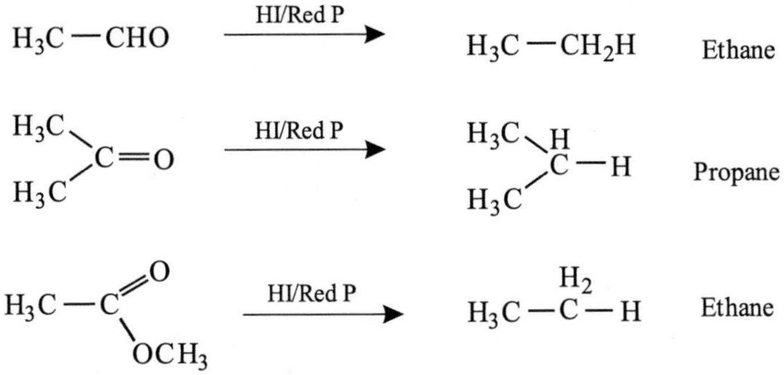

Clemmenson's Reduction: It is a reduction of a carbonyl group using Zn/Hg/HCl to given >CH_2 group.

Wolff-Kishner's Reduction: It is a reduction of hydrazone of a carbonyl group using $NaOC_2H_5$.

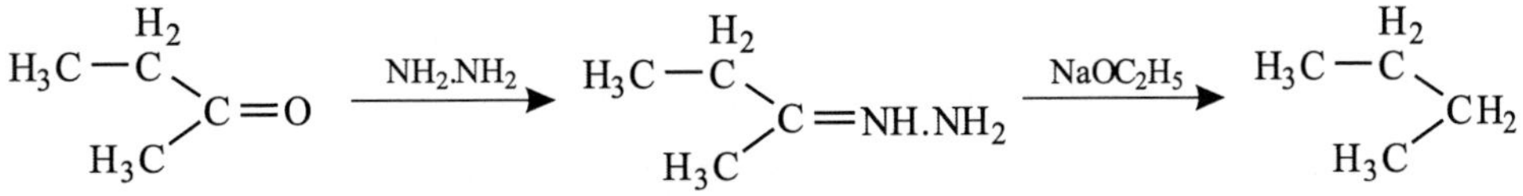

Ethyl methane ketone — Hydrazone of ethylmethyl ketone — Butane

Rosenmund's Reduction: It is the reduction of an acid chloride using Pd-C/$BaSO_4$/H_2 to form aldehydes. These can be further reduced to alkane.

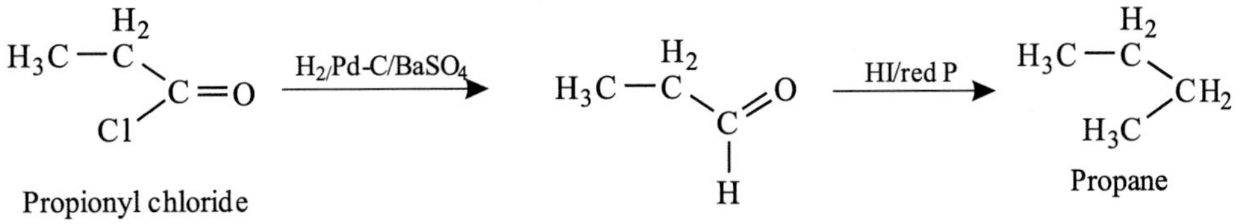

Propionyl chloride — Propane

Details of these reactions shall be discussed in the later chapters of the second volume.

(*v*) Reaction of Grignard reagents with reactive hydrogen

Grignard reagents are known to be very reactive for active hydrogens such as, in -COOH, water or NH_2. The alkyl group becomes the alkane and hydroxy magnesium halide is eliminated, when water is used.

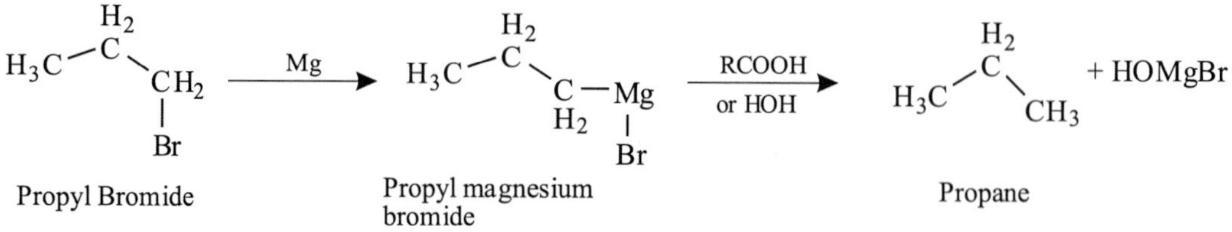

8.3 Properties of Alkanes

Normally alkanes containing 1 to 4 carbon atoms, are gases; those from C_5–C_{17} are colourless liquids; and from C_{18} onwards, are colourless solids. With the increase in number of carbon atoms, there is a regular increase of boiling point. It is regular because, we notice in the molecular formula, amongst each consecutive member of alkanes, there is a common difference of one –CH_2– group. Thus, the member of alkane series form a *homologous series*. Other physical properties like melting point, density and viscosity also increase with the increase in the number of carbon atoms in a chain.

8.3.1 General Physical Properties

Physical properties depend largely on two factors,

(*i*) Molecular mass or size

(*ii*) Intermolecular forces.

Alkanes, relative to other organic compounds containing functional groups, are inert. They do possess intermolecular forces - like hydrogen bonding, dipole moments but van der Waals forces, that exist are very weak. Still gradational variation in their physical properties has been observed.

Boiling and Melting Points: Boiling point of hydrocarbons increase with the increase in the number of carbon atoms. In case of alkanes it is molecular size and not the intermolecular forces that bring this change. Given below is the table of the boiling point and molecular mass of a few alkanes. There is a regular increase and graph between mass and boiling point is linear increase. Increase in melting point of alkanes follows a zig-zag path.

Table 8.2 Physical Properties of alkanes

Alkane	Structure of Alkane	Molecular Mass	Boiling Point °C, 1 Atm	Melting Point °C
Methane	CH_4	16	-161	-183
Ethane	CH_3CH_3	30	088.5	-172
Propane	$CH_3CH_2CH_3$	44	-42.5	-188
Butane	$CH_3CH_2CH_2CH_3$	58	-0.5	-135
Pentane	$CH_3CH_2CH_2CH_2CH_3$	72	36	-130
Hexane	$CH_3CH_2CH_2CH_2CH_2CH_3$	86	68.5	-95
Heptane	$CH_3CH_2CH_2CH_2CH_2CH_2CH_3$	100	98.5	-91
Octane	$CH_3CH_2CH_2CH_2CH_2CH_2CH_2CH_3$	114	125.6	-57
Nonane	$CH_3CH_2CH_2CH_2CH_2CH_2CH_2CH_2CH_3$	128	150.5	-54
Decane	$CH_3CH_2CH_2CH_2CH_2CH_2CH_2CH_2CH_2CH_3$	142	174	-30
Undecane	$CH_3CH_2CH_2CH_2CH_2CH_2CH_2CH_2CH_2CH_2CH_3$	156	196	-26
Dodecane	$CH_3CH_2CH_2CH_2CH_2CH_2CH_2CH_2CH_2CH_2CH_2CH_3$	170	216	-10

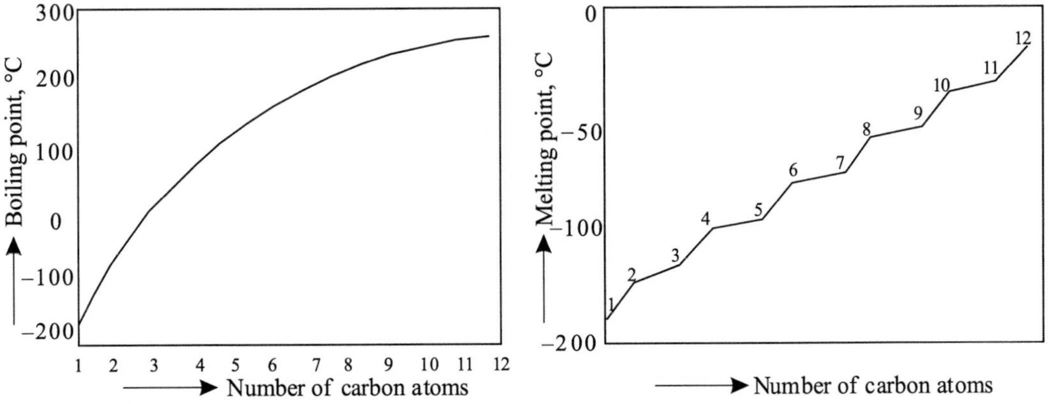

Fig. 8.1 Graph of variation of boiling points and melting points.

8.3.2 Chemical Reactions

Alkanes are inert molecules, so require strong reaction conditions. These undergo the following reactions:

8.3.2.1 Substitution Reactions

Substitution is a displacement of an atom or a group from a substrate molecule by another atom or group by the reagent. Hydrogens in alkanes can be replaced by halogeno- or nitro- or sulphonic acid groups by using appropriately strong reagent.

(*a*) Halogenation

Halogenation may be brought about by the reaction of any halogen in gas or liquid phase in presence of light, heat or catalyst. Most studied is the chlorination reaction. Alkanes that have primary, secondary or tertiary hydrogens have different reactivity towards chlorine to give a mixture of all kind of isomeric or multi-substituted products in different proportions. The order of substitution of hydrogen from alkane is tertiary > secondary > primary hydrogens. Considering the simplest alkane, methane, chlorination at 300 ^{0}C in presence of light, gives a mixture of methyl chloride, dichloromethane, chloroform and carbon tetrachloride.

$$H_3C{-}H + Cl_2 \xrightarrow{h\nu} H_3C{-}Cl \longrightarrow H_2CCl_2 \longrightarrow CHCl_3 \longrightarrow CCl_4$$

It follows the following free radical mechanism.

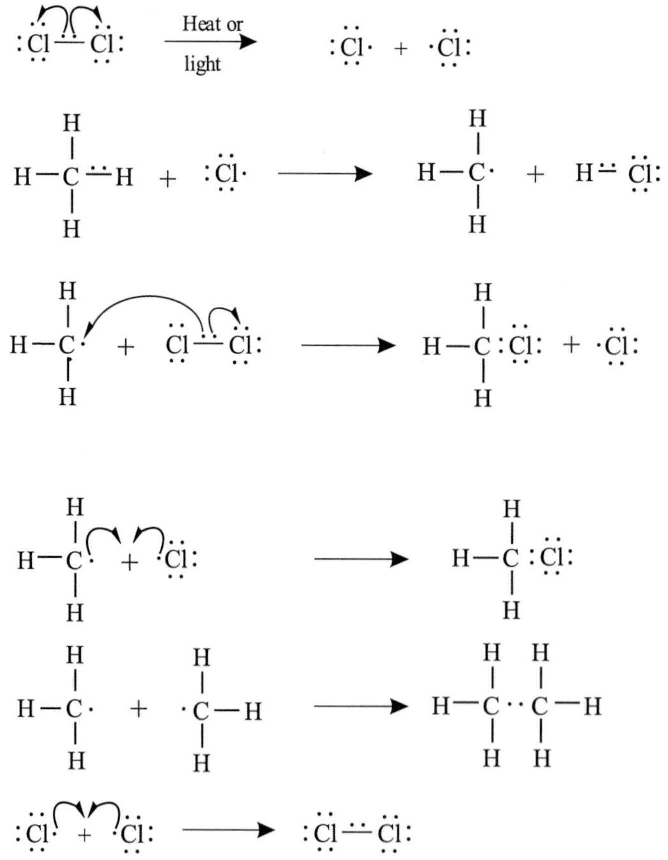

Step 1: Initiation step $\Delta H = -242.8$ kJ/mol

Step 2: This is repeated Propagation step $\Delta H = -4.186$ kJ/mol

This is repeated Propagation step $\Delta H = -106.78$ kJ/mol

Step 3: Termination step $\Delta H = -349.6$ kJ/mol

Termination step $\Delta H = -368.4$ kJ/mol

Termination step $\Delta H = -242.8$ kJ/mol

Similarly, when the number of carbon atoms and hydrogen is more, mixture result in varying proportions of the products. For example, chlorination of isobutane, gives 1-chloro-2-methyl propane and 2-chloro-2-methylpropane. The amount of tertiary substituted product is more than that of the primary susbtituted product.

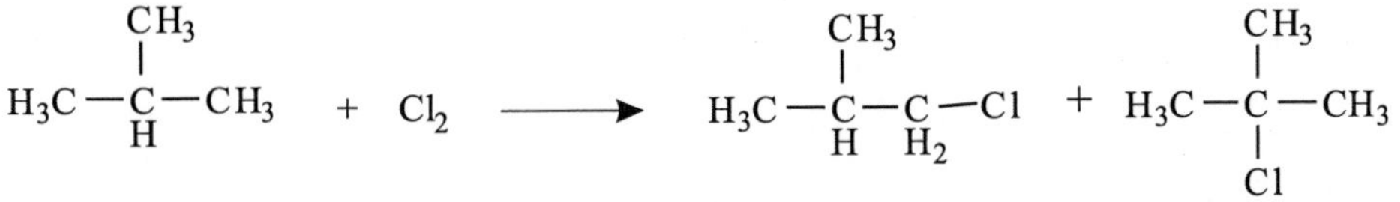

Bromination reactions is also very similar, but not so vigorous. Although, it is easy to form radical of bromine, still, may be because of its atomic size and electronegativity, its reaction is less volatile compared to that of chlorine. Iodination reaction is reversible, and iodinated compounds can be easily prepared by reaction of a chloro or bromo compound with sodium iodide in alcohol or acetone, in which it is soluble. In iodination reaction, hydroiodic acid formed makes it reversible, so it is advisable to add oxidising agents like, iodic acid, HIO_3, nitric acid, HNO_3 etc.

$$CH_4 + I_2 \rightleftharpoons CH_3I + HI$$

$$HI + HIO_3 \longrightarrow 3\,I_2 + 3\,H_2O$$

$$\text{or} \quad CH_3Cl + NaI \xrightarrow[\text{Acetone}]{} CH_3I + NaCl$$

(*b*) Nitration

Nitration is a reaction, where by we replace a hydrogen with a nitro group, and it can be carried out by reaction with nitric acid vapour at 300-475°C temperature. This reaction, like chlorination gives a mixture of mono-, di-or even trinitro-substituted products. It is because of the decomposition of reactive species.

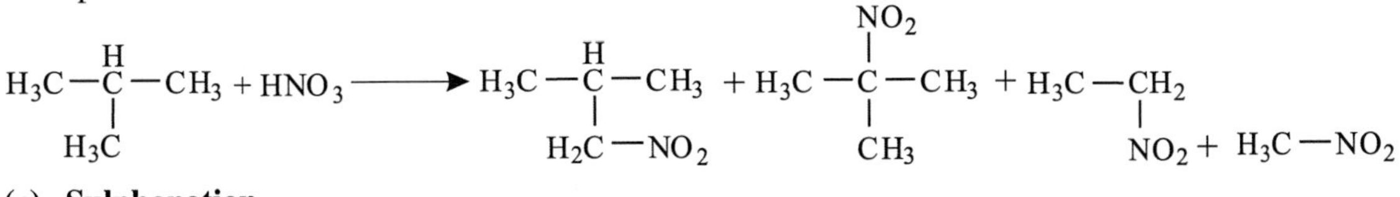

(*c*) Sulphonation

Sulphonation is a reaction of replacement of a hydrogen from a substrate with a sulphonic acid group. The reaction is carried out by reacting an alkane with fuming sulphuric acid. The reaction preferably takes place first at the tertiary carbon hydrogen, followed by a secondary carbon hydrogen and then the primary carbon. Reaction of tertiary butane gives t-butyl sulphonic acid.

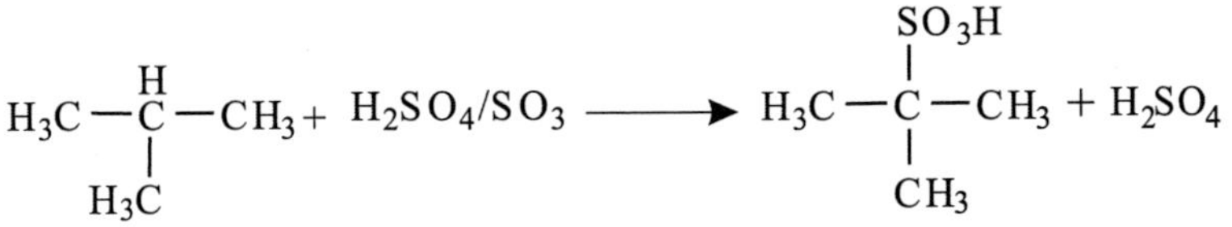

Mechanism:

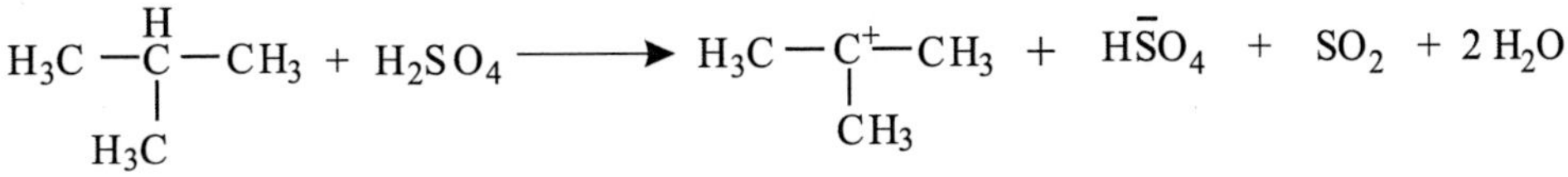

Formation of a carbonium ion is demonstrated by reaction of an optically active compound, such as 3-methyl hexane that led to racemisation in sulphuric acid.

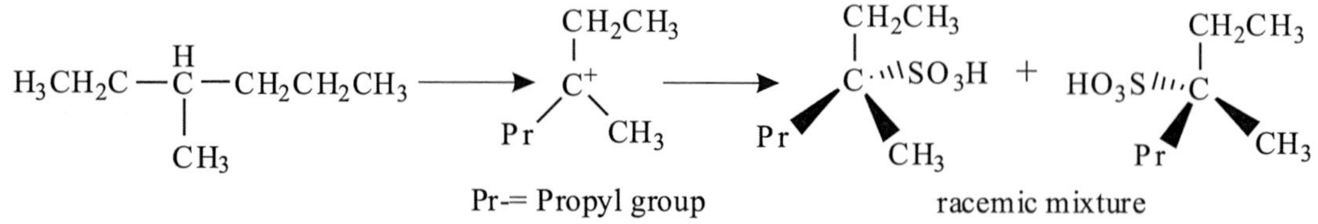

8.3.2.2 Oxidation Reactions

Oxidation reaction can be considered either as addition of oxygen or removal of hydrogen. Since hydrocarbons are the constitutent of petrol, these burn in excess of oxygen or air, to form carbon dioxide as an oxidation process. However, controlled oxidation reactions can also be carried. The nature of product depends upon the reagent used and the reaction conditions. Oxidation with potasssium permanganate, under boiling conditions, easily oxidises tertiary hydrogen to form an alcohol. The same example as in above cases, gives *t*-butanol.

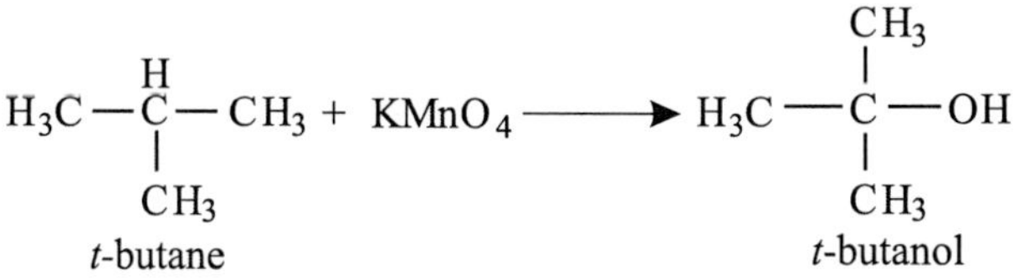

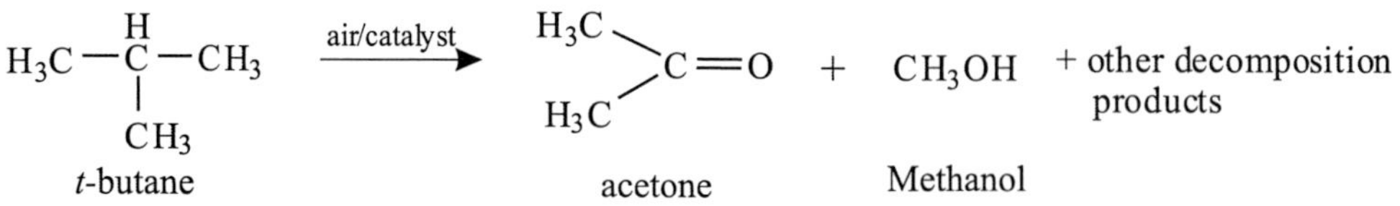

8.3.2.3 Isomerisation Reactions

When the vapours of an alkane are passed over aluminium chloride containing some HCl, isomerisation takes place. It is probably by formation of a carbonium ion, which isomerises to a more stable tertiary carbonium ion. Thus during isomerisation, tertiary product is formed in higher yield.

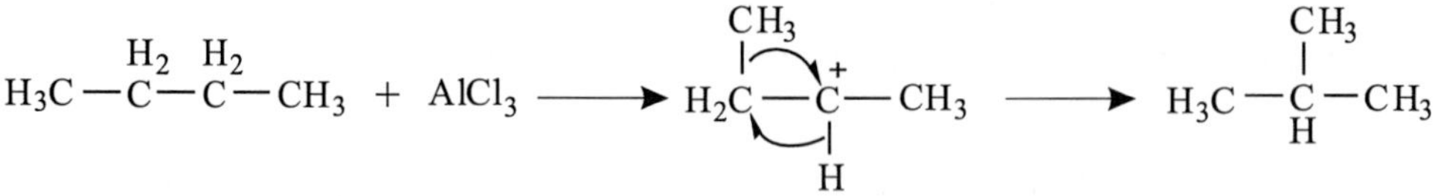

8.3.2.4 Petroleum and Cracking

Petroleum is a term used for a mixture of hydrocarbons obtained from the natural sources. Crude petroleum may contain hydrocarbon of varying carbon length, may be C_1-C_{40}. It also has natural gas as one of the components. It contains hydrocarbons with C_1-C_5 chains. The natural gas is thus used for manufacture of C_1-methanol, C_2- polyethylene, C_3-acetone, C_4-LPG, C_5-a component of refined petroleum.

From C_6-C_{11} are hydrocarbons called gasoline or petrol or petroleum, mostly used as motor fuel. Higher fractions, used as kerosene (C_{12}-C_{16}) and lubricating oils (C_{16}-C_{20}).

Cracking is a term used for pyrolysis of alkanes, which means, breakdown of longer carbon chains to smaller fragments. A few years earlier, petroleum was cracked in educational institutions to get laboratory gas. That has been now been replaced by LPG. The nature of product depends upon the temperature, pressure and catalyst used for cracking. Hydrocarbons containing six or more carbon atoms, can be cyclised to cyclohexane or higher members by using a suitable catalyst, such as chromic oxide, at a temperature of 500 ^{0}C.

The process of cracking involves, breaking down the carbon chain, isomerisation, dehydrogenation and cyclisation. All these processes produce hydrocarbon, that are smaller in size or cyclised or with branching of chain length. It has been observed that more branching lead to less knocking or these hydrocarbons act as anti-knock agents.

Cracking can be carried under two conditions

(*i*) **Liquid Phase Cracking:** Usually liquid phase cracking is carried out at 500 ^{0}C and under a pressure of about 200-1000 lb/sq.in. This process leads to formation of lower carbon atom containing hydrocarbons. Also in the process there is a decrease in the octane number.

(*ii*) **Vapour Phase Cracking:** In this cracking temperature is slightly higher, 600 ^{0}C and even higher number containing hydrocarbons remain in vapour phase.

Petroleum is a vast subject and cannot be further discussed in this chapter.

Points to Remember

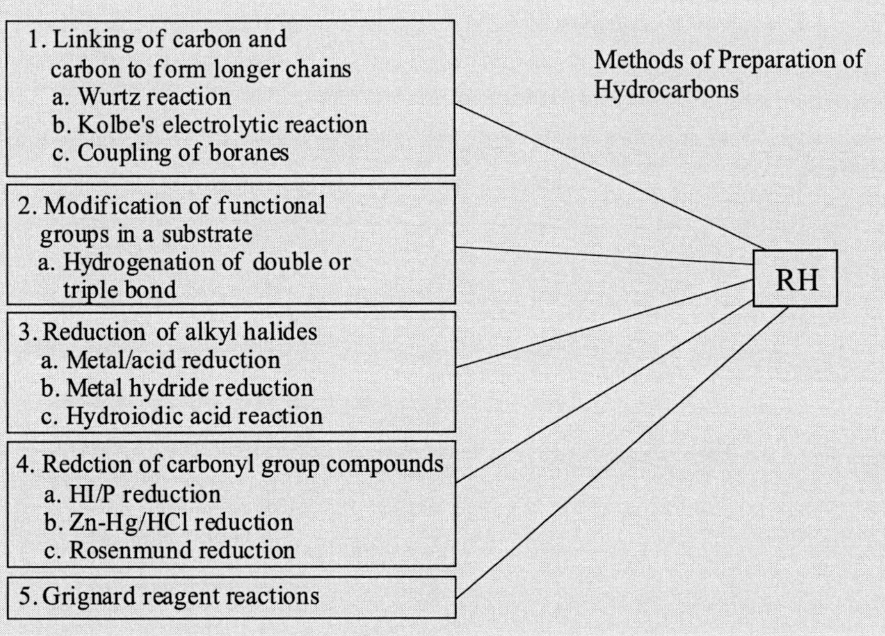

Properties of Hydrocarbons

RH

- If R is C_1-C_5, these are mostly **gas** if C_6-C_{11}, these are **liquids** if C_{12}-C_{18}, these are **heavy liquids** above this are **solids**
- Hydrocarbons undergo **substitution reactions**
  1. Halogenation
  2. Nitration
  3. Sulphonation
- **Oxidation reactions**, lead to CO_2 or alcohols or ketone
- Hydrocarbons undergo **isomerisation** reactions, leading to branching and products containing smaller number of carbon atoms in the product
- **Cracking** is a process of break down of hydrocarbon chains to ones of smaller number of carbon atoms in the product.

Exercises

1. What are the natural sources of alkanes? What use man, puts those in one's life?
2. Kerosene consists of C_{12} to C_{16} hydrocarbons, and on cracking forms lower hydrocarbons, the gas formed is used as a laboratory gas. What are the possible constitutents and write their structures?
3. Write the most stable conformation of ethane and give its energy diagram.
4. Give the IUPAC names of the following structures:

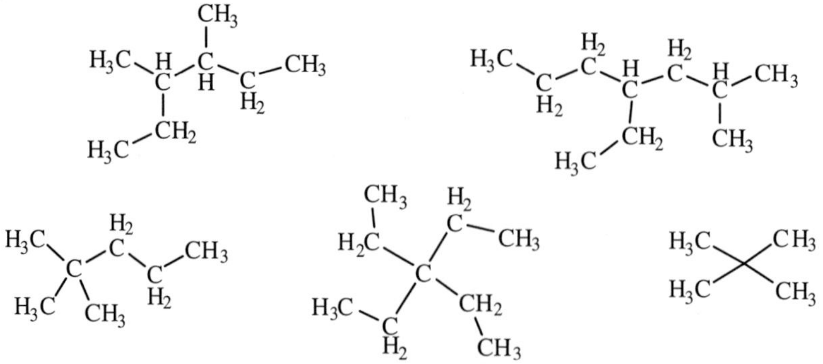

5. Write all the possible structures of compounds obtainable from C_6H_{14}. Give IUPAC names of the structures containing primary, secondary and tertiary hydrogens.

6. Write all the structures that can be formed from the molecular formula $C_5H_{11}Br$.and give their IUPAC names

   (*i*) How many structures have primary alkyl halides

   (*ii*) How many structures have secondary alkyl halides

   (*iii*) How many structures have tertiary alkyl halides .

7. In a hexane, C_6H_{14}, assuming a rotation around C_2-C_3 carbons, write the Newman's projections of the most stable conformation.

8. Write the structure of product of sulphonation of 2,2-dimethyl propane and 2-methylbutane.

9. Two cylinders contain *n*-hexane and 2,2-dimethylbutane, which can be used as a fuel. On the basis of energy considerations, which fuel would you like to select and why? What structural factors would you like to list to select a fuel that has maximum heat of combustion?

10. Arrange the following structures in order of their boiling points.

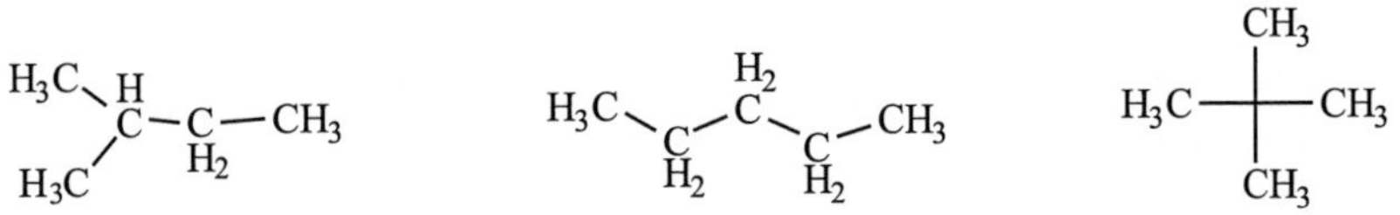

11. How would you carry out the following conversions?

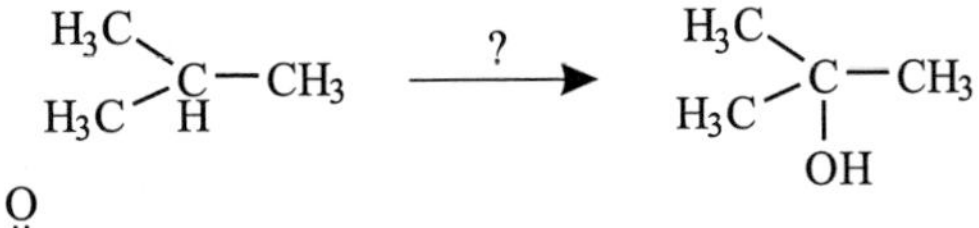

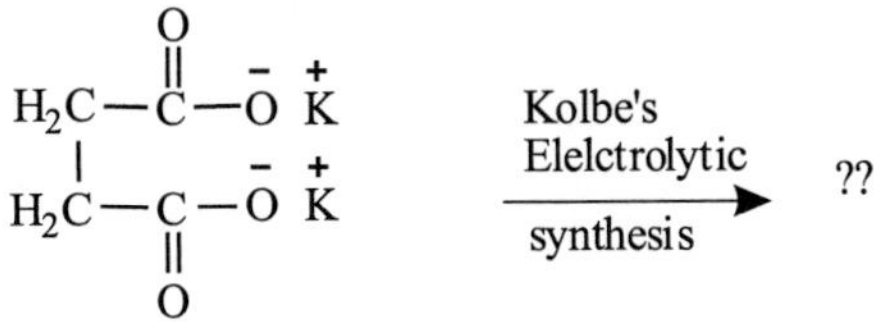

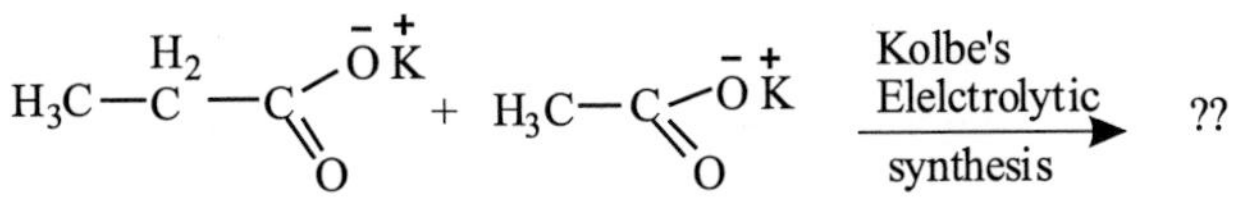

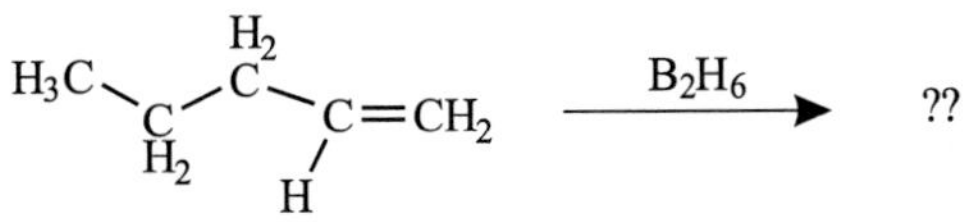

❑❑❑

Vladimir Vasilyevich Markovnikov
(1837–1904)

9

Aliphatic Hydrocarbons –Alkenes

IN THIS CHAPTER

He was born on December 22, 1837 at Shusha, Saint Petersburg, Russia. He studied at University of Kazan and University of Saint Petersburg. Markovnikov is best known for Markovnikov's rule, which he developed in 1869 to describe addition reactions of H-X to alkenes. According to this rule, the nucleophilic X- adds to the carbon atom with fewer hydrogen atoms, while the proton adds to the carbon atom with more hydrogen atoms bonded to it. The rule is useful in predicting the molecular structures of products of addition reactions. Why hydrogen bromide exhibited both Markovnikov as well as reversed-order, or anti-Markovnikov, addition, however, was not understood until Morris S. Kharasch offered an explanation in 1933. Hughes has discussed the reasons for Markovnikov's lack of recognition during his lifetime. Hughes concludes that the rule was an inspired guess, unjustified by the evidence of the time, but which turned out later to be correct (in most cases). Markovnikov also showed that butyric and isobutyric acids have the same chemical formula ($C_4H_8O_2$) but different structures; *i.e.*, they are isomers.

CHAPTER 9

Aliphatic Hydrocarbons –Alkenes

Introduction

Alkenes are unsaturated compounds, that contain double bond or bonds between carbon and carbon. Two carbons involved in double bond are sp^2 hybridised and consequently, the geometry of the molecule is such that all the substituents and carbon atoms lie in one plane, the *pi*-bond is perpendicular to the plane of sigma bonds. The bond angle between the substituents at the carbon-carbon double bond is usually around 120° as shown below:

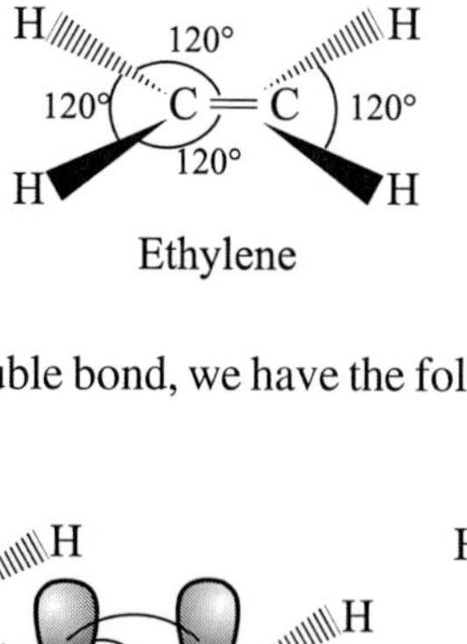

Ethylene

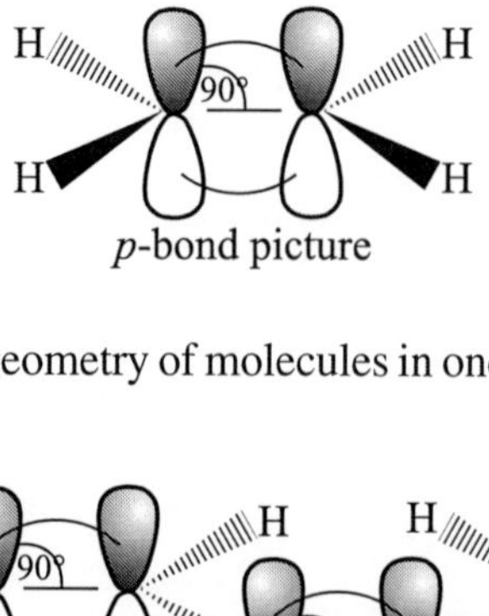

p-bond picture

For more than one double bond, we have the following geometry of molecules in one of the conformation.

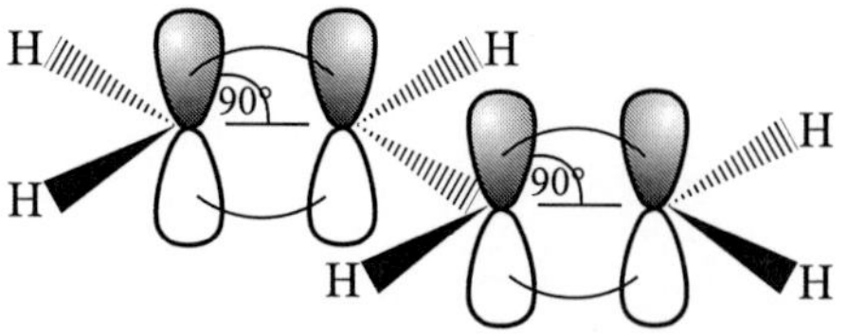

p-bond picture in buta-1,3-diene

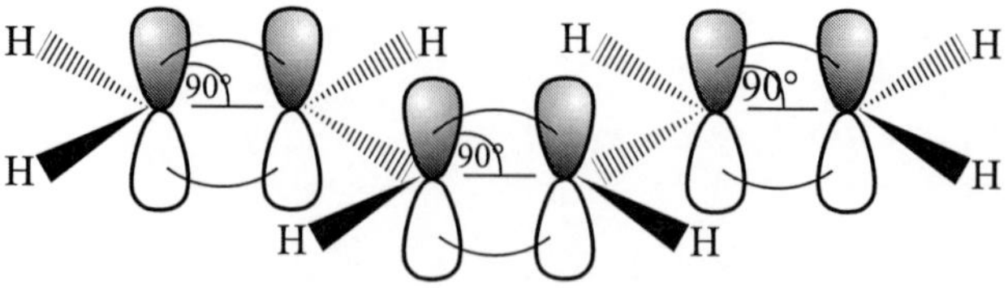

p-bond picture in hexa-1,5,3-triene

Shape or Molecular Geometry

Like covalent single bonds, double bonds can be described in terms of overlapping atomic orbitals, except that, a carbon-carbon double bond consists of one sigma bond and one *pi*-bond.

This double bond is stronger than a single covalent bond, bond energy for >C=C< = 611 kJ/mol . Also bond length is shorter, average bond length = 1.33 Angstroms (133 pm). The bond angle of sigma bonds is approximately 120°.

Each carbon of the double bond uses its three sp^2 hybrid orbitals to form three sigma bonds to three monovalent atoms. The unhybridized $2p$ atomic orbitals lie perpendicular to the plane created by the axes of the three sp^2 hybrid orbitals, that combine to form the pi bond.

Isomeric Structures of Substituted Alkenes:

Rotation about the carbon-carbon double bond is restricted because it involves breaking the π-bond, which requires a larger amount of energy, that in ethylene is equal to 264 kJ/mol. The consequence of this restricted rotation, substituted alkenes may exist as two isomers, called *cis or trans* isomers. Molecules having three or four different substituents, are named by (*E*), (*Z*) notation. Minimum necessary condition for a substituted alkene to show such an isomerism has been given in chapter 7.

Butene, for example, with two methyl groups facing the same side, is called (*Z*)-but-2-ene (as *cis*-2-butene), the other when two methyl groups lie on opposite face, it is called (*E*)-but-2-ene (as *trans*-2-butene). These two isomers of butene are slightly different in their chemical and physical properties.

Physical properties:

The physical properties of alkenes are comparable with those of alkanes. The physical state depends on molecular mass (gases from ethene to butene - liquids from pentene onwards). The simplest alkenes, ethene, propene and butene are gases. Linear alkenes of approximately five to sixteen carbons are liquids, and higher alkenes are waxy solids.

9.1 IUPAC Nomenclature of Alkene

Alkenes are compounds characterised by the presence of double bond in a carbon chain. The general formula for alkenes is C_nH_{2n}. The lowest member of the series has two carbons, there is a carbon-carbon double bond. The double bond is considered to be a functional group, compared to bromine or chlorine, that are regarded as substituents. Let us see some examples of compounds containing double bond.

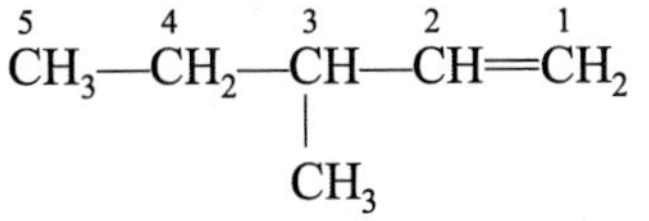

3-Methylpent-1-ene

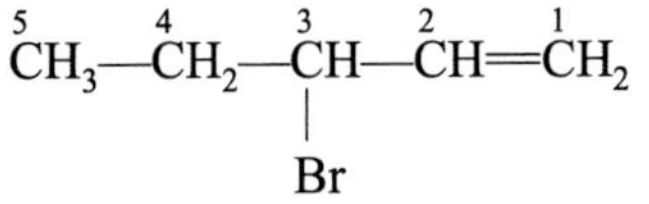

3-Bromo-pent-1-ene

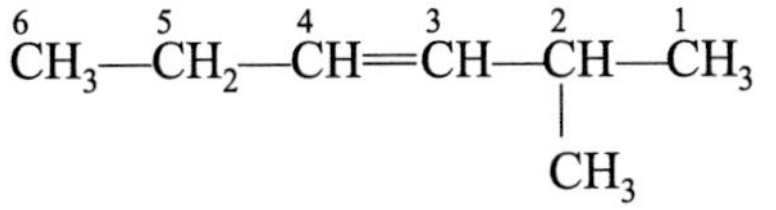

2-Methylhex-3-ene (correct name)
5-Methylhex-3-ene (wrong name)

(*a*) **Alkenyl Group:** It is a group obtained from alkene by reducing one hydrogen from alkene formula. The general formula of the alkenyl group is C_nH_{2n-1}. The point of linkage is also specified in the name and so also the position of the double bond. For example,

$CH_3—CH_2—CH{=}CH—$	But-1-enyl
$CH_3—CH(—)—CH{=}CH_2$	But-1-ene-3-yl
$—CH_2—CH(Br)—CH{=}CH_2$	3-Bromo-but-1-ene-4-yl
$—CH_2—CH{=}CH_2$	Prop-1-ene-3-yl (Allyl group)
$—CH{=}CH_2$	Ethenyl group (Vinyl group)

9.1.1 Isomeric Structure of C_5H_{10} and Their IUPAC Names

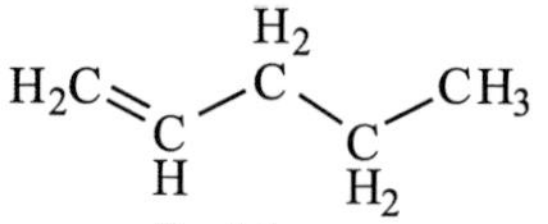
Pent-1-ene

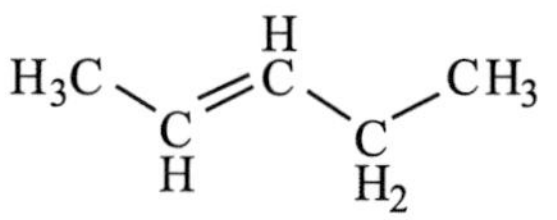
Pent-2-ene

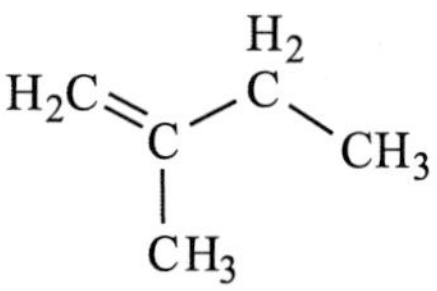
2-Methyl-but-1-ene

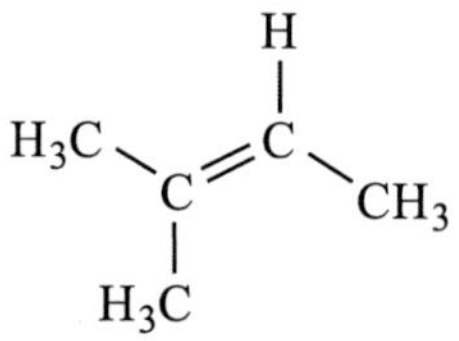
2-Methyl-but-2-ene

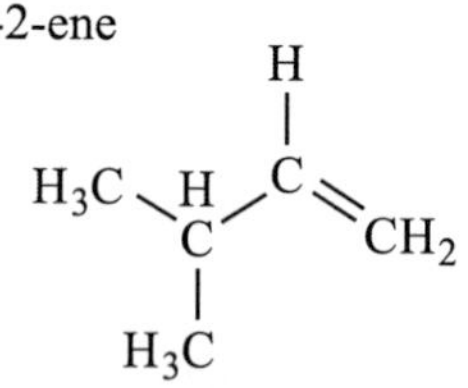
3-Methyl-but-1-ene

9.2 Methods of Preparation

Basically, there are two different types of methods that can be used for preparation of alkenes.

9.2.1 Creation of Double Bond by Elimination Reactions

(*i*) Dehydration reaction:

Primary alcohol can be dehydrated by acid catalysed dehydration to form alkenes. The most convenient method is by using conc. sulphuric acid,water is lost via the E1 mechanism. For example, the dehydration of ethanol produces ethene:

$$CH_3CH_2OH + H_2SO_4 \longrightarrow H_2C{=}CH_2 + H_3O^+ + HSO_4^-$$

Mechanism

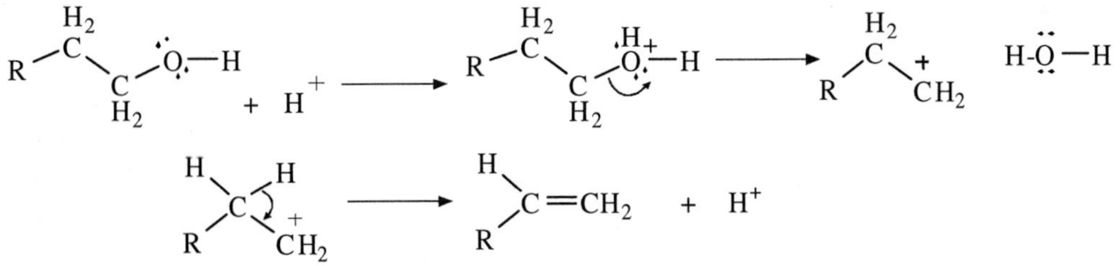

***(ii)* Dehydrohalogenation reaction:**

Halogen acid can be eliminated from the halide provided it has an α-hydrogen. To eliminate an acid, we make use of a strong base, such as sodium ethoxide in ethanol.

$$CH_3CH_2Br + NaOC_2H_5 \longrightarrow H_2C{=}CH_2 + NaBr + C_2H_5OH$$

Mechanism

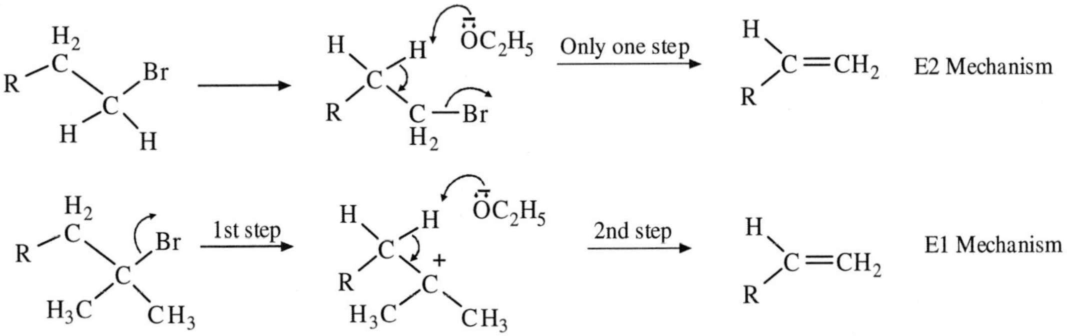

***(iii)* *vic* - Dehalogenation:**

Two halogen atoms on adjacent carbon atoms can be eliminated by reaction with a bivalent metal, *e.g.*, zinc.

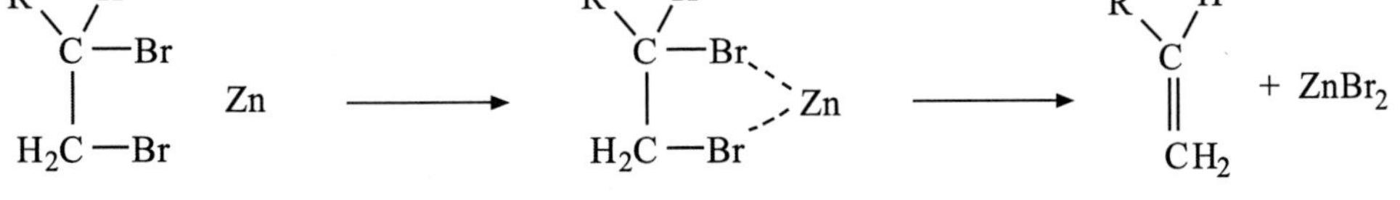

If instead of zinc, we use sodium, (like Wurtz's reaction), it gives dimerisation reaction product, but with double bond. It is due to two sodium atoms that react with two bromines on adjacent carbons in this substrate.

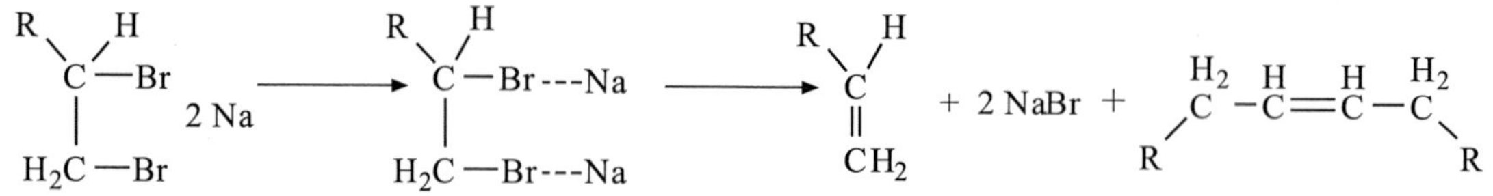

very little product   major product

Mechanism

Metal loses electrons and captures bromine anion.

$$Zn \longrightarrow Zn^{2+} + 2e$$

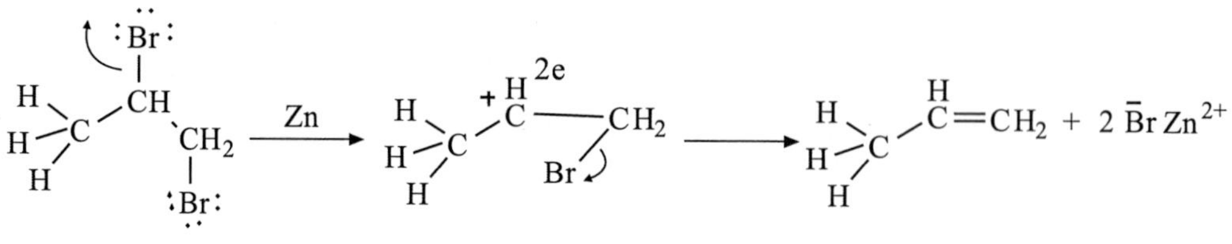

When iodine is used along with, the possible mechanism as given below.

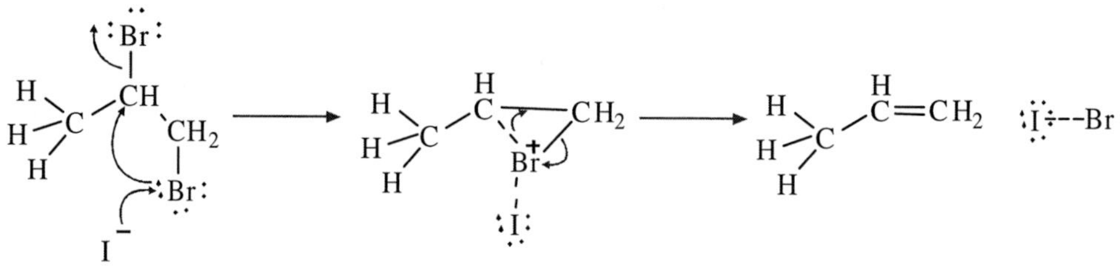

(*iv*) Hoffman Elimination:

Alkenes can be prepared indirectly from alkyl amines. The amine or ammonia is not a suitable leaving group, so the amine is first alkylated to form quaternary amine. When quaternary amine hydroxide is heated, it undergoes β-elimination of water molecule to form a double bond, along with formation of a tertiary amine. Hoffman elimination is unusual in that the less substituted alkene is usually the major product (opposite to Saytzeff elimination).

$$CH_3{-}CH_2{-}CH_2{-}\underset{\bar{O}H}{N^+}(CH_3)_3 \xrightarrow{\text{Heat}} H_3C{-}\underset{H}{C}{=}CH_2 + N(CH_3)_3 + H_2O$$

Mechanism

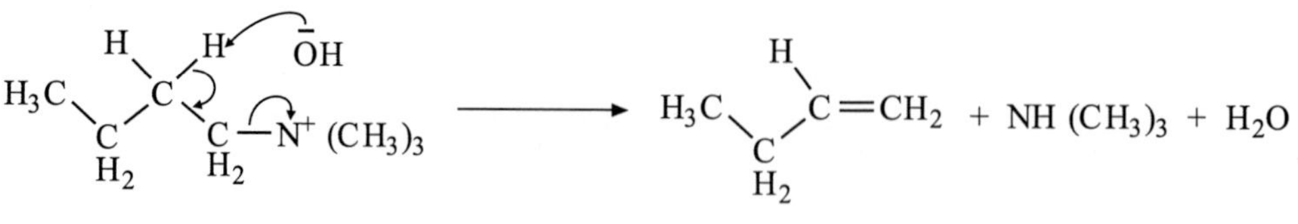

(*v*) Cope Reaction:

The Cope reaction is a syn-elimination that occurs when an amine is oxidized to an amine oxide as quaternary nitrogen (the Cope reaction), which makes it have a smooth elimination of β-hydrogen. Thus, giving a terminal double bond. It is a reaction that takes place at or below 150°C. For example, N,N-dimethyl propyl amine N-oxide on heating eliminates dimethylamine hydroxide. The reaction can be carried out at room temperature, when dimethyl sulphoxide or tetrahydrofuran is used as solvent.

$$CH_3.CH_2.CH_2{-}\underset{\displaystyle O^-}{\underset{|}{N^+}}(CH_3)_2 \xrightarrow{\text{Heat}} H_3C{-}\underset{H}{C}{=}CH_2 + HON(CH_3)_2$$

Mechanism

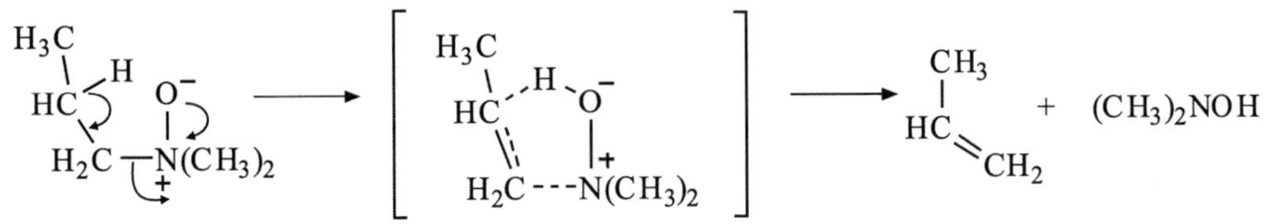

9.2.2 Modification of Functional Groups to Form a Double Bond

(*i*) Triple bond reduction to double bond

A controlled reduction of triple bond by using only one mole of hydrogen gives a double bond, an alkene. The most convenient method is to use Raney-Ni/H_2 at an elevated temperature.

$$H_3C{-}C{\equiv}CH + H_2 \xrightarrow{\text{Raney-Ni}} H_3C{-}\overset{H}{C}{=}CH_2$$

In case of unsymmetrical alkyne, two products are possible, the *cis*-and *trans*-, depending upon the catalyst or reagent used.

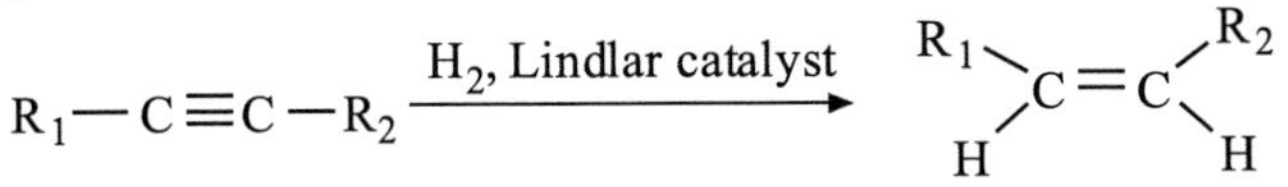

cis-product

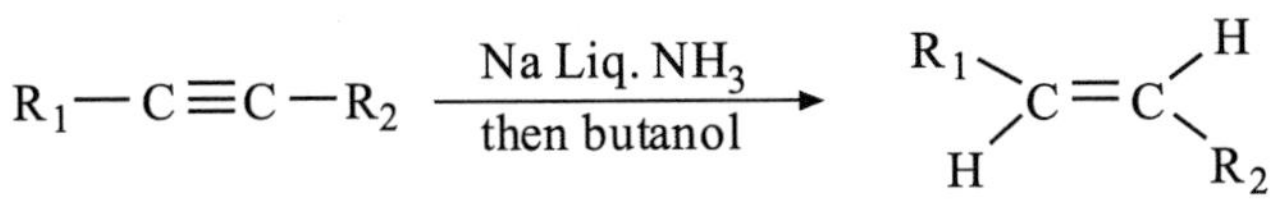

trans-product

(*ii*) Wittig reaction

Synthesis from carbonyl compounds:

Another important method for alkene synthesis involves construction of a new carbon-carbon double bond by coupling a phosphorus ylide with a carbonyl compound (such as an aldehyde or ketone). The ylide acts as carbanion equivalent. Such reactions are sometimes called olefinations. The most well-known of these methods is the Wittig reaction. Wittig's reaction is a method of converting an aldehyde or a ketone to an alkene by using an appropriate phosphorus ylide obtainable from triphenyl phosphine and alkyl halide. The dipolar intermediate leads to stable cyclic structure and physical methods also support this mechanism.

$$Ph_3P + H_3C{-}Br \xrightarrow{Ph\,Li} Ph_3P{=}CH_2 + PhH + LiBr$$

$$Ph_3P{=}CHR + CH_3CHO \longrightarrow H_2C{=}CH_2 + Ph_3PO$$

Mechanism

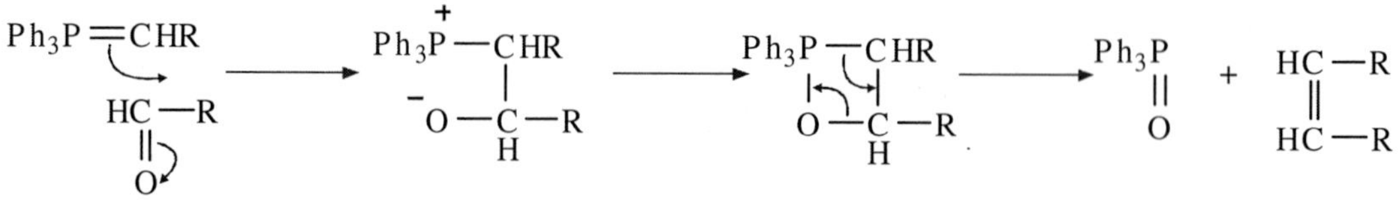

(*iii*) Pyrolysis of esters

Alcohol group in an ester or a xanthate form can abstract beta-proton leading to elimination of an acid part and formation of an alkene. This is just a method to convert an alcohol in to a better leaving group.

$$H_3C{-}\overset{H_2}{C}{-}\overset{H_2}{C}{-}O{-}COCH_3 \xrightarrow{\Delta} H_3C{-}\underset{H}{C}{=}CH_2 + CH_3.COOH$$

Mechanism

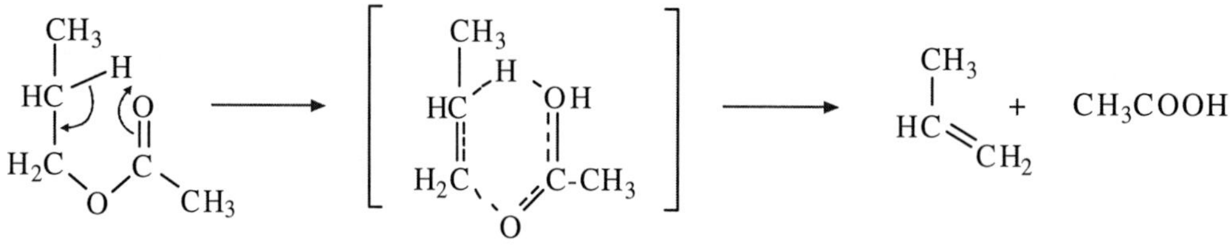

Similar elimination reactions that involve elimination of beta proton to give alkene are :

(*iv*) **Chugaev elimination:** It is reaction of breaking up of the xanthate

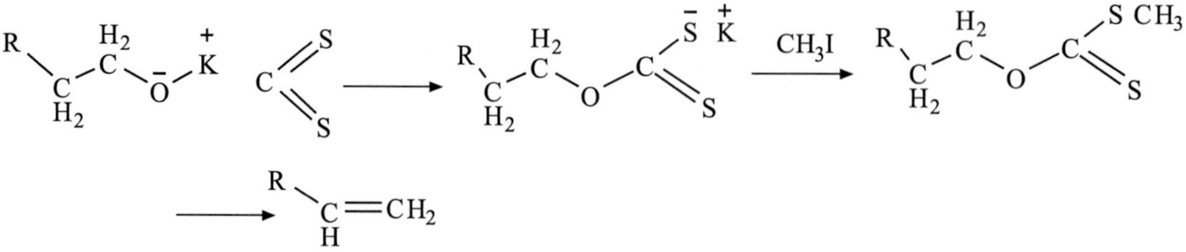

(*v*) **Ramberg-Bäcklund reaction,** involves halogen elimination from **sulfone:**

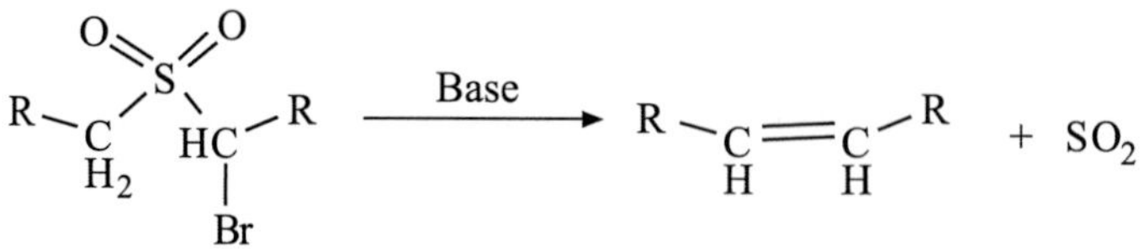

(*vi*) **Grieco elimination:**

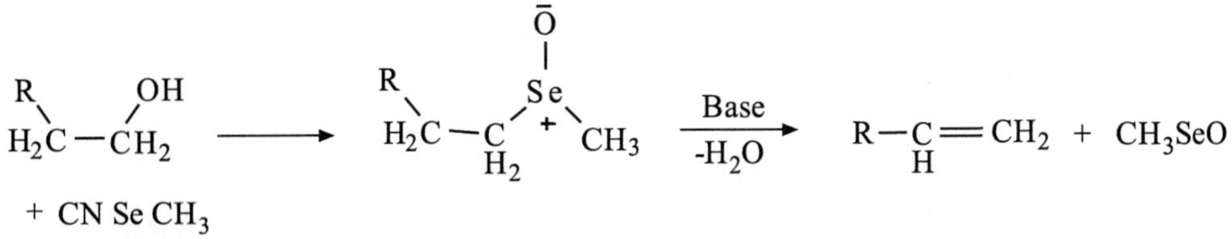

(*vii*) **Boord olefin synthesis:** β-Haloethers, under catalysis with magnesium or zinc metals, eliminate ROMX, to form an alkene.

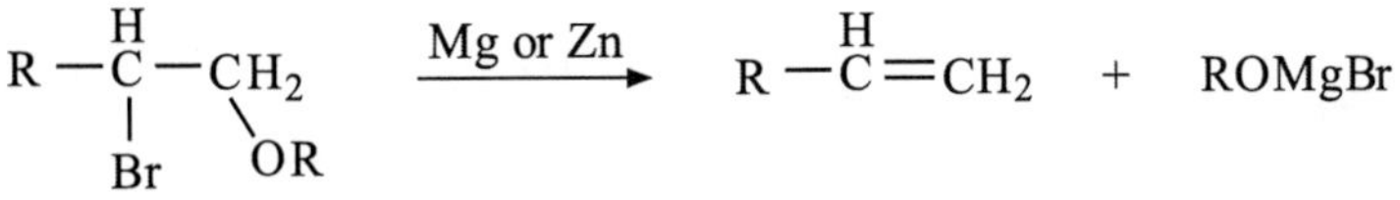

9.3 Chemical Reactions of Double Bond

Alkenes are relatively less stable compounds, but are more reactive than alkanes due to the presence of a carbon-carbon π-bond. The majority of the reactions of alkenes involve the rupture of this π-bond, forming new single bonds. Alkenes undergo many addition reactions, which occur by opening up the double-bond. Most addition reactions to alkenes follow the electrophilic addition mechanism.

9.3.1 Addition Reactions

Alkenes contain double bond, the valency of two carbons gets saturated by addition reactions. As we have already studied, that π-bond electrons can electromerise to form a polar species. Thus, two entities of opposite charge can add on to double bond. When electrophile portion adds first, this addition reaction is called electrophilic addition reaction, and when nucleophilic portion adds first, it is called nucleophilic addition reaction. Addition of an addendum A-B to a substituted carbon - carbon double can be represented in two ways as follows .

A. Electrophilic addition reaction

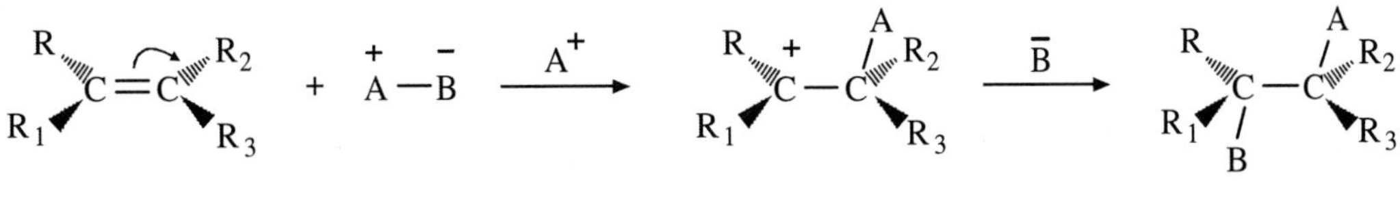

B. Nucleophilic addition reaction

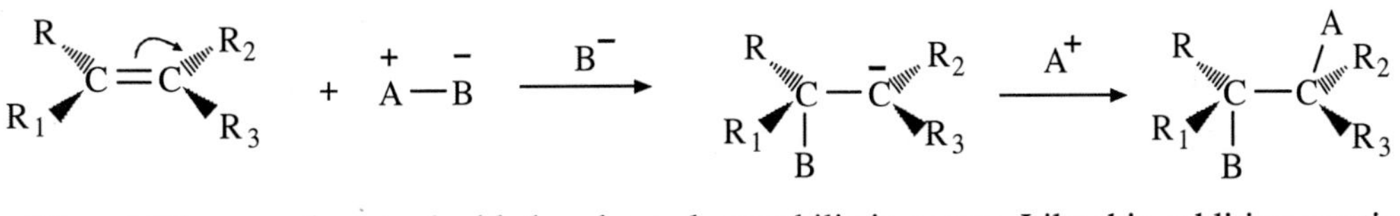

Most of the addition reactions to double bond are electrophilic in nature. Like this addition reaction, all other reagents can be added, that are given ahead.

9.3.1.1 Br_2 Addition:

Bromine addition to an alkene forms a vicinal dibromo compound. It is an ionic reaction, so the addition takes place by trans-addition mechanism.

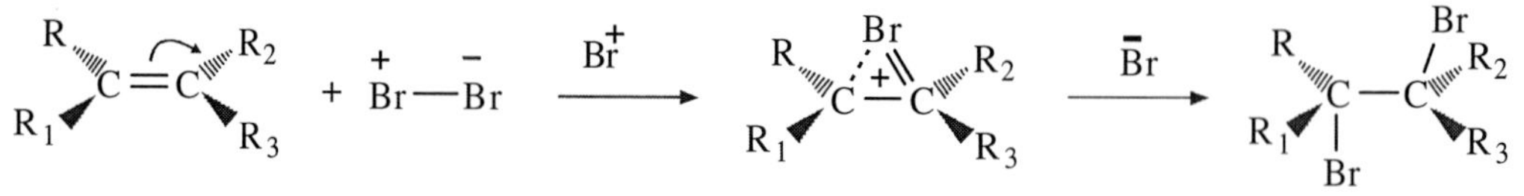

9.3.1.2 Cl_2 Addition:

Chlorine addition to an alkene forms a vicinal dichloro compound. It is an ionic reaction, so the addition takes place by trans-addition mechanism.

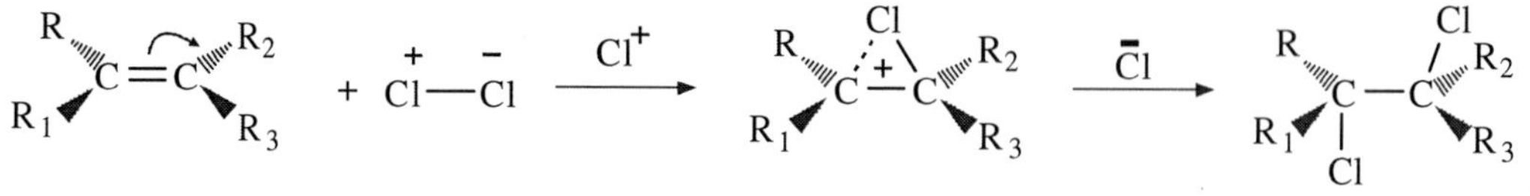

9.3.1.3 HBr Addition:

Hydrogen bromide addition to an alkene forms a bromo compound. It is an ionic reaction, so the addition takes place by trans-addition mechanism. Prior to full understanding of mechanisms, certain rules were made to explain such reactions, which could explain certain cases, but not all. One such rules that explained the addition reaction of halogen acids is called Markownikoff rule of addition of halogen acids.

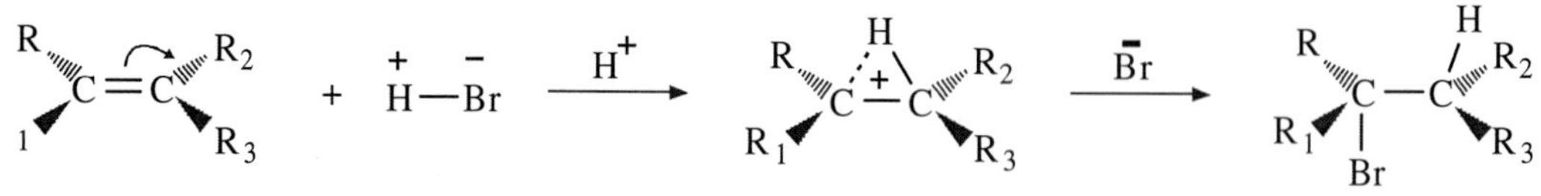

Markownikoff Rule

Markownikoff rule, was developed in 1869 to describe addition reactions of H-X to alkenes. According to this rule, the nucleophilic X^- adds to the carbon atom with fewer hydrogen atoms, while the proton adds to the carbon atom with more hydrogen atoms bonded to it. The rule is useful in predicting the molecular structures of products of addition reactions, but does not hold good in all cases.

HBr is an unsymmetrical addendum, so shall give unsymmetrical product. This type of reaction has been a matter of long discussion, as to which of the two carbons of double bond, the electrophile shall go. During early days of development, the Markownikoff rule was used to predict, the product. Later, systematic **electronic explanation** was offered. For example, addition of HBr to propene, takes place to form 2-bromo-propane and not 1-bromo-propane, as shown next.

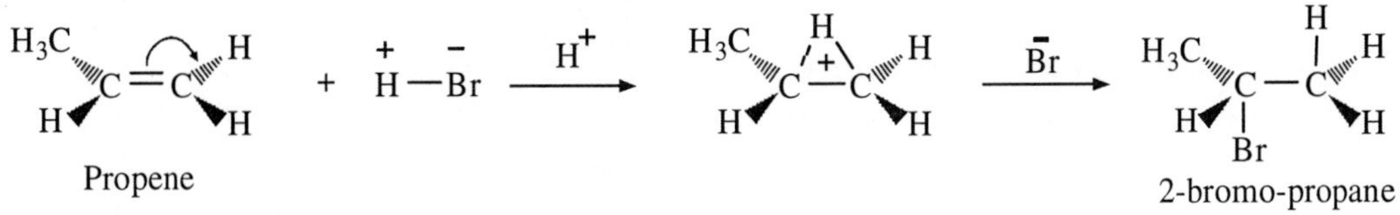

1-nitro-2-methylpropene addition of HBr can be visualised as to give two products, as shown. Infact only 1-Nitro-2-bromopropane is formed.

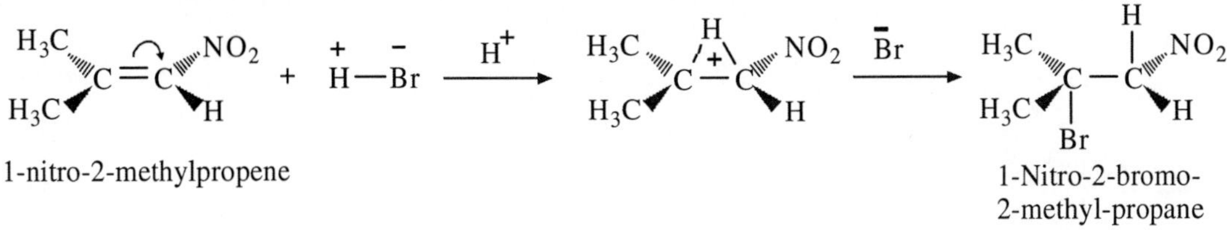

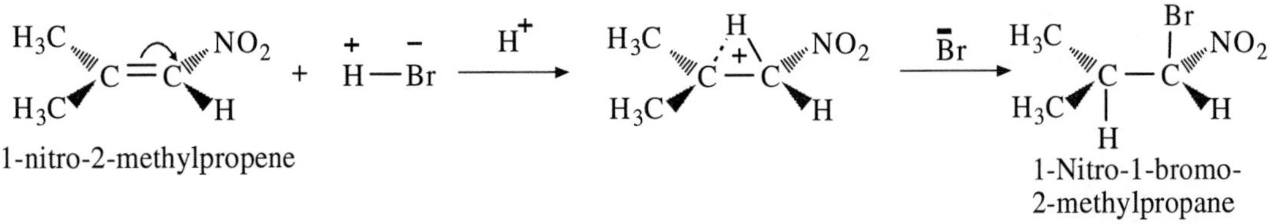

Which product is actually formed, can be predicted by electronic mechanism. The inductive effect of methyl groups as electron repelling group and nitro group as electron attracting group shall make the pi-bond electrons move in one direction only. Thus, creating positive and negative centres at carbon for the smooth addition reaction to occur. In this case it could be explained by Markownikoff rules also. There are certain examples, that Markownikoff rule is not followed.

Proof of mechanism

In case of unsymmetrical alkenes, the above given mechanism is justified by using a competetive anion like chloride anion. That is, when addition reaction of HBr to alkene is carried in presence of sodium chloride, two products get formed.

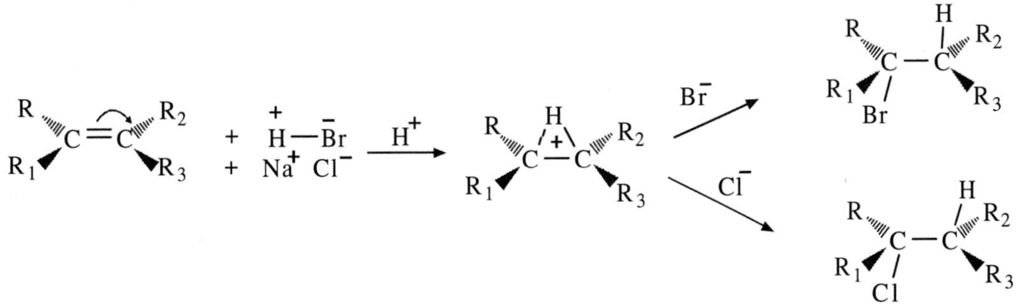

It was observed that addition of HBr to an alkene takes place according to Markownkoff rule, but in presence of a peroxide, addition took place in opposite fashion. Why hydrogen bromide exhibited both Markownkoff as well as anti-Markownkoff addition was not understood until Morris S. Kharasch offered an explanation in 1933.

Kharasch Peroxide effect or Anti-Markownikoff addition

Addition reactions that take place in presence of a peroxide, take place via radical intermediate. These are regioselective, and against what Markownkoff's rule predicts. Mechanisms which avoid the carbocation intermediate, react through free radical addition. Such reactions are said to be anti-Markownkoff, since the halogen adds to the less substituted carbon, exactly the opposite of Markownkoff reaction. Explanation of this effect is given by the fact that, like the positive charge, the radical is more stable in the more substituted position.

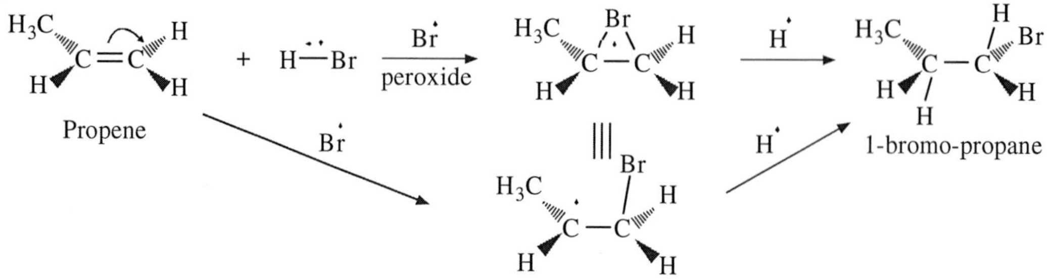

A secondary radical is more stable, so is formed in presence to the primary radical, that could get formed when addition of Br radical take place at the second carbon

9.3.1.4 Addition of Hydroiodic Acid

Addition of HI to an alkene is also in accordance to the Markownikoff rule, but it all depends upon the substituents of the alkene. A more plausible and alternate explanation to Markownikoff rule is the stability of carbonium ion that are formed as an intermediate in the addition reaction. It has been known that tertiary carbonium ion is more stable than secondary and primary carbonium, in the order, tertiary carbonium ion > secondary carbonium ion > primary carbonium ion. Thus, that reaction shall take place which involves the most stable carbonium ion. For example, HI adds to propene to form 2-iodopropane.

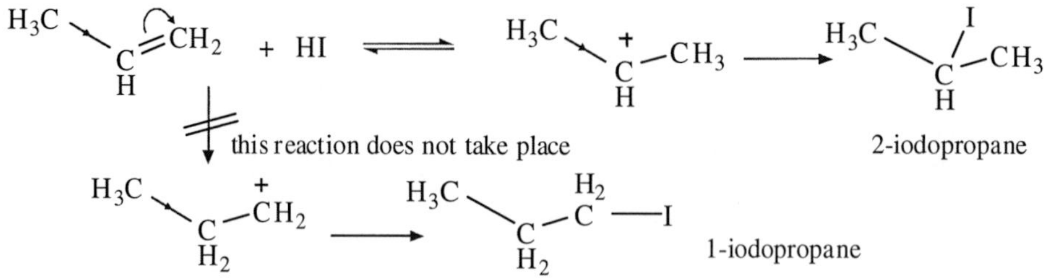

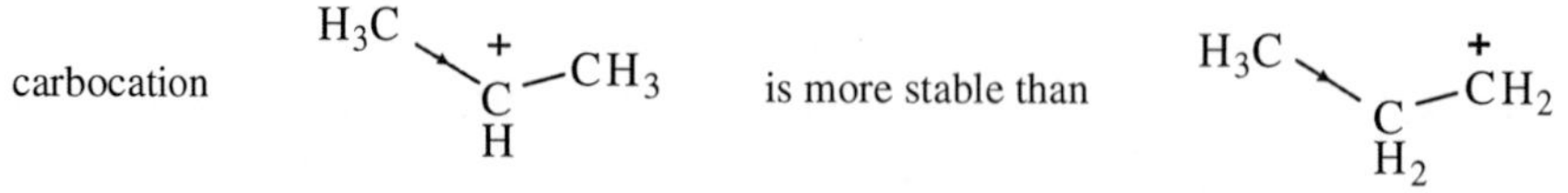

9.3.1.5 Addition of IBr and Br Cl

This addition reaction is strictly in accordance to direction of movement of electrons in the substrate. It follows the electronegativity order of halogens as, Cl > Br > I.

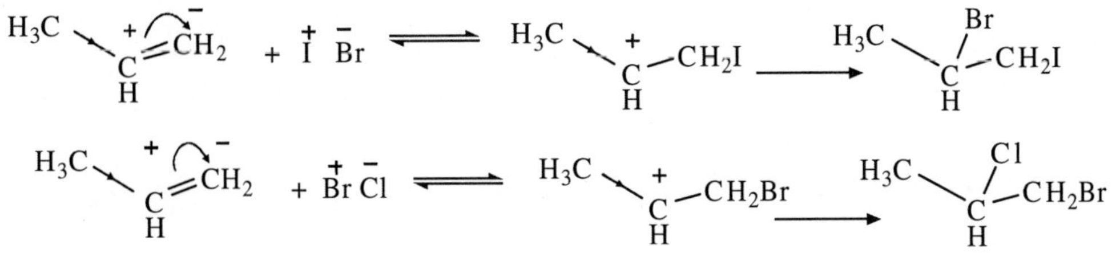

9.3.1.6 Addition of water molecule using sulphuric acid

Addition of elements of water using a small quantity of conc. sulphuric acid, takes place, in accordance to the Markownikoff rules to form alkyl hydrogen sulphate. However, in presence of water, hydrolysis of hydrogen sulphate takes place to yield alcohol. For which electronic explanations are also given. It is very similar to HBr addition, replace Br^- with HSO_4^- anion. That is a product, where we find, virtually, addition of water has taken place.

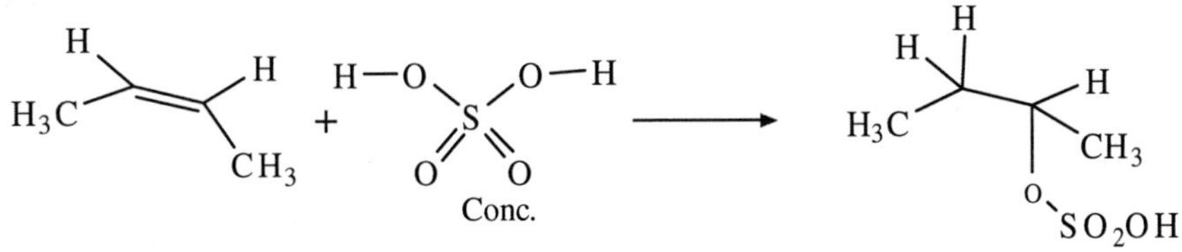

sec-Butyl hydrogen sulphate

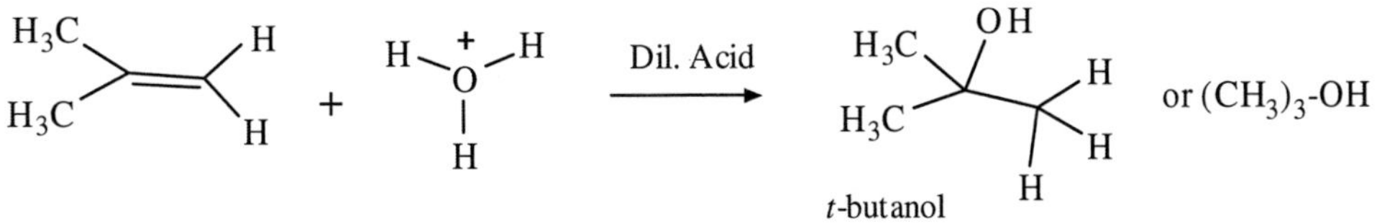

Mechanism

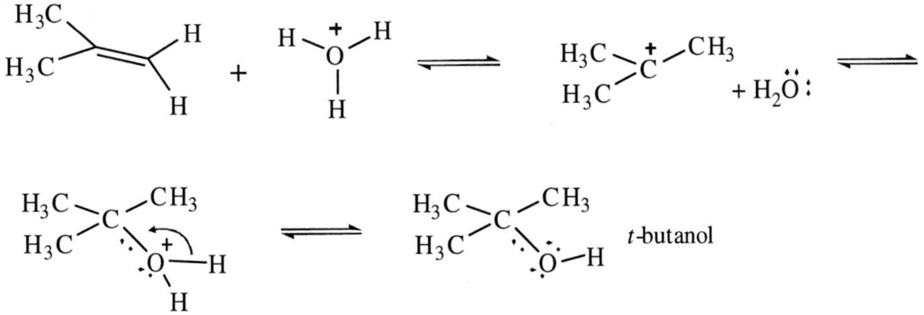

9.3.1.7 Mercuric acetate addition

Mercuric acetate, is $Hg^{++}(OCOCH_3)_2$, acetate is $^{-}OCOCH_3$, this negative part shall go to positive carbon and $Hg^{+}(OCOCH_3)$ shall go to negative carbon.

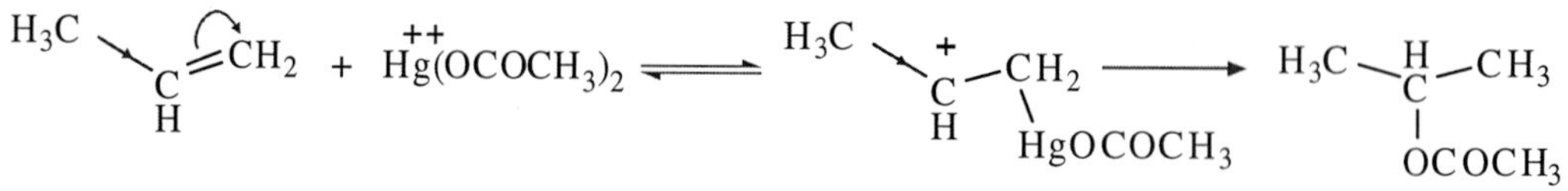

9.3.1.8 Nitrosyl chloride addition

Nitrosyl chloride is, O = N–Cl, oxygen is more electronegative than chlorine, so O=N, shall become negative addendum and Cl shall be positive addendum, addition takes place as follows

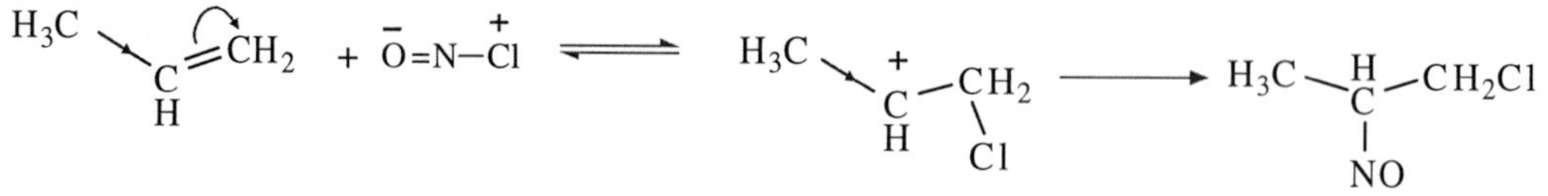

9.3.1.9 Ozone addition and double bond cleavage

Ozonolysis is a reaction for cleaving a C=C, to form two carbonyl compounds. It is a kind of oxidation of carbon atoms, so is considered under oxidation reactions, 9.5.6.

9.4 Reduction Reactions

9.4.1 Addition of Hydrogen

Addition of hydrogen to a double bond is also called hydrogenation. It is a reaction that takes place, by molecular hydrogen (H_2), in the presence of a catalyst. The process results in saturating the organic compounds. Hydrogen adds to double and triple bonds in hydrocarbons. The versatile catalyst for reducing a >C=C< double bond is Raney Ni/H_2. Platinum group metals, particularly platinum, palladium, rhodium, and ruthenium, form highly active catalysts, which operate at lower temperatures and lower pressures of H_2. An important characteristic of alkene and alkyne hydrogenations, is that hydrogen addition occurs with "syn addition," where hydrogen enters from the least hindered side. The mechanism of metal-catalyzed hydrogenation of alkenes and alkynes has been extensively studied. Isotope labelling using deuterium confirms the regiochemistry of the addition.
The addition of H_2 to an alkene affords an alkane in the protypical reaction

$RCH{=}CH_2 + H_2 \longrightarrow RCH_2CH_3$

alkene, $R_2C{=}CR'_2 \longrightarrow R_2CHCHR_2$, alkane,

alkyne, , $R{-}C{\equiv}C{-}R \longrightarrow$ *cis*-RHC=CHR , alkene,

An illustrative example of a hydrogenation reaction is the addition of hydrogen to maleic acid to succinic acid. Hydrogenation of unsaturated fats produces saturated fats and, in some cases, trans fats.

The reverse reaction, removal of hydrogen from a molecule, is called dehydrogenation. A reaction where bonds are broken while hydrogen is added is called hydrogenolysis. Hydrogenation differs from protonation or hydride addition. In hydrogenation, the products have the same charge as the reactants.

Thermodynamically, hydrogenation is a strongly exothermic reaction. In the hydrogenation of vegetable oils and fatty acids, for example, the heat released is about 25 kcal per mole (105 kJ/mol).

9.5 Oxidation Reactions

9.5.1 Baeyer Oxidation by $KMnO_4$ [Johann Friedrich Wilhelm Adolf von Baeyer (1835-1917)]

He was born on 31 October, 1835, in Berlin. He came from a family distinguished both in literature and the natural sciences. Baeyer devoted his first two years at the University of Berlin (1853-1855) chiefly to physics and mathematics. By 1856, however, his old love for chemistry re-awakened and drew him to Bunsen's Heidelberg and was associated with his ingenious structure theory. A study of uric acid, barbituric acid, provided the thesis by which he qualified as a university teacher in 1860. In 1865, he started his work on indigo - the blue dye had fascinated him since his youth-and this soon led to the discovery of indole and to the partial synthesis of indigotin. Baeyer, clarified the structure of alizarin and other dye-stuff-phthalein dyes and worked out the synthesis used industrially. On the death of Justus von Liebig in 1873, Baeyer was called to his Chair in the University of Munich. His on acetylene and polyacetylene, and from this derived the famous Baeyer strain theory of the carbon rings; there were studies of the constitution of benzene as well as comprehensive investigations into cyclic terpene. In this connexion the Baeyer-Villiger oxidation of ketones by means of per-acids was discovered. Special interest was aroused on the connection between constitution and colour."I have never set up an experiment to see whether I was right, but to see how the materials behave". Even in old age his views did not become fixed, and his mind remained open to new developments in chemical science. Honours were heaped upon him, including the Nobel Prize for Chemistry in 1905. He was still young in spirit when he succumbed to a seizure at his country house on 20 August, 1917.

Baeyer's oxidation is the reaction of an alkene with potassium permanganate to form cis-dihydroxy compound. During the process, $KMnO_4$ gets decolourised, so this reaction is also used as a test for the presence of double bond. Ethylene by reaction with dilute alkaline permanganate is converted into ethylene glycol.

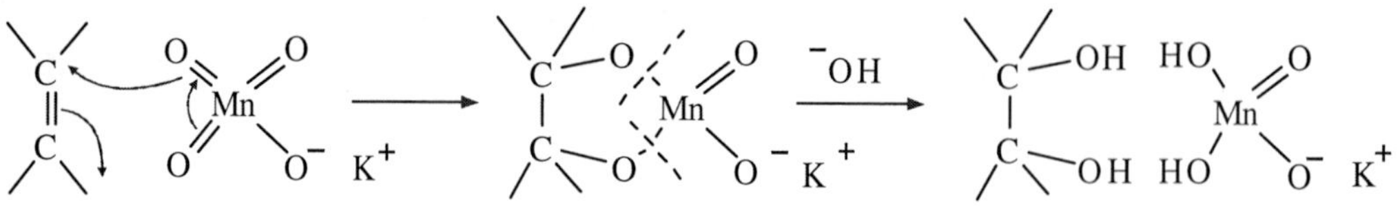

9.5.2 Hydrogen Peroxide Oxidation

Hydrogen peroxide (90 per cent), on reaction with an alkene in an acidic medium forms a dihydroxy compound. It is a reaction very similar to permanagnate oxidation reaction.

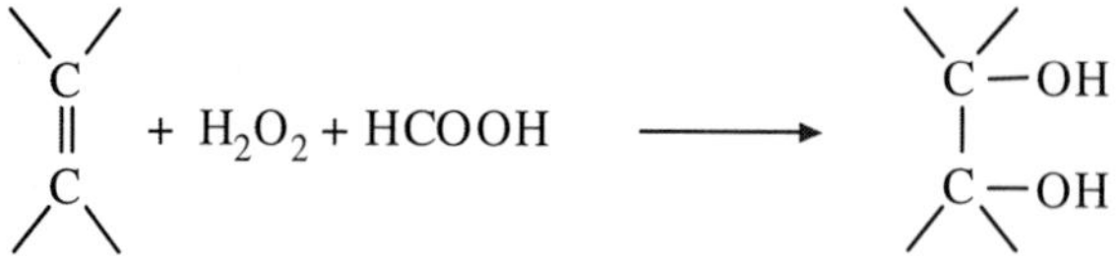

9.5.3 Osmium Tetroxide Oxidation

Osmium tetroxide like permanganate gives *cis*-hydroxylation product. It has been observed that rhuthenium tetroxide is more convenient and less toxic than osmium tetroxide.

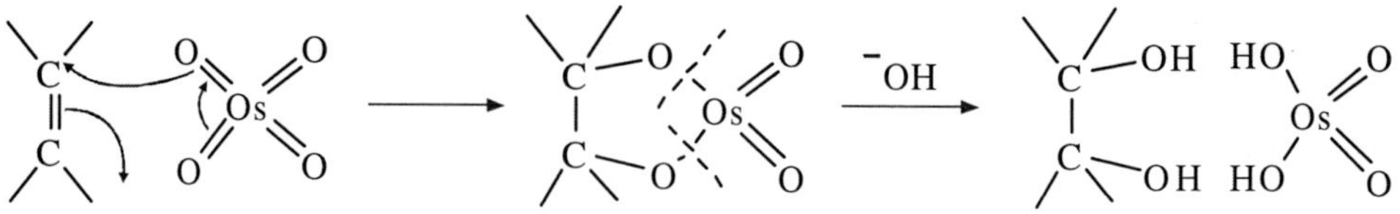

9.5.4 Iodine/silver Acetate Oxidation

Iodine/silver acetate is another reagent for *cis*-hydroxylation, where hydrolysis of mono- or diacetate occurs under alkaline conditions.

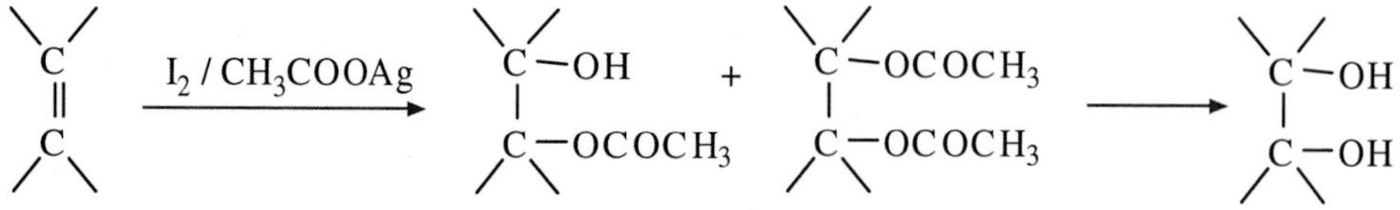

9.5.5 Peracid Oxidation to Epoxide

Alkenes get oxidised to epoxide by a peracid. The reaction is known as Prileschaiev's reaction. Recent years, Sharpless epoxidations have become very famous, because of specific stereochemistry of the product. Hydrolysis of epoxide gives a diol, like the above reactions.

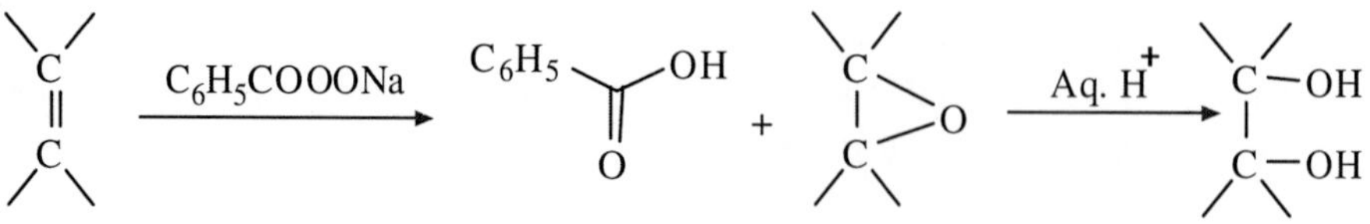

Mechanism

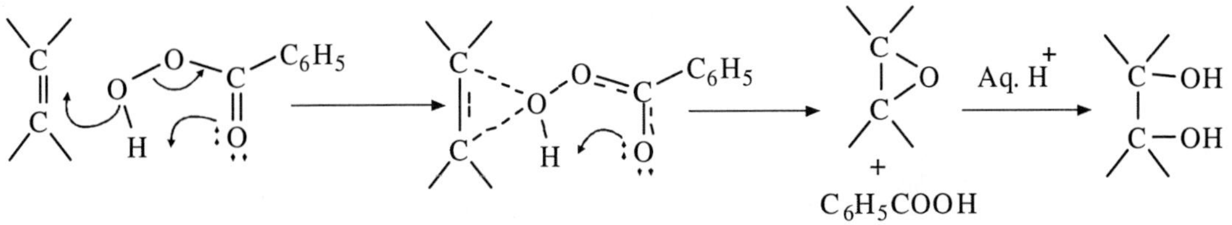

The **Sharpless epoxidation** reaction is an enantioselective chemical reaction to prepare 2,3-epoxyalcohols from primary and secondary allylic alcohols. The stereochemistry of the resulting epoxide is determined by which diastereomer of diethyl tartrate is employed in the reaction. This reaction gives good yields and diastereoselectivities over a broad range of substrates.

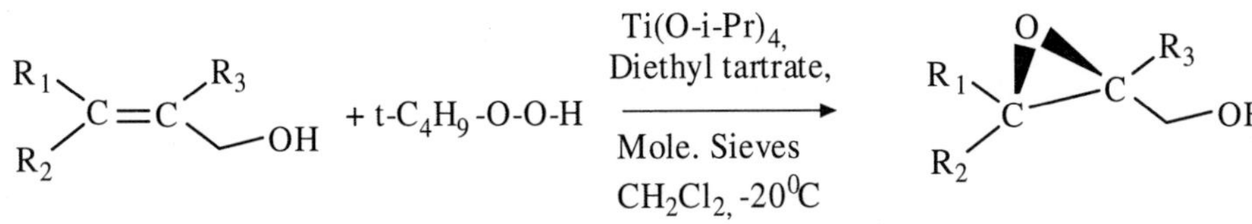

The oxidizing agent is tert-butyl hydroperoxide. Enantioselectivity is achieved by a catalyst formed from titanium tetra(isopropoxide) and diethyl tartrate. Only 5-10 mol% of the catalyst in the presence of 3Å molecular sieves (3Å MS) is required. K. Barry Sharpless shared the 2001 Nobel prize in Chemistry for his work on asymmetric oxidations. The prize was shared with William S. Knowles and Ryoji Noyori.

9.5.6 Ozone Addition and Ozonolysis

Addition of ozone to alkene gives ozonides, that on cleavage with silver oxide or any peracid, gives two molecules of carbonyl compounds. The net result is the cleavage of a molecule with double bond, at its double bond position, carbonyl group carbon are the ones involved in carbon-carbon double bond.

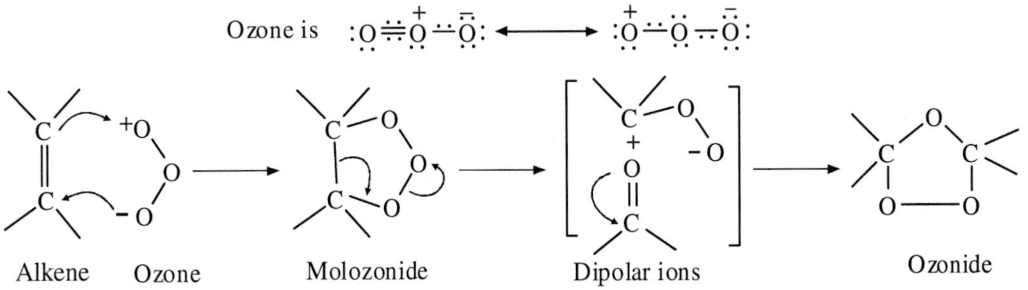

For example, pent-2-ene on ozonolysis gives propanal and ethanal, two aldehydes. However, when the alkene is cyclic, only one compound is formed, as shown in case of 3-methyl-5-chloro-hexandial. Similarly, other substituted alkenes give aldehydes or ketones, as the substitution might be. For example, ozonolysis of 2,3-dimethyl pent-2-ene gives butanone and acetone, as shown ahead.

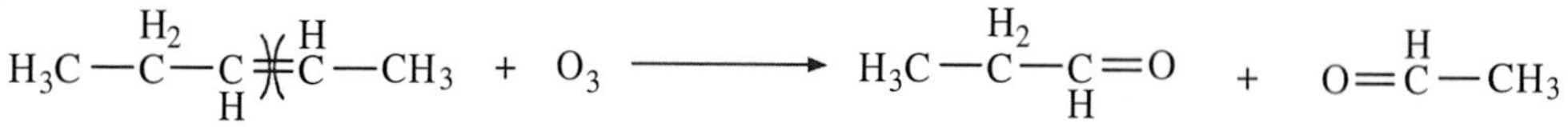

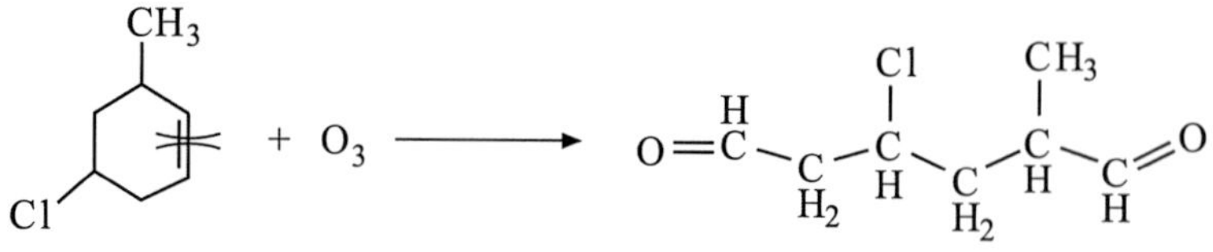

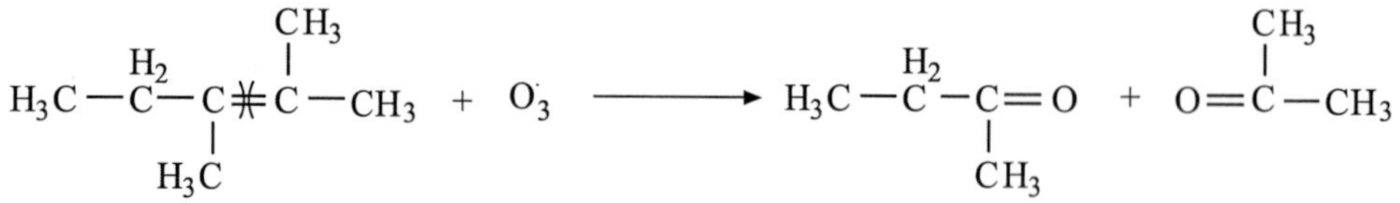

9.5.7 Hydroboration and Oxidation

Hydroboration is another famous example of anti-Markovnikov addition. Addition of diborane to a terminal double bond is very specific. For each mole of borane, three moles of alkene are required to form a trialkylborane. That is hydrolysed in presence of hydrogen peroxide to form terminal alcohol.

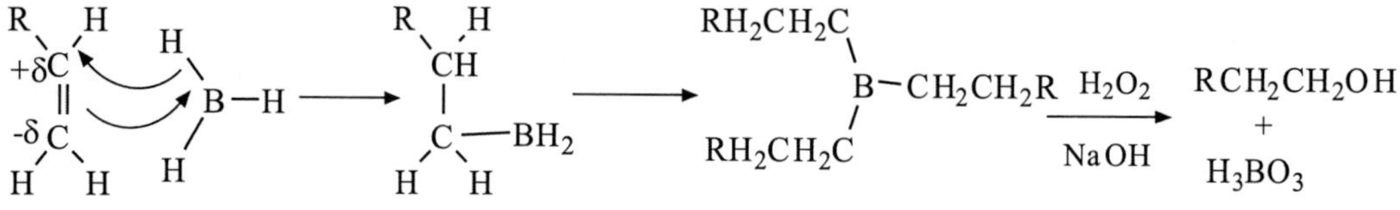

9.6 Friedel Craft Reaction in Presence of Acid

Friedel Craft reaction is used for alkylation of aromatic systems. Alkene or alkyl halides are used for the reaction, in presence of a Lewis acid, such as, aluminium trichloride or sulphuric acid.

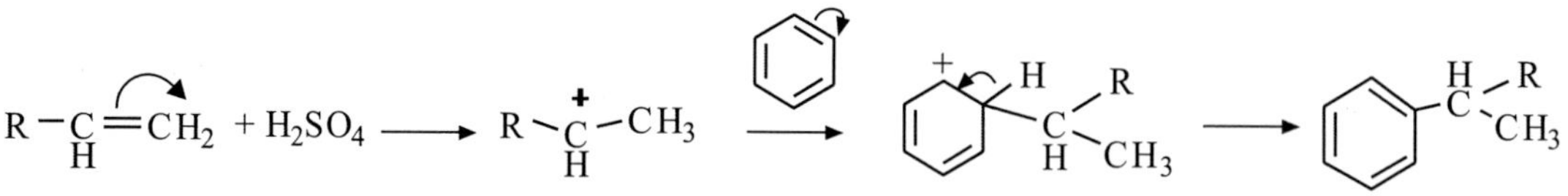

9.7 Polymerisation Reactions

Polymerisaton is a process of linking a unit of a molecule with another molecule of same or different compound. Nature carries out this process in a large number of synthesis of natural products, for example, synthesis of cellulose from glucose, starch from glucose and rubber form isoprene etc. The polymerisation process occurs by two methods, called addition polymerisation and the ether is called condensation polymerisation.

9.7.1 Addition Polymerisation

Addition polymerisation is just addition and linkage of many molecules of a compound, without loss of any entity. The molecular formula of a polymer is a simple whole number of the monomer. For example, **ethylene molecule** polymerises between say a thousand molecules by addition process to form a long carbon chain, then the molecular formula of the polymer shall be $(C_2H_2)_{1000}$. The linking can take place, by what is called head to tail or head to head or tail to tail linkages. It shall give rise to different types of units in the polymer.

Addition polymerisation reaction takes place by **ionic mechanism** with a suitable catalyst or by a **radical mechanism** by using a free radical generator, as given below.

polymerisation with only one monomer, *e.g.*, ethylene, we get polyethylene.

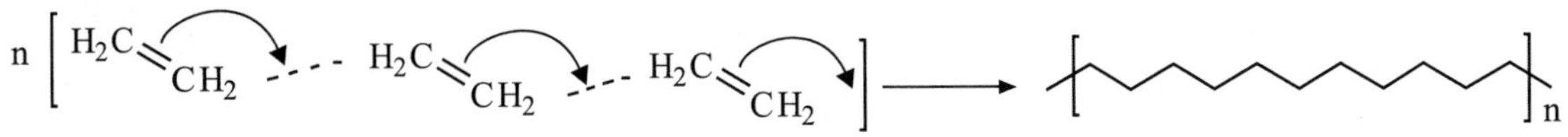

With a mixture of two ethylenes, as propylene and ethylene, we can get four type of polymers with different type of condensation. It can give different kind of products, either by regular repitition of monomers (I) or by irregular repitition of monomers (II), as shown below.

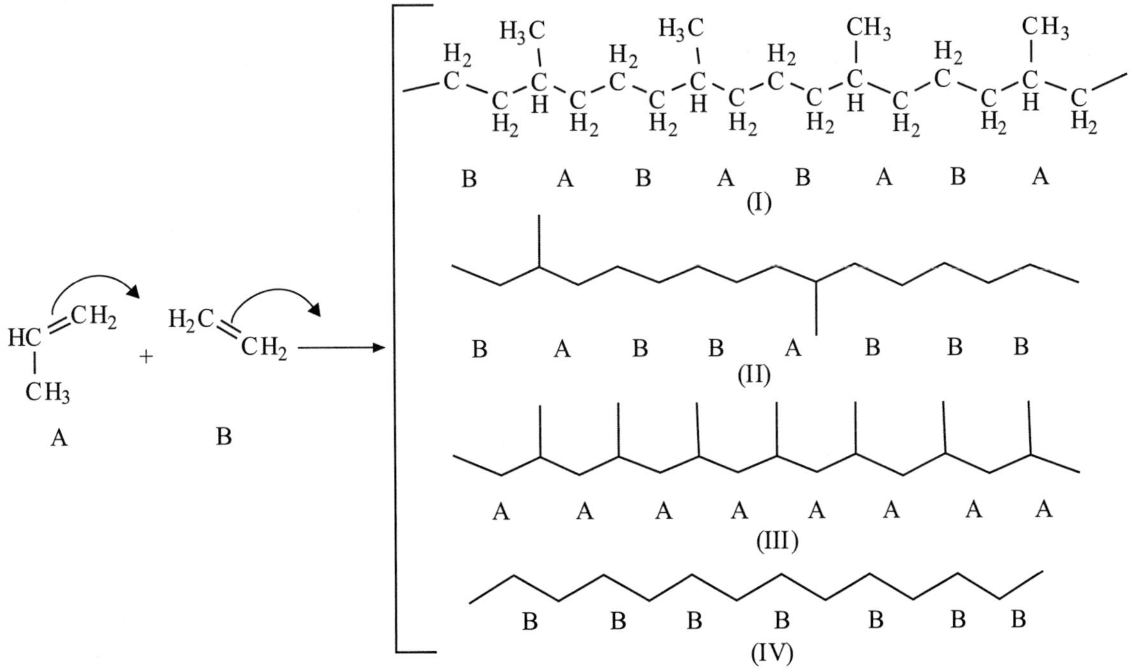

Mechanism: Ionic mechanism ususally requires the presence of a catalyst, such as, Lewis acid.

Initiation step: We need a radical generator for the initiation step.

Initiation Step

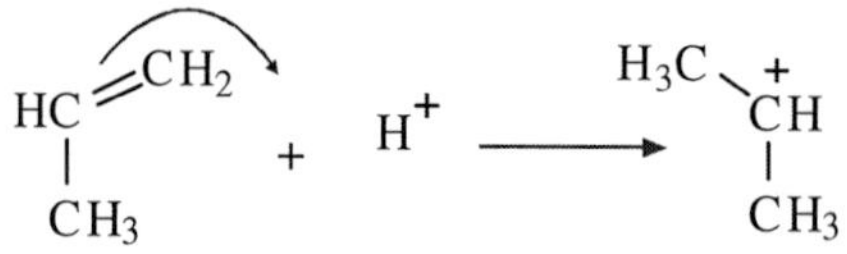

Propagation Step

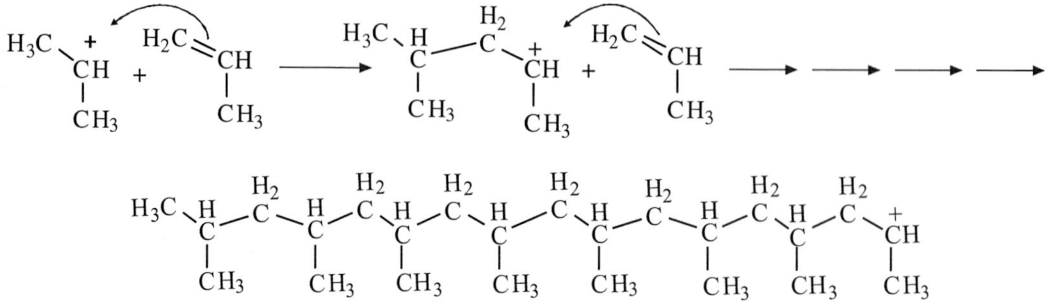

Termination Step

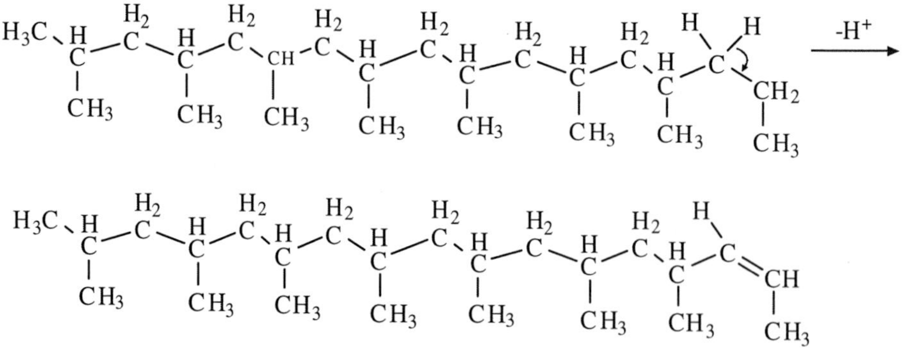

Radical Mechanism

Addition polymerisation may take place through radical mechanism, in presence of catalyst, such as, peracids, known to generate radicals. The radical formed from a peroxide on reaction with an alkene gives a radical, which interacts with other alkene to form further radicals. The termination of this takes place by coupling of radicals or by adding small amount of other radical quenchers.

Initiation Step

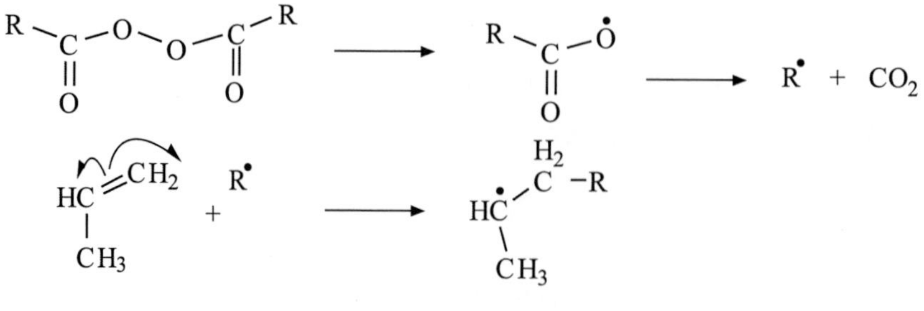

Propagation Step

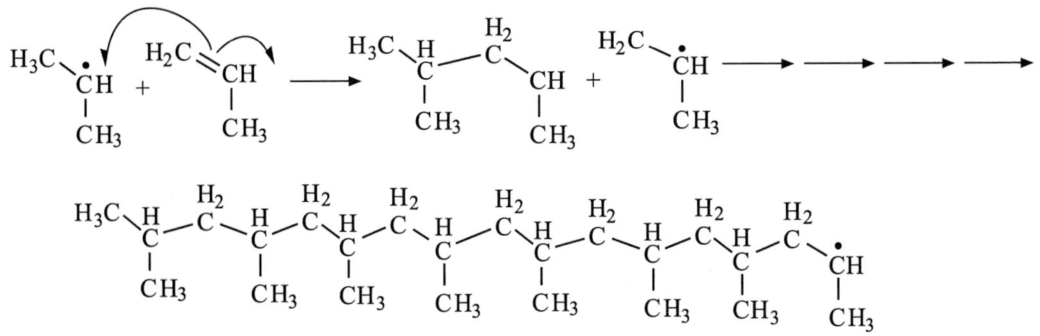

Termination Step

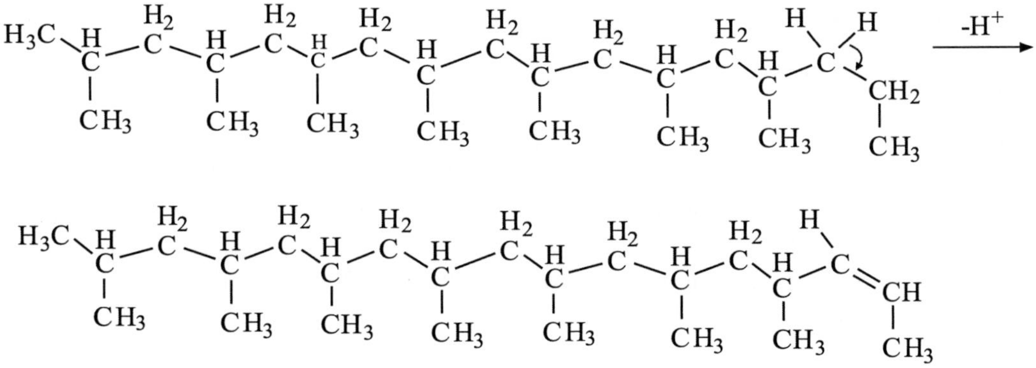

9.7.2 Condensation Polymerisation

The second type of polymerisation reactions are known as condensation reactions, although these are not the reactions of alkenes, but are considered here as related topic. In addition reaction, we have seen, there is no elimination of any moiety or a group when two or more monomers combine to form a polymer chain. In condensation reactions, usually some small group or moiety is eliminated at the juncture of two or more monomers. A reaction between a carboxyl group and a amino group containing compounds shall give us an amide group in the polymer chain. Thus, we see that it does not require the presence of double bonds, but the active functional groups. Numerous polymers of this group are known.

Nylon-66

Nylon is a polymer obtained by condensation of -COOH group and -NH_2 between two compounds. That is, an amino group at one end of the carbon chain, reacts with carboxyl group of another molecule to form an amide linkage, like that in proteins. Caproic acid (HOOC-CH_2-CH_2-CH_2-CH_2-CH_2-COOH), reacts with hexamethylene diamine ($H_2N(CH_2)_6NH_2$) to form nylon-66.

$$HOOC(CH_2)_4COOH + H_2N(CH_2)_6NH_2 + HOOC(CH_2)_4COOH \longrightarrow$$

$$HOOC(CH_2)_4CO\left[HN(CH_2)_6NHCO(CH_2)_4CO\right]_n NH(CH_2)_6NH_2$$

Nylon-66

Nylon-6

Another compound, called caprolactum, a cyclic amide, undergoes condensation is presence of a catalyst to form nylon-6, Caprolactum can be prepared from cyclohexane *via* cyclohexanone and Baeyer Villiger oxidation to form a cyclic ester, which on reaction with ammonia forms caprolactum.

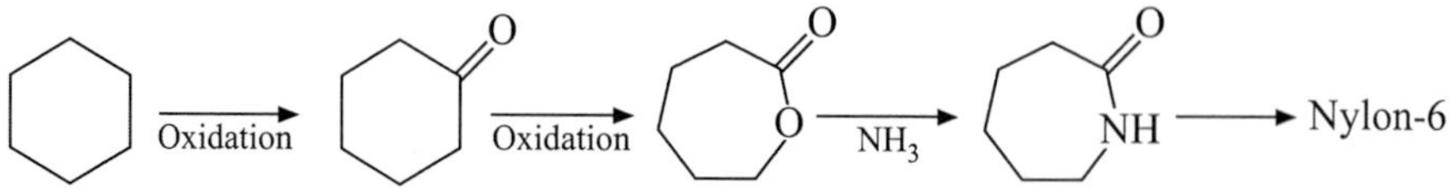

Proteins

Nylon functional unit is also present in life molecules, called proteins. These are formed by reaction between α-amino group of one acid and COOH group of another α-amino acid, H_2N - CH(R) - COOH. It is the variation of the value of R group to about 24 variants, that gives total life protein molecules. It may be seen that the difference in nylon and protein molecules is the position of -NH_2 group. Proteins have α-amino group, where as nylon has omega position of the -NH_2 group.

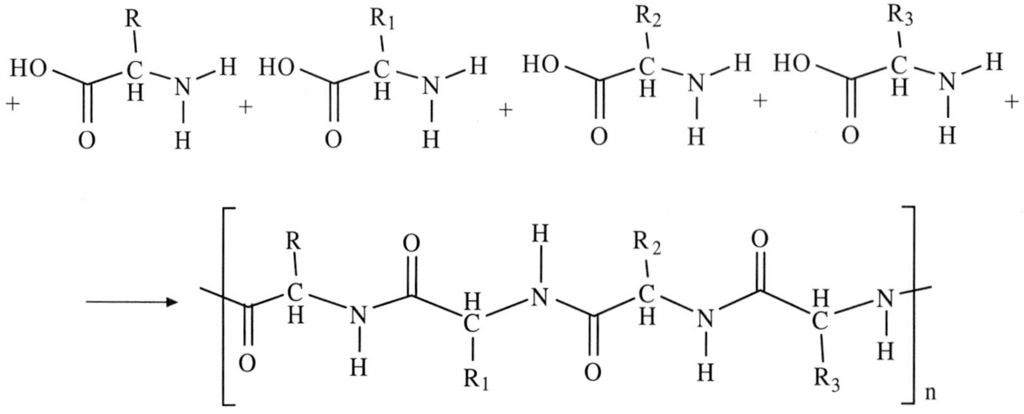

a Protein chain

9.8 Alkadienes

Alkenes are characterised by the properties of double bonds, such as addition, oxidation and reduction reactions. These properties of double bond hold good in case of all alkene and alkadienes, triene or polyenes. The difference is the number of times the addendum shall add to the substrate. In addition to these, some special properties of alkadienes are also observed.

In case of conjugated dienes (alternate double and single bond in a carbon chain), the special properties are observed due to the resonance property of conjugated compounds. For example, buta-1,3-diene undergoes mono-addition reaction at 1,2 as well as 1,4-carbons, besides di-addition at 1,2,3,4 - carbons. They do undergo addition, reduction and oxidation or poymerisation reactions but in different way and need be discussed in futher details.

Alkadienes are compounds that contain two double bonds in the carbon chain. In a five carbon chain, different position of double can be there. The isomeric structures of a five carbon atom molecule, C_5H_8 are:

Structural isomers of C_5H_8

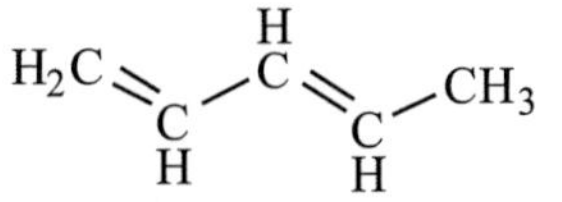

Pent-1,3-diene
Conjugated double bonds

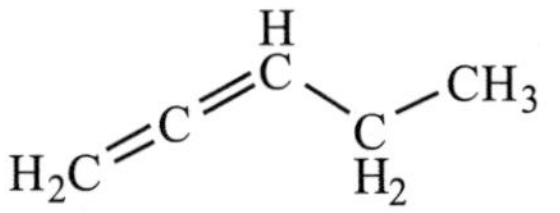

Pent-1,2-diene
Cumulative double bonds

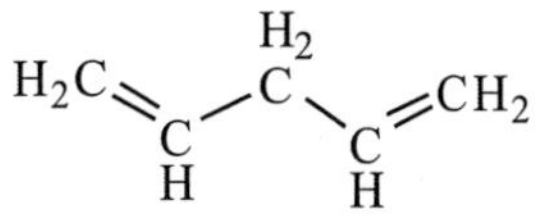

Pent-1,4-diene
Isolated double bonds

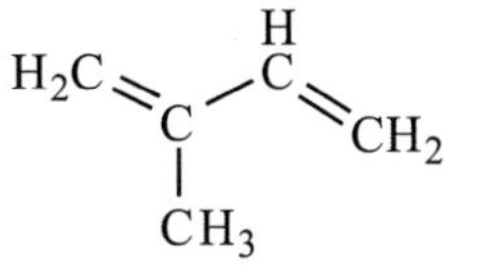

2-Methyl-but-1,3-diene
(Isoprene)

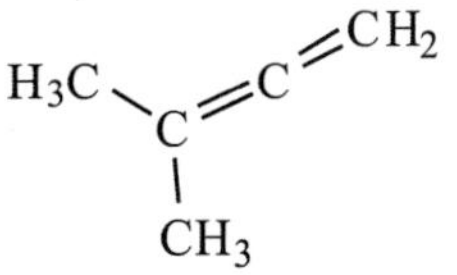

3-Methylbut-1,2-diene
(Allene)

Cis-, *trans*-isomers of C_5H_8

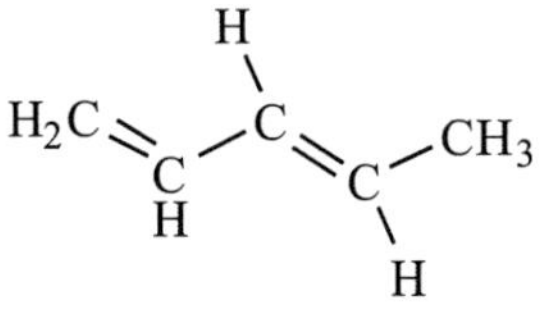

trans-Pent-1,3-diene

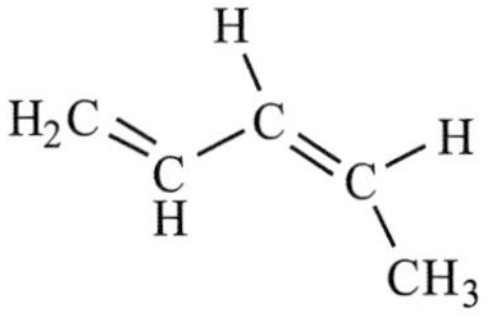

cis-Pent-1,3-diene

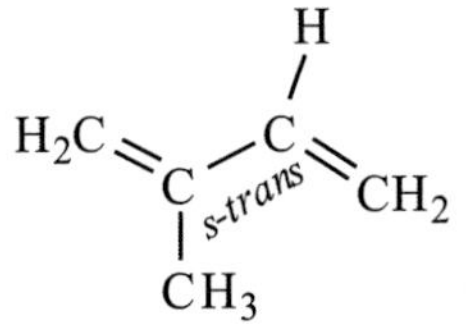

s-trans-2-Methyl-but-1,3-diene
(Isoprene)

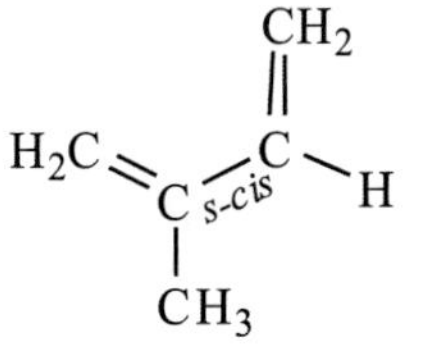

s-cis-2-Methyl-but-1,3-diene
(Isoprene)

s-stands for isomerism around single bond in alkene

9.8.1 Synthesis of Allenes

1. A simplest method of synthesis is dehydrobromination opf 1,2,3-tribromo-propane.

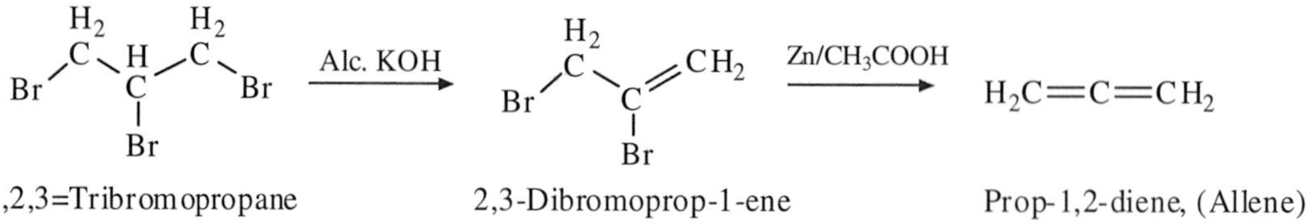

9.8.2 Synthesis of Butadiene

Butadiene can prepared in different ways.

(*i*) Dehydrogenation of butane or butene

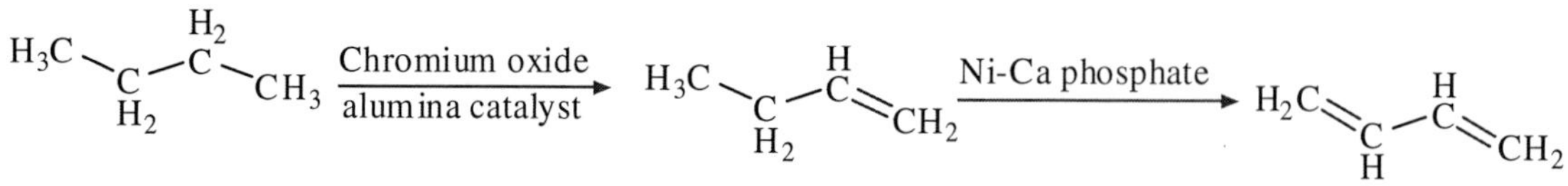

(*ii*) Heating cyclohexene, in presence of heated nichrome wire, undergoes retro-Diels Alder reaction.

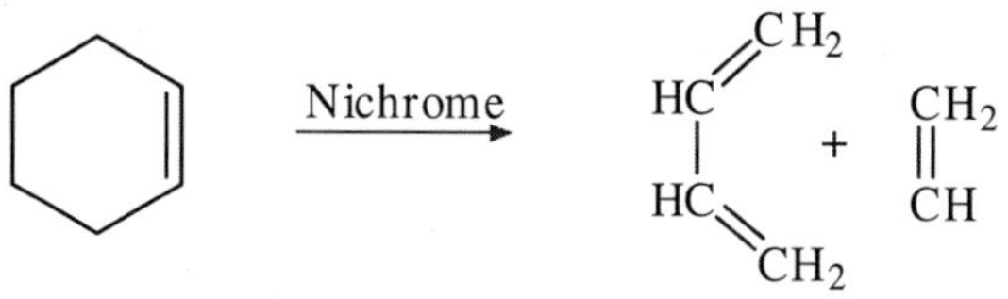

9.8.3 Synthesis of 2-methylbuta-1,3-diene (Isoprene)

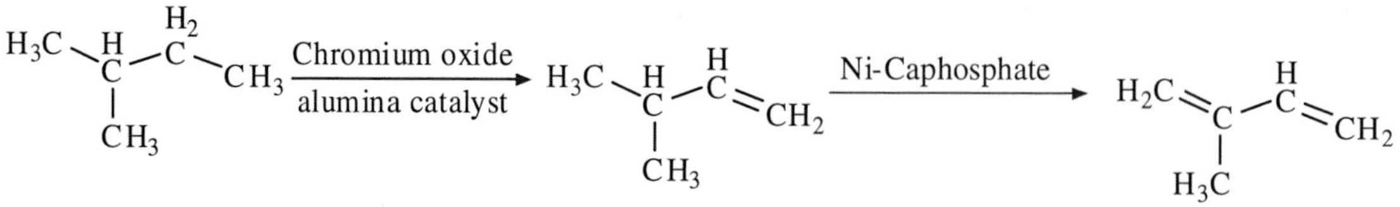

9.8.4 Chemical Reactions of Conjugated Dienes

9.8.4.1 Polymerisation reactions

Dienes like alkenes polymerise by addition condensation reactions. Butadiene polymerises to form butadiene rubber.

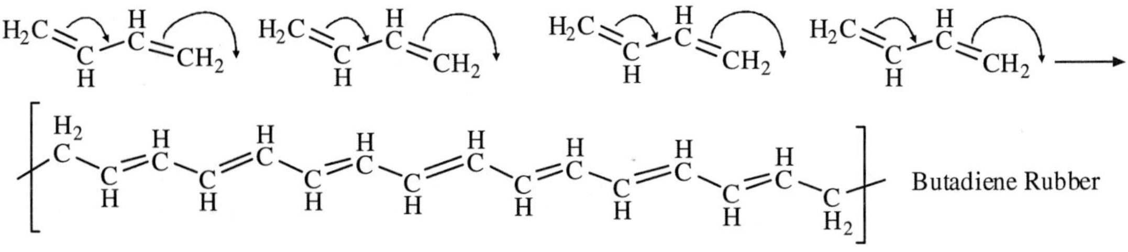

Butadiene Rubber

It can also be polymerised by different alkenes, such as styene. It is a synthetic rubber called BUNA-S**BUNA-S** rubber is a copolymer of butadiene and styrene

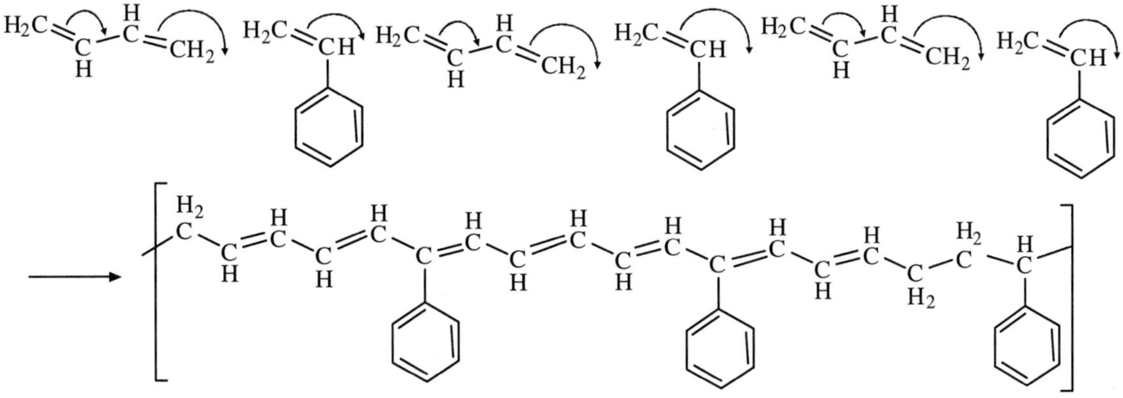

BUNA-S Rubber

Natural rubber is polymer of isoprene, 2-methylbuta-1,3-diene.

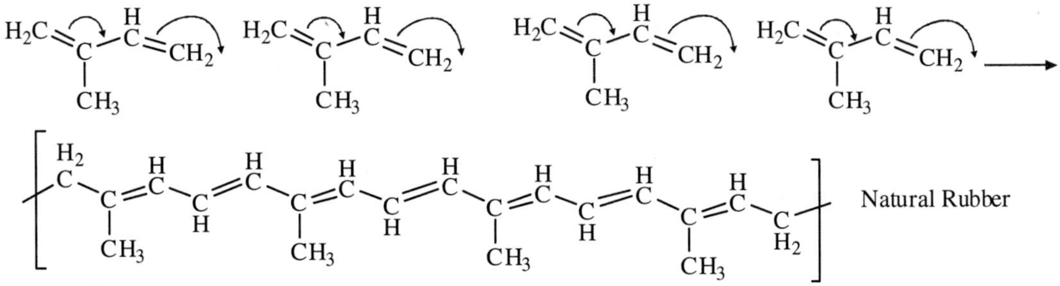

Natural Rubber

9.8.4.2 Addition of Bromine

Addition of bromine to a diene takes place in two ways, to form 3,4-dibromobut-1-ene and 1,4-dibromobut-2-ene, as shown.

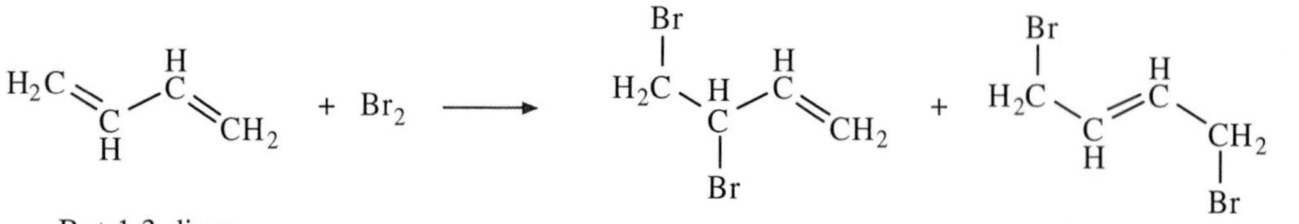

But-1,3-diene 3,4-Dibromo-but-1-ene 1,4-Dibromo-but-2-ene

Mechanism:

Step 1: Addition of brominium ion

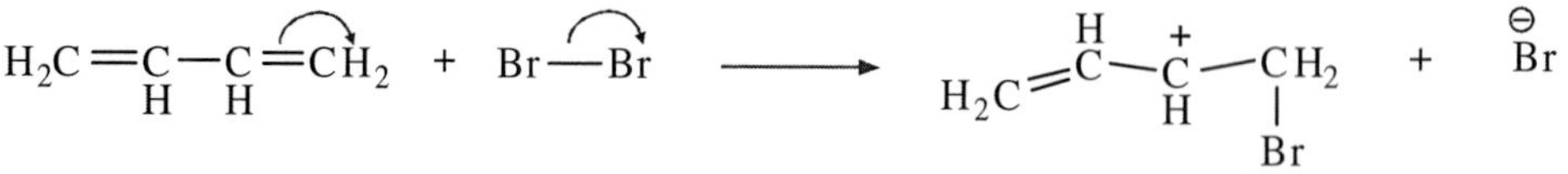

Step 2: Resonance in carbonium iontakes place in accordance to the predominance of carbocation species. The secondary carbocation is more stable, so predominates in the mixture of two possible ways of addition reaction.

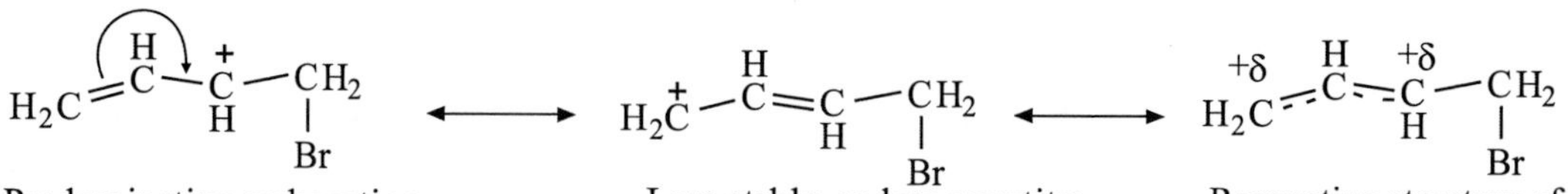

Predominating carbocation Less stable, so less quantity Resonating structure of two cation centre

Step 3: Addition of bromide ion at 2 or 4 position.

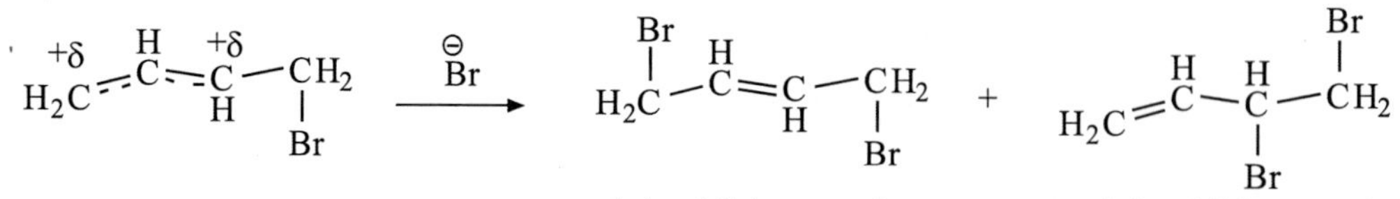

1,4- addition reaction 1,2- addition reaction

9.8.4.3 Addition of Halogen Acids

Addition of hydrogen bromide to a diene takes place in two ways, to form 3-bromobut-1-ene and 1-bromobut-2-ene, as shown.

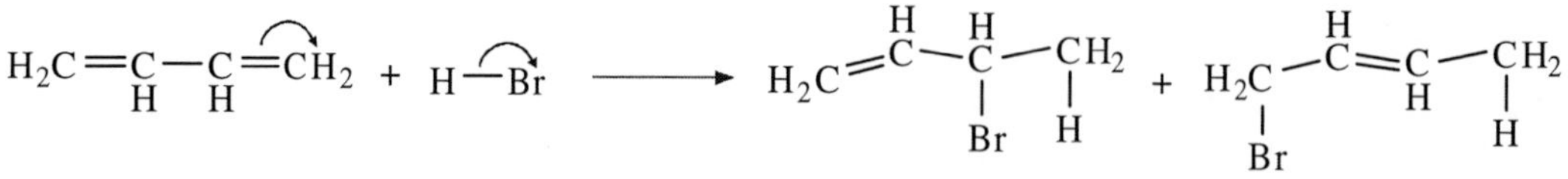

Mechanism:

Step 1: Addition of hydrogen cation

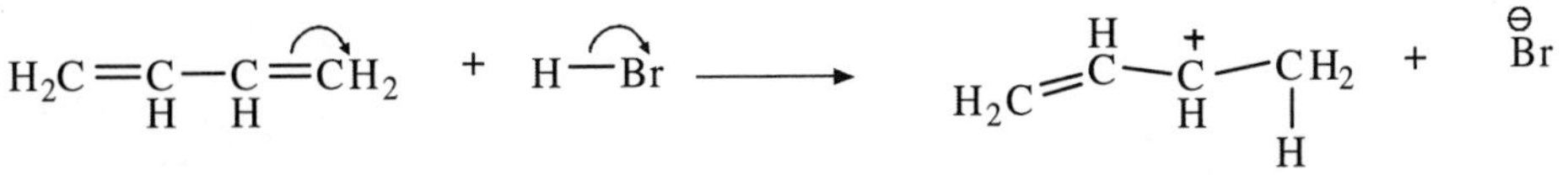

Step 2: Resonance in carbonium ion takes place in accordance to the predominance of carbocation species. The secondary carbocation is more stable, so predominates in the mixture of two possible ways of addition reaction.

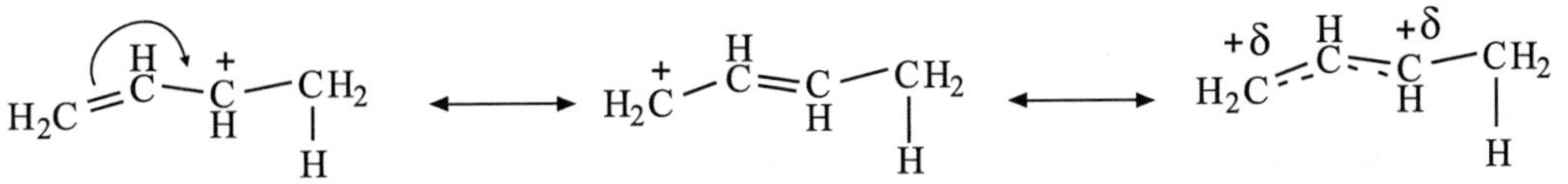

Predominating carbocation — Less stable, so less quantity — resonating structure of two cation centre

Step 3: Addition of bromide ion at 2 or 4 position.

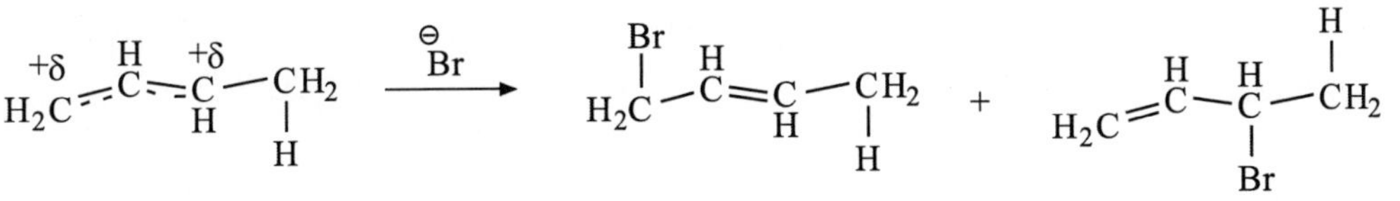

1,4-addition reaction 1-Bromo-but-2-ene — 1,2- addition reaction 3-Bromo-but-1-ene

Uses of Alkenes:

1. Ethylene is a plant (fruit and vegetable) hormone. It helps in ripening of fruits and vegetables, like tomato and others. However, due to its difficult handling in a gas cylinder and easy handling and generation of acetylene in situ from a packet of CaC_2 by slow reaction of atmospheric water, it makes acetylene a more often usable for ripening of fruits; but acetylene is not advisable.
2. Ethylene is the main component of synthesis of polyethylene and other polymers by changing to other suitable monomers. Substituted ethylenes, like cyanoethylene, propylene and aromatic alkenes form the starting meterials for different polymers or plastics.
3. 2-Methylbutadiene (isoprene) forms the starting material for the synthetic rubber. Likewise, chloroprene and styrene are used for the manufacture of different kind of rubbers.

Points to Remember

Synthesis of Alkene

Reaction	Substrate	Reagent	Product
1. Dehydration Reactions	An alcohol	Dil. Sulphuric acid	ALKENE
2. Dehydrohalogenation reaction	a halide with beta hydrogen	Alcoholic Potash	ALKENE
3. gem-Didehalogenation	gem-dihaloalkane	Zn	ALKENE
4. Hoffman elimination	amine quaternary salt with β-H	Heating the amine quaternary salt	ALKENE
5. Cope Reaction	N-oxide of an amine	Heating N-oxide of an amine	ALKENE
6. Functional group modification triple bond reduction to double bond	A triple bond compount	H_2/Pd-c	ALKENE
7. Witting Reaction	an ylide reaction	Phosphorus Ylides	ALKENE
8. Pyrolysis of Esters	an ester with a leaving grp	Heating the ester	ALKENE
9. Chugaev Elimination	A Xanthate on with CH_3I	Breaking up the Xanthate	ALKENE
10. Ramberg-Backlund reaction	A sulphone and a base	Sulphone elimination base	ALKENE

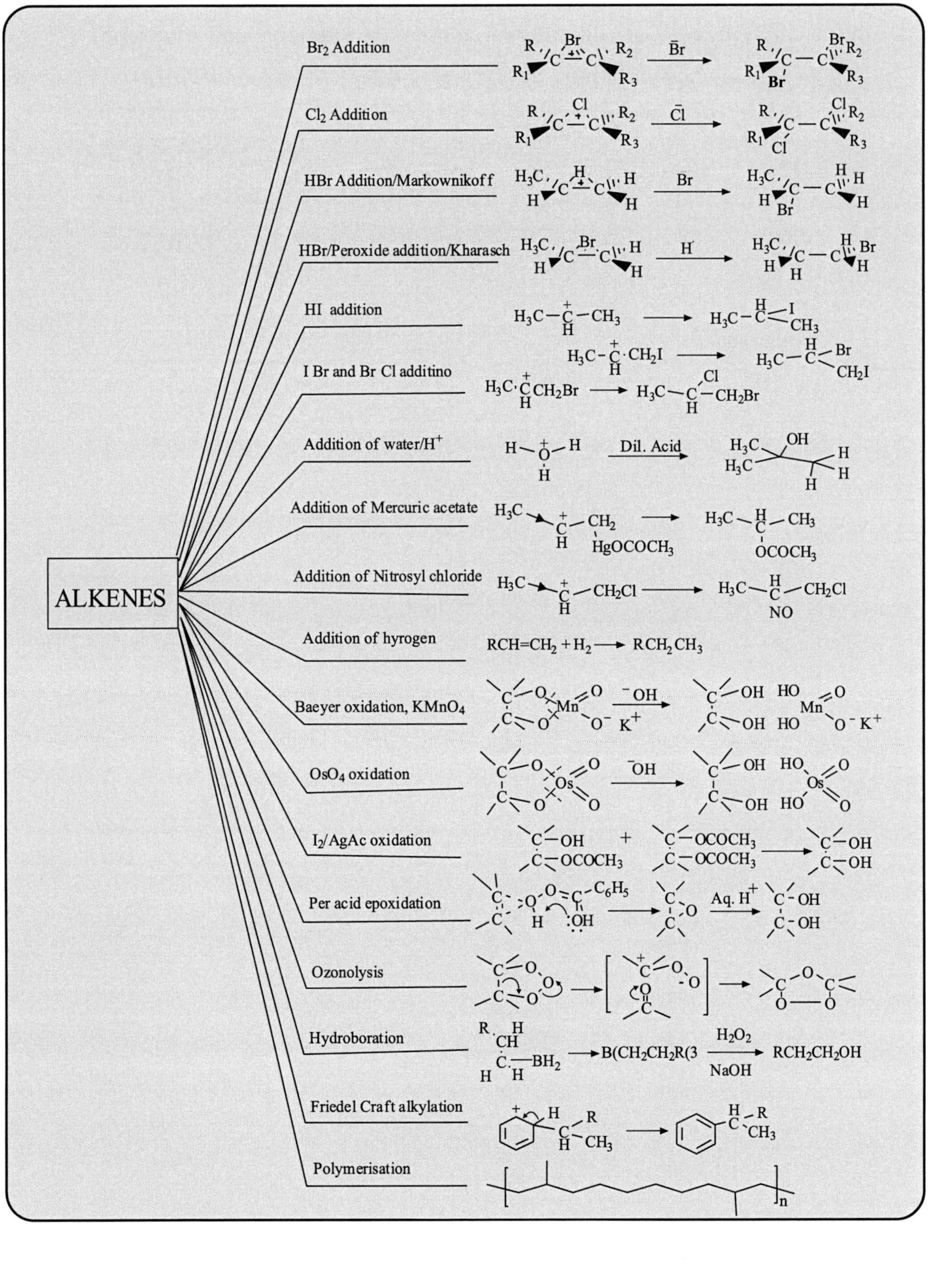
ALKENES
Br2 Addition
Cl2 Addition
HBr Addition/Markownikoff
HBr/Peroxide addition/Kharasch
HI addition
I Br and Br Cl additino
Addition of water/H+
Dil. Acid
Addition of Mercuric acetate
Addition of Nitrosyl chloride
Addition of hyrogen
RCH=CH2 + H2 → RCH2CH3
Baeyer oxidation, KMnO4
OsO4 oxidation
I2/AgAc oxidation
Per acid epoxidation
Aq. H+
Ozonolysis
Hydroboration
H2O2
NaOH
RCH2CH2OH
Friedel Craft alkylation
Polymerisation

Methods of Synthesis of Alkadienes (Allenes, Butadiene and Isoprene)

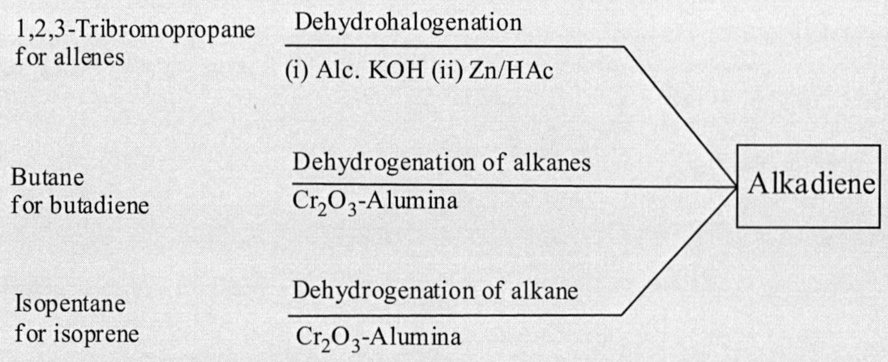

Chemical Reactions of Alkadienes (Allenes, Butaidene and Isoprene)

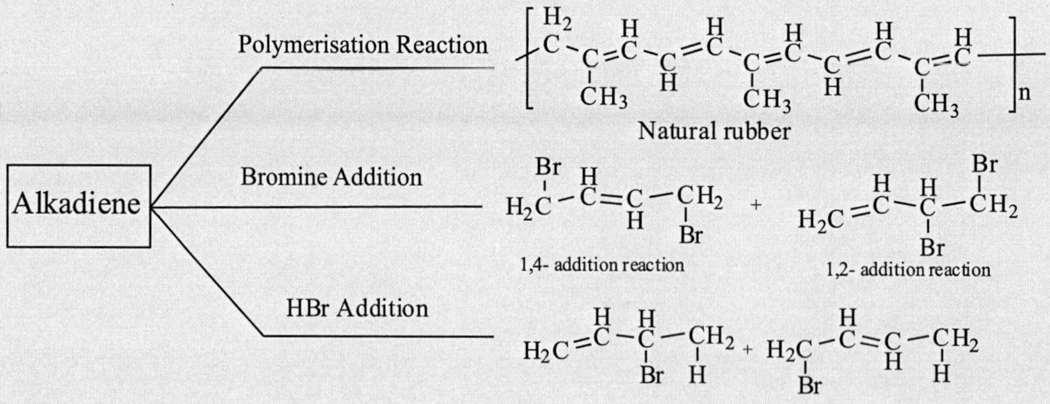

Exercises

1. Giving suitable substituents, write examples of structure of alkenes and the nature of product in case of all the reaction given in summary, p. 318.
2. How will you synthesise the following alkenes?

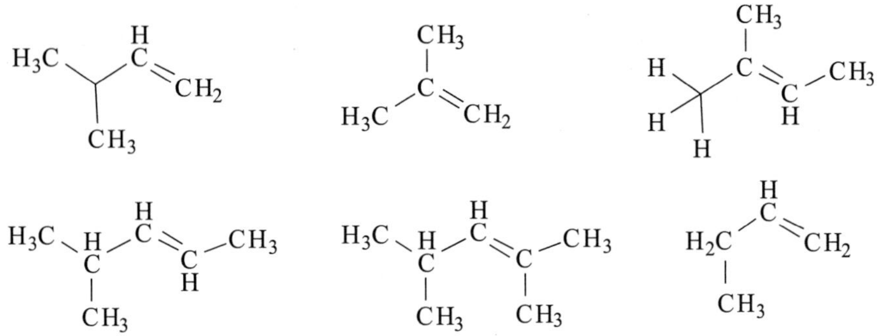

3. How will you carry out the following conversion?

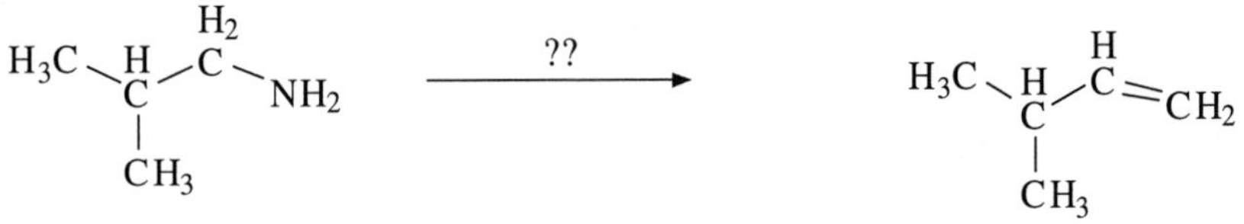

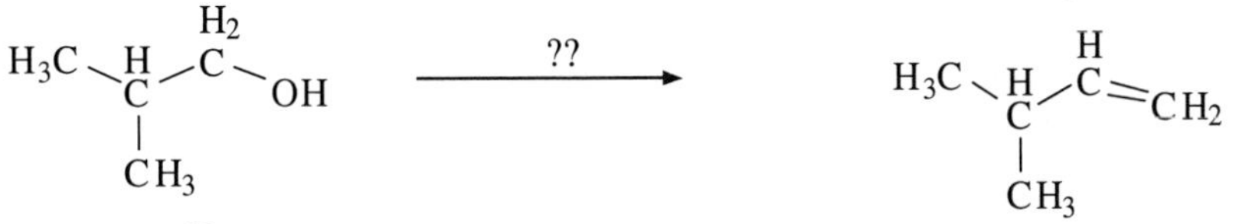

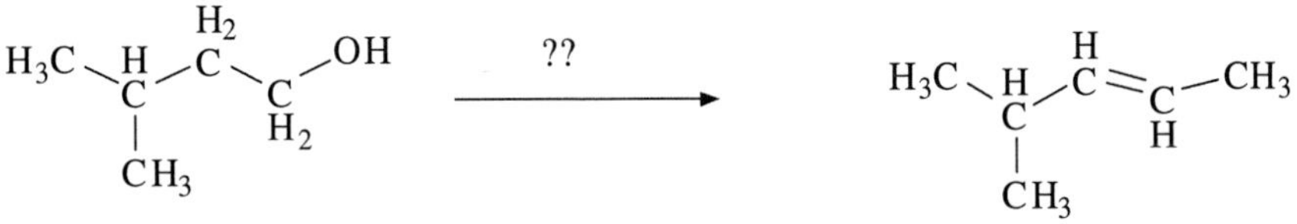

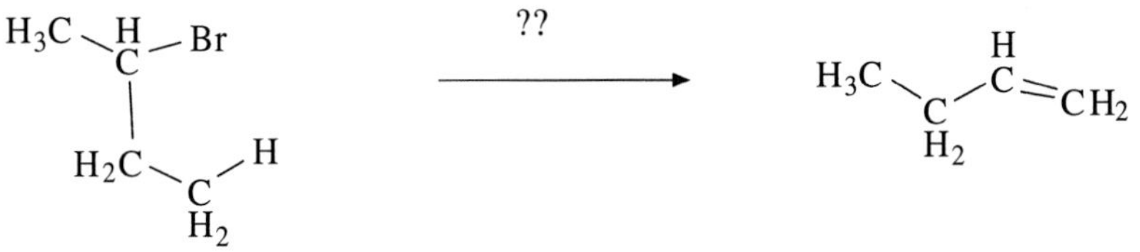

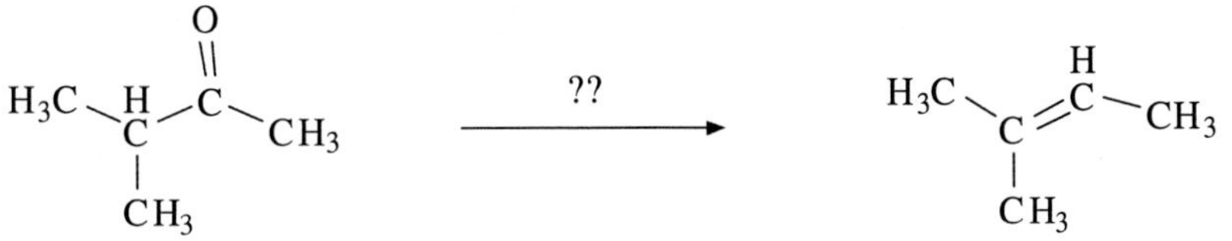

4. What is the structure of alkene formed from the following alkyl halide by dehydrobromination?

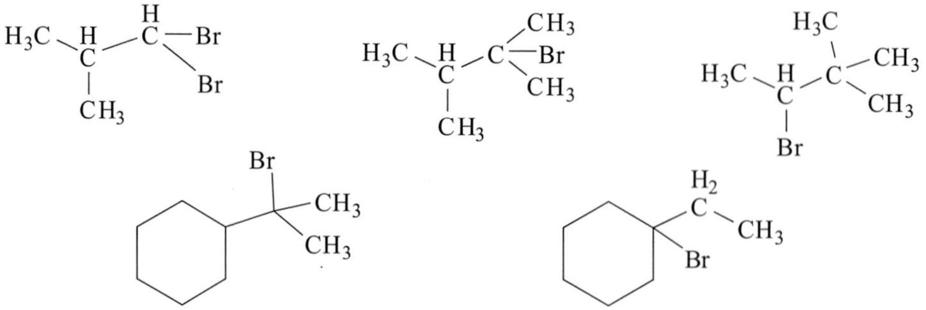

5. Deduce the structure of alkene that on ozonolysis gives the following product.

(*i*) 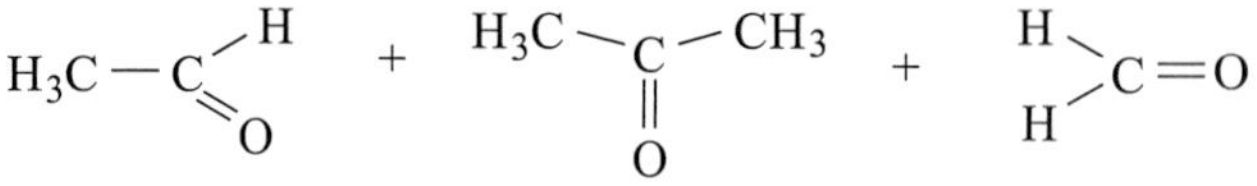

(*ii*) 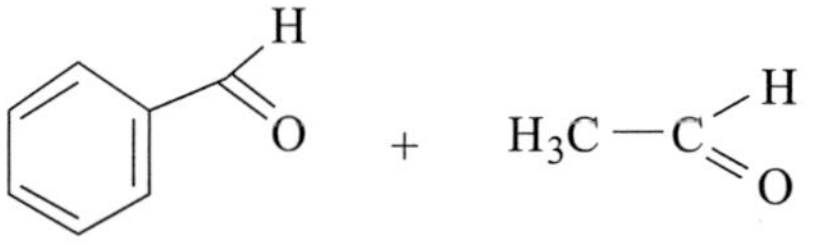

(*iii*) 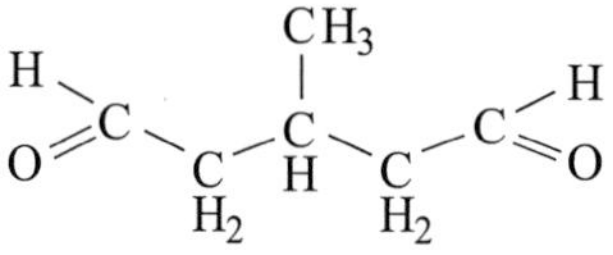

(*iv*) 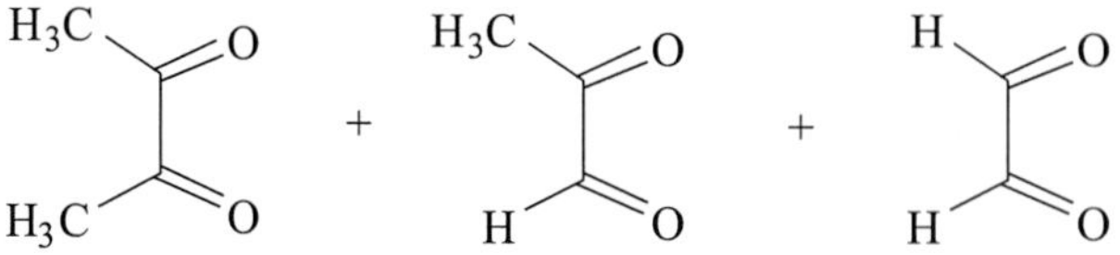

(*v*) 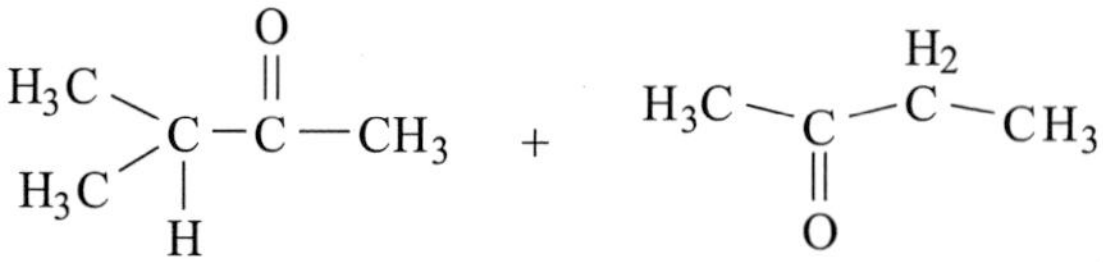

6. Addition of HBr to the following gives which as the major and minor products?

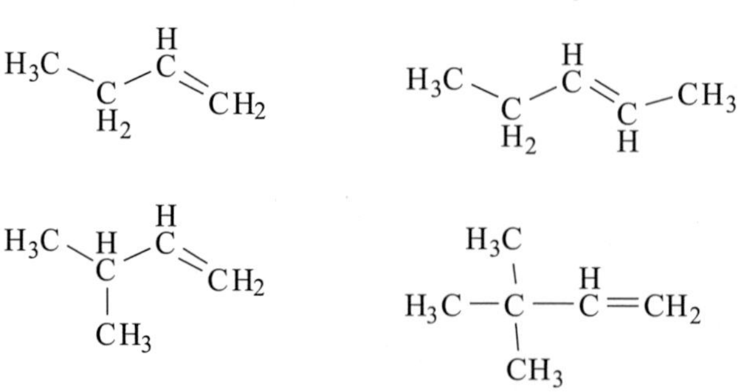

7. How will you prepare the following compounds by starting from a suitable alkene?

(*i*) 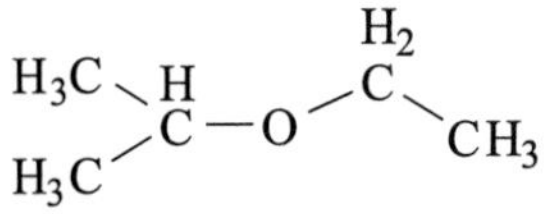

(*ii*) 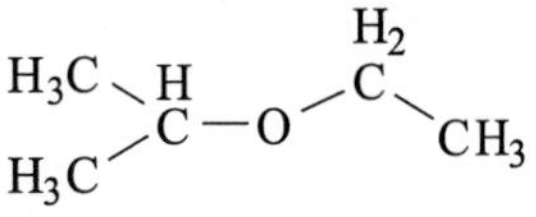

(*iii*) 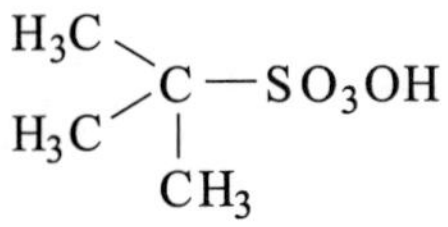

(*iv*) 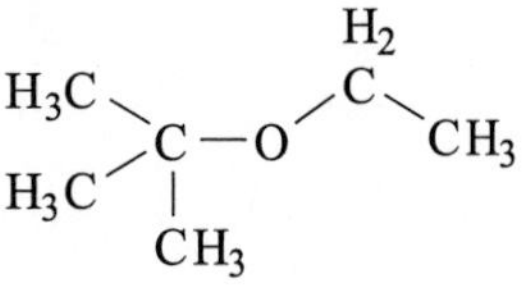

8. What shall be the product of addition of HBr to isopent-2-ene?

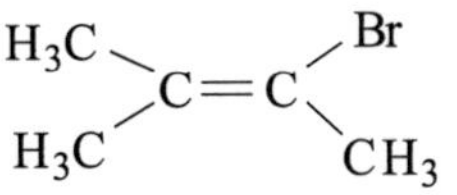

or

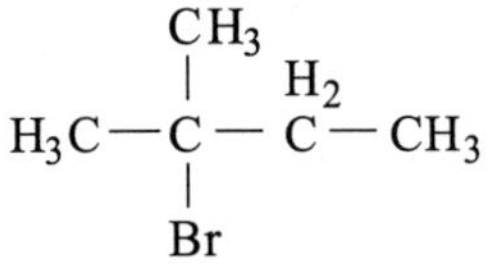

9. What could be the possible product of addition of HBr to the following compounds?

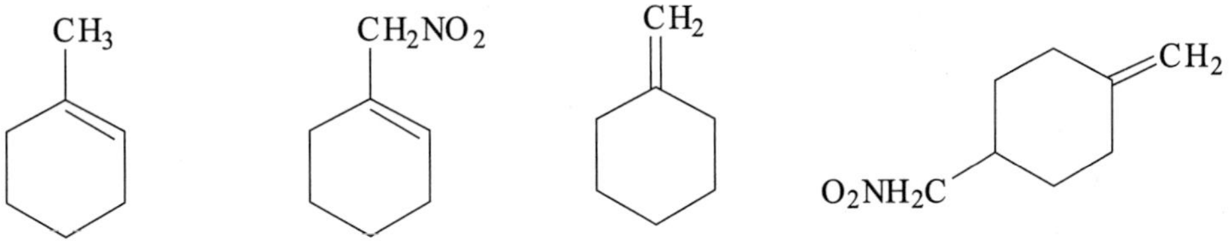

10. Addition of HBr to isobutene in presence of NaI takes place competitively. What reaction product shall predominate in this competition between I^- and Br^- during the course of addition?

11. Addition of methoxyl anion to an alkene in presence of mercuric acetate shall take place to which carbon of the double bond of the following substrates. Write the structure and IUPAC name of the product.

(*i*) 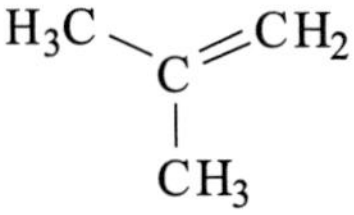

(*ii*) 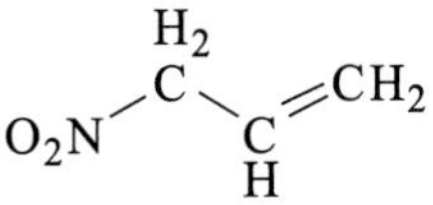

Kenneth B Wiberg (1930-   )

10

Aliphatic Hydrocarbons —Alkynes

He had his BS from MIT, Massachusetts, Ph.D. (1950) from Columbia University and worked for a long time at University of Yale, USA, as Eugene Higgins Professor of Chemistry and later, as Emeritus Member of Yale faculty since 1962

He has formed a great group of scientists working on - Studies of conformational equilibria and solvent effects on these equilibria, Studies of NMR chemical shifts, Studies of the effect of geometrical distortion on the properties of organic compounds., Studies of solvent effects on reactions, Studies of electronically excited states of molecules, He wrote a book on Computational theory of organic chemistry. His studies on the NMR shifts of acetyelenes and allenes revealed the structure of such molecules.

He has been honoured with following awards. A.P. Sloan Foundation Fellow, 1958, Guggenheim Fellow, 1961, J. F. Norris Award in Physical Organic Chemistry, 1974, Cope Scholar, 1986 Arthur C. Cope Award, 1988, Pauling Medal, 1992.

IN THIS CHAPTER

CHAPTER **10**

Aliphatic Hydrocarbons —Alkynes

Introduction

Alkynes are compounds that contain, at least one carbon - carbon triple bond. The first member of the series is called acetylene, that is why the triple bond containing compopund are called acetylenic compounds and the triple bond is called acetylenic bond. The dominant alkyne is acetylene itself, which is used as a fuel and for welding purposes.

The general formula of this class of compounds is C_2H_{2n-2}. The three dimensional structure is revealed from the consideration of bonding of atomic orbitals. The *sp*- hybridised orbitals of acetylene are shown below and it shows that acetylene has a linear structure. It has also been proved by NMR chemical shifts. .

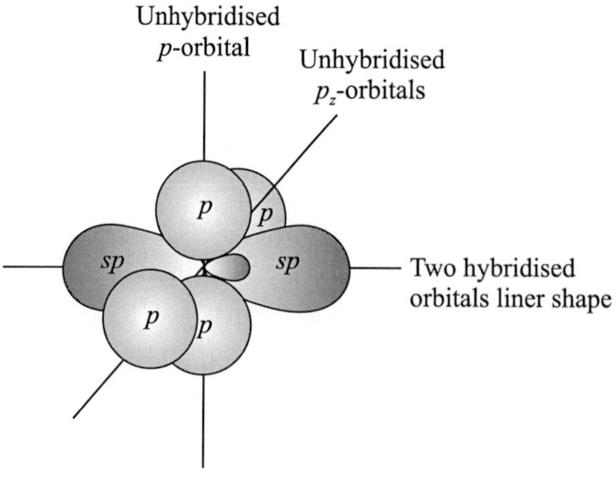

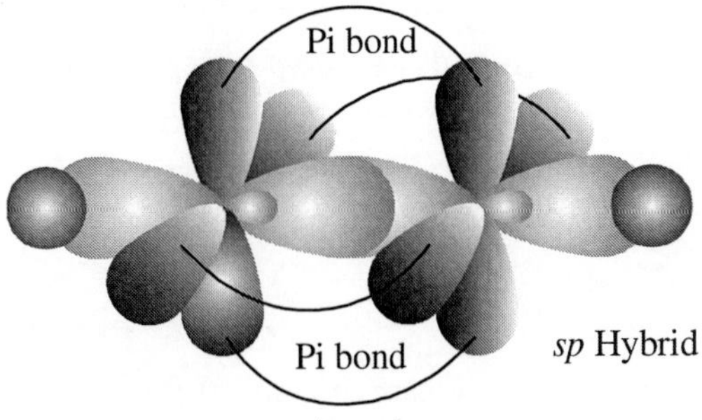

Acetylene

sp Hybrid orbitals of acetylene

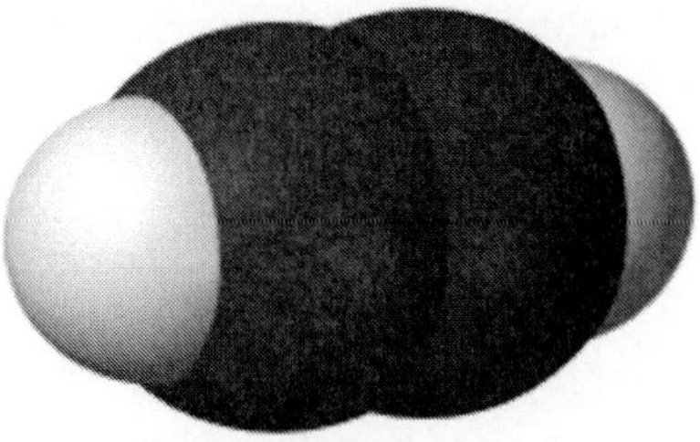

Bond Angles of Acetylene:

In acetylene, the H–C–C–H bond angles are 180°. By virtue of this bond angle, alkynes tend to be rod-like. Correspondingly, cyclic alkynes are rare. Benzyne is highly unstable.

Bond Length of Acetylene:

The C≡C bond distance of 121 picometers that is much shorter than the C=C distance in alkenes (134 pm) or the C–C bond in alkanes (153 pm).

Bond Strength of Acetylene:

The triple bond is very strong with a bond strength of 839 kJ/mol. The sigma bond contributes 369 kJ/mol, the first pi bond contributes 268 kJ/mol and the second pi bond of 202 kJ/mol bond strength.

Some examples of homologous series of the acetylenes are:

$H-C\equiv C-H$	$H_3C-C\equiv C-H$	$H_3C-C\equiv C-CH_3$	$H-C\equiv C-CH_2CH_3$
Ethyne	Propyne	But-2-yne	But-1-yne

Acetylenes can contain triple bond at the terminal carbon, such as:

$H-C\equiv C-CH_2CH_3$	$H-C\equiv C-CH_2CH_2CH_3$	$H-C\equiv C-CH_2CH_2CH_2CH_3$
But-1-yne	Pent-1-yne	Hex-1-yne

There are symmetrically substituted acetylenes, such as:

$Ph-C\equiv C-Ph$	$H_3C-C\equiv C-CH_3$	$H_3CH_2C-C\equiv C-CH_2CH_3$
Diphenyl acetylene	But-2-yne	Hex-3-yne

Acetylenic compounds are rare in nature, still some very important and biologically active acetylenic compounds are known. Some examples of such naturally occurring compounds are:

1-Phenylhepta-1,3,5-triyne is illustrative of a naturally occurring triyne.These compounds are highly bioactive as nematocides. Alkynes occur in some pharmaceuticals, including the contraceptive nor-ethynodrel. A carbon-carbon triple bond is also present in marketed drugs such as the antiretroviral- 'Efavirenz' and the antifungal- 'Terbinafine'.

10.1 IUPAC Nomenclature of Alkynes

Alkynes are characterised by presence of triple bond. In numbering the carbon atoms, triple bond is given preference as in case of double bond. The following examples shall help in learning the nomenclature of alkynes.

(*a*) Alkynyl Groups

It is a group obtained from alkyne by reducing one hydrogen from its formula. The general formula of the alkynyl group is C_nH_{2n-3}. The point of linkage is also specified in the name and so also the position of the triple bond. For example,

$CH_3—CH_2—C\equiv C—$	But-1-yne-1-yl
$CH_3—\underset{\vert}{C}H—C\equiv CH$	But-1-yne-3-yl
$—CH_2—\underset{\underset{Br}{\vert}}{C}H—C\equiv CH$	3-Bromo-but-1-yne-4-yl
$—CH_2—C\equiv CH$	Prop-1-yne-3-yl (propargyl group)
$—CH_3—C\equiv C—$	Prop-l-yne-l-yl
$—C\equiv CH$	Ethynyl group

(*b*) Some Acetylenes

Some examples of nomenclature of compounds containing acetyelenic bond are as below.

1. $\overset{5}{HC}\equiv\overset{4}{C}—\overset{3}{C}H_2—\overset{2}{C}H=\overset{1}{C}H_2$ **Pent-1-ene-4-yne**

When the double bond and the triple bond appear the identical position in the carbon chain, the preference is given to double bond in numbering but the writing of the suffix is -ene first then -yne.

2. $H_3\overset{5}{C}—\overset{4}{C}=\overset{3}{C}H—\overset{2}{C}\equiv\overset{1}{C}—H$ **Pent-3-ene-1-yne**

It is an example, where triple bond appears at number 1 carbon in the chain and double bond appears at the number 2 carbon from the other side of the chain. Preference is given to triple bond in numbering the carbon chain but writing of the -ene and -yne follows the suffix of -yne, as shown,

3. $\overset{5}{H_3C}-\overset{4}{C}\equiv\overset{3}{C}-\overset{2}{CH}=\overset{1}{CH_2}$ **Pent-1-ene-3-yne**

In this example, preference in numbering is given to double bond. Name shall have the suffix –yne as above.

4. 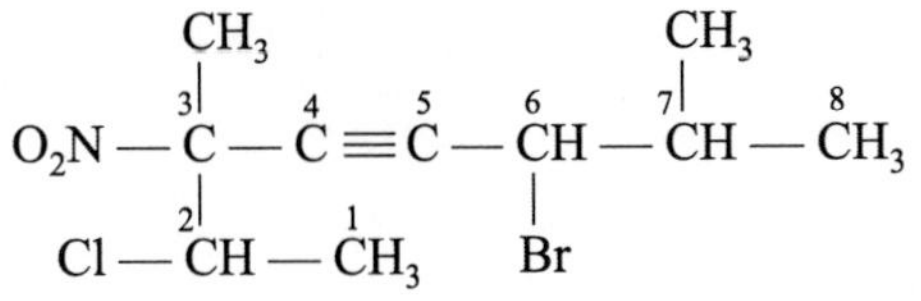 **6-Bromo -2-chloro- 3,7-dimethyl-3-nitro-oct-4-yne**

5. 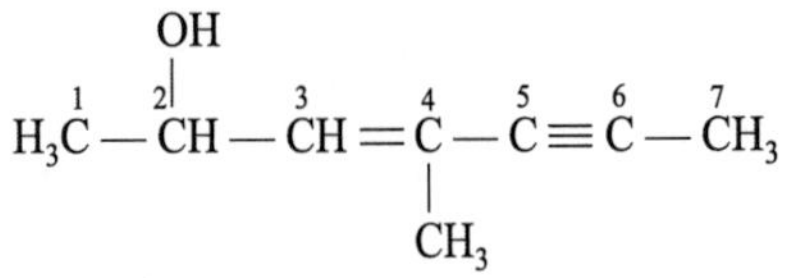 **4-Methyl-hept-3-ene- 5-yne-2-ol**

10.2 Methods of Preparation

1. **Commercial Method:**

Hundreds of millions of kilograms of acetylene is produced annually by dehydrogenation of natural gas:

$$2\,CH_4 \xrightarrow{\text{Catalyst}} \underset{\text{Ethyne}}{H-C\equiv C-H} + 3H_2$$

Propyne, is industrially useful, and is prepared by thermal cracking of hydrocarbons.

2. Small scale commercial preparation of acetylene is carried by the reaction of water with calcium carbide.

$$CaC_2 + H_2O \longrightarrow \underset{\text{Ethyne}}{H-C\equiv C-H} + 3H_2$$

3. Alkynes are prepared by dehydrohalogenation of vicinal alkyl dihalides or vinyl halides by the action of ethanolic potassium hydroxide or sodamide /liquid ammonia, for example, 1,2-dibromoethane gives acetylene.

$$\underset{vic\text{-dibromo ethane}}{Br-\overset{H_2}{C}-\overset{H_2}{C}-Br} + KOH \longrightarrow \underset{\text{Vinyl bromide}}{H_2C=\underset{H}{C}-Br} + KBr + H_2O$$

$$\underset{\text{Vinyl bromide}}{H_2C=\underset{H}{C}-Br} + KOH \longrightarrow \underset{\text{Acetylene}}{H-C\equiv C-H} + KBr + H_2O$$

4. *gem*-Dihalogeno compound *e.g.*, ethylidene chloride can also form the starting material, that on reaction with strong base-ethanolic potassium hydroxide, gives acetylene.

$$H_3C-CH(Br)_2 + KOH \longrightarrow H_2C=CH-Br + KBr + H_2O$$

gem-dibromo ethane or ethylidene bromide → Vinyl bromide

$$H_2C=CH-Br + KOH \longrightarrow H-C\equiv C-H + KBr + H_2O$$

Vinyl bromide → Acetylene

6. Via the Fritsch-Buttenberg-Wiechell Rearrangement:

It is a chemical reaction whereby a 1,1-diaryl-2-bromo-alkene rearranges to a 1,2-diaryl alkyne by reaction with a strong base such as an alkoxide.

$$(C_6H_5)_2C=CH-Br + K^+ \ {}^-O-C(CH_3)_3 \longrightarrow C_6H_5-C\equiv C-C_6H_5$$

1,1-Diphenyl-2-bromo-ethene → Diphenylacetylene

7. Higher Homologous of Acetylenes:

Metal acetylides can be coupled with primary alkyl halides. Alkynes are also prepared from vinyl bromides.

$$H-C\equiv C-H + NaNH_2 \longrightarrow Na-C\equiv C-Na + NH_3$$

Acetylene → Disodium acetylide

$$Na-C\equiv C-Na + CH_3Br \longrightarrow H_3C-C\equiv C-CH_3 + NaBr$$

But-2-yne

$$Na-C\equiv C-Na + CH_3CH_2Br \longrightarrow H_3CH_2C-C\equiv C-CH_2CH_3 + NaBr$$

Hex-3-yne

Some of reactions are carried for sake of their chemistry, that are applied in special compounds.

8. Alkynes can be Prepared from Aldehydes Using the Corey-Fuchs Reaction:

It is a reaction between *gem*-dihalogeno-ethenes with butyl lithium and then treating with water. Corey and Fuchs method involves preparation of *gem*-dihalogeno ethene by reaction of an aldehyde with phosphine and tetra-bromomethane.

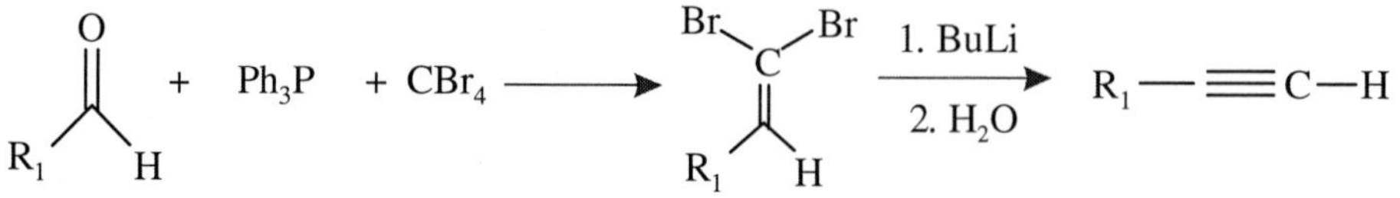

9. From aldehydes or ketones by the Seyferth–Gilbert homologation:

It is the chemical reaction of an aryl ketone (or aldehyde) with dimethyl (diazomethyl) phosphonate and potassium tert-butoxide to give substituted alkynes. Diazomethyl phosphanate is called Seyferth–Gilbert reagent.

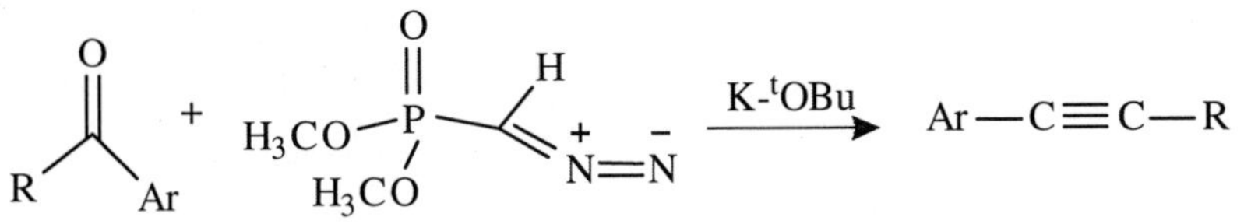

10. Iodoform reaction with silver powder

$$2CHI_3 + 6Ag \longrightarrow H{-}C{\equiv}C{-}H + 6AgI$$

11. Kolbe's electrolysis of potassium salt of maleic or fumaric acid (like synthesis of alkane from a monocarboxylic acid and, alkene from saturated dicarboxylic acid).

$$KOOC{-}\underset{H}{C}{=}\underset{H}{C}{-}COOK \xrightarrow[H_2O]{\text{Electrolysis}} C_2H_2 + 2\,CO_2 + 2\,NaOH + H_2$$

10.3 Physical Properties

Density	1.097 kg/m^3
Melting point	–80.8 °C (189 K, subl)
Boiling point	–84 °C
Acidity	(p*Ka*) 25
Structure	Molecular shape Linear
Standard enthalpy of formation	$\Delta_f H_0^{298}$ = +226.88 kJ/mol

10.4 Chemical Reactions

Like alkenes, acetylenes also undergo addition reactions, based on the stability of the product, the addendum can add two times, where ever possible,

10.4.1 Addition Reactions

Acetylenes undergo additon reaction like alkenes, but addition due to the presence of two pi-bonds, can take place twice. If we want a mono-addition to take place, we need to add limited quantity of the reagent.

10.4.1.1 Addition of Hydrogen

Hydrogen addition results in saturation of the triple bond stepwise, to double and then to single bond. The reduction is best carried out in presence of Lindlar catalyst, Pd-C/ $CaCO_3$. Even Raney Nickel reduction is also a very convenient method to reduce the double or triple bond.

$$R_1-C\equiv C-R_2 + H_2 \xrightarrow[CaCO_3]{Pd\text{-}C/} R_1-\underset{H}{C}=\underset{H}{C}-R_2 \xrightarrow[CaCO_3]{Pd\text{-}C/} R_1-\overset{H_2}{C}-\overset{H_2}{C}-R_2$$

10.4.1.2 Addition of Chlorine

Chlorine is most electronegative element, adds to pi-electrons to form sigma bond. Thus forming dichloro or tetra-chloro compound.

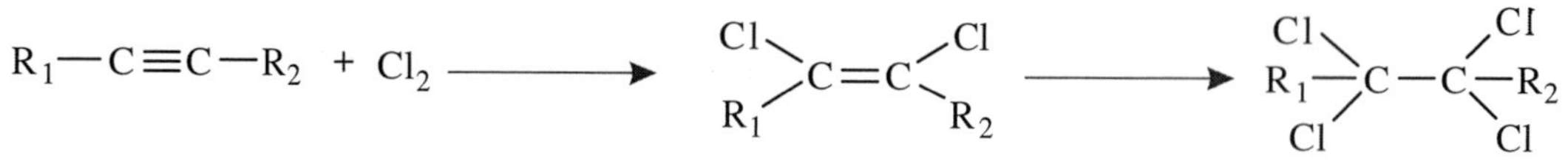

10.4.1.3 Addition of Bromine

Addition of bromine is very similar to that of chlorine, to give a dibromo or tetrabromo alkane.

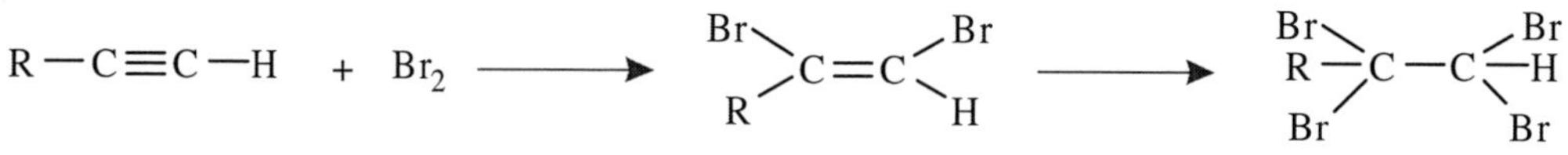

10.4.1.4 Addition of Iodine

Addition of Iodine is very similar to that of chlorine, to give a di-iodo or tetra-iodo alkane, however due to low electronegativity of iodine, reaction may be slow. Also since, iodine is a good leaving group, it may have some kind of equilibrium and the yield of the product may be less.

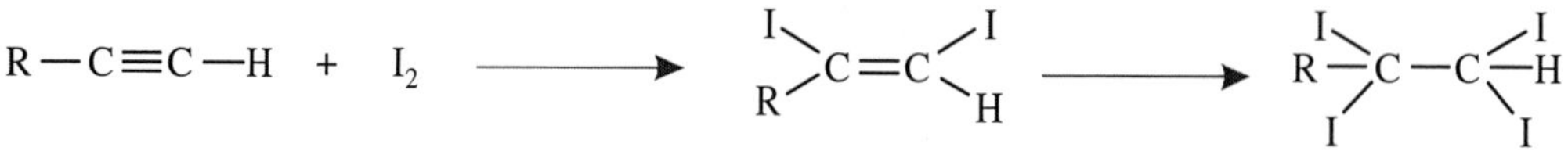

10.4.1.5 Addition of Hydrogen Chloride

Addition of HCl, takes place in accordance to the inductive effect, as well as, the electronegativity effect. If the alkyne is substituted by an alkyl group, both the inductive effect as well as the electronegativity effect operates, causing a dipole in opposite direction (after addition of HCl). More so, due the resonance, the addition becones restricted. The two possibilities are shown below. However, in case of acetylene, the first addition of HCl gives vinyl chloride, which undergoes resonance. Due to this effect, the second addition becomes difficult. Infact, the *pi*-bond of vinyl chloride is less reactive.

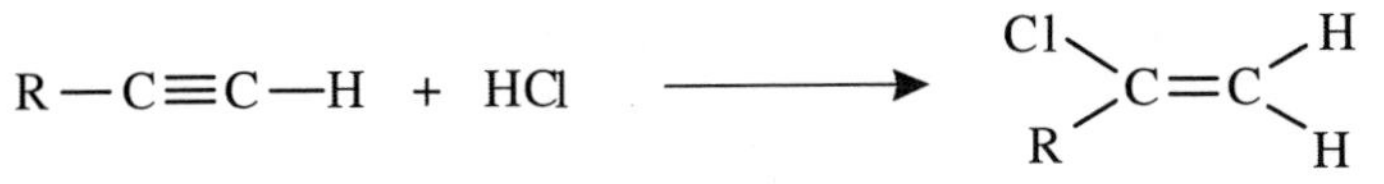

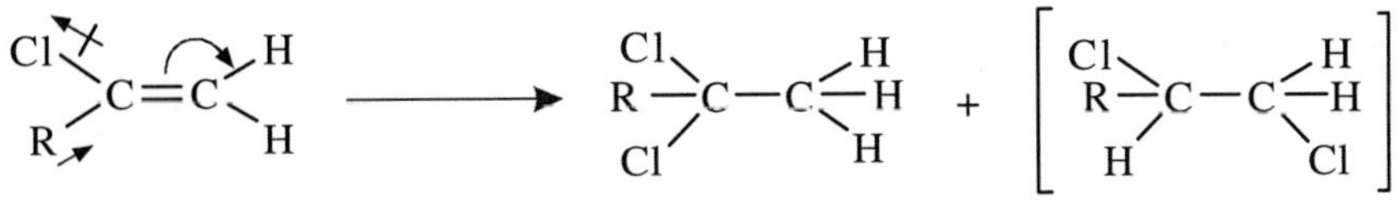

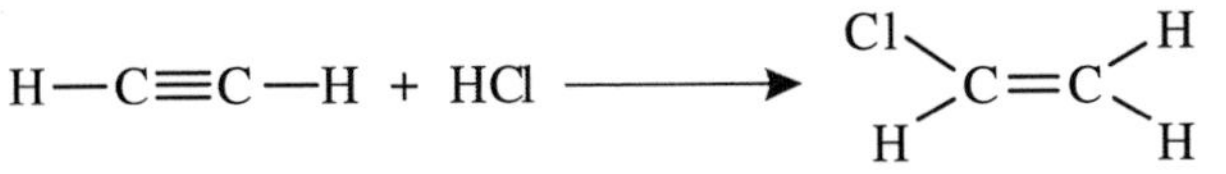

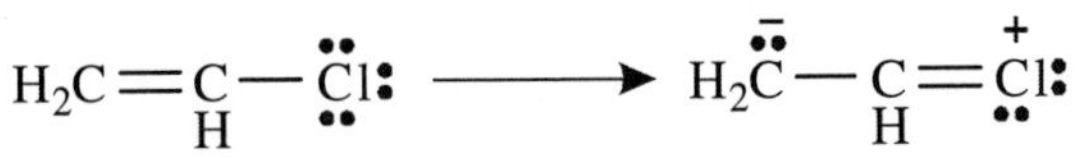

10.4.1.6 Addition of Hydrogen Bromide

Additon of HBr, takes place slightly differently than that of HCl. It is due to less electronegativity of bromine. Addition of one molecule of HBr, gives vinyl bromide. The C-Br bond dipole, no doubt shall be towards bromine, but it is weak. The second addition can take place, by first adding proton, to form two possible, carbocations as intermediates, CH_3-CH^+-Br or $^+CH_2$ CH_2-Br . The first carbocation being secondary is more likely to get formed, so the addition of bromide anion shall take place to form ethylidene bromide, rather than 1,2-dibromoethane.

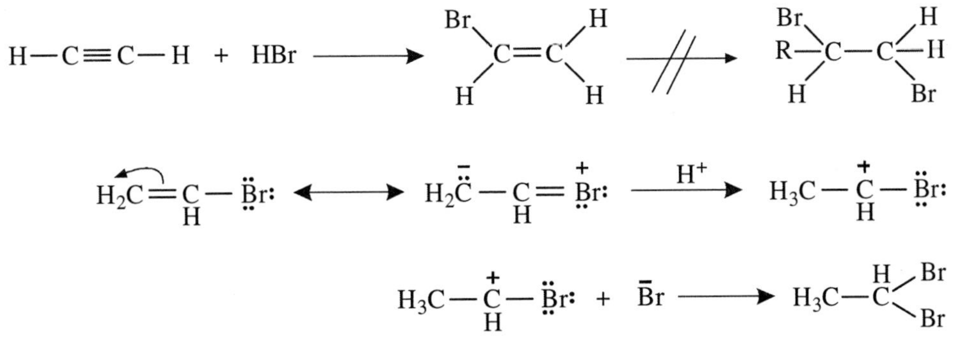

10.4.1.7 Addition of Hydrocyanic acid

In case of hydrocyanic acid, vinyl cyanide is the end product. It is very useful product in industry for preparing useful polymers- polyvinyl cyanides.

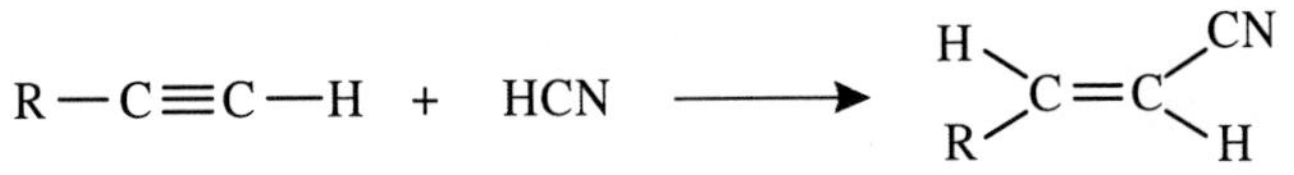

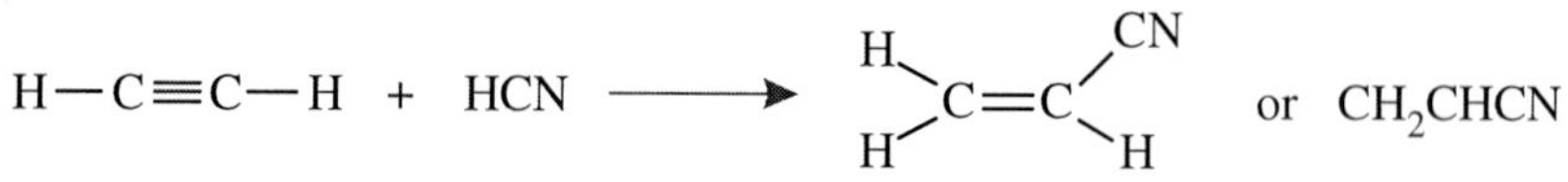

10.4.1.8 Addition of Acetic acid in presence of Hg^{2+} ions

Addition of acetic acid to triple bond is very similar to HBr addition, to form a *gem*-diacetoalkane.

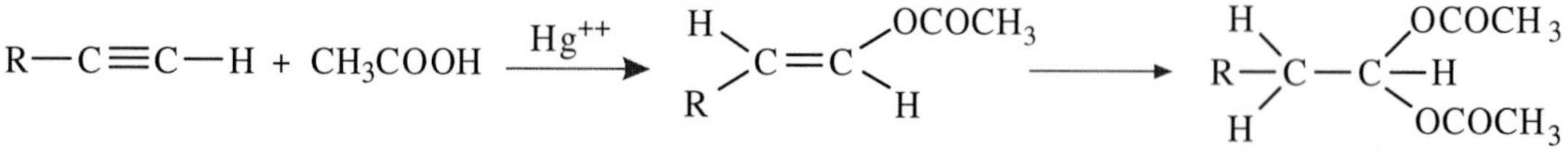

10.4.1.9 Addition of water in presence of Hg^{++} ions

Addition of water to an acetylene molecule forms an aldehyde, whereas, alkyl acetylenes form ketones as the end product. The intermediate, can enolise to carbonyl compound or add another molecule of water, followed by loss of water again.

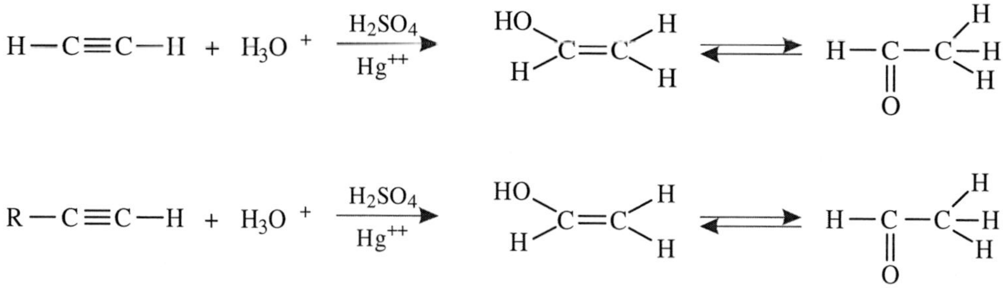

10.4.1.9 Addition of methoxide in methanol

Addition of methoxide ion to acetylene takes place as expected to give vinyl methyl ether–a very useful for synthesis of polymers.

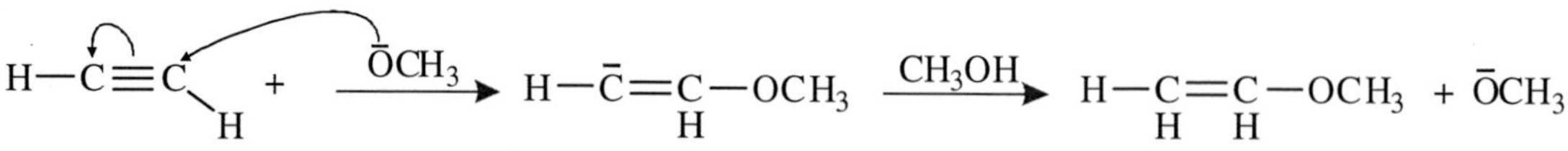

10.4.1.10 Addition of hypochlorous acid

Oxygen in hypochlorous acid is more electronegative than chlorine, thus OH becomes negative addendum and chlorine is positive addendum. After first addition, enolisation takes place to form alkyl chloromethyl ketone. For R= CH_3, it forms chloroacetone.

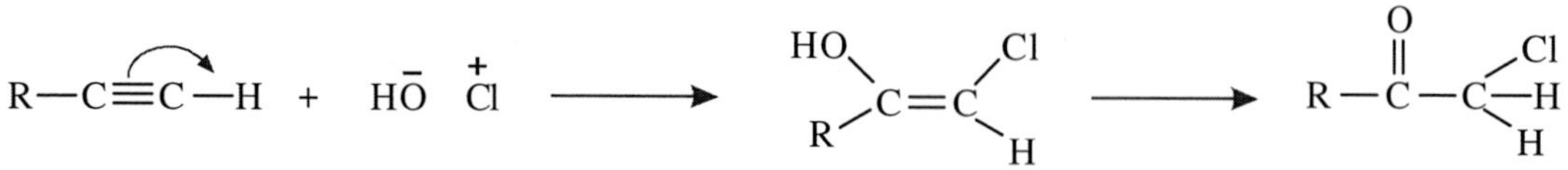

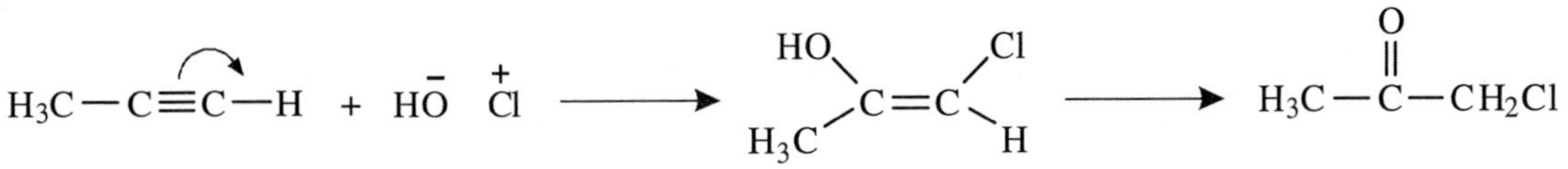

10.4.2 Addition of Ozone and Cleavage

Like alkenes, ozone can cleave the double bond, leaving one bond to give a diketone. This could be oxidised further by hydrogen peroxide to give two molecules of carboxylic acids.

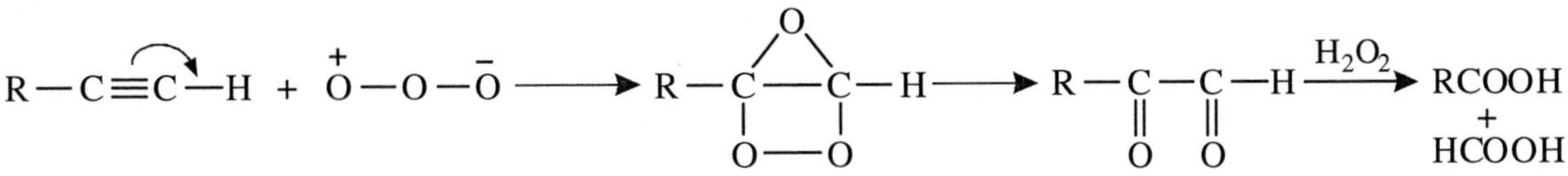

10.4.3 Addition of Potassium Permanganate

Double bond and triple bond can be effectively differentiated by action of oxidising agents like chromic acid and potassium permanganate. $KMnO_4$ is known to oxidise the triple bond, as well as the double bond, but chromic acid, does not oxidise triple bond, but the double bond.

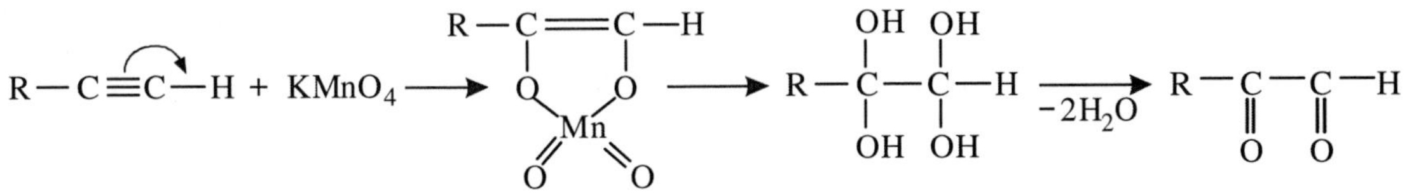

$$H_3C-C\equiv C-CH_2-CH=CH_2 \xrightarrow[HOAc]{CrO_3} H_3C-C\equiv C-CH_2-COOH + HCOOH$$

10.4.4 Reaction with Carbonyl Compounds

Carbonyl has a sigma bond, as well as, a pi-bond. It is a reactive group and the direction of electromeric change, makes it a suitable substrate for nucleophilic attack at the carbon atom. In presence of a base, the acidic proton of an acetylene leave a negative carbon, that attaches with carbon of carbonyl group. Thus, it helps in preparing the higher acetylenes by addition of carbon.

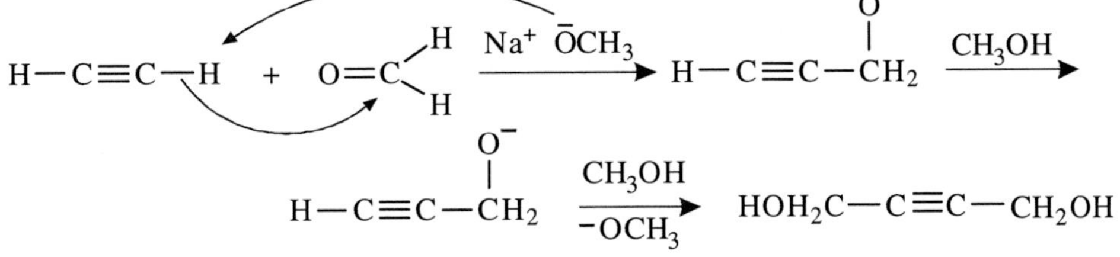

Condensation with formaldehyde gives acetylenic alcohol, that can be reduced to form higher acetylenes, Most of the industrially useful alkyne derivatives are prepared from acetylene *via* such condensation with carbonyl compounds.

10.4.5 Polymerisation Reactions

Unlike alkenes, polymerisation of acetylenes give tri- or tetramerisation product, that are very useful in synthetic organic chemistry. Thus, three molecules of acetylene on passing through a heated tube, form benzene. or four molecules of acetylene, produce cyclooctatetraene. In presence of catalyst like copper chloride, linear polymerisation also takes place to produce, but-1-ene-3-yne or hexa-1,5-diene-3-yne.

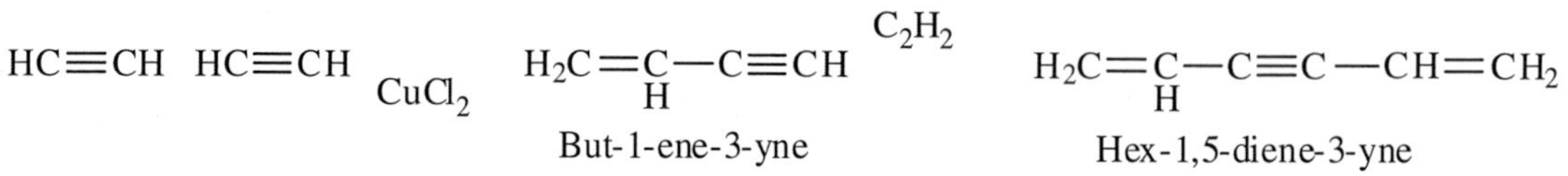

In case of substituted acetylenes, homologues of benzene are obtained. For example, propyne on trimerisation, gives 1,3,5-trimethyl benzene. Three molecules of but-2-yne gives, 1,2,3,4,5,6-hexamethyl benzene.

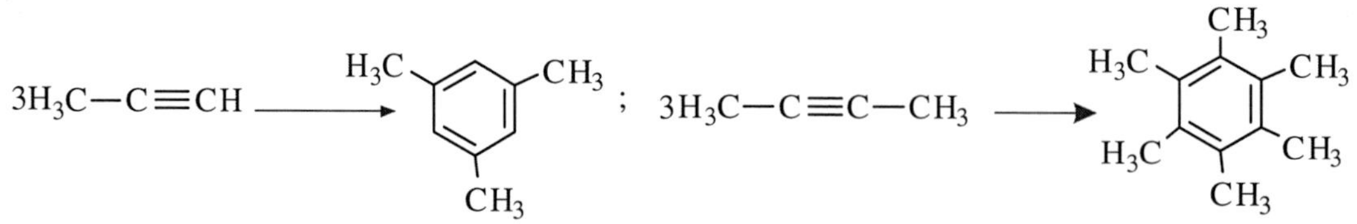

10.5 Uses of Acetylene

Acetylene is used

1. for oxyacetylene torch for welding
2. for synthesis of variety of organic compounds in industry. Walter Reppe discovered that in presence of metal ions, acetylene is useful for preparing a variety of industrial chemicals, like HCN, alcohols, vinyl compounds and carboxylic acids.
3. It is used in ripening of fruits, but is not desirable, only ethylene should be used.

Points to Remember

Methods of Synthesis of Acetylenic Compounds

- CH_4 Dehydrogenation Catalytically
- *vic*-Dibromo-alkane + alc. KOH
- *gem*-dibromo-alkane + alc. KOH
- Acetylenes + $NaNH_2$ then RBr
- a carboyl compound + Ph_3P + CBr_4
- a carbonyl compound + dimethyl diazophsophonate + K *t*-butaoxide
- CHI_3 + 6 Ag
- Ethylenic dicarboxylic potassium salt kolbe's electrolysis method

Acetylenes

Chemical Reactions of Acetylenes

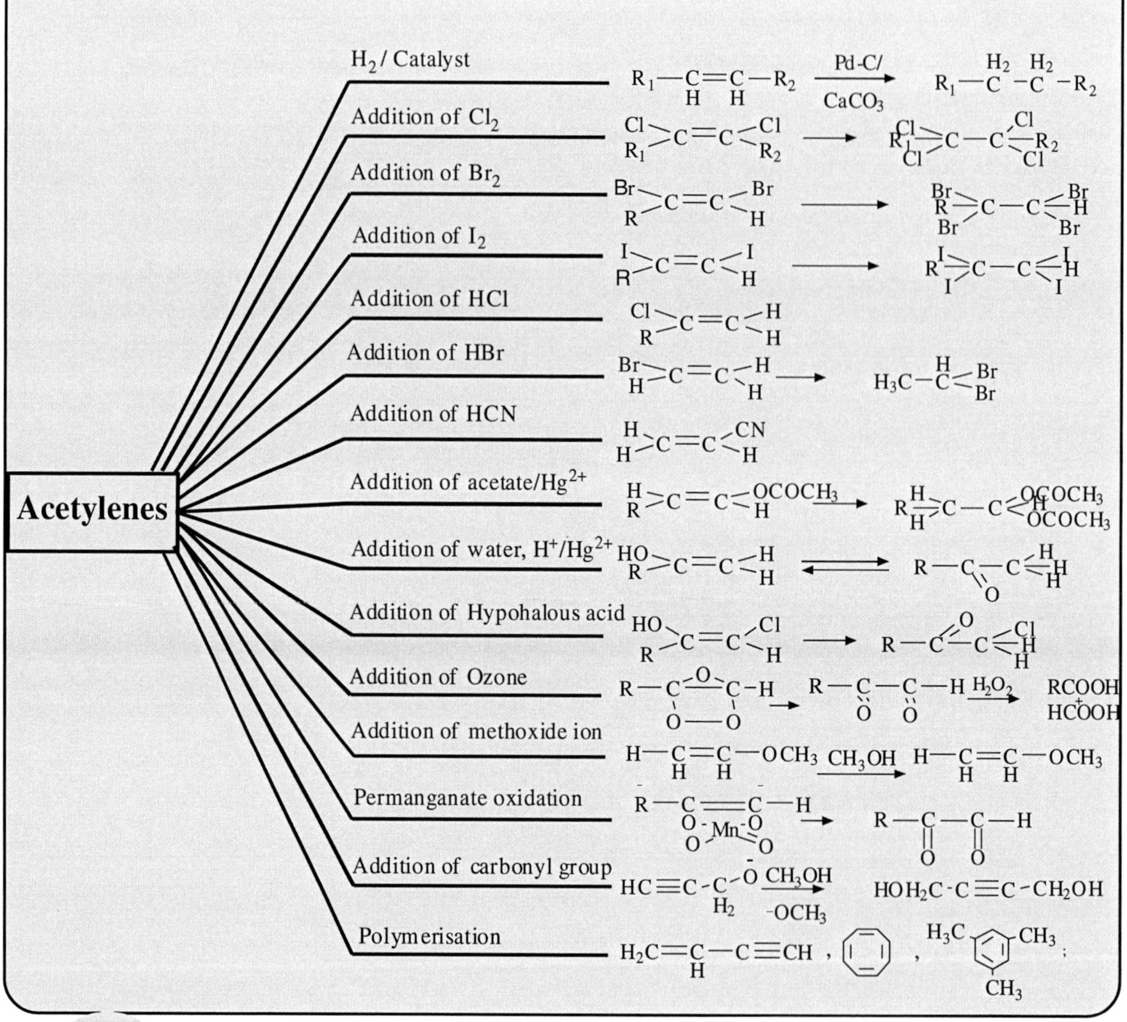

Exercises

1. The number of hydrogen atoms in butane are 10, that in butene are 8 and in butyne, it is 6, which of these compounds is most acidic. The structures are given below.

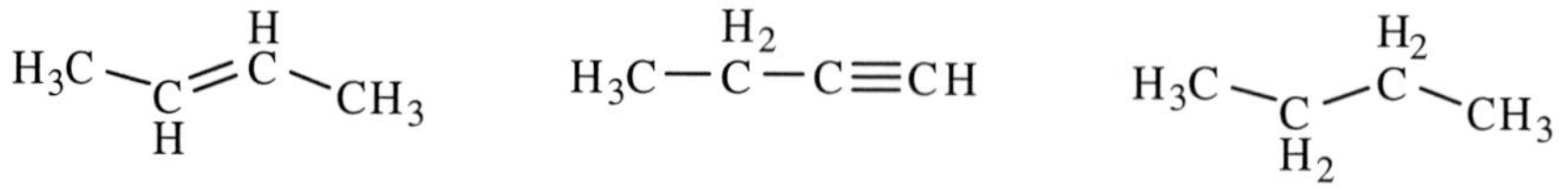

2. How will you convert propyne in to propanal and propanone?

3. What are the products of reaction of the following alkynes with Hg^{2+} / H_2SO_4?

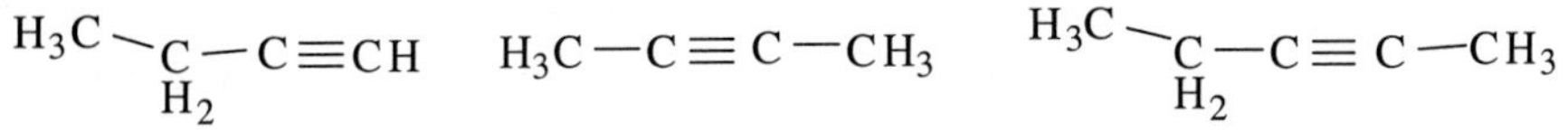

4. How will you convert pent-2-yne into *cis*-pent-2ene and *trans*-pent-2-ene?

5. Complete the following reactions:

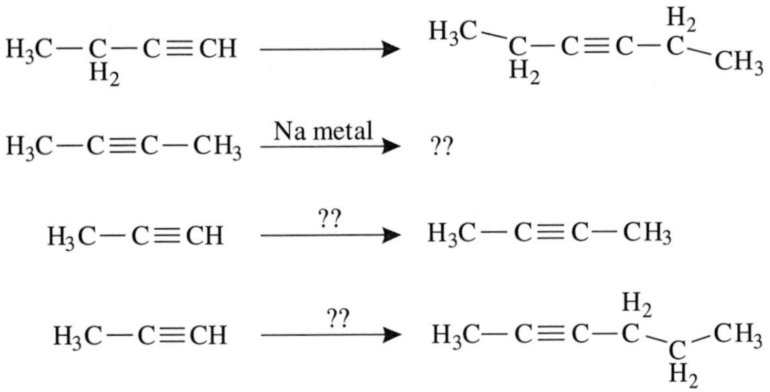

6. Which of the following compounds are more stable within the pairs given below?

(*i*) $H_3C-\overset{H}{\underset{+}{C}}-CH_3$ or $H_3C-\overset{+}{C}=CH_2$

(*ii*) $H_3C-\overset{H}{\underset{+}{C}}-CH_3$ or $H_3C-C\equiv\overset{+}{C}$

(*iii*) $H_3C-\overset{+}{C}=CH_2$ or $H_3C-C\equiv\overset{+}{C}$

7. How will you carry out the following conversions?

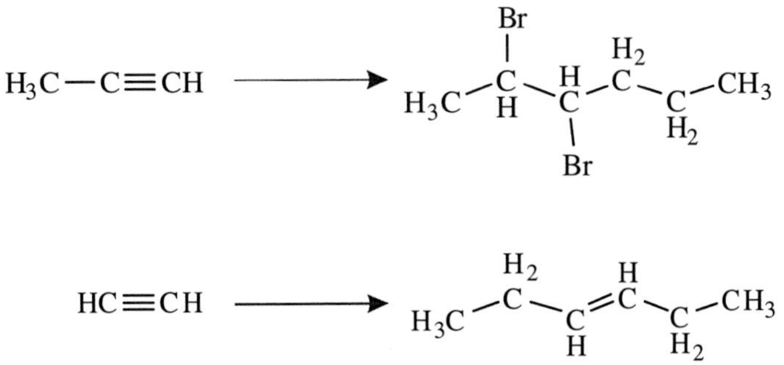

8. Using 2,2-dichloropropane, how will you make but-1-yne, but-2-yne and methyl allene?

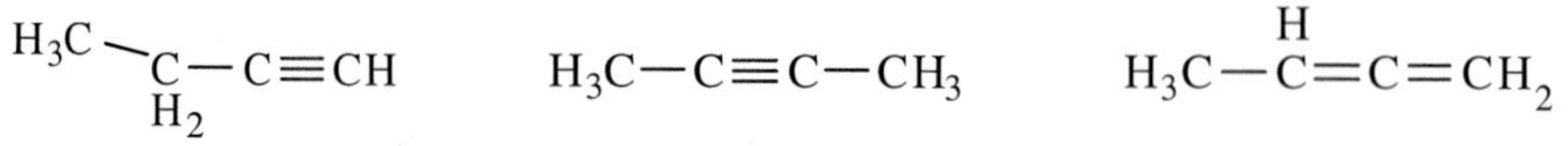

9. How will you convert acetylene in to

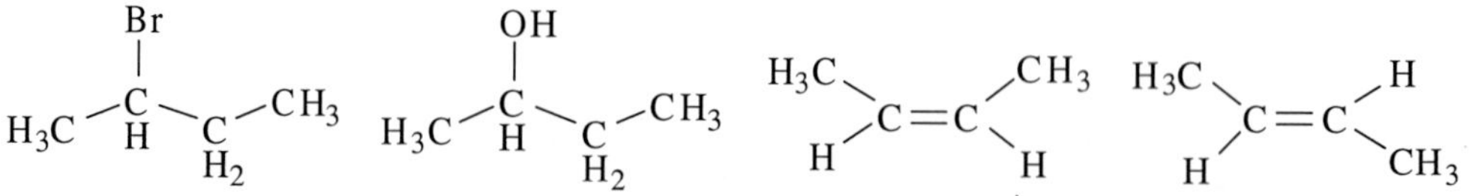

2-bromobutane | *sec.*-butyl alcohol | *cis*-but-2-ene | *trans*-but-2-ene

10. What is the nature of product that shall be obtained by hydroxylation-oxidation of but-2-yne and while the same reaction with pent-2-yne gives more than one product?

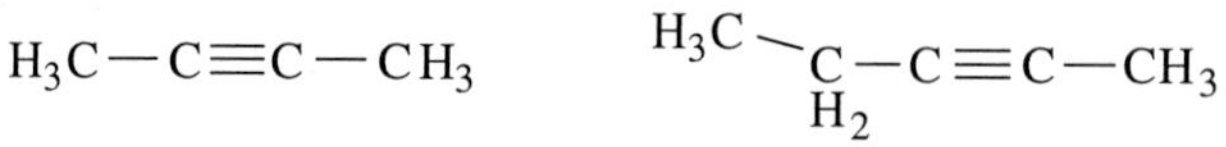

❑❑❑

PRACTICALS

UNITS

Practicals in Chemistry

What do we do in practicals

1. Preparation of Compounds
2. Analysis of Compounds
   A. Qualitative Analysis
   B. Quantitative Analysis
      (*a*) Volumetric Analysis
      (*b*) Gravimetric Analysis
      (*c*) Instrumental Analysis

Syllabus of Semester I – PRACTICALS

Section A:

Inorganic Chemistry - Volumetric Analysis

1. Estimation of sodium carbonate and sodium hydrogen carbonate present in a mixture.
2. Estimation of oxalic acid by titrating it with $KMnO_4$.
3. Estimation of water of crystallization in Mohr's salt by titrating with $KMnO_4$.
4. Estimation of Fe(II) ions by titrating it with $K_2Cr_2O_7$ using internal indicator.
5. Estimation of Cu(II) ions iodometrically using $Na_2S_2O_3$.

Section B:

Organic Chemistry

1. Detection of extra elements (N,S,Cl,Br,I) in organic compounds (containing upto two extra elements).
2. Separation of mixtures by Chromatography: Measure the R_f value in each case (combination of two compounds to be given).
   (*a*) Identify and separate the components of a given mixture of 2 **amino acids** (glycine, aspartic acid, glutamic acid, tyrosine or any other amino acid) by paper chromatography.
   (*b*) Identify and separate the **sugars** present in the given mixture by paper chromatography.

CHAPTER 11

Inorganic Practicals (Volumetric Analysis)

11.1 Volumetric Analysis

It is one of the chemical analytic technique, that is used for quantitative determination of chemical compounds in solution. The method involves reaction of materials in solution and the process is quantised by measuring volumes of the solutions. The reactions take place in equivalent amounts of materials present in solution, irrespective of their volumes. The calculations of quantities of materials involved can be done by using the famous normality equation, $N_1V_1 = N_2V_2$ and their equivalent weights. The other quantitative method is the gravimetric method, that is not the subject of the present chapter.

Experimentally, Volumetric analysis is carried out by taking a known volume of one solution in a flask (the solution contains some unknown quantity of the compound) and reacting with another solution of a compound of known concentration (meaning standard solution) by repeatedly putting in small measured volumes from the burette. The exact volume of the solution from burette that will bring the end point of the reaction, meaning equivalent amounts of each component has reacted is determined by the use of indicators or instruments. Indicators are reagents, the type of which depends upon the type of the reactions involved. At the end points, the indicators produce definite colour, that helps us to know that no further quantity of the material from the burette need to be added. Detailed procedure of different titrations are given as experiments that are included in this chapter giving individual experimental estimations.

11.2 Terms Used in Volumetric Analysis

Volumetric Analysis involves certain terms and it may be better to understand them. These terms are:

1. **Titration:** It is a procedure in volumetric analysis in which a solution of known concentration is added to the solution of unknown concentration from a burette until the equivalent point or end point is reached.

2. **Titrant:** The solution used in volumetric analysis whose concentration is known. It may also be called a standard solution.

3. **Titrand:** The solution used in volumetric analysis whose concentration is to be determined.

4. **Indicator:** A reagent that is added in small quantity of the reaction flask, produces colour change by addition of little excess quantity of either of the solutions taking part in the volumetric analysis.

5. **Equivalence Point or End Point:** While adding one compound into the other solution, the end point is the stage of reaction, where the equivalent amounts of the reactants have reacted. The point is determined by use of indicators bringing about the observable physical or colour change.

6. **Strength of Solution:** It is a parameter, that determines the amount of solute present in a solution. There are different ways of expressing strength of a solution, like molarity, molality etc.

7. **Normality:** It is the number of gram equivalents of the substance dissolved per liter per cubic decimeter of solution. It is denoted by the letter *N*, and is equal to

$$N = \frac{\text{grams of solute dissolved in one litre solution}}{\text{gram equivalent weight of the substance.}}$$

8. **Molarity:** It is the number of moles of solute per litre of solution. It is denoted by the letter, *M*.

$$M = \frac{\text{grams of solute dissolved in one litre solution}}{\text{gram moleular mass of the substance.}}$$

9. **Molality:** It is the number of moles of a substance that is dissolved in 1000g or 1 kg of solvent and is denoted by *m*.

$$m = \frac{\text{grams of solute dissolved in 1000g of solvent}}{\text{gram moleular mass of the solute}}$$

10. **Formality:** It is the number of gram formula weight of the substance dissolved in one litre of the solution. It is denoted by *F*, where *F* is equal to:

$$F = \frac{\text{grams of solute dissolved in one litre solution}}{\text{formula weight of the substance.}}$$

11. **Equivalent Weight:** It is a measure of 'combining power' that is used in volumetric analysis for finding equivalent reactions. The equivalent weight of an element is number of grams of the element that would react with or displace 1 gram of hydrogen or 8 grams by weight of oxygen or 35.5 grams by weight of chlorine. It is also equal to the atomic weight of an element divided by it valence. It is not a fixed number. It changes according to the reaction. For example, one equivalent of sodium hydroxide reacts with half equivalent of sulphuric acid. Thus, the equivalent weight of sulphuric acid is half of molecular weight of sulphuric acid.

12. **Standard Solution:** A standard solution is one, whose concentration is known. Standard solutions are of two types,

    (*a*) **Primary Standard Solution:** A primary standard solution of compond is one, which can be prepared by directly weighing the solute accurately and making it upto a known volume in a standard flask. The solute has exact composition and is stable under normal conditions of storage. That is, with time, its composition does not change. For example, oxalic acid crystallises with two water molecules of crystallisation and the compound is very stable. Its solution can be prepared as a primary standard solution. If we want ferrous ion solution, we cannot use ferrous sulphate salt to prepare the solution, because, it changes to ferric ion by aerial oxidation. Thus, the stable salt, $FeSO_4,(NH_4)_2SO_4,6\ H_2O$ is the best salt that is used for preparing the standard solution of ferrous ions.

(*b*) **Secondary Standard Solution:** Compounds that are not stable or do not have fixed composition, are not good for preparing standard solution. In such cases, the solution prepared though of known concentration, will not have exactly the concentration of different ions, as prepared. Such compounds do not give standard solutions good for to be used as primary standards.

The exact strength of such solution is determined by titrating against a primary standard solution. This method is called standardization. The examples of secondary standards are; Sodium hydroxide, $KMnO_4$ HCl, H_2SO_4 etc.

11.3 Types of Volumetric Titrations

Volumetric analysis can be carried out by different types of titrations,

1. Acid - base titrations
2. Redox titrations
3. Precipitation titrations
4. Complexometric titrations
5. Iodometric titrations

11.3.1 Acid Base Titrations

These titrations are based on the reaction between hydrogen and hydroxyl ions of an acid and a base to form unionized water to a small extent *i.e.*

$$H^+ + OH^- \rightleftharpoons H_2O$$

If hydronium ion (H_3O^+) is used in place of hydrogen ion, the reaction is,

$$H_3O^+ + OH^- \rightleftharpoons 2H_2O$$

The most common example is the titration of sodium hydroxide against hydrochloric acid, the reactions involved are:

$$NaOH + HCl \longrightarrow NaCl + H_2O$$

$$Na^+ + OH^- + H^+ + Cl^- \longrightarrow Na^+ + Cl^- + H_2O$$

Acid base titrations are also known as acidimetry and alkalimetry. Acidimetry involves the determination of the strength of an unknown acid with the help of a base, whose strength is known.

Alkalimetry involves the determination of the strength of an unknown base with the help of an acid whose strength is known.

The indicators generally employed are (*i*) phenolpthalein and (*ii*) methyl orange. These indicators are complex organic compounds, which are themselves weak acids or bases. These associate or dissociate, according to the pH of the solution.

Phenolpthalein is a diprotic acid and is colourless in acidic medium, having a benzenoid structure. But, in basic medium, it changes to pink, having a quinonoid form.

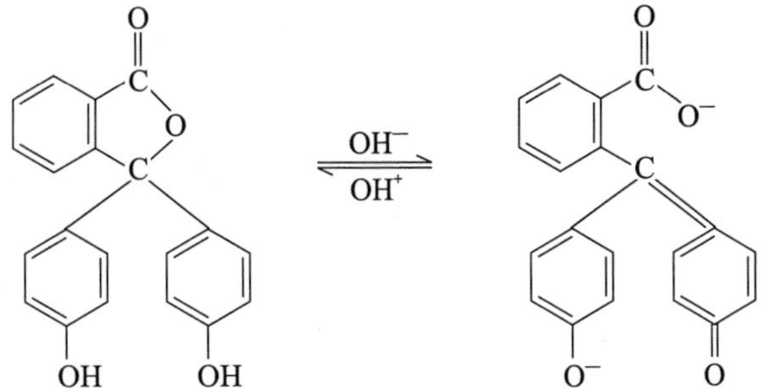

(Colourless) (Pink coloured)

Methyl orange is a weak base, and possesses yellow colour in the alkaline medium and red colour in the acidic medium.

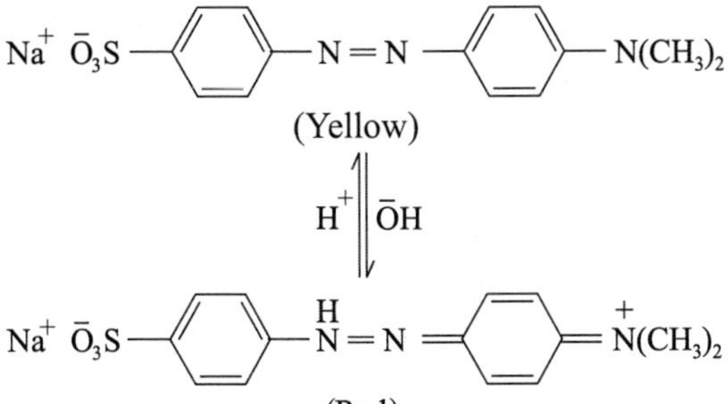

11.3.2 Redox Titrations

Redox titrations involve oxidation reduction reactions *i.e.*, the transfer of electrons from a substance which releases electrons to another substance, which accepts electrons. Examples include:

$$Fe^{2+} \longrightarrow Fe^{3+} + \bar{e} \qquad \text{(Oxidation)}$$

$$Ce^{4+} + \bar{e} \longrightarrow Ce^{3+} \qquad \text{(Reduction)}$$

$$Fe^{2+} + Ce^{4+} \longrightarrow Fe^{3+} + Ce^{3+} \qquad \text{(Redox reaction)}$$

The indicators used in redox titrations are redox indicators. Redox indicator is one whose colour in the oxidised form is different from the colour in the reduced form.

Redox indicators can be classified as:

(*i*) Self indicator

(*ii*) Internal indicator

(*iii*) External indicator

(*i*) **Self Indicator:** The most commonly employed self indicator is potassium permanganate. ($KMnO_4$). It is intensely coloured but when added to a reducing agent in solution, it becomes colourless. $KMnO_4$ is purple in its oxidized form and colourless in the reduced form (while in solution).

$$\underset{\text{(Purple)}}{MnO_4^-} + 8H^+ + 5\bar{e} \longrightarrow \underset{\text{(colourless)}}{Mn^{2+}} + 4H_2O$$

(*ii*) **Internal Indicator:** These are the indicators which are added to the volumetric flask in which titration is carried out *e.g.*, Diphenylamine is the most common internal indicater used in the titration of Fe (II) with potassium dichromate.

Diphenylamine has different colours in oxidized and reduced form, when all the Fe (II) ions are oxidized to Fe (III) ions by $K_2Cr_2O_7$ then an extra drop of $K_2Cr_2O_7$ added in the solution, will oxidize the indicator (diphenylamine), which changes to deep blue violet colour and hence indicates the completion of reaction.

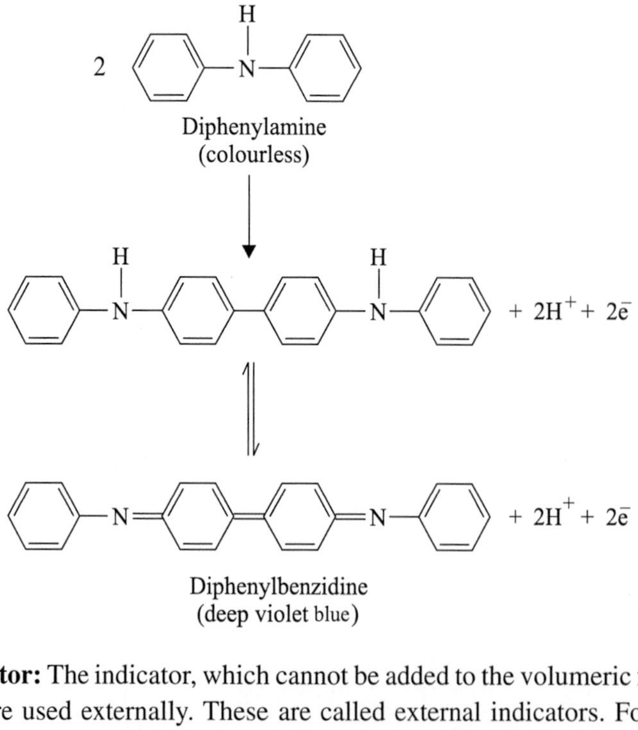

(*iii*) **External Indicator:** The indicator, which cannot be added to the volumeric flask in which titration is carried out, are used externally. These are called external indicators. For example, potassium ferricyanide is an external indicator in the titration of $K_2Cr_2O_7$ with Mohr's salt.

Potassium ferricyanide ($K_3[Fe(CN)_6]$) gives a blue colour with Fe^2 ions due to formation of the complex ferrous ferricyanide.

$$2K_3Fe(CN)_6 + 2FeSO_4 \longrightarrow \underset{\substack{\text{Ferrous ferricyanide} \\ \text{(Dark blue complex)}}}{Fe_3[Fe(CN)_6]_2} + 3K_2SO_4$$

11.3.3 Precipitation Titrations

These are the titrations in which there is formation of a sparingly souble salt, when the solutions of two reacting substances are brought in contact with each other. For example, a white precipitate of silver chloride is formed, when a solution of silver nitrate is added to a solution of sodium chloride.

$$AgNO_3 + NaCl \longrightarrow AgCl\downarrow + NaNO_3$$

Formation of AgCl, marks the completion of the reaction.

11.3.4 Complexometric Titrations

These titrations involve the formation of a soluble complex between a metal ion and a complexing agent. Here, the metal ion is called the central atom and the complexing agent is called the ligand. The number of bonds formed by the central atom is taken as the co-ordination number of the atom.

Ligands generally contain an electronegative atom, such as nitrogen, oxygen etc. The ligand or a complexing agent can be unidentate, bidentate or polydentate having two or more donor atoms respectively. Though, the polydentate ligands are effective in forming stable complexes with a wide varieties of metals, as they form ring structure called chelates. Most commonly used ligand in the complexometric titrations is Ethylene diaminetetraacetic acid (EDTA). Indicators used in these titrations are the organic substances, that change colour when they form metal complex. The indicator forms a less stable complex with the metal atom than EDTA–metal Complex. The most widely used indicator is Eriochrome Black T (EBT). Eriochrome Black T chemically is :

1– (hydroxy 2–napthylazo) 6 nitro-2napthol–4 sulphonate

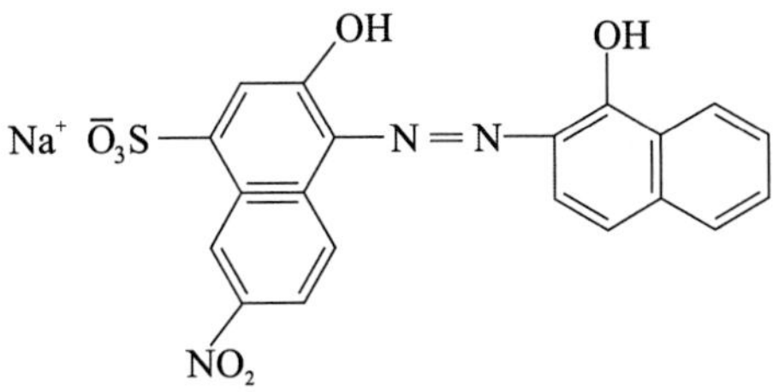

(Structure of EBT)

These titrations are performed in buffers of pH 8 – 10 as the stability of the metal –EDTA complex and the colour change of the indicator are sensitive to pH changes.

11.3.5 Iodometric Titrations

Iodine is a mild oxidising agent and is used in volumetric estimations. Iodine can act as an oxidizing agent in neutral aqueous medium *i.e.*

$$I_2 + 2\bar{e} \longrightarrow 2I^- \text{ (gets reduced)}$$

On the orther hand, an oxidising agent can oxidise I^- to I_2

$$2I^- \longrightarrow I_2 + 2\bar{e} \text{ (gets oxidised)}$$

In these titrations, iodine is liberated by the action of an oxidising agent on potassium Iodide (source of iodine) in neutral or acidic medium. This liberated iodine is then titrated with a reducing agent.

Iodometric titrations are used to determine the strength of oxidizing agents like the estimation of Cu^{2+} using sodium thiosulphate.

$$2Cu^{2+} + 4I^- \longrightarrow Cu_2I_2 + 2I_2$$

$$I_2 + 2S_2O_3^{2-} \longrightarrow S_4O_6^{2-} + 2I^-$$

Indicator used in these titrations is freshly prepared starch solution, which forms a dark blue complex with triiodate ions.

Freshly prepared starch solution should be used as the starch solution is decomposed due to bacterial action.

Free iodine forms complex with strach and takes time to release it, that is why, starch should be added to the titration flask only when most of free iodine has been used upto the pale yellow stage.

Iodometric titrations are performed in cold conditions. Since, iodine is volatile, these titrations are perfomed in cold.

These titrations are not performed in alkaline medium since, alkali react with iodine to form hypoiodite, which is a strong oxidising agent and oxidises thiosulphate ions.

$$I_2 + 2NaOH \longrightarrow NaIO + NaI + H_2O$$

or

$$I_2 + 2OH^- \longrightarrow IO^- + I^- + H_2O$$

$$4IO^- + S_2O_3^{2-} + 2OH^- \longrightarrow 4I^- + 2SO_4^{2-} + H_2O$$

In these titrations, always excess of KI is used as the solubility of I_2 in water is very low. In presence of KI, it forms a complex KI_3, which releases iodine slowly during the titration due to reversible reaction.

$$KI + I_2 \rightleftharpoons KI_3$$

$$I^- + I_2 \rightleftharpoons I_3^-$$

When free iodine is used up in the reaction with thiosulphate, the equilibrium shifts to the left releasing more I_2.

11.4 Some Volumetric Experiments

I. Determine the amount of Na_2CO_3 and $NaHCO_3$ in the given mixture of Na_2CO_3 and $NaHCO_3$

Theory: The determination of the strength of Na_2CO_3 and $NaHCO_3$ present in the solution is based on the selective use of indicators. The indicators used are phenolphthalein and methyl orange.

The reactions may be represented as:

(*a*) $CO_3^{2-} + H^+ \longrightarrow HCO_3^-$

(*b*) $HCO_3^- + H^+ \longrightarrow H_2CO_3$

When phenolpthalein is used as an indicator, the volume of acid will correspond to half neutralisation of Na_2CO_3 or we can say, neutralization upto half stage *i.e.*

$$Na_2CO_3 + HCl \longrightarrow NaHCO_3 + NaCl$$

Let the volume of acid used for neutralization be y mL, when phenolphalein is used as an indicator.

Therefore, the volume used for complete neutralization of Na_2CO_3 will be 2y mL.

Similarly, when methyl orange is used as an indicator, the volume of acid will neutralize both, Na_2CO_3 and $NaHCO_3$ respectively *i.e.*

$$Na_2CO_3 + 2HCl \longrightarrow 2NaCl + H_2O$$

$$NaHCO_3 + HCl \longrightarrow NaCl + H_2O + CO_2$$

Let the volume of acid used is *x* mL when methyl orange is used as an indicator. The volume used for neutralization of $NaHCO_3$ will be *x*-2*y* mL.

Procedure

(*i*) Preparation of standard (N/10) Na_2CO_3 solution for calculating the exact normality of HCl solution.

Weigh out nearly 1.325 g of the dried Na_2CO_3 and transfer it to 250 mL of measuring flask. Add about 30-40 mL of distilled water and stir gently to disiolve the solid. Now, make up the volume to 250 mL with distilled water.

(*ii*) Rinse and fill the burette with given HCl solution. Pipette out 10 mL of standard Na_2CO_3 solution and add 2-3 drops of methyl orange. Titrate with HCl till red colour is obtained. Repeat until three concordant readings are obtained.

(*iii*) Pipette out 10 mL of the given mixture and add two drops of phenolpthalein indicator. Add 2 drops of phenolpthalein. Now, add standard solution of HCl till the pink colour of the solution changes to colourless. Note this end point carefully. Now, add two diops of methyl orange and titrate with HCl till light pink colour appears. Also, note this reading. (This reading will also include previous volume used with first indicator).

Repeat the experiment, till three conccordant readings are obtained.

Observations

Weight of the weighing bottle = w_1 g

Weight of the weighing bottle + sodium carbonate = w_2 g.

Weight of the weighing bottle after transferring the salt = w_3 g.

Weight of sodium carbonate transferred to the standard flask = $(w_2 - w_3)$g

$$\text{Normality of } Na_2CO_3 \text{ solution} = \frac{4 \times (w_2 - w_3)}{53} \text{ N}$$

Burette Reading:

(*i*) Titration of std. Na_2CO_3 *vs* HCl

S.N.	Initial Reading	Final Reading	Vol. of titrant used (mL)
1.			
2.			
3			
4.			

(*ii*) Titration of Given Mixture *vs* HCl

S.N.	Initial Reading	Final Reading	Vol. of HCl used upto phonolpthalein end point (mL)	Vol. of HCl used upto Methyl orange end point (mL)
1.				
2.				
3				
4.				

The volume of acid required for neutralizing Na_2CO_3 upto half stage *i.e.*, upto phenolpthalein end point be y mL.

The volume of acid required for neutralizing Na_2CO_3 and $NaHCO_3$ *i.e.*, upto methyl orange end point by x mL.

Calculations

(*i*)
$$N_{HCl} \times V_{HCl} = N_{Std\ Na_2CO_3} \times V_{Std\ Na_2CO_3}$$

$$N_{HCl} = \frac{N_{Std\ nNa_2CO_3} \times V_{Std\ Na_2CO_3}}{V_{HCl}}$$

$$N_{HCl} = \frac{N_{Std\ Na_2CO_3} \times 10}{V_{HCl}}$$

(*ii*) Now, the volume of acid required for neutralizing Na_2CO_3 completely = $2y$ mL

$$\underset{\text{(acid)}}{N_1 V_1} = \underset{(Na_2CO_3)}{N_2 V_2}$$

$$N_{HCl} \times 2y = N_2 \times 10$$

$$N_2 = \frac{N_{HCl} \times 2y}{10}$$

Hence, the strength of Na_2CO_3 in the solution = Normality × eq. wt.

$$= \left(\frac{N_{HCl} \times 2y}{10}\right) \times 53 \text{ g / L}$$

Volume of acid required for neutralization of $NaHCO_3$ only=$(x - 2y)$ mL

$$\underset{\text{(acid)}}{N_1 V_1} = \underset{(NaHCO_3)}{N_2 V_2}$$

$$N_{HCl} \times (x - 2y) = N_2 \times 10$$

$$N_2 = \frac{N_{HCl} \times (x - 2y)}{10}$$

Hence, the strength of $NaHCO_3$ in the solution = $\dfrac{(x-2y) \times N_{HCl}}{10} \times 84$ g/L

Precautions are as given in text portion.

II. Determination of the strength of given oxalic acid solution by titrating it against approximately N/20 $KMnO_4$ solution.

Theory: $KMnO_4$ solution reacts with oxalic acid in the presence of H_2SO_4.

$$2KMnO_4 + 3H_2SO_4 \longrightarrow K_2SO_4 + 2MnSO_4 + 3H_2O + 5[O]$$

$$\begin{matrix} COOH \\ | \\ COOH \end{matrix} + [O] \xrightarrow[\Delta]{60-70^\circ} 2CO_2 + H_2O] \times 5$$

$$2KMnO_4 + 3H_2SO_4 + 5H_2C_2O_4 \longrightarrow K_2SO_4 + 2MnSO_4 + 8H_2O + IOCO_2$$

Ionic eqn:

$$2\,[MnO_4]^- + 5[C_2O_4]^{2-} + 16\,H^+ \longrightarrow Mn^{2+} + 8H_2O + 10CO_2$$

The rate of reaction of oxalic acid with $KMnO_4$ is very slow, therefore, oxalic acid solution is heated to 60–70°C initially. Once the reaction is started the rate increases due to the formation of Mn^{2+} ions which catalyzes the reaction.

Procedure

(*i*) Preparation of standard N/20 oxalic acid solution

Dissolve about $\frac{63 \times 250}{1000} \times 0.05 = 0.79$g of oxalic acid crystals in a 250 mL measuring flask and then make up the volume by adding distilled water.

(*ii*) Rinse and fill the burette with N/20 $KMnO_4$ solution and note the initial reading. Pipette out 20ml of standard oxalic acid solution in a conical flask and about a test tube full of dil H_2SO_4. Heat the contents gently to 60-80°, till it is really hot to touch. Run in $KMnO_4$ from the burette, until light pink colour just appears and persist for about 30 seconds. Repeat the titration to get three concordant readings.

(*iii*) Repeat the procedure, with the given oxalic acid solution.

Observations

Weight of the weighing bottle = w_1 g.

Weight of the weighing bottle + oxalic acid crystals = w_2 g

Weight of the weighing bottle after transferring oxalic acid crystals, = w_3 g

Weight of oxalic acid crystals transferred to prepare 250 mL of standard solution = $(w_2 - w_3)$ g

w is the mass of oxalic acid dissolved in 250 ml of distilled water, the normality of oxalic acid solution = $\frac{4w}{63}$ N

Burette Readings:

(I) Standard Oxalic Acid solution *vs* $KMnO_4$ solution

S.N.	Initial Reading	Final Reading	Vol. of titrant used
1.			
2.			
3			

(II) Given Oxalic Acid solution *vs* $KMnO_4$ solution

S.N.	Initial Reading	Final Reading	Vol. of titrant used (mL)
1.			
2.			
3			
4.			

Calculations

(*i*) For finding the normality of $KMnO_4$ solution

$$N_1 V_1 = N_2 V_2$$

(Std oxalic Acid) ($KMnO_4$)

$$\frac{N}{20} \times 20 = N_2 \times V_2$$

$$N_2 = \frac{N}{20} \times \frac{20}{V_2}$$

∴ Normality of $KMnO_4 = N_2$

(*ii*) For finding the normality and strength of given oxalic acid solution.

$$N_1 V_1 = N_2 V_2$$

(Given oxalic acid) ($KMnO_4$)

$$N_1 = \frac{N_2 \times V_2}{20}$$

Strength of given oxalic acid solution

$$= \frac{N_2 \times V_2}{20} \times \text{Eq. wt of oxalic acid}$$

$$= \frac{N_2 \times V_2}{20} \times 63 \text{ g/L}$$

Result: The strength of given oxalic acid solution is __________g/L

Precautions

(*i*) Since, $KMnO_4$ is dark in colour, the burette reading is noted at the upper meniscus.

(*ii*) A brown precipitate of MnO_2 is formed, if H_2SO_4 is not added in sufficient amount.

(*iii*) $KMnO_4$ solution attacks rubber, hence it should be taken in a glass stoppered burette.

(*iv*) At the end point, light pink colour must persist for 30 seconds. Add a drop of oxalic acid to it, if pink colour disappears, the end point is correct.

III. Determination of the number of water molecules of crystallization in Mohr's salt, provided with approximately N/20 $KMnO_4$ solution

Theory: Acidified $KMnO_4$ acts as an oxidising agent and converts Fe^{2+} ions to Fe^{3+} ion, according to the following equation:

$$2MnO_4^- \ 16H^+ + 10\bar{e} \longrightarrow 2Mn^{2+} + 8H_2O$$

$$10Fe^{2+} \longrightarrow 10\,Fe^{3+} + 10\bar{e}$$

$$2MnO_4^- \ 16H^+ + 10Fe^{2+} \longrightarrow 2Mn^{2+} + 10Fe^{3+} + 8H_2O$$

Since $KMnO_4$ is a self-indicator, no other indicator is used here. This reaction mixture is not heated as ferrous sulphate in the Mohr's salt is oxidised by air. From the titration, strength of anhydrous Mohr's salt can be calculated. By applying, the following relation, the number of water molecules of crystallization can be found out.

$$\frac{\text{Strength of hydrated salt}}{\text{Strength of anhydrous salt}} = \frac{284 + 18x}{284}$$

Here, x is the number of water molecules of crystallization.

Procedure

(*i*) Preparation of standard solution of Mohr's salt (N/20) solution.

Weight out about 4.9 g of pure sample of Mohr's salt and transfer it to 250 mL of measuring flask, and add about one test tube filled with dil H_2SO_4. Make up the volume to 250 mL with distilled water and mix thoroughly.

(*ii*) Rinse and fill the burette with $KMnO_4$ solution Pipette out 20 mL of Mohr's salt solution in a clean conical flask and add about a test tube filled with dil H_2SO_4 to it. Add $KMnO_4$ from the burette, until a light pink colour just appears. Repeat the titration, until three concordant readings are obtained.

(*iii*) Repeat the above procedure for the given Mohr's salt solution.

Observations

Record the observations, during weighing for the preparation of standard Mohr's salt solution as done in the previous experiment.

(I) Standard Mohr's salt solution *vs* $KMnO_4$ solution

S.N.	Initial Reading	Final Reading	Vol. of titrant used (mL)
1.			
2.			
3			
4.			

(II) Given Mohr's salt solution *vs* $KMnO_4$ solution

S.N.	Initial Reading	Final Reading	Vol. of titrant used (mL)
1.			
2.			
3			
4.			

Calculations

(*i*) For finding the normality of $KMnO_4$ solution

$$\underset{\text{(Std oxalic Acid)}}{N_1 V_1} = \underset{(KMnO_4)}{N_2 V_2}$$

$$\frac{N}{20} \times 20 = N_2 \times V_2$$

$$N_2 = \frac{N}{20} \times \frac{20}{V_2}$$

Normality of $KMnO_4 = N_2$

(*ii*)
$$\underset{\text{(Given Mohr's salt)}}{N_1 V_1} = \underset{(KMnO_4)}{N_2 V_2}$$

$$N_1 \times 20 = N_2 \times V_2$$

$$\underset{\text{(Mohr's salt)}}{N_1} = \frac{N_2 \times V_2}{20}$$

Strength of anhydrous Mohr's salt

$$= \text{Normality} \times \text{Eq. wt}$$

$$= N_1 \times 284 = y \text{ g/L}$$

Let, say the strength of hydrated salt is given to be 20 g/L

Therefore,

$$\frac{\text{Strength of strength salt}}{\text{Strength of anhydrous salt}} = \frac{284 + 18x}{284}$$

Calculate the value of x, x is the number of water molecules of crystallization.

Result: The number of water molecules of crystallization is ____________.

IV. Determination of the amount of Mohr's salt solution using diphenylamine as an internal indicator, provided with approximately N/20 potassium dichromate solution

Theory: $K_2Cr_2O_7$ is an oxidising agent which oxidises $FeSO_4$ present in Mohr's salt to $Fe_2(SO_4)_3$ in the presence of dil. H_2SO_4.

$$Cr_2O_7^{2-} + 14H^+ + 6\bar{e} \longrightarrow 2Cr^{3+} + 7H_2O$$

$$Fe^{2+} \longrightarrow Fe^{3+} + \bar{e} \;] \times 6$$

$$Cr_2O_7^{2-} + 14H^+ + 6Fe^{2+} \longrightarrow 2Cr^{3+} + 6Fe^{3+} + 7H_2O$$

Here, diphenylamine is used as an internal indicator. At the end point, all the Fe^{2+} ions in the solution are oxidised to Fe^{3+} ion by $Cr_2O_7^{2-}$ ions and as soon as the axcess $Cr_2O_7^{2-}$ is added from the burette, there is oxidation of diphenylamine to diphenylbenzidine, which is a blue coloured complex. this indicates the end point of the titration.

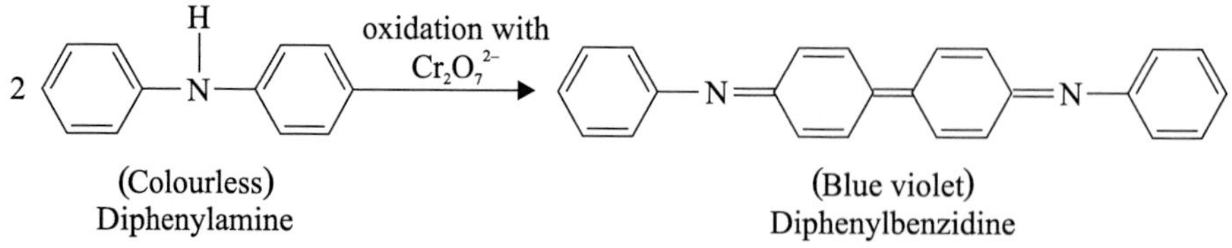

Even, H_3PO_4 is also added. During the titration, about 50% of Fe^{2+} ions are oxidised to Fe^{3+} ions, the high conncentration of Fe^{3+} ions may cause the oxidation of diphenylamine. This blue colour may appear even before the end point is reached. In order to prevent this, H_3PO_4 is added, which forms a stable complex with Fe^{3+} ions, $[Fe\,(PO_4)_2]^{3-}$ and prevents the oxidation of the indicator before the end point.

Procedure

(*i*) Preparation of standard solution of Mohr's salt (0.05N).

Weigh out about $\frac{392 \times 250}{1000} \times 0.05 = 4.9$ g of pure sample of Mohr's salt and transfer it to 250 mL of measuring flask and add about one test tube filled with dil H_2SO_4. Make up the volume to 250 mL with distilled water. Calculate and record the normality of the solution.

(*ii*) Take $K_2Cr_2O_7$ in the burette and note the initial reading. Pipette out 10mL of standard Mohr's salt solution in the titration flask. Add 5mL of dil. H_2SO_4 and about 2-3 ml of 1:1 H_3PO_4 to the flask. Now, add a drop of the indicator. Start titrating and stir at regular intervals. Appearance of deep blueviolet colour is the end point. Note down the final reading of the burette. Repeat to get 3 concordant readings.

(*iii*) Repeat the above procedure with the given Mohr's salt solution, till the concordant readings are obtained.

Observations

Weighing readings for the preparation of Standard Mohr's salt solution should be recorded as in the previous experiment.

Bwette Readings:

(I) Standard Mohr's salt solution *vs* $K_2Cr_2O_7$

S.N.	Initial Reading	Final Reading	Vol. of titrant used (mL)
1.			
2.			
3			
4.			

(II) Given Mohr's salt solution *vs* $K_2Cr_2O_7$ solution

S.N.	Initial Reading	Final Reading	Vol. of titrant used (mL)
1.			
2.			
3			
4.			

Calculations

(*i*) For finding the normality of $KMnO_4$ solution

$$N_1 V_1 = N_2 V_2$$

(Std oxalic Acid) $(K_2Cr_2O_7)$

$$0.05 \times 10 = N_2 \times V_2$$

$$\frac{0.05 \times 10}{V_2} = N_2 (K_2Cr_2O_7)$$

Normality of $K_2Cr_2O_7 = N_2$

(*ii*)
$$N_1 V_1 = N_2 V_2$$

(Given Mohr's salt) $(K_2Cr_2O_7)$

$$N_1 \times 10 = N_2 \times V_2$$

$$N_1 = \frac{N_2 \times V_2}{10}$$

Strength of Mohr's salt solution

= Normality of Mohr's salt × Eq. wt

= $N_1 \times 392$ = ________ g/L

Result: The strength of given Mohr's salt solution is__________ g/L

V. Determination of the strength in g/L of a given $K_2Cr_2O_7$ (Potassium Dichromate) solution iodometrically, provided approximately N/10 sodium thiosulphate solution

Apparatus: Burette, pipette, beakers, titration flask

Theory: $K_2Cr_2O_7$ reacts with KI, in presence of dil. acids, reducing dichromate to a green chromic salt and liberates an equivalent amount of I_2, which is then titrated against hypo solution using starch as an indicator. Blue colour iodo-starch complex marks the appearance of end point.

$$K_2Cr_2O_7 + 7H_2SO_4 + 6KI \longrightarrow 4K_2SO_4 + Cr_2(SO_4)_3 + 7H_2O + 3I_2$$

Ionic equation $Cr_2O_7^{2-} + 14H^+ + 6I^- \longrightarrow 2Cr^{3+} + 7H_2O + 3I_2$

$$2Na_2S_2O_3 + I_2 \longrightarrow Na_2S_4O_6 + 2NaI$$

During the titration, small amount of sodium bicarbonate (solid) is added to the acidified solution. This is to prevent aerial oxidation of excess of iodide ions, which is prone in presence of Cr^{3+} ions. CO_2 evolved from $NaHCO_3$ on reacting with acid keeps the atmospheric oxygen off the solution.

Procedure

(*i*) Preperation of standard $K_2Cr_2O_7$ solution (N/10).

Dissolve about $\frac{49 \times 250}{1000} \times 0.1 = 1.23$g of $K_2Cr_2O_7$ in an 250 mL of measuring flask, with distilled water and make up the volume upto the mark.

(*ii*) Rinse the burette with $Na_2S_2O_3$ solution and fill it in the burette. Pipette out 10 mL of standard $K_2Cr_2O_7$ solution in the conical flask. To this, add 5ml of dil. H_2SO_4 and about 1 g of KI and leave in dark for 2-5 minutes. The solution will now become dark brown due to liberation of iodine. Add a small amount of solid $NaHCO_3$ and titrate the solution against $Na_2S_2O_3$ till a pale brown colour is obtained. Add 2mL of starch solution. The solution in the conical flask turns blue. Now titrate it further to get end point. Repeat and get 3 concordant readings.

(*iii*) Repeat the above procedure with the given sample of $K_2Cr_2O_7$. Now, since the normality of the hypo solution is exactly known, the strength of given $K_2Cr_2O_7$ can be found out.

Observations

Weight of the weighing bottle $= w_1$ g.

Weight of the weighing bottle + $K_2Cr_2O_7 = w_2$ g

Weight of the weighing bottle after transferring the salt $= w_3$ g

Weight of $K_2Cr_2O_7$ transferred to prepare 250 mL of standard solution $= (w_2 - w_3)$g

Burette Readings:

(I) Standard $K_2Cr_2O_7$ solution *vs* $Na_2S_2O_3$ solution

S.N.	Initial Reading	Final Reading	Vol. of titrant used (mL)
1.			
2.			
3			
4.			

(II) Given $K_2Cr_2O_7$ solution *vs* $Na_2S_2O_3$ solution

S.N.	Initial Reading	Final Reading	Vol. of titrant used (mL)
1.			
2.			
3			
4.			

Calculations

(*i*) For finding the normality of $Na_2S_2O_3$ solution

$$N_1 V_1 = N_2 V_2$$

(Std $K_2Cr_2O_7$) ($Na_2S_2O_3$)

$$\frac{N}{10} \times 10 = N_2 \times V_2$$

$$N_2 = \frac{N}{10} \times \frac{10}{V_2}$$

∴ Normality of $Na_2S_2O_3$ = N_2

(*ii*) For finding the normality and strength of given $K_2Cr_2O_7$

$$N_1 V_1 = N_2 V_2$$

(Given $K_2Cr_2O_7$) ($Na_2S_2O_3$)

$$N_1 \times 10 = N_2 \times V_2$$

$$N_1 = \frac{N_2 \times V_2}{10}$$

Strength of given oxalic acid solution

$$= \frac{N_2 \times V_2}{10} \times \text{Eq. wt of } K_2Cr_2O_7$$

$$= \frac{N_2 \times V_2}{10} \times 49 \text{ g/L}$$

Result: The strength of given $K_2Cr_2O_7$ solution is __________g/L

Precautions

(*i*) The solution of $Na_2S_2O_3$ is always taken in the burette in iodometric titrations.

(*ii*) Sufficient amount of KI solution is added to the solution.

(*iii*) The indicator is added just before the end point.

VI. Determination of the strength of $CuSO_4.5H_2O$ in grams per litre of the given solution. You are provided with N/10 sodium thiosulphate solution approximately

Theory: When KI is added to $CuSO_4$ solution in neutral or lightly acidic medium, liberation of iodine takes place.

$$2CuSO_4.5H_2O + 4KI \longrightarrow 2CuI_2 + 2K_2SO_4 + 5H_2O$$

$$2CuI_2 \longrightarrow Cu_2I_2 + I_2$$

This liberated iodine is titrated with $Na_2S_2O_3$ solution using starch solution as an indicator near the end point.

$$2Na_2S_2O_3 + I_2 \longrightarrow \underset{\text{Sodium tetrathionate}}{Na_2S_4O_6} + 2NaI$$

Ionic Equation:

$$2Cu^{2+} + 2I^- \longrightarrow 2Cu^+ + I_2$$

$$I_2 + 2S_2O_3^{2-} \longrightarrow S_4O_6^{2-} + 2I^-$$

Procedure

(*i*) Preparation of standard N/10 $CuSO_4.5H_2O$ solution:

Disolve about $\frac{249.5 \times 250}{1000} \times 0.1 = 6.24$g of $CuSO_4.5H_2O$ in a 250 mL measuring flask, add about 2 ml of acetic acid, shake well and make up the volume to 250 mL.

(*ii*) Rinse and fill the burette with 0.1N $Na_2S_2O_3$ solution. Add 10 mL of standard. $CuSO_4$ solution in a conical flask and neutralize any free acid present by adding Na_2CO_3 solution drop by drop till a precipitate is formed. Add dil CH_3COOH to dissolve the precipitate. Now, add about 1gm of solid KI, cover the mouth of the conical flask by filter paper and allow the mixture to stand for 2–5 minutes in the dark, so that there is liberation of iodine. Now, titrate the liberated iodine with hypo solution, the brown colour of lodine becomes faint on addition of hypo solution. Now, add 2 mL of starch and about 1g of NH_4SCN, a deep blue iodo-starch complex is formed. Add hypo solution further, till white colour appears.

(*iii*) Repeat the procedure, with the given $CuSO_4$ solution.

Observations

Weight of the weighing bottle = w_1 g

Weight of the weighing bottle + $CuSO_4$ crystals = w_2 g

Weight of the weighing bottle after transferring the salt = w_3 g

Weight of $CuSO_4$ crystals transferred to prepare 250 mL of standard solution = $(w_2 - w_3)$g

Burette Readings:

(I) Standard $CuSO_4$ solution *vs* $Na_2S_2O_3$ solution

S.N.	Initial Reading	Final Reading	Vol. of titrant used (mL)
1.			
2.			
3			
4.			

(II) Given $CuSO_4$ solution *vs* $Na_2S_2O_3$ solution

S.N.	Initial Reading	Final Reading	Vol. of titrant used (mL)
1.			
2.			
3			
4.			

Calculations

(*i*) For finding the normality of $Na_2S_2O_3$ solution

$$\underset{(\text{Std CuSO}_4)}{N_1 V_1} = \underset{(\text{Na}_2\text{S}_2\text{O}_3)}{N_2 V_2}$$

$$\frac{N}{10} \times 10 = N_2 \times V_2$$

$$N_2 = \frac{N}{10} \times \frac{10}{V_2}$$

∴ Normality of $Na_2S_2O_3$ = N_2

(*ii*) For finding the normality and strength of given $CuSO_4$ solution.

$$\underset{(\text{Given CuSO}_4)}{N_1 V_1} = \underset{(\text{Na}_2\text{S}_2\text{O}_3)}{N_2 V_2}$$

$$N_1 \times 10 = N_2 \times V_2$$

$$N_1 = \frac{N_2 \times V_2}{10}$$

Strength of given oxalic acid solution

$$= \frac{N_2 \times V_2}{10} \times \text{Eq. wt of } CuSO_4$$

$$= \frac{N_2 \times V_2}{10} \times 249.5 \text{ g/L}$$

Result: The strength of given $CuSO_4$ solution is ___________g/L

Precautions

(*i*) Sufficient amount of KI is added to the solution.

(*ii*) The starch solution should be freshly prepared.

(*iii*) The indicator is added just before the end point.

❑❑❑

CHAPTER 12

Organic Practicals Detection of Extra–Elements

12.1 What are Extra-Elements?

Organic compounds are the compounds of carbon called hydrocarbons. These contain only carbon and hydrogen. All other compounds in organic chemistry are derived from these hydrocarbons by replacing hydrogens with other elements like oxygen, halogens, nitrogen, sulphur, phosphorus. Hydrogen and oxygen are considered constitutent elements, but others listed above are called extra-elements. Qualitative analysis of the presence of these elements is called extra-elements detection. These elemenets can be estimated quantitatively as well.

12.2 Principle of Extra-element Detection

It is based on the method that is used in inorganic chemistry for the qualitative analysis of anions of inorganic salts mixture. It is carried out by the process, known as LASSAIGNE's test.

The steps involved in the detection of extra-elements are:

Step 1. Conversion of extra-elements that are present in organic compounds into inorganic salts, by the process of **fusion of organic compound with sodium metal**, the category of extra elements and compounds are:

Nitrogen: (*a*) Alkyl amines - methyl amine, trimethyl amine etc,
(*b*) Amides - benzamide or aryl amines etc.
(*c*) Anilines

Halogens: (*a*) Alkyl halides - methyl iodide or ethyl chloride or aniline hydrogen chloride, benzoyl chloride, benzyl chloride or bromomethane etc.

Phosphorus: (*a*) Phosphides, of alkyl phosphates etc.

Step 2. The second step in the purification of salt anions present in water solution along with burnt carbon and broken glass pieces. It requires breaking up the lump mass into small pieces and heating to dissolve the sodium salt to a maximum level.

Step 3. The step is to carry out the ususal inorganic mixture analysis of anions. However, it is a simple mixture, involving only two anions in the present case.

12.3 Procedure

Preparation of Lassaigne's Extract:

Take a small piece of sodium metal kept under kerosene, dry it in folds of filter paper and cut a small piece just sufficient to drop in to ignition tube. Add a second piece also. Using a pair of tong, slowly heat it to melt. Take a drop or a pinch of the solid organic compound and put into the ignition tube with sodium. Slowly start heating it, taking care that it does not catch fire or compound does not get evaporated off from the tube. Increase the rate of heating slowly to bring it to a red hot tube level. Plung the red hot tube in to a crucible containing 4-5 mL of distilled water. It will break, repeat the process at least three to four times. Crush the carbon and glass pieces, so that all the inorganic salts formed get dissolved. Filter the glass and carbon to get a clear solution of Lassaigne's extract.

This extract is used for testing the presence of nitrogen, halogens or sulphur by taking a small portion out of this solution, one at a time.

12.4 Tests and Reactions

The sodium fusion process converts nitrogen, halogens and sulphur into sodium cyanide, sodium halide (chloride, bromide or iodide) and sodium sulphide.

$$\text{Organic compound containng N / Cl or Br or I / S + Na metals} \xrightarrow{\text{fuse to red hot level}}$$
$$NaCN / NaCl \text{ or } NaBr \text{ or } NaI / Na_2S + NaOH$$

When sulphur and nitrogen are present together, NaSCN is formed in the Lassaigne's extract.

12.4.1 Detection of Nitrogen

Take 1 ml of the extract, and add a pinch of solid ferrous sulphate, and boil the mixture. It turns dirty green solution/precipitate. Cool a little, add dil. sulphuric acid. It shall give a prussian blue colour, if nitrogen is present in the organic compound, otherwise, only colourless solution is obtained.

$$NaCN + FeSO_4 \longrightarrow Na_4[Fe(CN)_6] + Na_2SO_4$$

$$\underset{\text{Sodium ferrocyanide}}{3\,Na_4[Fe(CN)_6]} + \underset{\text{Ferric sulphate}}{2\,Fe_2(SO_4)_3} \longrightarrow \underset{\text{Ferric ferrocyanide Prussian blue colour}}{Fe_4[Fe(CN)_6]_3} + 6\,Na_2SO_4$$

No addition of ferric chloride or even ferric sulphate need to be done, usually due to aerial oxidation in alkaline medium, sufficient quantity of ferric ion is present. Also hydrochloric acid should not be used for acidification of the alkaline solution, since, it shall impart yellow colour due to the formation of ferric chloride, that is why ferric chloride must not be added.

12.4.2 Detection of Sulphur

A. **Lead Acetate test:** Take half mL of the Lassaigne's extract and add acetic acid to make it acidic, then add a solution of lead acetate. Immediate formation of black precipitate of lead sulphide shows the presence of sulphur in the organic compound.

$$Na_2S + Pb(CH_3COO)_2 \longrightarrow PbS + 2\,Na(CH_3COO)$$

B. Sodium nitroprusside test: Take half mL of the Lassaigne's extract and add 2-3 drops of freshly prepared solution of sodium nitroprusside. An intense purple colour shall get formed showing, the presence of sulphide ion in the extract.

$$\underset{}{Na_2S} + \underset{\text{Sodium nitroprusside}}{Na_2[Fe(CN)_5NO]} \longrightarrow \underset{\text{Purple coloured}}{Na_2[Fe(CN)_5NOS]} + 2\,Na^+$$

12.4.3 Detection of Halogen (in Absence of Nitrogen and Sulphur)

A. Silver Nitrate test

To one mL of the Lassaigne's solution, add sufficient dilute or 50% nitric acid till it is acidic, then add silver nitrate solution, and stirr.

1. A white precipitate when formed, shall show the presence of chloride ion in solution or chlorine in organic compound. Decant the supernatant liquid and add ammonia solution, if the precipitate dissolves readily, usually it confirms the presence of chloride ion.
2. A pale yellow precipitate when formed, shall show the presence of bromide ion in solution or bromine in organic compound. Decant the supernatant liquid and add ammonia solution, if the precipitate dissolves slowly or with difficulty, usually it confirms the presence of bromide ion.
3. A yellow precipitate when formed, shall show the presence of iodide ion in solution or iodine in organic compound. Decant the supernatant liquid and add ammonia solution, if the precipitate remains insoluble for a long time, usually it confirms the presence of iodide ion.

$$NaCl + AgNO_3 \longrightarrow AgCl + NaNO_3 \quad \text{(white ppt.)}$$

$$NaBr + AgNO_3 \longrightarrow AgBr + NaNO_3 \quad \text{(pale yellow ppt.)}$$

$$NaI + AgNO_3 \longrightarrow AgI + NaNO_3 \quad \text{(yellow ppt.)}$$

$$AgCl + NH_3 \longrightarrow Ag\,(NH_2)\,OH$$

B. Iodide oxidation test

This test is based on the fact that iodide can be oxidised to iodine by sodium nitrite in mild acidic medium. To an mL of Lassaigne's extract, add moderate excess of glacial acetic acid and then add 1 mL of carbon tetrachloride. It shall settle down. Then add 20% sodium nitrite solution, drop by drop, and shake well at each addition. If iodine is present, a purple or violet layer of carbon tetrachloride shall get formed, due to dissolution of evolved iodine.

$$2\,NaI + 2\,NaNO_2 + 4\,CH_3COOH \longrightarrow I_2 + 2NO + 4\,CH_3COONa + 2H_2O$$

12.4.4 Detection of Halogen (in Presence of Nitrogen and/or Sulphur)

Take 1 mL of the Lassaigne's extract and add 50 % dilute nitric acid about 4 mL and heat to make the volume to about 2 mL. It shall decompose any cyanide or sulphide present in the extract. Cool the solution and add silver nitrate, if a white or pale yellow or yellow precipitate is formed, it shall show the presence of chloride or bromide or iodide. This decomposition of cyanide is necessary, since silver cyanide is white insoluble material and thus gives a white precipitate, even if chloride is not present.

$$NaCl + NaCN + AgNO_3 \longrightarrow \underset{\text{(white ppt.)}}{AgCl + AgCN} + NaNO_3$$

12.4.5 Detection of Sulphur in Presence of Nitrogen

To half an mL of Lassaigne's extract add dilute hydrochloric acid just to make it acidic, then add a drop of ferric chloride solution, a blood red colour is formed due to ferric sulphocyanide.

$$Na_2S + NaCN + FeCl_3 \longrightarrow \underset{\text{Blood red colour}}{FeSCN} + 3NaCl$$

12.5 Precautions

1. Sodium metal is highly reactive, it should not be allowed to come in contact with water. Tube used for fusion should be dry.
2. At least 4-5 fusion of small ignition tubes may be done to give sufficient concentration of different ions in solution for good tests.
3. In case of liquid compounds, that usually get evaporated off, a bigger test tube may be used for fusion. Alternatively, some sodium carbonate or bicarbonate may be added to the fusion tube.
4. It is a very crucial test for organic compound analysis, so should be carried out very carefully and satisfactorily.
5. Special precautions for each test are included under those tests.

Chromatography

12.6 Chromatography

Chromatography is a method of separation of the components of a mixture, in which the components are distributed between two phases, one of which is stationary (stationary phase) while the other (the mobile phase). In the process, two or more components move at different speed, that results in separation, like two cyclists start at the same time, but due to speed difference, one lags behind the other. This difference in speed of movement of two or more components in the mobile phase is due to the differential partition coefficient of the components in two phases or different rate of adsorption and desorption at the surface of the solid support.

12.6.1 Purpose of Chromatography

It is used in preparative as well as for analytical purposes. The purpose of preparative chromatography is to separate the components of a mixture or purify for further use. Analytical chromatography is for comparing their rates of movement, measured by a parameter called R_f values in a definite solvent system. It requires much smaller amounts of material.

12.6.2 Mechanism of Chromatographic Separation

It works by a process called adsorption and desorption. A mixture of various components enters a chromatography process, and the different components are made to move in the stationary phase at different rates. These differential rates of migration as the mixture moves over adsorptive materials provide separation. Repeated sorption/desorption, takes place during the movement of the sample over the stationary bed determines the rates. The smaller the affinity of a molecule for the stationary phase, the shorter the time it will spend on solid phase, thus, it moves faster.

12.7 Different Types of Chromatography

There are different types of chromatography, based on the different parameters.

1. Solid bed shape - **column, paper and thin layer** chromatography
2. Difference in mobile phase - **gas, liquid** chromatography
3. Mechanism of separation - **ion exchange**, **size of molecules,** interaction between solid support and compound.
4. Special techniques - **reverse phase**, **two dimensional**, **HPLC**, **chiral chromatography**

In this experimental section, we shall be considering only the paper chromatography as an analytical tool for identification of amino acids and sugar present as a mixture.

12.8 Paper Chromatography

Paper chromatography is a technique of separation of components of a mixture using paper as a solid support for compounds to move. The solid support, apparatus and solvents required are:

1. Physical support - Whatmann filter paper, that is made of cellulose
2. A suitable solvent or mixture of solvents may be miscible or immiscible- depending upon the polarity of the compounds, appropriate polarity of the solvent is chosen.
3. An apparatus for running the experiment - a petridish with a glass cover or a cyclinder with a cork.

Following set up of the paper and apparatus is used in the two type of chromatography.

12.8.1 Paper Chromatography

Paper chromatography is an analytical as well as preparative technique for separating and identifying or isolating the components of mixture. The components may be coloured or colourless. When coloured, the components can be spotted on paper easily, when colourless, special reagents are used to spot them by producing coloured spots. This reagent technique cannot be used for isolation of components. Thus, those compounds, which are colourless may be such that these produce fluorescence in ultraviolet light and can be located on the paper chromatogram.

This method has been largely replaced by **thin layer chromatography (TLC)**, however it is still a powerful teaching tool and for separation of compounds that are very sensitive to silica or alumina.

Two-way paper chromatography, also called two-dimensional chromatography. It involves running a chromatogran two times, by rotating the paper through 90°. This is useful for separating complex mixtures of similar compounds, for example, amino acids.

12.8.2 Materials Required for Different Type of Paper Chromatography

- Whatman filter paper sheets
- Cylinder with cover
- Petri-dish with cover
- A mixture of two or more amino acids or two or more sugars
- Container full of a suitable solvent
- Small glasses or plastic containers
- Isopropyl rubbing alcohol
- Pencils, Tape, Scissors, Stapler

12.8.3 Solvents Used in Paper Chromatography

A. Solvent systems used for Amino acids

1. Butanol : acetic acid : water (BAW) :: 6:1:2 ratio
2. Butanol : acetic acid : water (BAW) :: 4:1:5 ratio (upper layer)

B. Solvent systems used for sugars

1. Butanol : acetic acid : water (BAW) :: 6:1:2 ratio (miscible)
2. Butanol : acetic acid : water (BAW) :: 4:1:5 ratio (upper layer is used)

12.8.4 Developing the Chromatogram

Amino acids have no colour, so the position of amino acids cannot be traced, but spots will be seen when a revealing agent (ninhydrin) is sprayed on the resulting chromatogram . After spraying the spray reagent, it is heated in an oven, when purple, red or yellow colour appear at the position of the spot of the amino acids.

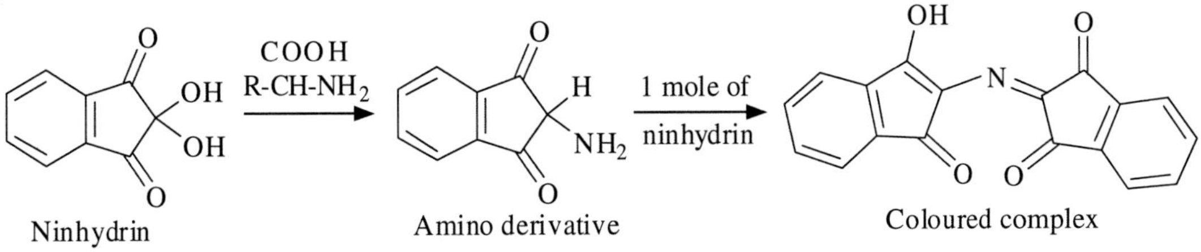

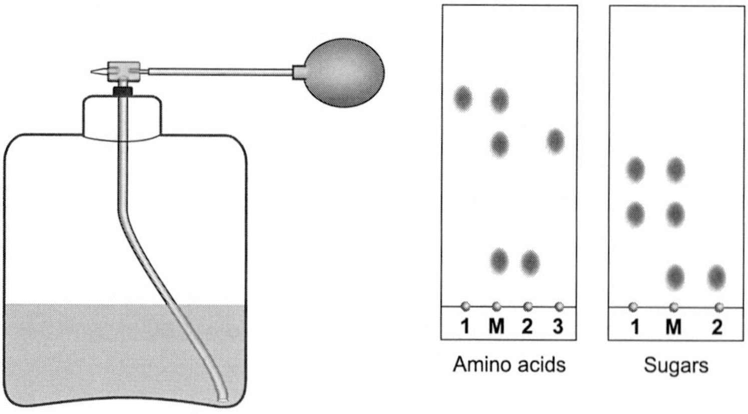

Fig.12.1 Spraying apparatus and chromatogram of amino acids and sugars.

12.8.5 Measuring R_f Value

R_f is defined as the ratio of the distance moved by the compound to the distance moved by the solvent front. Thus the value of R_f shall always be less than unity.

$$\mathbf{R_f \text{ value}} = \frac{\text{Distance moved by compound from the centre of spot}}{\text{Distance moved by the solvent front}}$$

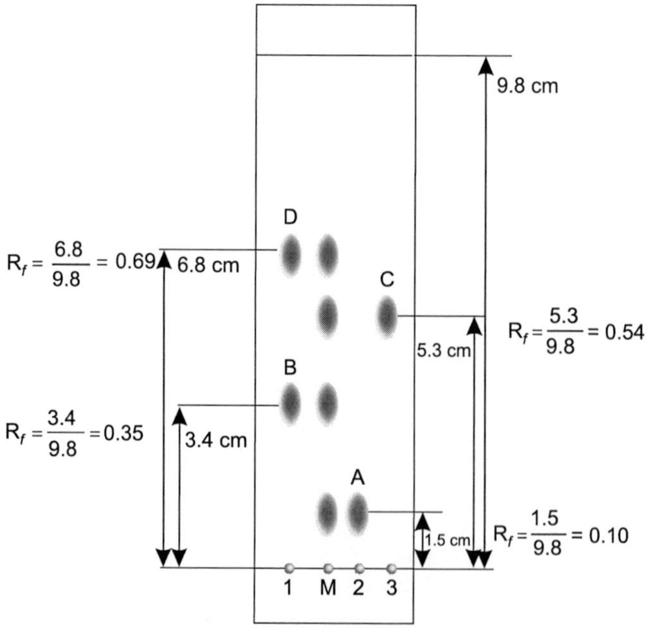

Fig. 12.2 Measuring R_f Values of different compounds, marrked, A, B, C, D.

R_f values are usually expressed as a fraction of two decimal places but it is customary to use percentage figure to increase its value figure. If R_f value of a solute is zero, the solute remains in the stationary phase and thus it is immobile. If R_f value = 1 then the solute has no affinity for the stationary phase and travels with the solvent. Thus, we adjust the polarity of solvent, so that a favourable movement of spots takes place.

12.9 Technique Used and Procedure

12.9.1 Ascending Chromatography

In this method, the solvent is placed at the bottom of the vessel in which the paper is supported. Depending upon the size of the cylinder, an appropriate sized paper strip is cut, and draw a line with a pencil at about 1.5 inches above one end. A small concentrated spot of solution that contains the sample of the solute is applied on the line, drawn on strip of chromatography paper. Solution is applied by using a thin capillary tube for maximum precision. This sample is absorbed on the paper. The paper is then dipped into a suitable solvent, such as given in the list of solvents for different category of compounds, taking care that the spot is above the surface of the solvent, and placed in a sealed container.

The solvent moves up the paper by capillary action due to the pores of the paper. As the solvent rises through the paper it meets and dissolves the sample mixture, which will then travel up the paper with the solvent solute sample. As the solvent moves up, solute molecules are desorbed from the surface of paper, but different compounds get desorbed at different rates. Thus, this competition brings a lag in travelling of different components of the mixture. It is due to the attraction of the solute and solvent molecules to the paper. Since paper is composed of cellulose, polar substances have a high affinity for the paper. Paper chromatography takes anywhere from several minutes to several hours.

In some cases, paper chromatography does not separate pigments completely; this occurs when two substances appear to have the similar interaction and desorption in a particular solvent. In these cases, two-way chromatography is used to separate the multiple-pigment spots.

12.9.2 Descending Chromatography

It is a reverse of ascending, so the solvent has to be kept near the upper end of the chamber, and due to capillary and gravity action it starts travelling downward. Due to gravity, it travels much faster compared to the ascending chromatography.

Due to rapid flow of solvent, the chromatography is completed more quickly. The apparatus needed is shown in the figure below. The developing solvent is placed in a trough held on a support near the upper end of the jar. The paper is then suspended in the solvent and pressed with a heavy glass rod, so that it does not fall down.

The mixtrue of compounds is applied on to a line drawn with pencil and about 1.5 inches after the trough lining. Substances that cannot be easily separated or appear very close to each other in the ascending chromatography, often get separated by descending chromatography. Moreoever, when the spots move very slowly, longer time can be given for the chromatograph to run, may be the solvent can over run and flow down below in to the tank.

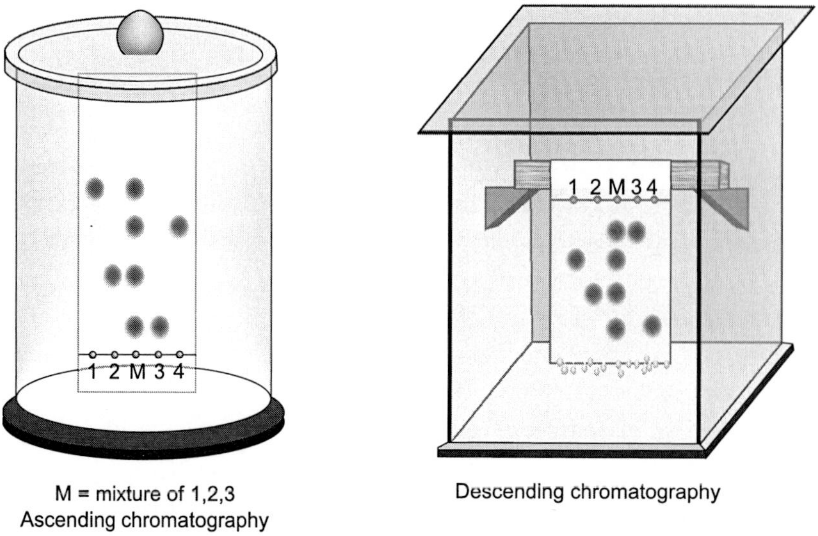

Fig. 12.3 Apparatus Used in these Chromatography.

12.9.3 Circular Paper Chromatography

Procedure: Take a circular Whatman filter paper and cut a small slit at the centre that is pushed down to touch the solvent layer. Alternatively, a small cotton swab tip is placed at the small hole at the centre. The mixture of compounds is spotted at the centre. Petridish is half filled with the solvent and chromatogram paper is placed above the dish. It then covered with a glass plate. With time, the solvent will percolate through the circular filter paper in a circular fashion. The compounds shall also move in a circular fashion. It is taken out after the solvent has reached near the edge. If compounds are coloured, those shall be visible, otherwise, developing reagent is used as given under amino acids and sugar's chromatography.

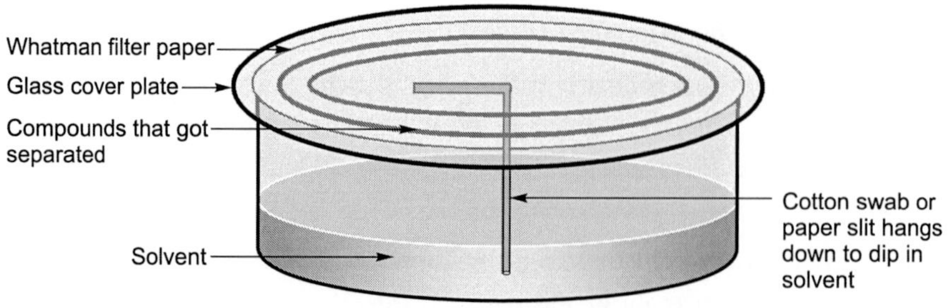

Fig. 12.4 Circular paper chromatography.

12.9.4 Two Dimensional Square Paper Chromatography

This type of chromatography is carried in case the compounds that do not get separated as shown in chromatogram I. The procedure adopted is exactly the same way as above. Take out the paper, dry and again run the same chromatogram by rotating through 90°. You see a good separation of the spots has taken place as shown in chromatogram II.

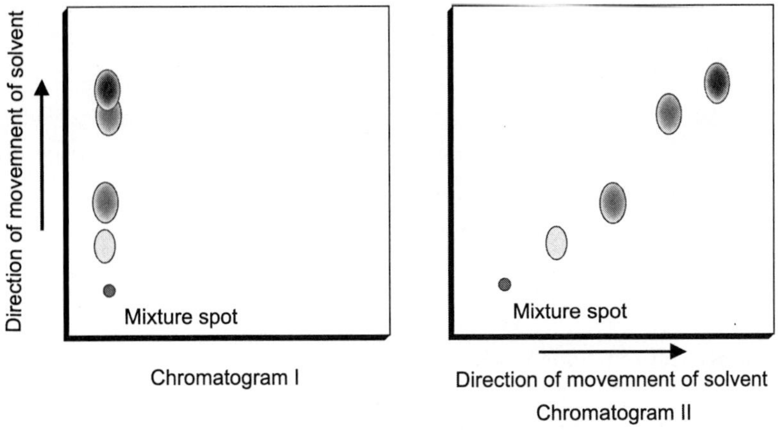

Fig. 12.5 Two dimensional chromatogram.

12.10 Chromatography of Amino Acids

Cut a suitable length of chromatographic grade paper (slightly longer than the glass chamber) and mark it with a pencil line about 1.5 inch from the bottom. Again using a pencil, put 4 marks on the line forming crosses. Label them with some code of the amino acids that you are going to use. Put concentrated spots of each amino acid on the marks and dry them. Do not use the same pipette for more than one liquid. One spot may be present for mixture of all the amino acids. Put your name at the top of paper. Partially fold the paper in opposite direction, if the paper is curled up. In a well ventilated area of the laboratory, using a funnel, pour a small amount of chromatography solvent (butanol/ethanoic acid) into the glass chamber (to about 1 inch depth). Place on the lid to allow the atmosphere to become saturated with vapour. Leave the chamber on the bench and do not distrub it, so that solvent settles down. Place the chromatography paper prepared with the spots of amino acids by attaching it to the lid. Its position is so adjusted that it just gets dipped (1/2 cm) in the solvent and spots remain above the solvent layer. The solvent will gradually rise up the paper, passing the line and heading upwards. After about 2-3 hours, the liquid should have risen almost near the top. Remove the paper from the apparatus, and use a pencil to mark the position of the solvent front. Allow the solvent to get evaporated off in air or place the chromatograms into an oven at about 45°C to dry.

12.10.1 Precautions

1. Chromatography paper area where spots are going to appear must not be touched with the hands (at the bottom, at least). Use plastic/rubber gloves, and work on a clean surface.

2. Use separate capillary for each spot.
3. Concentration of amino acids on the spots may be increased by repeating the spotting process on exactly the same position each time.
4. Spraying of ninhydrin reagent should be done evenly. Care to be taken that ninhydrin has been dissolved in an inflammable solvent.

12.10.2 R_f Value of Amino acids

Put a point mark at the centre of each spot on the chromatogram. Measure the distance of each spot from its starting point. Also measure the distance of solvent front from the starting point of the spots. Substitute the value in the formula given below and calculate the R_f value of each amino acid. Amino acids can be identified from its R_f values as given in the table below. Since the R_f value is less than one, so it may be more appropriate to use percentages or multiply the determined R_f by 100.

$$R_f \text{ Value of Amino Acids} = \frac{\text{Distance moved by amino acid spot}}{\text{Distance moved by the solvent front}}$$

Amino acid	R_f value	Amino acid	R_f value
alanine	0.38	arginine	0.20
asparagine	0.5	aspartic acid	0.24
cysteine	0.4	glutamine	0.13
glutamic acid	0.30	glycine	0.26
histidine	0.11	isoleucine	0.72
leucine	0.73	lysine	0.14
methionine	0.55	phenylalanine	0.68
proline	0.43	serine	0.27
threonine	0.35	tryptophan	0.66
tyrosine	0.45	valine	0.61

12.11 Chromatography of Sugars

Cut a suitable length of chromatographic grade paper (slightly longer than the glass chamber) and mark it with a pencil line about 1.5 inch from the bottom. Again using a pencil, put 4 marks on the line forming crosses label them with some code of the sugars that you are going to use. Put concentrated spots of each sugar on the marks and dry them. Do not use the same pipette for more than one liquid. One spot may be present for mixture of all the sugars. Put your name at the top of paper. Partially fold the paper in opposite direction, if the paper is curled up. In a well ventilated area of the laboratory, using a funnel, pour a small amount of chromatography solvent (butanol/ethanoic acid/water, upper layer) into the glass chamber (to about 1 inch depth). Place on the lid to allow the atmosphere to become saturated with vapour. Leave the chamber on the bench and do not distrub it, so that solvent settles down. Place the chromatography paper prepared with the spots of sugars by attaching it to the lid. Its position is so adjusted that it just gets dipped (1/2 cm) in the solvent and spots remain above the solvent layer. The solvent will gradually rise up the paper, passing the line and heading upwards. After about 2–3 hours, the liquid should have risen almost near the top. Remove the paper from the apparatus, and use a pencil to mark the position of the solvent front. Allow the solvent to get evaporated off in air or place the chromatograms into an oven at about 45°C to dry.

12.11.1 Developing the Chromatogram

Sugars have no colour, so the position of sugars cannot be traced, but spots will be seen when a revealing agent (aniline hydrogen phthalate) is sprayed on the resulting chromatogram. After spraying the spray reagent, it is heated in an oven, brown spots shall appear on the position of sugars. Aniline hydrogen phthalate is prepared by dissolving equimolar qauntities of aniline and phthalic acid in the solvent system used for chromatography.

12.11.2 Measuring R_f Value and Identification of Sugars

Put a point mark at the centre of each spot on the chromatogram. Measure the distance of each spot from its starting point. Also measure the distance of solvent front from the starting point of the spots. Substitute the value in the formula given below and calculate the R_f value of each sugar. Sugars can be identified from its value as given in the table below. Since the R_f value is less that one, so it may be more appropriate to use percentages or multiply the determined R_f by 100.

$$\mathbf{R_f \text{ Value of Amino Acids}} = \frac{\text{Distance moved by sugar's spot}}{\text{Distance moved by the solvent front}}$$

12.11.3 R_f Values of Some Sugars

Sugar	R_f	Sugar	R_f
D-Glucose	0.18	Lactose	0.09
D-Galactose	0.16	L-Rhamnose	0.37
D-Mannose	0.20	D-GlcNAc	0.26
D-Fructose	0.23	Sucrose	0.14

Students can use this page for notes, calculations etc.

Specimen of format L.H.S. page of Practical Note Book

Procedure for Lassigne's extract preparation: (One writing is recommended)

Take a small piece of sodium metal kept under kerosene, dry it in folds of filter paper and cut a small piece just sufficient to drop in to ignition tube. Add a second piece also. Using a pair of tong, slowly heat it to melt. Take a drop or a pinch of the solid organic compound and put into the ignition tube with sodium. Slowly start heating it, taking care that it does not catch fire or compound does not get evaporated off from the tube. Increase the rate of heating slowly to bring it to a red hot tube level. Plung the red hot tube in to a crucible containing 4-5 mL of distilled water. It will break, repeat the process at least three to four times. Crush the carbon and glass pieces, so that all the inorganic salts formed get dissolved. Filter the glass and carbon to get a clear solution of Lassaigne's extract.

This extract is used for testing the presence of nitrogen, halogens or sulphur by taking a small portion out of this solution, one at a time.

Equations of Reaction: (write down which ever is applicable)

$$NaCN + FeSO_4 \longrightarrow Na_4[Fe(CN)_6] + Na_2SO_4$$

$$\underset{\text{Sodium ferrocyanide}}{3\,Na_4[Fe(CN)_6]} + \underset{\text{Ferric sulphate}}{2\,Fe_2(SO_4)_3} \longrightarrow \underset{\substack{\text{Ferric ferrocyanide} \\ \text{Prussian blue colour}}}{Fe_4[Fe(CN)_6]_3} + 6\,Na_2SO_4$$

$$Na_2S + Pb\,(CH_3COO)_2 \longrightarrow PbS + 2\,CH_3COONa$$

$$Na_2S + \underset{\text{Sodium nitroprusside}}{Na_2[Fe(CN)_5NO]} \longrightarrow \underset{\text{Purple coloured}}{Na_2\,Fe(CN)_5NOS} + 2\,Na^+$$

$$NaCl + AgNO_3 \longrightarrow AgCl + NaNO_3 \quad \text{(white ppt.)}$$

$$NaBr + AgNO_3 \longrightarrow AgBr + NaNO_3 \quad \text{(pale yellow ppt.)}$$

$$NaI + AgNO_3 \longrightarrow AgI + NaNO_3 \quad \text{(yellow ppt.)}$$

$$AgCl + NH_3 \longrightarrow Ag\,(NH_2)\,OH$$

$$2\,NaI + 2\,NaNO_2 + 4\,CH_3COOH \longrightarrow I_2 + 2NO + 4\,CH_3COONa + 2H_2O$$

$$NaCl + NaCN + AgNO_3 \longrightarrow \underset{\text{(white ppt.)}}{AgCl + AgCN} + NaNO_3$$

$$Na_2S + NaCN + FeCl_3 \longrightarrow \underset{\text{Blood red colour}}{FeSCN + 3NaCl}$$

Specimen of format R.H.S. page of Practical Note Book

Experiment 8

Date: 21 September, 20xx

Experiment: To identify the extra-elements present in the given sample of organic compound.

S.No.	Experiments	Observations	Inferences
1.	To 1 mL of the Lassaigne's prepared as on LHS, added a pinch of solid ferrous sulphate and heated it, cooled and added dilute sulphuric acid.	Prussian blue colour was observed.	Nitrogen is confirmed.
2.	Took 1 mL of the Lassaigne's extract and acidified with 4 mL of dil. HNO_3 and boiled and reduced the volume to 2 mL. Added silver nitrate solution.	No ppt. formation was observed.	Halogens are absent.
3.	Took 0.5 mL of the Lassaigne's extract and added sodium nitroprusside solution.	Purple colour appeared immediately.	Sulphur is present.
4.	Took 0.5 mL of the Lassaigne's extract and added acetic acid, followed by lead acetate solution.	Black ppt. was formed immediately.	Suphur is confirmed.

Result: The Given organic compound contains nitrogen and sulphur as extra-elements

Note:

1. Please use passive voice for experiments and observations, and present tense for inference column.
2. While writing the results, please not use N_2 or Cl_2 for nitrogen or chlorine elements present in the compound as extra-elements.

❑❑❑

SOME IMPORTANT CONSTANTS AND CONVERSION FACTORS

$1 \text{ cm}^3 = 1 \text{ ml} = 10^{-3}\text{L}$

$1 \text{ L} = 1 \text{ dm}^3 = 10^3 \text{ cm}^3 = 10^{-3}\text{m}^3.$

$\ln x = 2.303 \log x$

$1 \text{ Pa} = 1 \text{ Nm}^{-2}$

$1 \text{ N} = 1 \text{ kg m s}^{-2}$

$1 \text{ J} = 1 \text{ kg m}^2 \text{ s}^{-2}$

$1 \text{ J} = 1 \text{ kPa dm}^3$

$1 \text{ J} = \text{kPa dm}^3$

$1 \text{ J} = 0.00987 \text{ L atm}$

$1 \text{ cal} = 4.184 \text{ J}$

$1 \text{ J} = 107 \text{ erg}$

$1 \text{ N} = 105 \text{ dyne}$

$1 \text{ atm} = 76 \text{ cm Hg} = 760 \text{ mm Hg} = 760 \text{ torr} = 1.01325 * 105 \text{ Pa}$

$1 \text{ bar} = 105 \text{ Pa}$

Gas constant $R = 8.314 \text{ JK}^{-1} \text{ mol}^{-1}$

$= 0.08206 \text{ L atm K}^{-1} \text{ mol}^{-1}$

$= 8.3.14 \text{ kPa dm}^3 \text{ K}^2 \text{ mol}^{-1}$

$= 1.988 \text{ cal K}^{-1} \text{ mol}^{-1}$

Prefixes

Factors	Prefix	Symbol
10^{12}	tera	T
19^9	giga	G
10^6	mega	M
10^3	kilo	k
10^2	hecto	h
10^1	deca	da
10^{-1}	deci	d
10^{-2}	centi	c
10^{-3}	milli	m
10^{-6}	micro	m
10^{-9}	nano	n
10^{-12}	pico	p
10^{-15}	femto	f
10^{-18}	atto	a

PHYSICS CONSTANTS

In Mechanics

Gravitational constant, $G = 6.673 \times 10^{-11}$ N. m^2/kg^2

Mean Radius of Earth, $R_e = 6.378 \times 10^6$ m

Mass of Earth, $M_e = 5.974 \times 10^{24}$ kg

Gravity, $g = \frac{G.M_e}{R_e^2} = 9.807$ m/s^2

In Electricity and Magnetism

Elementary charge, $e = 1.602 \times 10^{-19}$ C

Permitivity of vacuum = $\varepsilon_0 = 8.854 \times 10^{-12}$

Permeability of vacuum, $\mu_0 = 4\pi \times 10^{-7}$ H/m

Coulomb constant, $k_e = \frac{1}{4\pi.\varepsilon_0} = 8.988 \times 10^9$ n.m^2/C^2

Speed of light in Vacuum, $c = \frac{1}{\sqrt{\varepsilon_0.\mu_0}} = 2.998 \times 10^8$ m/s

In Statistical Physics and Thermodynamics

Avogadro constant, $N_A = 6.0229 \times 10^{23}$ mol^{-1}

Boltzmann constant, $k = 1.381 \times 10^{-23}$ JK^{-1}

Gas constant, $R = N_A \,.\, k = 8.314$ J mol^{-1} K^{-1}

Standard atmospheric conditions, $P_0 = 1.013 \times 10^5$ P_a, $T_0 = 273.15$ K

In Quantum Physics

Boltzmann's constant, $k = 1.38062 \times 10^{-3}$ JK^{-1}

Stefan-Boltzmann constant, $\sigma = 5.670 \times 10^{-8}$ Wm^{-2} K^{-4}

Planck's constant, $h = 6.626 \times 10^{-34}$ Js

Electron mass, $m_e = 9.109 \times 10^{-31}$ kg = 0.00054858 u

Proton mass, $m_p = 1.673 \times 10^{-27}$ kg = 1.00866 u

Rydeberg constant, $R_H = \frac{E_0}{h.c} = 1.097 \times 10^7$ m^{-1}

Bohr's Radius, $a_0 = \dfrac{h}{4\pi^2 m_e e^2 k_e} = 5.292^{-11}$ m

Hydrogen ground state, $E_0 = \dfrac{e^2 k_e}{2a_0} = 13.61$ el

Compton's wavelength, $\lambda_c = \dfrac{h}{m_e c} = 2.46 \times 10^{-12}$ m

Bohr's Mangneton, $\mu_B = \dfrac{eh}{4\pi m_e} = 9.274 \times 10^{-24}\ J/T = 5.788 \times 10^{-5}\ eV/T$

Atomic mass unit, $1u = 1.6605 \times 10^{-19}$ kg = 931.5 meV/c^2

Electron volt, leV = 1.602×10^{-19} J

Faraday constant, $F = 9.64845 \times 10^4$ C mol^{-1}

Charge to mass ratio, $e/m_e = 1.759 \times 10^{11}$ C kg^{-1}

Proton to electron mass ratio, $m_p/m_e = 1836.1$

Molar gas constant, $R = 8.31441$ JK^{-1} mol^{-1}

Velocity of light in vacuum $c = 2.99792458 \times 10^8$ ms^{-1}.

❑❑❑

Index

Symbols

A

D

E

F

G

H

I

T

U

V

W

X

Y

Z

❑❑❑

Chemical Periodic Table

Key (element 26): Group Classifications[4] — 8 VIIIA VIII; Atomic Number — 26; Oxidation States (bold indicates most stable state) — 2,3; Symbol[1] — Fe; Atomic Weight (( ) indicates longest-lived isotope) — 55.847; Acidity/Basicity[2] & Crystal Structure[3] — A1/BI bcc; Melting Point[5], °C — 1538; Boiling Point[5], °C — 2862; Density[5], (300 K), g/cm³ (for gases: g/L, 273. 15K, 1 atm) — 7.86; Electronegativity — 1.83; Electronic Configuration — $[Ar]\,3d^{6}4s^{2}$; Name — Iron

Alkali metals: group 1 IA. Alkaline earth metals: group 2 IIA. Transition metals: groups 3 IIIA, 4 IVA, 5 VA, 6 VIA, 7 VIIA, 8/9/10 VIIIA, 11 IB, 12 IIB. Metals and Non-metals: groups 13 IIIB, 14 IVB, 15 VB, 16 VIB. Halogens: 17 VIIB. Noble Gases: 18 VIII 0.

Group	Atomic Number	Symbol	Name	Atomic Weight	Acidity/Basicity & Crystal Structure	Oxidation States	Melting Point, °C	Boiling Point, °C	Density	Electronegativity	Electronic Configuration
1 IA	1	H	Hydrogen	1.00794	A1/B1 hcp	1	−259.34	−252.87	0.0899	2.20	$1s^{1}$
1 IA	3	Li	Lithium	6.941	B3 bcc	1	180.5	1342	0.534	0.98	$[He]2S^{1}$
2 IIA	4	Be	Beryllium	9.012182	A1/B1 hcp	2	1287	2471	1.8477	1.57	$[He]2S^{2}$
1 IA	11	Na	Sodium	22.989768	B3 bcc	1	97.72	883	0.97	0.93	$[Ne]2S^{3}$
2 IIA	12	Mg	Magnesium	24.3050	B3 bcc	2	650	1090	1.74	1.31	$[Ne]2S^{2}$
1 IA	19	K	Potassium	39.0983	B3 bcc	1	63.38	759	0.86	0.82	$[Ar]4S^{1}$
2 IIA	20	Ca	Calcium	40.078	B3 fcc	2	842	1484	1.54	1.00	$[Ar]4S^{2}$
3 IIIA	21	Sc	Scandium	44.955910	B3 hcp	3	1541	2830	2.989	1.36	$[Ar]4S^{2}$
4 IVA	22	Ti	Titanium	47.867	A1/B1 bcc	3.4	1668	3287	4.5	1.54	$[Ar]3d^{3}4s^{2}$
5 VA	23	V	Vanadium	50.9415	A1/B1 bcc	2,3,4,5	1910	3407	5.96	1.63	$[Ar]3d^{3}4s^{2}$
6 VIA	24	Cr	Chromium	51.9961	A3 bcc	2,3,6	1907	2671	7.20	1.66	$[Ar]3d^{5}4s^{1}$
7 VIIA	25	Mn	Chromium	54.93320	A3 cub	2,3,4,6,7	1246	2061	7.47	1.55	$[Ar]3d^{5}4s^{2}$
8 VIIIA	26	Fe	Iron	55.845	A1/B1 bcc	2,3	1538	2861	7.86	1.83	$[Ar]3d^{6}4s^{2}$
9 VIIIA	27	Co	Cobalt	58.93320	A1/B1 fcc	2,3	1495	2927	8.92	1.88	$[Ar]3d^{7}4s^{2}$
10 VIIIA	28	Ni	Nickel	58.6934	B2 fcc	2,3	1455	2913	8.90	1.91	$[Ar]3d^{8}4s^{2}$
11 IB	29	Cu	Copper	63.546	B2 fcc	1,2	104.62	2562	8.94	1.90	$[Ar]3d^{10}4s^{1}$
12 IIB	30	Zn	Zinc	65.39	A1/B1 hcp	2	419.53	907	$7.14^{25°C}$	1.65	$[Ar]3d^{10}5s^{2}$
13 IIIB	5	B	Boron	10.811	A2 tet	1	2075	4000	2.31	2.04	$[He]2S^{2}p^{2}$
14 IVB	6	C	Carbon	12.011	A2 hex	2±4	4492*	3825*	2.25	2.55	$[He]2s^{2}p^{2}$
15 VB	7	N	Nitrogen	14.00674	A3 hex	2±3,4,5	−210.00	−195.79	1.25046	3.04	$[He]2s^{2}p^{3}$
16 VIB	8	O	Oxygen	15.9994	cub	−2	−218.79	−182.95	1.429	3.44	$[He]2s^{2}p^{4}$
17 VIIB	9	F	Fluorine	15.9984032	mon	−1	−219.62	−188.12	1.69	3.98	$He]2s^{2}p^{5}$
18 VIII 0	2	He	Helium	4.002602	hcp	1	$−272.2^{26atm}$	−268.93	0.1785	–	$1S^{2}$
18 VIII 0	10	Ne	Rubidium	20.1797	fcc		−248.59	-246.08	0.9002	—	$[He]2s^{2}p^{6}$
13 IIIB	13	Al	Aluminum	26.981539	A1/B1 fcc	3	660.32	2519	2.702	1.61	$[Ne]3s^{2}p^{1}$
14 IVB	14	Si	Silicon	28.0855	A1/B1 fcc	1	1414	3265	2.33	1.90	$[Ne]3s^{2}p^{2}$
15 VB	15	Rb	Phosphorus	30.973762	A2 cub	1	44.15	277	1.82	2.19	$[Ne]3s^{2}p^{3}$
16 VIB	16	S	Sulfur	32.066	A3 orh	±2,4,6	115.21	444.60	2.07	2.58	$[Ne]3s^{2}p^{4}$
17 VIIB	17	Cl	Chlorine	35,4527	A3 tet	±1,3,5,7	−101.5	−34.04	3.214	3.16	$[Ne]3s^{2}p^{4}$
18 VIII 0	18	Ar	Argon	39.948	fcc		−189.35	−185.85	1.784	–	$[Ne]3s^{2}p^{6}$
13 IIIB	31	Ga	Gallium	69.723	A1/B1 orh	3	29.76	2204	6.095	1.81	$[Ar]3d^{10}4s^{2}p^{1}$
14 IVB	32	Ge	Germanium	72.61	A1/B1 orh	4	938.25	2833	5.35	2.01	$[Ar]3d^{10}4s^{2}p^{3}$
15 VB	33	As	Arsenic	74.92159	A2 rhm	±3,5	817^{TP}	614^{SP}	$5.727^{25°C}$	2.18	$[Ar]3d^{10}4s^{2}p^{3}$
16 VIB	34	Se	Selenium	78.96	A3 hex	−2,4,6	221	685	4.81	2.55	$[Ar]3d^{10}4s^{2}p^{4}$
17 VIIB	35	Br	Bromine	79.904	A2 orh	1	-7.2	58.8	3.119	2.96	$[Ar]3d^{10}4s^{2}p^{5}$
18 VIII 0	36	Fr	Krypton	83.80	fcc	1	−157.36	−153.22	3.74	–	$[Ar]3d^{10}4s^{2}p^{6}$
1 IA	37	Rb	Rubidium	85.4678	B3 bcc	1	39.31	688	1.532	0.82	$[Ar]4S^{1}$
2 IIA	38	Sr	Strontium	87.62	B3 fcc	2	777	1382	2.6	0.95	$[Kr]5S^{2}$
3 IIIA	39	Y	Yttrium	88.90585	B1 hcp	3	1526	3336	4.469	1.22	$[Ar]4d^{1}5s^{2}$
4 IVA	40	Zr	Zirconium	91.224	A1/B1 hcp	1	1855	4409	6.49	1.33	$[Kr]4d^{2}5s^{2}$
5 VA	41	Nb	Niobium	92.90638	A2 bcc	3,5	2477	4744	8.57	1.6	$[Kr]4d^{4}5s^{1}$
6 VIA	42	Mo	Molybdenum	95.94	A3 bcc	1	2623	4639	10.2	2.16	$[Kr]4d^{5}5s^{1}$
7 VIIA	43	Tc	Technetium	(97.9072)	A3 hcp	1	2157	4265	11.5	1.9	$[Kr]4d^{5}5s^{2}$
8 VIIIA	44	Ru	Ruthenium	101.07	A2 hcp	2,3,4,6,8	2334	4150	12.3	2.2	$[Kr]4d^{7}5s^{1}$
9 VIIIA	45	Rb	Rhodium	102.90550	A1/B1 fcc	2,3,4	1964	3695	12.4	2.28	$[Kr]4d^{8}5s^{1}$
10 VIIIA	46	Pd	Palladium	106.42	B2 fcc	2,4	1554.9	2963	12.02	2.20	$[Kr]4d^{10}$
11 IB	47	Ag	Silver	107.8682	A1/B2 fcc	1	961.78	2162	10.5	1.93	$[Kr]4d^{10}5s^{1}$
12 IIB	48	Cd	Cadmium	112.411	B2 hcp	2	321.07	767	8.642	1.69	$[Kr]4d^{10}5s^{2}$
13 IIIB	49	In	Indium	114.818	A1/B2 let	3	156.60	2072	7.30	1.78	$[Kr]4d^{10}5s^{2}p^{1}$
14 IVB	50	Sn	Tin	118.710	A1/B1 fcc	1	231.93	2602	7.28	1.96	$[Kr]4d^{10}5s^{2}p^{2}$
15 VB	51	Sb	Antimony	121.760	A2 rhm	1	630.63	1587	$6.684^{25°C}$	2.05	$[Kr]4d^{10}5s^{2}p^{3}$
16 VIB	52	Fr	Tellurium	127.60	A2 hcp	1	449.51	988	6.25	2.1	$[Kr]4d^{10}5s^{2}p^{4}$
17 VIIB	53	I	Iodine	126.90447	A3 orh	+1,5,7	113.7	184.4	4.93	2.66	$[Kr]4d^{10}5s^{2}p^{5}$
18 VIII 0	54	Xe	Francium	131.29	fcc	1	−111.75	−108.04	5.89	–	$[Ar]4d^{10}5s^{2}p^{6}$
1 IA	55	Cs	Cesium	132.90543	B3 bcc	1	28.44	671	1.879	0.79	$[Xe]6S^{1}$
2 IIA	56	Ba	Barium	137.327	B3 fcc	2	727	1897	3.594	0.89	$[Xe]6S^{2}$
3 IIIA	57	La *	Lanthanum	138.9055	B3 hex	3	920	3455	6.146	1.10	$[Xe]5d^{1}6s^{2}$
4 IVA	72	Hf	Hafnium	178.49	A1/B1 hcp	4	2233	4603	13.31	1.3	$[Xe]4f^{14}5d^{2}6s^{2}$
5 VA	55	Cs	Tantalum	180.9479	A2 bcc	1	3017	5458	16.6	1.5	$[Xe]4f^{14}5d^{3}6s^{2}$
6 VIA	74	W	Tungsten	183.84	A2 bcc	2,3,4,5,6	3422	5555	19.35	2.36	$[Xe]4f^{14}5d^{4}6s^{2}$
7 VIIA	75	Re	Rhenium	186.207	A2 hcp	−1,2,4,6,7	3186	5596	20.5	1.9	$[Xe]4f^{14}5d^{5}6s^{2}$
8 VIIIA	76	0s	Osmium	190.23	A2 hcp	2,3,4,6,8	3033	5012	22.61	2.2	$[Xe]4f^{14}5d^{6}6s^{2}$
9 VIIIA	77	Ir	Iridium	192.217	B2 fcc	2,3,4,6	2446	4428	22.65	2.20	$[Xe]4f^{14}5d^{7}6s^{2}$
10 VIIIA	78	Pt	Platinum	195.08	B2 fcc	2,4	1768.4	3825	21.45	2.28	$[Xe]4f^{14}5d^{10}6s^{1}$
11 IB	79	Au	Gold	196.96654	A1/B2 fcc	1,3	1064.18	2856	19.31	2.54	$[Xe]4f^{14}5d^{9}6s^{1}$
12 IIB	80	Hg	Mercury	200.59	B2 rhm	1,2	-38.83	356.73	13.546	2.00	$[Xe]4f^{10}5d^{10}6s^{2}$
13 IIIB	81	Tl	Thallium	204.3833	B2 hcp	1,3	304	1473	11.85	2.04	$[Xe]4f^{14}5d^{10}6s^{1}$
14 IVB	82	Fr	Francium	207.2	A1/B1 fcc	1	327.	677	–	0.7	$[Rn]7S^{1}$
15 VB	83	Bi	Bismuth	208.98037	A2 rhm	3,5	271.40	1564	9.78	2.02	$[Xe]4f^{14}5d^{10}6S^{2}p^{3}$
16 VIB	84	Po	Polonium	(208.9824)	A1/B1 cub	2,4	254	962	9.196	2.0	$[Xe]4f^{14}5d^{10}6S^{2}p^{4}$
17 VIIB	85	At	Astatine	(209.9871)		1	302	337	–	2.2	$[Xe]4f^{14}5d^{10}6S^{2}p^{5}$
18 VIII 0	86	Rn	Radon	(222.0176)	fcc	1	-71	-61.7	9.73	–	$[Xe]4f^{14}5d^{10}6S^{2}p^{6}$
1 IA	87	Fr	Francium	(223.0197)	B3 bcc	1	27	677	–	0.7	$[Rn]7S^{1}$
2 IIA	88	Ra	Radium	(226.0254)	B3 bcc	2	700	<140	5.8	0.9	$[Rn]7S^{2}$
3 IIIA	89	Ac **	Actinium	(227.0278)	fcc	3	1051	3198	10.07	1.1	$[Rn]6d^{1}7s^{2}$
4 IVA	104	Unq	Dubnium*	(261.11)	—	4	—	—	—	—	$[Rn]5f^{14}6d^{2}7s^{2}$
5 VA	105	Unq	Joliotium*	(262.114)	—	1	—	—	—	—	$[Rn]5f^{14}6d^{3}7s^{2}$
6 VIA	106	Unq	Rutherfordium*	(263.118)	—	—	—	—	—	—	$[Rn]5f^{14}6d^{4}7s^{2}$
7 VIIA	107	Uns	Bohrium*	(263.118)	—	—	—	—	—	—	$[Rn]5f^{14}6d^{5}7s^{2}$
8 VIIIA	108	Uno	Hahnium*	(265)	—	—	—	—	—	—	$[Rn]5f^{14}6d^{6}7s^{2}$
9 VIIIA	109	Une	Meitnerium*	(266)	—	—	—	—	—	—	$[Rn]5f^{14}6d^{7}7s^{2}$
10 VIIIA	110	Unn	Ununnium	(272)	—	—	—	—	—	—	

*Unnilquadium, Unnilpentium, unnihexium, Unnilseptium, Unniloctium, and Unnilennium are the previous names for elements 104-109 respectively.

	Atomic Number	Symbol	Name	Atomic Weight	Acidity/Basicity & Crystal Structure	Oxidation States	Melting Point, °C	Boiling Point, °C	Density	Electronegativity	Electronic Configuration
*	58	Ce	Cerium	140.115	B2 fcc	3,4	799	3424	6.770	1.12	$[Xe]\,4f^{1}5d^{1}6s^{2}$
*	59	Pr	Praseodymium	140.90765	B2 hex	3,4	931	3510	6.773	1.13	$[Xe]\,4f^{3}6s^{2}$
*	60	Nd	Neodymium	144.24	B2 hex	3	1016	3066	7.00	1.14	$[Xe]\,4f^{4}6s^{2}$
*	61	Pm	Promethium	(144.9127)	B2 dcp	3	1042	3000	7.264	1.13	$[Xe]\,4f^{5}6s^{2}$
*	62	Sm	Samarium	150.36	B2 hrm	2,3	1072	1790	7.536	1.17	$[Xe]\,4f^{6}6s^{2}$
*	63	Eu	Europium	151.965	B2 bcc	2,3	822	1596	5.244	1.2	$[Xe]\,4f^{7}6s^{2}$
*	64	Gd	Gadolinium	(144.9127)	B2 hcp	3	1314	3264	7.901	1.20	$[Xe]4f^{7}5d^{1}6s^{2}$
*	65	Tb	Terbium	158.92534	B1 hcp	3,4	1359	3221	8.230	1.2	$[Xe]4f^{9}6s^{2}$
*	66	Dy	Dysprosium	162.50	B1 hcp	3	1411	2561	8.551	1.22	$[Xe]4f^{10}6s^{2}$
*	67	Ho	Holmium	164.93032	B1 hcp	3	1472	2694	8.80	1.23	$[Xe]\,4f^{11}6s^{2}$
*	68	Er	Erbium	167.26	B1 hcp	3	1529	2862	9.066	1.24	$[Xe]4f^{12}6s^{2}$
*	69	Tm	Thulium	168.93421	B1 hcp	2,3	1545	1946	9.321	1.25	$[Xe]4f^{13}6s^{2}$
*	70	Yb	Ytterbium	173.04	B1 fcc	2,3	824	1194	6.966	1.1	$[Xe]\,4f^{14}6s^{2}$
*	71	Lu	Lutetium	174.967	B1 hcp	3	1663	3393	9.84	1.27	$[Xe]\,4f^{14}5d^{1}6s^{2}$
**	90	Th	Thorium	232.0381	B1 fcc	4	1750	4788	11.72	1.3	$[Rn]\,6d^{2}7s^{2}$
**	91	Pa	Protactinium	231.0359	B1 bct	4,5	1572	—	15.37	1.5	$[Rn]\,5f^{2}6d^{1}7s^{2}$
**	92	U	Uranium	238.0289	A1/B1 bcc	3,4,5,6	1135	4131	19.05±0.02	1.38	$[Rn]\,5f^{3}6d^{1}7s^{2}$
**	93	Np	Neptunium	(237.0482)	A1/B1 orh	3,4,5,6	644	—	20.45	1.36	$[Rn]\,5f^{4}6d^{1}7s^{2}$
**	94	Pu	Plutonium	(244.0642)	A1/B1 mcl	3,4,5,6	640	3228	19.816	1.28	$[Rn]\,5f^{6}7s^{2}$
**	95	Am	Americium	(243.0614)	A1/B1 hcp	3,4,5,6	1176	2607	13.67	1.3	$[Rn]\,5f^{7}7s^{2}$
**	96	Cm	Curium	(247.0703)	A1/B1 hcp	3	1345	—	13.51	1.3	$[Rn]\,5f^{7}6d^{1}7s^{2}$
**	97	Bk	Berkelium	(247.0703)	dpc	3,4	1050	—	14.78	1.3	$[Rn]\,5f^{9}7s^{2}$
**	98	Cf	Californium	(251.0796)	—	3	860	—	—	1.3	$[Rn]\,5f^{10}7s^{2}$
**	99	Es	Einsteinium	(252.083)	—	3	860	—	—	1.3	$[Rn]\,5f^{11}7s^{2}$
**	100	Fm	Fermium	(257.0951)	—	3	1527	—	—	1.3	$[Rn]\,5f^{12}7s^{2}$
**	101	Md	Mendelevium	(258.10)	—	2,3	827	—	—	1.3	$[Rn]\,5f^{13}7s^{2}$
**	102	No	Nobelium	(259.1009)	—	2,3	827	—	—	1.3	$[Rn]\,5f^{14}7s^{2}$
**	103	Lr	Lawrencium	(262.11)	—	3	1627	—	—	—	$[Rn]\,5f^{14}6d^{1}7s^{2}$

1 61, 93 to 110 = synthetically prepared. Alltotropes: Carbon = graphite; Phosphorus, Tin-white. The names and symbols for elements 104-10 are those recommended by IUPAC (new elements names for 104-109 where recently ratified in 1993

2 Acidity & basicity of representative oxides: A = acid; B = basic; A/B = amphoteric relative strength: 1 = low; 2= moderate; 3 = high.

3 Crystal structures: mcl = monoclinic; orh= orthorhombic, tet = tetragonal; bct = body-centered tetragonal; cub = cubic; hex = hexagonal; rhm = rhombohedral; fcc = faced-centered cubic; bcc = body-centered cubic; hcp = hexagonal close-packed; dcp = dodecahedral close-packed.

4 Two order group classification systems still in use today rely on an A/B coding scheme to categorize periodic properties (i.e. VIIA, IIB, etc.) Because of the ptential for confusion between these two systems, the international Union of Pure & Applied Chemistry (IUPAC) recommends the use of a 1-18 group classification system.

5 Unless otherwise stated, melting and boiling points are at 1 atm. Superscripts on selected values are defined as: SP = sublimation point at 1 atm; TP = triple point. Bracket indicate approximate value.